CHECK LIST OF THE LEPIDOPTERA
OF AMERICA NORTH OF MEXICO

Check List of the Lepidoptera of America North of Mexico

INCLUDING GREENLAND

EDITED BY

RONALD W. HODGES

AND

TATIANA DOMINICK

DONALD R. DAVIS

DOUGLAS C. FERGUSON

JOHN G. FRANCLEMONT

EUGENE G. MUNROE

JERRY A. POWELL

1983

LONDON

E. W. CLASSEY LIMITED AND
THE WEDGE ENTOMOLOGICAL RESEARCH FOUNDATION

This work is to be cited as
Hodges, Ronald W., *et al.*, 1983,
*Check List of the Lepidoptera
of America North of Mexico*

DISTRIBUTORS

THE WEDGE ENTOMOLOGICAL RESEARCH FOUNDATION
c/o National Museum of Natural History,
MRC–127, Washington, D.C. 20560, U.S.A.

E. W. CLASSEY LTD
P.O. Box 93, Faringdon, Oxfordshire SN7 7DR, England

BIOQUIP PRODUCTS
P.O. Box 61, Santa Monica, California 90406, U.S.A.

ENTOMOLOGICAL REPRINT SPECIALISTS
P.O. Box 77224, Dockweiler Station,
Los Angeles, California 90007, U.S.A.

The text has been computer set in Monotype Ehrhardt on the Lasercomp and printed
on GB Fine Book Wove Vol. 16, 90 gm²

Designed by Gertrude Temkin and Basil Harley
printed in Great Britain by the University Press, Cambridge

CONTENTS

To the memory of

RICHARD B. DOMINICK

His enthusiasm for
The Moths of America North of Mexico
was infectious and enduring

PREFACE

Forty years have passed since the names of North American (North of Mexico) moths and butterflies have been detailed serially. During this period much taxonomic information has been learned about many of the species, genera, families and superfamilies; however, our knowledge of others remains as it was when McDunnough (1938, 1939) made his list. All such lists are out-of-date by the time they are published, and this one is no exception. It represents the state of published and unpublished knowledge available to the authors to the end of 1978.

Because the last catalog of the North American Lepidoptera was published in 1903 (Dyar), and records of synonyms, homonyms, misspellings and misidentifications are not generally available, we have included this information. However, no attempt was made to search for all misspellings. Those in common use, or that cause confusion, in the taxonomic literature are cited. No synonymy of categories above the genus is indicated.

New systematic information, such as new combinations, new synonymies and changes in status, is indicated throughout the list. It reflects the results of study of type specimens, taxonomic revisions in progress or incidental knowledge gained by the authors. Some infrasubspecific names are included, even though they do not have nomenclatural standing, to prevent concern or confusion over apparent omissions.

Authorship is indicated throughout the list, usually in association with familial names. The senior editor was responsible for developing and maintaining the format, editing all parts, writing the introduction, interacting with the printer–publisher, and resolving questions and problems on a daily basis. Review of the text by the board of editors was extensive. Special thanks go to John D. Bradley, Richard L. Brown, Michael Carter, Harry Clench (deceased), Julian Donahue, D. S. Fletcher, J. Donald Lafontaine, William Miller, I. F. B. Nye, Paul Opler, Frederick Rindge, Gaden Robinson, Klaus Sattler, Gerhard Tarmann, Norman Tindale, Alan Watson and Paul E. S. Whalley who reviewed parts of the list. Their comments materially improved its quality. Errors of commission or omission are those of the individual authors.

Errors occur in the list; they have been found by individuals working with proof copies and by the authors. However, some of the required corrections have not been included because to do so would mean that the list never would appear. We plan to publish sheets of errata, as needed, so the list will be as useful and nearly correct as possible. To help achieve this goal I would appreciate receiving notice of any errors that you find together with a citation to the correct information. These should be sent to Ronald W. Hodges; Systematic Entomology Laboratory, USDA; c/o U.S. National Museum of Natural History, NHB 127; Washington, D.C. 20560

INTRODUCTION

EXTRALIMITAL NAMES

The following names, cited by McDunnough or published subsequently, are excluded from this list because they are extralimital. They are indicated here to prevent concern for their omission.

Utetheisa idae Clarke, 1940, is not American. The type-locality, Swain's Island, Newfoundland, probably refers to Swain's Island, about 3 degrees north of Pago Pago, Samoa. It is a junior synonym of *Utetheisa pulchelloides marshallorum* Roths., 1910.

Lichnoptera illudens (Wlk., 1856), and synonym *L. python* (Druce, 1889), is from the montane pine forests of Mexico and Central America. The Florida record is based on dealer's specimens. Because the moth has not been taken in Florida by any collector, it is assumed that the locality was incorrectly recorded. (McD., 1139)

Amathes ditrapezium (D. & S., 1775) is a *nomen nudum*. The record of this species from Labrador by Hampson (1903) apparently is based on an incorrectly labelled or misidentified specimen. (McD., 1514)

Polia passa (Morr., 1874) is from New Zealand. It is a junior synonym of *Graphania mutans* (Wlk., 1857). (McD., 1676)

Nephelodes minians (Cram., 1780) is a junior synonym of *Agrochola helvola* (L., 1758), a European species. (McD., 1895)

Leucania pertracta (Morr., 1875) is palearctic. Franclemont (1951) indicated that it is a junior synonym of *Aletia littoralis* (Curt., 1827). (McD., 1997)

Agriopodes corticosa (Gn., 1852) is thought to be Asiatic or South American. (McD., 2583)

Achaea ablunaris (Gn., 1852) is a misidentification of *Mimophisma delunaris* (Gn., 1852), and it is South American. (McD., 3417)

Zale sexplagiata (Wlk., 1858) is South American. As treated by McDunnough (1938), it is a misidentification of *Zale smithi* Haim., 1928. (McD., 3473)

Eupithecia arceuthata (Freyer, 1842) is a European subspecies of *E. intricata* (Zett., 1839). It does not belong to the North American fauna as treated by McDunnough (1949).

Ecliptopera silaceata deflavata (Stgr., 1871) was described from Europe. The widespread North American subspecies is *albolineata* (Pack., 1873). (McD., 4410b)

"Semiothisa" sabularia (Gn., 1857) is a junior synonym of *Tephrina catalaunaria* (Gn., 1857), an Old World moth that occurs from Spain to South Africa. The type female of *sabularia* is in the USNM. It supposedly came from North America. (McD., 4741)

"Itame" nervata (Gn., 1857), described from Brazil, was listed because its synonym, *tractata* (Wlk., 1861), was based on a specimen that supposedly came from North America. There is no further evidence that this species, properly *Semiothisa vernata*, occurs in the United States or Canada. (McD., 4796)

Apolema carata (Hulst, 1887) is a species of *Pseudopanthera* Hbn., 1823. Study of the type specimen in the American Museum of Natural History showed it probably to be equal to *P. ennomosaria* (Wlk., 1862) from India or Pakistan. *Apolema* Hulst, 1896, with *carata* as type-species, is a junior synonym of *Pseudopanthera*, **new synonymy**, and is removed from our fauna. (McD., 5018)

Lycimna Wlk., 1860 is an Old World genus. *Lycimna peccataria* B. & McD., 1916 belongs to the neotropical genus *Phyllodonta* Warr., 1894. (McD., 5060)

Patalene hamulata (Gn., 1857) apparently has not been collected in America north of Mexico. Two other species of *Patalene*, not listed by McDunnough (1938), have been collected in Texas. (McD., 5185)

Eudule Hbn., 1823, as revised by Fletcher (1954), applies to a group of neotropical species. *E. mendica* and its allies are placed in the genus *Eubaphe*.

Dysodia vitrina (Bdv., 1829) is African. (McD., 5316)

Acrolepia assectella (Zell., 1839) is a palearctic species, and it is the type-species of *Acrolepiopsis* Gaedike, 1970. The common name, leek moth, is correctly associated with *Acrolepiopsis incertella* (Cham., 1872).

Tineola uterella Wlsm., 1897 is a neotropical species, and it is the type-species of *Phereoeca* Hinton & Bradley, 1956. The common name, household casebearer, is correctly associated with *Phereoeca walsinghami* (Bsk., 1934).

UNPLACED NAMES

Four names have been placed in the Noctuoidea, but none of them has been identified satisfactorily. The types of the Behr names were destroyed during the San Francisco earthquake and ensuing fire; that of the Guenée name appears to be lost.

Anarta mimuli Behr, 1885 probably is a species of *Annaphila*. This is suggested by the association of the types with the plant genera *Antirrhinum* Tournefort and *Mimulus* L. It also could be a species of the noctuidlike pyralid genus *Gyros* or some similar appearing pyralid. (McD., 1793)

Epunda onychina Gn., 1852 may be an earlier name for *Egira alternans* (Wlk.) The description of the larva, probably based on an Abbot drawing, agrees in most essentials with the larva of *alternans*. (McD., 2206)

Capnodes californica Behr, 1870, could be any of several small moths. If it is a noctuid, it probably is a species of *Hemeroplanis*. (McD., 3607)

Agassizia urbicola Behr, 1870, if a noctuid, may be a species of *Mycterophora* and a senior synonym of one of the western species. Because *Agassizia* Behr, 1870 is a junior homonym, it would not take precedence over *Mycterophora* should *urbicola* be a senior synonym of a species of *Mycterophora*. (McD., 3715)

ABBREVIATIONS

Abbreviations of many authors' names and descriptors of taxa are used to conserve space. They are the following.

AUTHORS' NAMES

Auriv.	Aurivillius
B.-H.	Bang-Haas
B. & Benj.	Barnes and Benjamin
B. & Bsk.	Barnes and Busck
B. & L.	Barnes and Lindsey
B. & McD.	Barnes and McDunnough
Benj.	Benjamin
Brgstr.	Bergsträsser
Blkmre.	Blackmore
Bdv.	Boisduval
Bkh.	Borkhausen
Bott.	Bottimer
Bsk.	Busck
Cass.	Cassino
C. & S.	Cassino and Swett
Cham.	Chambers
Clarke	J. F. G. Clarke
Clem.	Clemens
Cl.	Clerck
Ckll.	T. D. A. Cockerell
Comst.	J. A. Comstock
Cram.	Cramer
Curt.	Curtis
Dalm.	Dalman
D. & S.	Denis & Schiffermüller
Don.	Donovan
Dup.	Duponchel
Hy. Edw.	Henry Edwards
Edw.	W. H. Edwards
Engelh.	Engelhardt
Esp.	Esper
Evers.	Eversmann
F.	Fabricius
Fgn.	Ferguson
Fern.	C. H. Fernald
Fbs.	W. T. M. Forbes
Franc.	Franclemont
Free.	T. N. Freeman
F. & B.	Frey and Boll
Germ.	Germar
Gey.	Geyer
Gibs.	Gibson
Gmel.	Gmelen
Godt.	Godart
Godm.	Godman
Grossb.	Grossbeck
Grt.	Grote
G. & R.	Grote and Robinson
Gn.	Guenée
Guér.	Guérin
Gump.	Gumppenberg
Haim.	Haimbach
Hamp.	Hampson
Hdwk.	Hardwick
Harr.	Harris
Harv.	Harvey
Haw.	Haworth
Heinr.	Heinrich
H.-S.	Herrich-Schäffer
Hochen.	Hochenwarth
Holl.	Holland
Hbn.	Hübner

<table>
<tr><td>

Hufn.   Hufnagel
Kft.   Kearfott
Keif.   Keifer
Kby.   W. F. Kirby
Koll.   Kollar
Latr.   Latreille
Led.   Lederer
Linds.   Lindsey
L.   Linnaeus
Lint.   Lintner
Luc.   Lucas
McD.   McDunnough
Mén.   Ménétriés
Meyr.   Meyrick
Morr.   Morrison
Mösch.   Möschler
Mun.   Munroe
Murt.   Murtfeldt
Neum.   Neumoegen
N. & D.   Neumoegen and Dyar
Oberth.   Oberthür
Obr.   Obraztsov
Ochs.   Ochsenheimer
Ottol.   Ottolengui
Pack.   Packard
Pears.   Pearsall
Prt.   L. B. Prout
Rag.   Ragonot
Reak.   Reakirt
Rob.   Robinson
Roths.   Rothschild
R. & J.   Rothschild and Jordan
Saund.   Saunders
Schiff.   Schiffermüller
Schr.   Schrank
Scop.   Scopoli
Skin.   Skinner
Sm.   J. B. Smith
Snell.   Snellen
Staint.   Stainton
Stgr.   Staudinger
Steph.   Stephens
Stkr.   Strecker
Sulz.   Sulzer
Tayl.   Taylor
Thunb.   Thunberg
Tr.   Treitschke
Wlk.   Walker
Wallgr.   Wallengren
Wlsm.   Walsingham
Warr.   Warren
Woll.   Wollaston
Wgt.   W. G. Wright
Zell.   Zeller
Zett.   Zetterstedt
Zinck.   Zincken

</td><td>

DESCRIPTORS

</td></tr>
</table>

ab.	aberration
ab. infrasubsp.	aberration, infrasubspecific
auth.	of authors = misidentification
auth., not Rag., 1878	of authors, not of Ragonot, 1878 = misidentification
auth., part, not Zell., 1873	of authors, in part, not of Zeller 1878 = misidentification
emend.	emendation, incorrect emendation
extralim.	extralimital, used when nominate subspecies does not occur in America north of Mexico
ident. uncert.	identity uncertain = unrecognized species
incorr. orig. spell.	incorrect original spelling
mispl.	generic placement incorrect or uncertain
missp.	misspelling, incorrect subsequent spelling
n. comb.	new combination
n. syn.	new synonymy
nom. nud.	*nomen nudum* = undescribed name
preocc. by Wlk., 1864	preoccupied
quest. occur.	questionable occurrence in America north of Mexico
quest. syn.	doubtful or questioned synonymy
repl. name	replacement name
rev. stat.	revised status
sec. hom.	secondary homonym
spr. f.	spring form
sum. f.	summer form
suppr.	suppressed, or in a work that has been suppressed, example: [suppr. (ICZN Op. 97)]
unassoc. ♀; ♂	unassociated ♀, or unassociated ♂
unavail., hybr. name	unavailable name, based on a hybrid
unavail., publ. in syn.	unavailable name, published in synonymy
uncert. stat.	status uncertain
unused sr. syn.	unused senior synonym = *nomen oblitum*

CLASSIFICATION

The classification of the Lepidoptera to the subfamily level is outlined for those taxa that occur in America north of Mexico. The number following each taxon shows the species recognized in this list.

OUTLINE

LEPIDOPTERA – 11283

Zeugloptera – 2
  Micropterigoidea – 2
    Micropterigidae – 2

Dacnonypha – 15
  Eriocranioidea – 15
    Eriocraniidae – 12
    Acanthopteroctetidae – 3

Exoporia – 20
  Hepialoidea – 20
    Hepialidae – 20

Monotrysia – 224
  Nepticuloidea – 137
    Nepticulidae – 82
    Opostegidae – 7
    Tischeriidae – 48
  Incurvarioidea – 87
    Incurvariidae – 56
      Incurvariinae – 22
      Prodoxinae – 16
      Adelinae – 18
    Heliozelidae – 31

Ditrysia – 11022
  Tineoidea – 598
    Tineidae – 174
    Psychidae – 26
      Taleporiinae – 2
      Psychinae – 16
      Oiketicinae – 8
    Ochsenheimeriidae – 1
    Lyonetiidae – 122
    Gracillariidae – 275
      Gracillariinae – 138
      Lithocolletinae – 126
      Phyllocnistinae – 11
  Gelechioidea – 1460
    Oecophoridae – 225
      Depressariinae – 108
      Ethmiinae – 50
      Peleopodinae – 2
      Stenomatinae – 23
      Oecophorinae – 41
      Chimabachinae – 1

    Elachistidae – 57
      Coelopoetinae – 1
      Elachistinae – 56
    Blastobasidae – 121
      Symmocinae – 12
      Blastobasinae – 109
    Coleophoridae – 169
      Coleophorinae – 145
      Batrachedrinae – 24
    Momphidae – 37
    Agonoxenidae – 6
      Blastodacninae – 6
    Cosmopterigidae – 180
      Antequerinae – 3
      Cosmopteriginae – 91
      Chrysopeleiinae – 86
    Scythrididae – 35
    Gelechiidae – 630
      Anomologinae – 44
      Gelechiinae – 491
      Anacampsinae – 39
      Chelariinae – 7
      Dichomeridinae – 48
      Lecithocerinae – 1
  Copromorphoidea – 35
    Copromorphidae – 1
    Alucitidae – 1
    Carposinidae – 11
    Epermeniidae – 11
      Epermeniinae – 9
      Ochromolopinae – 2
    Glyphipterigidae – 11
  Yponomeutoidea – 166
    Plutellidae – 54
    Yponomeutidae – 32
    Argyresthiidae – 52
    Douglasiidae – 5
    Acrolepiidae – 3
    Heliodinidae – 20
  Sesioidea – 144
    Sesiidae – 115
      Tinthiinae – 8
      Paranthreninae – 15
      Sesiinae – 92
    Choreutidae – 29
  Cossoidea – 45
    Cossidae – 45
      Hypoptinae – 19
      Cossulinae – 1
      Cossinae – 21
      Zeuzerinae – 4

NOTES ON CLASSIFICATION

Classification of the Lepidoptera reflects widely differing states of knowledge in different taxa. Some groups are relatively well known and perhaps have a relatively stable classification. Others are extremely poorly known, and the study of type specimens and additional material will result in development of new concepts of taxa. A short statement is made about each family or superfamily to indicate the status of current knowledge about it.

The higher classification of the Lepidoptera, particularly above the superfamily level, probably has demonstrated more confusion than that of any arthropod order. The classification of this list still contains some uncertainty, but recent efforts have begun to establish a foundation for a phylogenetically sound and, we hope, stable arrangement.

The system followed in the current list represents a conservative approach, based on information published prior to 1978. The use of the term "suborder" was known to be inconsistent, but because any attempt to follow a more natural system would require the introduction of new names, the more traditional system was followed. The relationships of the Nepticuloidea and Incurvarioidea present some of the greatest difficulties in the higher classification of the order. These two groups, although diverging more than superfamily status indicates, have been retained in this list under Monotrysia with the Ditrysia considered as a coordinate group.

Since 1978 some publications have appeared that further elucidate the subordinal relationships in the Lepidoptera. Viette (1979) reviewed the numerous subordinal names proposed for the Lepidoptera and synonymized some under the oldest available name for the group. Although the *International Code of Zoological Nomenclature* does not require the use of priority in the case of names above the superfamily level, Viette's paper attempts to stabilize these names using a consistent principle. It raises problems, such as the recognition of Protolepidoptera Packard (1895), a seldom-used name, instead of the younger but more familiar Zeugloptera Chapman (1917).

The subordinal classification of the Lepidoptera may be demonstrated best in a cladistic sequence that was first outlined by Hennig (1953). Several papers following, or attempting to follow, his principles have appeared, the most recent of which has been by Kristensen and Nielsen (1980). They present a preliminary outline of nearly all the major groups currently recognized (excluding Ditrysia), but they do not supply category names for the levels of the hierarchy. The divisional sequence of the groups recognized by them generally agrees with that currently followed (e.g. Davis, 1978), with the relative position of only a few families (the Tischeriidae in particular) remaining in doubt. Perhaps the most serious criticism of this classification, particularly in the light of Viette's paper, is their selection and use of some new or unfamiliar group names for previously named categories. Also, the primary division of the order is presented as an unresolved trichotomy consisting of the groups (suborders?) Zeugloptera, Aglossata and Glossata. Although not shown in their outline of the non-ditrysian Lepidoptera, it is assumed that the Ditrysia would be placed as a primary division (another trichotomy) of the Heteroneura together with the Nepticulina and the Incurvariina.

To supplement the foregoing remarks as well as the "subordinal" classification used in this list, an outline of Kristensen and Nielsen's table with representative superfamilies follows.

LEPIDOPTERA

Zeugloptera	Micropterigoidea
Aglossata	Agathiphagoidea
Glossata	
Dacnonypha	Eriocranioidea
Myoglossata	
Neopseustina	Neopseustoidea
Neolepidoptera	
Exoporia	Hepialoidea
Heteroneura	
Nepticulina	Nepticuloidea
Incurvariina	Incurvarioidea
[Ditrysia]	[Tineoidea – Noctuoidea]

MICROPTERIGIDAE: The family is being revised by Davis with one new western species to be added to the list.

ERIOCRANIOIDEA: The arrangement of taxa is based upon the revision of Davis (1978).

HEPIALIDAE: The arrangement of species is essentially the same as that followed by McDunnough (1939). The family needs study and is being revised by Tindale for *The Moths of America North of Mexico.*

NEPTICULIDAE: The arrangement of these species is tentative, pending the completion of a revision by Wilkinson and Davis of the North American members of the family. Although only about 40 additional species will be proposed in this revision, the size of this family in North America is probably at least two times greater than now realized. (DRD)

OPOSTEGIDAE: This family is being revised by

Davis for North America, and the arrangement partially reflects that study.

TISCHERIIDAE: The arrangement of species follows that of Braun (1972).

INCURVARIIDAE: The family is being revised by Davis for the New World, and the arrangement of species follows the results of this study. The Prodoxinae are arranged according to Davis (1967).

HELIOZELIDAE: The family is being revised by Lafontaine for North America, and the arrangement of taxa partially reflects this study.

TINEIDAE: American members of this family are being studied by Davis, and the arrangement of genera partially reflects this study. Several subfamilies have been proposed for the world, but many of our genera cannot be placed in them with any confidence. Consequently, subfamilies have not been recognized in this list. Pending the completion of the North American part of the revision, wherein approximately 15 new genera and 70 new species will be proposed, the species in this list have been arranged alphabetically. (DRD)

PSYCHIDAE: The arrangement of species follows that of Davis (1964).

OCHSENHEIMERIIDAE: The single species was discussed by Davis (1975).

LYONETIIDAE: This family needs revision, particularly at the generic level for the world. Several subfamilies have been proposed (Kuroko, 1964), but nearly half of our North American genera cannot be placed in that classification. For this reason subfamilies have not been recognized. The Bucculatricinae appear to be a well-defined group; it was revised by Braun (1963) for North America. The arrangement of taxa follows that of McDunnough (1939) and Braun (1963). (DRD)

GRACILLARIIDAE: The family is being revised by Davis for *The Moths of America North of Mexico.* The phylogenetic arrangement of genera follows that of Vari (1961). Pending the completion of the North American study, the species have been arranged alphabetically. (DRD)

OECOPHORIDAE: The arrangement of taxa follows Hodges (1974, 1978), Powell (1973) and Duckworth (1964).

ELACHISTIDAE: The arrangement of taxa follows that of Braun (1948). The fauna is much larger than recognized; many species remain to be described.

BLASTOBASIDAE: This primarily New World group is extremely poorly known. Type specimens have not been studied. More taxa probably remain to be described than are recognized. Species are listed alphabetically.

COLEOPHORIDAE: The Coleophorinae are arranged according to the results of an ongoing revision by Wright; the Batrachedrinae according to Hodges (1966).

MOMPHIDAE: The taxa are extremely poorly known, and the species are listed alphabetically.

AGONOXENIDAE: Relatively few species occur in North America; they are listed alphabetically.

COSMOPTERIGIDAE: The taxa are listed according to Hodges (1978). Further collecting will result in additional undescribed species and genera.

SCYTHRIDIDAE: The species and genera are very poorly known. Type specimens have not been studied. It appears that less than one-tenth of the North American genera and species have been described. The species are listed alphabetically.

GELECHIIDAE: Busck (1939) defined 14 genera in the *Gelechia* complex and assigned hundreds of species to them based on genital characters. Povolný (1967) associated species in the *Gnorimoschema* group with several genera. Sattler (1973) cataloged the generic names to provide much useful information. The family is being revised by Hodges for *The Moths of America North of Mexico.* The described fauna is less than half that recognized. New taxonomic information is given to the extent possible; it is based on examination of type specimens of nearly all names. Species are listed alphabetically pending publication of the results.

COPROMORPHIDAE: One species is known from North America in this largely tropical family. The tropical taxa need study.

ALUCITIDAE: North America has a depauperate fauna of this primarily palearctic group. A second, undescribed species is known to occur in North America.

CARPOSINIDAE: The taxa are arranged according to Davis (1967). A few new species have been discovered since and need to be incorporated.

EPERMENIIDAE: The arrangement follows Gaedike's (1977) revision of the North American fauna.

GLYPHIPTERIGIDAE: The family is being revised by Heppner for *The Moths of America North of Mexico.* The described fauna is less than half that known.

PLUTELLIDAE: The family is poorly known in North America except for the few economically important species. The arrangement is essentially that of McDunnough (1939).

YPONOMEUTIDAE: Some species are relatively well known, but many are very poorly known and need study.

ARGYRESTHIIDAE: This family is being revised by Opler for *The Moths of America North of Mexico.*

DOUGLASIIDAE: The only recent revisionary work in this family has been by Gaedike (1974) on the palearctic fauna. The North American species are poorly known.

ACROLEPIIDAE: The species need study.

HELIODINIDAE: Species of this family are little collected, probably as a result of the adults' diurnal habits. The family needs revision.

SESIIDAE: This family is being revised by Eichlin and Duckworth for *The Moths of America North of Mexico.* The North American fauna is relatively well known, but several species are undescribed. The arrangement follows the results of their studies.

CHOREUTIDAE: The family was separated by Heppner (1977) from the unrelated Glyphipterigidae, and he is revising it for *The Moths of America North of Mexico.* Nearly half the species are undescribed.

COSSIDAE: The Barnes and McDunnough (1911) revision has been followed with some modification based on work by Clench (in manuscript).

TORTRICIDAE: The higher classification is a consensus arrangement based on review of the world fauna, primarily following Obraztsov (1954–1968), Common (1958, 1963, 1965), Diakonoff (1961, 1977), MacKay (1962) and Dugdale (1966). Recent classifications by Kuznetsov and Stekol'nikov (1973, 1977) and Razowski (1976) are derived by cladistic inference, based exclusively on male genitalia and correlated functional musculature. These authors propose massive shifts in relationships of tribal level groupings from those used here, including treatment of the Cochylidae as a tribe within Tortricinae, synonymy of Cnephasiini within the Archipini, and reduction of the Sparganothidini to a subtribe of

Archipini. The Tortricinae and Olethreutinae were considered to be separate families at the time of the McDunnough list, but they are regarded as subfamilies by most contemporary authors. In the Olethreutinae there have been descriptions and reassignments of genera and species by W. E. Miller, R. L. Brown and others, but generally the arrangement of higher taxa and genera follows that of Heinrich (1923, 1926) used in McDunnough (1939), with nomenclatural modifications according to Bradley (1972) and in a few cases to Diakonoff (1973). The Tortricinae, by contrast, have undergone comprehensive taxonomic revision during the past 40 years, and the treatment here represents the work of Obraztsov (1954–1957) for the palearctic and several subsequent papers on nearctic genera, Freeman (1958), Powell (1964) and Bradley et al. (1973). Thus, present nomenclatural and taxonomic placements differ greatly from those of the McDunnough list. (JAP)

COCHYLIDAE: Little modern research has been carried out with the nearctic cochylids. The arrangement follows Busck (1940) and Razowski (1970) as far as possible, but most species have not been studied adequately to enable their placement accurately within existing or needed new genera. (JAP)

HESPERIOIDEA: The arrangement follows the catalog of the butterflies and skippers of America north of Mexico by Miller and Brown (1981).

PAPILIONOIDEA: The arrangement follows the catalog of Miller and Brown (1981).

ZYGAENOIDEA: The arrangement of the taxa in this superfamily closely follows that of McDunnough (1939). Tarmann is studying the genera on a world basis.

PYRALIDAE: Classification of the subfamilies varies in quality and in the amount of new information presented. The subfamilies Scopariinae, Nymphulinae, Odontiinae, Glaphyriinae and Evergestinae and the tribe Pyraustini of the Pyraustinae follow the arrangement of Munroe (1972–1976) with some corrections and updating. In the tribe Spilomelini of the Pyraustinae there are extensive changes in arrangement, generic classification and synonymy, anticipating in part the treatment in fascicle 13.3 of *The Moths of America North of Mexico* and also bringing together material already published in many scattered papers. Many species are left in inappropriate genera, or transferred to genera where

they fit better than in their former positions, but in which they will not remain permanently. The more incongruous combinations are noted as 'mispl.'. The Crambinae have been treated by A. B. Klots in much the same way in relation to his projected revision of the subfamily. The Cybalomiinae have been separated from the Pyraustinae and the Ancylolomiinae from the Crambinae, the latter following Forbes (1923) on Klots' recommendation. The treatment of the Schoenobiinae, Pyralinae and Epipaschiinae is conventional, though the Endotrichinae have been united with the Pyralinae, following Whalley's (1961) arrangement. The Chrysauginae are arranged to incorporate the results published in the abstract of E. C. Cashatt's thesis, but not those recorded only in the unpublished full text. The classification of the Galleriinae follows that of Whalley (1964), except that the Macrothecini have been added as a fourth tribe instead of treating them as a separate subfamily. The Phycitinae basically follow the system of Heinrich (1956), though it is largely a classification of convenience rather than a natural one. Not enough of the new concepts of phycitine taxonomy being developed by Ulrich Roesler have been published to make a satisfactory correlation with the American fauna possible. The Peoriinae and non-peoriine "Anerastiinae" follow the revision of Shaffer (1968) with the non-peoriine elements fitted into the Phycitinae on a somewhat arbitrary basis. (EGM)

THYRIDIDAE: The species have been distributed into the subfamilies recognized by Whalley (1964), though one name change was necessary.

HYBLAEIDAE: The family needs revision, but this will not affect our single species.

PTEROPHORIDAE: Generic transfers were made as the required genera had been described by Zimmerman (1958) for the species introduced to Hawaii. Further rearrangement will probably be needed, but so far as possible the classification is harmonized with that now current for the palearctic species. (EGM)

THYATIRIDAE: Changes in this group reflect the revisionary works of Clarke (1938) and Werny (1966). However, two of Werny's taxonomic decisions to elevate infrasubspecific forms to the rank of species are reversed. One of his changes is known to be incorrect on the basis of personal experimental rearing in which both forms were obtained from a single female parent; the other is a somewhat parallel case believed to be similarly incorrect but not proved. (DCF)

DREPANIDAE: The nomenclature for the five species of Drepanidae remains essentially unchanged. (DCF)

GEOMETRIDAE: Publications initiating taxonomic changes in the Geometridae since 1938 are too numerous to list individually. Major contributions include (a) the many revisions by F. H. Rindge, particularly in the tribes Boarmiini, Melanolophiini, Bistonini, Caberini and Nacophorini published in the *Bulletin* and *Novitates* series of the American Museum of Natural History between 1949 and the present; (b) revisions of *Eupithecia* (1949) and *Hydriomena* (1954) by McDunnough; (c) revision of the Geometrinae by Ferguson (1969); (d) of *Scopula* by Covell (1970); (e) and of *Pero* by Poole (in press). Other important works that have served as sources of information for the present list are McGuffin (1972, 1977), Fletcher (1972) and Lerault (1980). Fletcher's (1979) *The Generic Names of Moths of the World, Geometridae* is particularly important for any catalog or list; pertinent information from it was freely made available before publication. For the Sterrhinae classification follows Covell (1970) for *Scopula* and taxonomic decisions to be reflected in his revision of the subfamily in *The Moths of America North of Mexico.* The arrangement of *Pero* follows Poole (in press) but with the reinstatement of subspecies as treated by Rindge (1955). The arrangement of *Metarranthis* is a summation of Rupert (1943), Franclemont (unpublished) and Ferguson (unpublished); that for *Plagodis* follows Rupert (1949) and Munroe (1959). Tribal groupings are mainly derived from Forbes (1948); but the Cingiliini could not be differentiated from the Ourapterygini, so they were combined under the latter name. Forbes treated 129 of the 253 genera; several others were classified to tribe by Rindge or McGuffin; but the remaining nearly 120 genera had never been assigned to any tribe. These were investigated and distributed within the existing tribal framework. Transfer of the anomalous genus *Hypagyrtis* from the Boarmiini to the Bistonini was indicated in a manuscript revision of geometrid classification by the late L. R. Rupert. The genera *Alsophila, Ametris* and *Almodes* are placed in the Oenochrominae, but this is not done without misgiving. It seems unlikely that these genera should belong to a subfamily with a curious Australian genus as its type, or that the three genera belong in one subfamily, but it is expedient to leave them there until their affinities are better understood. (DCF)

EPIPLEMIDAE: These remain unchanged except for the addition of three genera and three species recently recorded from the United States. (DCF)

SEMATURIDAE: This family is represented by one rare species from Arizona. Although the moth was described in 1919, it was accidentally omitted by McDunnough (1938). (DCF)

URANIIDAE: This family was not listed by McDunnough (1938), but 10 records of the occurrence in the United States of the well-known, neotropical, migrant species, *Urania fulgens*, have been documented by Kendall (1979). (DCF)

MIMALLONIDAE, APATELODIDAE, BOMBYCIDAE, LASIOCAMPIDAE: The arrangements follow those of Franclemont (1973).

SATURNIIDAE: The arrangement follows that of Ferguson (1971, 1972).

SPHINGIDAE: The arrangement follows that of Hodges (1971) with some modifications.

NOTODONTIDAE: The classification follows Forbes (1939, 1948). Geographical distribution has been used as a character, but more important characters have been the genitalia of both sexes. (JGF)

DIOPTIDAE: This family contains moths that are little more than dayflying, mimetic notodontids. One species is a true member of the fauna north of the Mexican border although one or two other species may stray north of the border on occasion. The family, or subfamily, needs to be revised. (JGF)

ARCTIIDAE: The classification is eclectic but based in great part on my studies. The arrangement in *Holomelina* is mostly from an unpublished thesis of R. T. Cardé, that for the genus *Halysidota* follows Allan Watson (1980). The Ctenuchidae are reduced to a subfamily, and there is some question as to the monophyletic origin of the group. Julian Donahue, Douglas Ferguson, and Allan Watson have offered comments based on personal observations. The Pericopinae are wholly New World. The Old World Hypsinae are probably an aberrant group of Noctuidae; the larvae are better placed in that family than in the Arctiidae (Gardner, 1941). The Lithosiinae are cosmopolitan and much in need of revisionary studies. The Afridini are of uncertain position and may be related to some of the odd elements placed in the Eustrotiini in the Noctuidae. The Arctiinae are divided into tribes of which the Callimorphini and *Holomelina* in the Arctiini seem to be the primitive elements in the subfamily. (JGF)

LYMANTRIIDAE: The arrangement follows that of Ferguson (1978)

NOCTUIDAE: The tendency has been to retain the traditional classification, but to try to rectify the many arbitrary decisions that the unswerving application of a rigidly construed set of character states forced upon Hampson (1903). The Hampsonian classification was based mainly on a series of character reductions, and it resulted in many misassociations and misalignments of genera and species. A classification, to have any possibility of lasting, must be developed by striking a balance between similarities and differences. Although the present arrangement begins with the Herminiinae, it is not thought that this is the most primitive subfamily, rather it is a specialized group of the quadrifids. The Sarrothripinae and Acontiinae–Eustrotiini complexes are thought to represent the modern elements of the ancestral stock from which both the quadrifid and trifid groups arose. Thus, the classification begins with a specialized group of the quadrifids, "descends" through the quadrifids to the groups most nearly akin to the stock from which the Noctuidae are thought to have evolved and then "ascends" to the specialized trifids. The six or seven recognized trifid subfamilies are a very close assemblage, and I am inclined to agree with those who advocate treating them as tribes of a single subfamily, Noctuinae, or of two, the Noctuinae and Acronictinae (including the Pantheinae); however, there have been shifts in the associations of many genera. The only group of the trifids that has been the subject of a contemporary revision in North America is the Heliothinae (Hardwick, 1970); with few modifications this revision is followed. In the other trifid groups the arrangement is, in part, based on the concurrence of opinions by the two authors and, in part, on the opinion of one or the other of the authors. The tribe Acontiini follows Todd's conclusions from his research on the group; the Eustrotiini are arranged to follow the conclusions of both authors, but a number of genera are "disposed of" in this tribe until the group is studied on a world basis. The quadrifid subfamilies with the exception of the Plusiinae (Eichlin & Cunningham, 1978) are in great need of revisional studies. Much misalignment and misassociation has resulted from the attempt to use the presence or absence of spines on the tibiae to divide the main body of the quadrifids into two subfamilies, Catocalinae and Ophiderinae. In the list, genera that are obviously related, whether with or without spines on the tibiae, have been grouped together. Many of the genera in the list are isolated members of large groups that are well developed in the tropics, and they have been placed so as not to interrupt the sequences of related genera and groups of genera. (JGF)

AUTHORSHIP

Authors and parts of the list that each prepared are the following.

F. Martin Brown (Research Associate, Allyn Museum of Entomology) – Apaturidae, Danaidae, Hesperiidae, Libytheidae, Lycaenidae, Nymphalidae, Papilionidae, Pieridae, Riodinidae, Satyridae (all with Miller).

Charles V. Covell, Jr. (Department of Biology, University of Louisville) – Sterrhinae.

Donald R. Davis (Department of Entomology, Smithsonian Institution) – Acanthopteroctetidae, Dalceridae, Epipyropidae, Eriocraniidae, Gracillariidae, Heliozelidae, Hepialidae, Incurvariidae, Limacodidae, Lyonetiidae, Megalopygidae, Micropterigidae, Nepticulidae (with Wilkinson), Ochsenheimeriidae, Opostegidae, Psychidae, Tineidea, Tischeriidae, Zygaenidae.

W. Donald Duckworth (Department of Entomology, Smithsonian Institution) – Acrolepiidae, Argyresthiidae, Douglasiidae, Heliodinidae, Plutellidae, Sesiidae, Yponomeutidae (all with Heppner).

Douglas C. Ferguson (Systematic Entomology Laboratory, Science and Education Administration, United States Department of Agriculture) – Drepanidae, Epiplemidae, Geometridae (except Sterrhinae), Lymantriidae, Saturniidae, Sematuridae, Thyatiridae, Uraniidae.

John G. Franclemont (Department of Entomology, Cornell University) – Arctiidae, Apatelodidae, Bombycidae, Dioptidae, Lasiocampidae, Mimallonidae, Noctuidae (with Todd and except *Euxoa, Papaipema*).

John B. Heppner (Department of Entomology, Smithsonian Institution) – Alucitidae, Acrolepiidae (with Duckworth), Argyresthiidae (with Duckworth), Carposinidae, Choreutidae, Copromorphidae, Douglasiidae (with Duckworth), Epermeniidae, Glyphipterigidae, Heliodinidae (with Duckworth), Plutellidae (with Duckworth), Sesiidae (with Duckworth), Yponomeutidae (with Duckworth).

Ronald W. Hodges (Systematic Entomology Laboratory, Science and Education Administration, United States Department of Agriculture) – Agonoxenidae, Batrachedrinae, Blastobasidae, Cosmopterigidae, Cossidae, Gelechiidae, Momphidae, Oecophoridae, Scythrididae, Sphingidae.

Alexander B. Klots (Research Associate, Department of Entomology, American Museum of Natural History) – Crambinae.

J. Donald Lafontaine (Biosystematics Research Institute, Canada Agriculture) – Genus *Euxoa* (Noctuidae).

Lee D. Miller (Allyn Museum of Entomology) – Apaturidae, Danaidae, Hesperiidae, Libytheidae, Lycaenidae, Nymphalidae, Papilionidae, Pieridae, Riodinidae, Satyridae (all with Brown).

Eugene Munroe (Research Association, Lyman Entomological Museum, McGill University) – Hyblaeidae, Pterophoridae, Pyralidae (except Crambinae), Thyrididae.

Jerry A. Powell (Division of Entomology, University of California, Berkeley) – Cochylidae, Tortricidae.

Eric L. Quinter (Department of Entomology, American Museum of Natural History) – Genus *Papaipema* (Noctuidae).

E. L. Todd (Systematic Entomology Laboratory, Science and Education Administration, United States Department of Agriculture) – Noctuidae (with Franclemont).

Christopher Wilkinson (Biological Laboratory, Vrije Universiteit, Amsterdam) – Nepticulidae (with Davis).

Barry Wright (Nova Scotia Museum) – Coleophorinae.

LITERATURE

BARNES, W. and MCDUNNOUGH, J. H. 1911. Revision of the Cossidae of North America. *Contrib. Nat. Hist. Lep. N. Am.*, 1 (1): 1–36, pl. 1–7.

BRADLEY, J. D. 1972. *In* G. S. KLOET and W. D. HINCKS. A check list of British insects, 2nd edition (rev.), part 2, Lepidoptera. *Roy. Ent. Soc. London Handbook*, 11(2): i–viii+1–153.

BRADLEY, J. D., TREMEWAN, W. G. and SMITH A. 1973. *British Tortricoid Moths – Cochylidae and Tortricidae: Tortricinae*. The Ray Society, London. viii+1–251, figs. 1–52, pl. 1–47.

BRAUN, A. F. 1948. Elachistidae of North America (Microlepidoptera). *Mem. Amer. Ent. Soc.*, 13: 1–110+i, ii, figs. 1–137.

BRAUN, A. F. 1963. The genus *Bucculatrix* in America north of Mexico (Microlepidoptera). *Mem. Amer. Ent. Soc.*, 18: 1–208+i–iii, i, ii, pl. 1–45.

BRAUN, A. F. 1972. Tischeriidae of America north of Mexico (Microlepidoptera). *Mem. Amer. Ent. Soc.*, 28: 1–148+i, pl. 1–24.

BUSCK, A. 1939. Restriction of the genus *Gelechia* (Lepidoptera: Gelechiidae), with descriptions of new genera. *Proc. U.S. Natl. Mus.*, 86: 563–593, figs. 1–64.

BUSCK, A. 1940. Notes on North American microlepidoptera with descriptions of new genera and species. *Bull. So. California Acad. Sci.*, 39: 87–98, pl. 13, 14.

CHAPMAN, T. A. 1917. *Micropteryx* entitled to ordinal rank. Order Zeugloptera. *Trans. Ent. Soc. London*, 1916: 310–314, 12 pl.

CLARKE, J. F. G. 1938. A study of some North American moths allied to the thyatirid genus *Bombycia* Hübner. *Bull. So. California Acad. Sci.*, 37(2): 55–73, figs. 1–21.

COMMON, I. F. B. 1958. The genera of the Australian Tortricidae (Lepidoptera). *Proc. Tenth International Congress Ent.*, 1: 289–295, figs. 1–7.

COMMON, I. F. B. 1963. A revision of the Australian Cnephasiini (Lepidoptera: Tortricidae: Tortricinae). *Australian Jour. Zool.*, 11: 81–151, figs. 1–14, pl. 1–3.

COMMON, I. F. B. 1965. A revision of the Australian Tortricini, Schoenotenini and Chlidanotini (Lepidoptera: Tortricidae: Tortricinae). *Australian Jour. Zool.*, 13: 613–726, figs. 1–27, pl. 1–3.

COVELL, C. V., JR. 1970. A revision of the North American species of the genus *Scopula* (Lepidoptera, Geometridae). *Trans. Amer. Ent. Soc.*, 96: 101–221, figs. 1–99.

DAVIS, D. R. 1964. Bagworm moths of the Western Hemisphere (Lepidoptera: Psychidae). *U.S. Natl. Mus. Bull.*, 244: v+1–233, maps 1–11, figs. 1–385.

DAVIS, D. R. 1967. A revision of the moths of the subfamily Prodoxinae (Lepidoptera: Incurvariidae). *U.S. Natl. Mus. Bull.*, 255: vii+1–170, maps 1–17, figs. 1–155.

DAVIS, D. R. 1968. A revision of the American moths of the family Carposinidae (Lepidoptera: Carposinoidea). *U.S. Natl. Mus. Bull.*, 289: v+1–105, maps 1–11, figs. 1–222.

DAVIS, D. R. 1975. Review of Ochsenheimeriidae and the introduction of the cereal stem moth *Ochsenheimeria vacculella* into the United States (Lepidoptera: Tineoidea). *Smithsonian Contrib. Zool.*, 192: iii+1–20, maps 1, 2, figs. 1–31.

DAVIS, D. R. 1978. A revision of the North American moths of the superfamily Eriocranioidea with the proposal of a new family, Acanthopteroctetidae (Lepidoptera). *Smithsonian Contrib. Zool.*, 251: iv+1–131, maps 1–6, figs. 1–344.

DIAKONOFF, A. 1961. Taxonomy of the higher groups of the Tortricoidea. *XI Internationaler Kongress für Ent. Wien., Verhandl.*, 1: 124–126.

DIAKONOFF, A. 1977. Description of Hilarographini, a new tribus in the Tortricidae (Lepidoptera). *Ent. Berichten*, 37: 76–77.

DUCKWORTH, W. D. 1964. North American Stenomidae (Lepidoptera: Gelechioidea). *Proc. U. S. Natl. Mus.*, 116: 23–71, pl. 1–4, maps 1–12, figs. 1–45.

DYAR, H. G., FERNALD, C. H., HULST, G. D. and BUSCK, A. 1902 [1903]. A list of the North American Lepidoptera and key to the literature of this order of insects. *Bull. U.S. Natl. Mus.*, 52: xix+1–723.

EICHLIN, T. D. and CUNNINGHAM, H. B. 1978. The Plusiinae (Lepidoptera: Noctuidae) of America north of

Mexico, emphasizing genitalic and larval morphology. *U.S. Dept. Agr. Tech. Bull.*, **1567**: i–vi+1–122, figs. 1–227.

FERGUSON, D. C. 1969. A revision of the moths of the subfamily Geometrinae of America north of Mexico (Insecta, Lepidoptera). *Bull. Peabody Mus. Nat. Hist.*, **29**: 1–251, pl. 1–49.

FERGUSON, D. C. 1978. *In* R. B. DOMINICK *et al.* Bombycoidea (Saturniidae). *The Moths of America North of Mexico*, fasc. **20.2**: xxi+1–277, text figs. 1–30, pl. 1–22.

FERGUSON, D. C. 1978. *In* R. B. DOMINICK *et al.* Noctuoidea: Lymantriidae. *The Moths of America North of Mexico*, fasc. **22.2**: x+1–110, text figs. 1–23, pl. A, 1–8.

FLETCHER, D. S. 1954. A revision of the genus *Eubaphe* (Lepidoptera: Geometridae). *Zoologica*, **39**: 153–166, figs. 1–63.

FLETCHER, D. S. 1972. *In* G. S. KLOET and W. D. HINCKS. A check list of British insects, 2nd edition (rev.), part 2, Lepidoptera. *Roy. Ent. Soc. London Handbook*, **11**(2): i–viii+1–153.

FLETCHER, D. S. 1979. *In* I. W. B. NYE. Geometridae. *The generic names of moths of the world*, **3**: i–xx+1–243.

FORBES, W. T. M. 1923. The Lepidoptera of New York and neighboring states. *Cornell University Agric. Experiment Station Mem.*, **68**: 1–729, figs. 1–439.

FORBES, W. T. M. 1939. The Lepidoptera of Barro Colorado Island, Panama. *Bull. Mus. Comparative Zool.*, **85**(4): vii+97–322, pl. 1–8.

FORBES, W. T. M. 1948. Lepidoptera of New York and neighboring states, part 2. *Cornell University Agric. Experiment Station Mem.*, **274**: 1–263, figs. 1–255.

FRANCLEMONT, J. G. 1951. The species of the *Leucania unipuncta* group, with a discussion of the generic names for the various segregates of *Leucania* in North America. *Proc. Ent. Soc. Washington*, **53**: 57–85, figs. 1–41.

FRANCLEMONT, J. G. 1973. *In* R. B. DOMINICK *et al.* Mimallonoidea: Bombycoidea (in part). *The Moths of America North of Mexico*, fasc. **20.1**: viii+1–86, text figs. 1–22, pl. 1–11.

FREEMAN, T. N. 1958. The Archipinae of North America (Lepidoptera: Tortricidae). *Can. Ent., Supplement*, **7**: 1–89, figs. 1–258.

GAEDIKE, R. 1974. Revision der paläarktischen Douglasiidae (Lepidoptera). *Acta Faunistica Entomologica Musei Nationalis Pragae*, **15**: 79–101, figs. 1–69.

GAEDIKE, R. 1977. Revision der nearktischen und neotropischen Epermeniidae (Lepidoptera). *Beiträge Ent., Berlin*, **27**: 301–312, figs. 1–62.

GARDNER, J. C. M. 1941. Immature stages of Indian Lepidoptera (2) Noctuidae, Hypsidae. *Indian Forest Records (new series) Ent.*, **6**: 253–298, figs. 1–22, pl. 1, 2.

HAMPSON, G. F. 1903. Catalogue of the Noctuidae in the collection of the British Museum. *Catalogue of the Lepidoptera Phalaenae in the British Museum*, **4**: xx+1–689, figs. 1–125, pl. 55–77.

HARDWICK, D. F. 1970. A generic revision of the North American Heliothidinae (Lepidoptera: Noctuidae). *Mem. Ent. Soc. Canada*, **73**: 1–59, figs. 1–114.

HEINRICH, C. 1923. Revision of the North American moths of the subfamily Eucosminae of the family Olethreutidae *U.S. Natl. Mus. Bull.*, **123**: iv+1–298, figs. 1–432.

HEINRICH, C. 1926. Revision of the North American moths of the subfamilies Laspeyresiinae and Olethreutinae *U.S. Natl. Mus. Bull.*, **132**: v+1–216, figs. 1–448.

HEINRICH, C. 1956. American moths of the subfamily Phycitinae *U.S. Natl. Mus. Bull.*, **207**: iii+1–581, figs. 1–1138.

HENNIG, W. 1953. Kritische Bemerkungen zum phylogenetischen System der Insekten. *Beiträge zur Ent.*, **3** (Sonderheft): 1–85, figs. 1–12.

HEPPNER, J. B. 1977. The status of the Glyphipterigidae and a reassessment of relationships in yponomeutoid families and ditrysian superfamilies. *Jour. Lep. Soc.*, **31**: 124–134, table 1, fig. 1.

HODGES, R. W. 1966. Review of the New World species of *Batrachedra* with description of three new genera (Lepidoptera: Gelechioidea). *Trans. Amer. Ent. Soc.*, **92**: 585–651, figs. 1–144.

HODGES, R. W. 1971. *In* R. B. DOMINICK *et al.* Sphingoidea. *The Moths of America North of Mexico*, fasc. **21**: xii+1–158, text figs. 1–19, pl. 1–14.

HODGES, R. W. 1974. *In* R. B. DOMINICK *et al.* Gelechioidea: Oecophoridae (in part). *The Moths of America*

North of Mexico, fasc. **6.2**: x+1–142, text figs. 1–32, pl. A, 1–7.

HODGES, R. W. 1978. *In* R. B. DOMINICK *et al.* Gelechioidea: Cosmopterigidae. *The Moths of America North of Mexico*, fasc. **6.1**: x+1–166, text figs. 1–53, pl. 1–6.

KENDALL, R. O. 1979. Periodic occurrence of *Urania fulgens* (Uraniidae) in the United States. *Jour. Lep. Soc.*, **32**: 307–309.

KRISTENSEN, N. P. and NIELSEN, E. S. 1980. The ventral diaphragm of primitive (non-ditrysian) Lepidoptera – A morphological and phylogenetic study. *Zeits. für Zoologische Systematik und Evolutionforschung*, **18**: 123–146, table 1, figs. 1–51.

KUROKO, H. 1964. Revisional studies on the family Lyonetiidae of Japan (Lepidoptera). *Esakia*, **4**: 1–61, figs. 1–86.

KUZNETZOV, V. I. and STEKOLNIKOV, A. 1973. Phylogenetic relationships in the family Tortricidae (Lepidoptera) treated on the base of study of functional morphology of genital apparatus. (in Russian) *Horae Soc. Ent. Union Sovieticae*, **56**, 18–43, figs. 1–14.

KUZNETZOV, V. I. and STEKOL'NIKOV, A. 1977. Functional morphology of the male genitalia and phylogenetic relationships of some tribes in the family Tortricidae (Lepidoptera) of the fauna of the far east. (in Russian) *Acad. Sci. USSR, Proc. Zool. Inst.*, **70**: 65–97, figs. 1–18.

LERAUT, P. 1980. List systématique et synonymique des Lépidoptères de France, Belgique et Corse. *Alexanor*, Suppl. (Soc. ent. de France). 334 pp.

MACKAY, M. R. 1962. Larvae of the North American Tortricinae (Lepidoptera: Tortricidae). *Can. Ent., Supplement*, **28**: 1–182, figs. 1–85.

MCDUNNOUGH, J. 1938. Check list of the Lepidoptera of Canada and the United States of America. Part I. Macrolepidoptera. *Mem. So. California Acad. Sci.*, **1**: 1–272.

MCDUNNOUGH, J. 1939. Check list of the Lepidoptera of Canada and the United States of America. Part II. Microlepidoptera. *Mem. So. California Acad. Sci.*, **2** (1): 1–171.

MCDUNNOUGH, J. H. 1949. Revision of the North American species of the genus *Eupithecia* (Lepidoptera, Geometridae). *Bull. Amer. Mus. Nat. Hist.*, **93**(8): 535–728, pl. 26–32.

MCDUNNOUGH, J. H. 1954. The species of the genus *Hydriomena* occurring in America north of Mexico (Geometridae, Larentiinae). *Bull. Amer. Mus. Nat. Hist.*, **104**(3): 239–358, pl. 1–3.

MCGUFFIN, W. C. 1972. Guide to the Geometridae of Canada, II(1). *Mem. Ent. Soc. Canada*, **86**: 1–159, figs. 1–239.

MCGUFFIN, W. C. 1977. Guide to the Geometridae of Canada, II(2). *Mem. Ent. Soc. Canada*, **101**: 1–191, figs. 1–206.

MILLER, L. D. and BROWN, F. M. In press. Synonymic catalogue/check list of the Rhopalocera of America north of Mexico. *Lep. Soc. Mem.* **2**.

MUNROE, E. G. 1959. The *phlogosaria* complex of the genus *Plagodis* (Lepidoptera: Geometridae). *Can. Ent.*, **91**: 193–208, figs. 1–63.

MUNROE, E. 1972–1973. *In* R. B. DOMINICK *et al.* Pyraloidea: Pyralidae (part). *The Moths of America North of Mexico*, fasc. **13.1**: xx+1–304, pl. 1–13, A–K.

MUNROE, E. 1976. *In* R. B. DOMINICK *et al.* Pyraloidea: Pyralidae (part). *The Moths of America North of Mexico*, fasc. **13.2**: xviii+1–150, pl. 1–9, A–U.

OBRAZTSOV, N. 1954–1968. Die Gattungen der palaearktischen Tortricidae. *Tijds. voor Ent.*, **97** (1954): 141–231, figs. 1–248; **98** (1955): 147–228, figs. 249–366; **99** (1956): 107–154; **100** (1957): 309–347; **101** (1958): 229–261, figs. 1–24, **102** (1959): 175–216, figs. 25–63, pl. 23–26; **103** (1960): 111–143, figs. 64–106, pl. 11–13; **104** (1961): 51–70, figs. 107–126; **104** (1961): 231–240; **107** (1964): 1–48, figs. 127–177, pl. 1–8; **108** (1965): 1–40, figs. 1–6, pl. 1–4; **108** (1965): 365–387, figs. 178–191, pl. 16–22; **110** (1967): 13–36, figs. 1–2, pl. 1–2; **110** (1967): 65–88, figs. 1–24, pl. 3–10; **111** (1968): 1–48, figs. 1–5, pl. 1–11.

OBRAZTSOV, N. S. 1959. Note on North American *Aphelia* species (Lepidoptera, Tortricidae). *Amer. Mus. Novitates*, **1964**: 1–9, figs. 1–15.

OBRAZTSOV, N. S. 1959. On the systematic position of *Tortrix nigrivelata* (Lepidoptera, Tortricidae). *Amer. Mus. Novitates*, **1959**: 1–6, figs. 1–5.

OBRAZTSOV, N. S. 1962. *Anopina*, a new genus of the Cnephasiini from the New World (Lepidoptera, Tortricidae). *Amer. Mus. Novitates*, **2082**: 1–39, figs. 1–71.

OBRAZTSOV, N. S. 1962. New species and subspecies of

North American Archipini, with notes on other species (Lepidoptera, Tortricidae). *Amer. Mus. Novitates*, **2101**: 1–26, figs. 1–49.

OBRAZTSOV, N. S. 1963. North American species of the genus *Eana*, with a general review of the genus, and descriptions of two new species (Tortricidae). *Jour. Lep. Soc.*, **16**: 175–192, figs. 1–7.

PACKARD, A. S. 1895. On the phylogeny of the Lepidoptera. *Zoologischer Anzeiger*, **18** (no. 477): 228–236, table 1.

PACKARD, A. S. 1895. On a new classification of the Lepidoptera. *Amer. Naturalist*, **29**: 636–647, figs. 1–6.

POOLE, R. W. In press. A taxonomic revision of the New World moth genus *Pero* (Lepidoptera: Geometridae). *U.S. Dept. Agric. Technical Bull.*

POVOLNÝ, D. 1967. Genitalia of some nearctic and neotropic members of the tribe Gnorimoschemini (Lepidoptera, Gelechiidae). *Acta Entomologica Musei Nationalis Pragae*, **37**: 51–127, figs. 1–131.

POWELL, J. A. 1964. Biological and taxonomic studies on tortricine moths, with reference to the species in California. *Univ. California Publications in Ent.*, **32**: iv+1–317, maps 1–15, figs. 1–108, pl. 1–8.

POWELL, J. A. 1973. A systematic monograph of New World ethmiid moths (Lepidoptera: Gelechioidea). *Smithsonian Contrib. Zool.*, **120**: iv+1–302, maps 1–68, figs. 1–294, pl. 1–22.

RAZOWSKI, J. 1970. *In* H. G. AMSEL *et al.* Cochylidae. *Microlepidoptera Palaearctica*, **3**: xiv+1–528, pl. 1–161, 132 figs.

RAZOWSKI, J. 1976. Phylogeny and system of Tortricidae (Lepidoptera). *Acta Zoologica Cracoviensia*, **21**: 73–118, fig. 1.

RINDGE, F. H. 1955. A revision of some species of *Pero* from the western United States (Lepidoptera, Geometridae). *Amer. Mus. Novitates*, **1750**: 1–33, figs. 1–20, table 1.

RUPERT, L. R. 1943. A specific revision of the genus *Metarranthis* (Lepidoptera, Geometridae, Ennominae). *Jour. New York Ent. Soc.*, **51**: 133–159, pl. 7–9.

RUPERT, L. R. 1949. A revision of the North American species of the genus *Plagodis* (Lepidoptera, Geometridae, Ennominae). *Jour. New York Ent. Soc.*, **57**: 19–49, pl. 1–5.

SATTLER, K. 1973. A catalogue of the family-group and genus-group names of the Gelechiidae, Holcopogonidae, Lecithoceridae and Symmocidae (Lepidoptera). *Bull. British Mus. (Nat. Hist.) (Ent.)*, **28**: 153–282.

SHAFFER, J. C. 1968. A revision of the Peoriinae and Anerastiinae (*auctorum*) of America north of Mexico (Lepidoptera: Pyralidae). *U.S. Natl. Mus. Bull.*, **280**: vi+1–124, table 1, figs. 1–178.

VÁRI, L. 1961. South African Lepidoptera, Lithocolletidae. *Transvaal Mus. Mem.*, **12**: xix+1–238, pl. 1–112.

VIETTE, P. 1979. Réflexions sur les classifications en sous-ordres de l'ordre des Lepidoptera. *Bull. Soc. Ent. France*, **84**: 68–78.

WATSON, A. 1980. A revision of the *Halysidota tessellaris* species-group (*Halysidota sensu stricto*) (Lepidoptera: Arctiidae). *Bull. British Mus. (Nat. Hist.) (Ent.)*, **40**: 1–65, figs. 1–106.

WERNY, K. 1966. Untersuchungen über die systematik der tribus Thyatirini, Macrothyatirini, Habrosynini und Tetheini (Lepidoptera: Thyatiridae). *Universität des Saarlandes* (Saarbrücken, Germany), *Inaugural-Dissertation*. 1–463, figs. 1–436, maps.

WHALLEY, P. E. S. 1961. A change in status and a redefinition of the subfamily Endotrichinae (Lep. Pyralidae), with the description of a new species. *Ann. Mag. Nat. Hist.*, (13) **3**: 733–736.

WHALLEY, P. E. S. 1964. Catalogue of the world genera of the Thyrididae (Lepidoptera) with type selection and synonymy. *Ann. Mag. Nat. Hist.*, (13) **7**: 115–127.

WHALLEY, P. E. S. 1964. Catalogue of the Galleriinae (Lepidoptera, Pyralidae) with descriptions of new genera and species. *Acta Zoologica Cracoviensia*, **9**: 561–618, pl. 14–44.

ZIMMERMAN, E. C. 1958. Lepidoptera: Pyraloidea. *Insects of Hawaii*, **8**: xii+1–456, figs. 1–347.

CHECK LIST OF THE LEPIDOPTERA
OF AMERICA NORTH OF MEXICO

ZEUGLOPTERA

Micropterigoidea

MICROPTERIGIDAE
by DONALD R. DAVIS

EPIMARTYRIA Wlsm., 1898
1 **auricrinella** Wlsm., 1898
2 **pardella** (Wlsm., 1880)

DACNONYPHA

Eriocranioidea

ERIOCRANIIDAE
by DONALD R. DAVIS

DYSERIOCRANIA Spuler, 1910
  MNEMONICA Meyr., 1912
3 **griseocapitella** (Wlsm., 1898)
  auricyanea; auth., not Wlsm., 1882
  auricrinella; Fracker, 1930
4 **auricyanea** (Wlsm., 1882)
  cyanosparsella (Williams, 1908)

ERIOCRANIA Zell., 1851
5 **semipurpurella** (Steph., 1834)
  amentella (Zell., 1850)
  inconspicuella (Wood, 1890)
  a. **pacifica** Davis, 1978
    semipurpurella; auth., not Steph., 1834
6 **breviapex** Davis, 1978

ERIOCRANIELLA Viette, 1949
  ERIOCRANIELLA Viette, 1949,
    subgenus
7 **aurosparsella** (Wlsm., 1880)
8 **xanthocara** Davis, 1978
  aurosparsella; auth., not Wlsm., 1880
9 **longifurcula** Davis, 1978
10 **platyptera** Davis, 1978

  DISFURCULA Davis, 1978,
    subgenus
11 **variegata** Davis, 1978
12 **trigona** Davis, 1978
13 **falcata** Davis, 1978

  NEOCRANIA Davis, 1978
14 **bifasciata** Davis, 1978

ACANTHOPTEROC-
TETIDAE
by DONALD R. DAVIS

ACANTHOPTEROCTETES Braun,
  1921
15 **tripunctata** Braun, 1921
  tripunctella Davis, 1969, missp.
16 **bimaculata** Davis, 1969
17 **unifascia** Davis, 1978

EXOPORIA

Hepialoidea

HEPIALIDAE
by DONALD R. DAVIS

STHENOPIS Pack., 1864
18 **argenteomaculatus** (Harr., 1842)
  argentata Pack., 1864
  alni (Kellicott, 1885)
19 **purpurascens** (Pack., 1863)
  los (Stkr., 1893)
  perdita (Dyar, 1893)
20 **quadriguttatus** (Grt., 1864)
  semiauratus N. & D., 1893
21 **thule** Stkr., 1875
22 **auratus** Grt., 1878

HEPIALUS F., 1775
  HEPIOLUS Illiger, 1801, emend.
  EPIALUS Led., 1852, emend.
23 **hyperboreus** Mösch., 1862
23.1 **sciophanes** Fgn., 1979
24 **confusus** Hy. Edw., 1884
25 **roseicaput** N. & D., 1893
  a. **mutatus** B. & Benj., 1925
  b. **demutatus** B. & Benj., 1925
26 **pulcher** Grt., 1864
27 **mcglashani** Hy. Edw., 1896
28 **mathewi** Hy. Edw., 1874
29 **novigannus** B. & Benj., 1925
  a. **mackiei** B. & Benj., 1925
30 **mustelinus** Pack., 1864
  labradoriensis Pack., 1864
31 **gracilis** Grt., 1864
  furcatus Grt., 1883
32 **lembertii** Dyar, 1894
33 **behrensi** (Stretch, 1872)
  mendocinolus Behrens, 1876
  a. **sequoiolus** Behrens, 1876
  b. **tacomae** Hy. Edw., 1874
34 **montanus** (Stretch, 1872)
  desolatus Stkr., 1875
  baroni Behrens, 1876
  anceps Hy. Edw., 1881
  rectus Hy. Edw., 1881
35 **californicus** Bdv., 1868
36 **hectoides** Bdv., 1868
  modestus Hy. Edw., 1873
  inutilis Hy. Edw., 1881
  a. **lenzi** Behrens, 1876
  sangaris Stkr., 1878

MONOTRYSIA

Nepticuloidea

NEPTICULIDAE
by DONALD R. DAVIS & CHRISTOPHER
  WILKINSON

OBRUSSA Braun, 1915
  ETAINIA Beirne, 1945
37 **ochrefasciella** (Cham., 1873)
38 **sericopeza** (Zell., 1839)

ECTOEDEMIA Bsk., 1907
  DECHTIRIA Beirne, 1945
39 **argyropeza** (Zell., 1839), extralim.
  a. **downesi** Wilkinson & Scoble,
    1979
40 **canutus** Wilkinson & Scoble, 1979
41 **trinotata** (Braun, 1914)
42 **marmaropa** (Braun, 1925), n. comb.
43 **platanella** (Clem., 1861)
  maximella (Cham., 1873)
44 **clemensella** (Cham., 1873)
45 **similella** (Braun, 1917), n. comb.
46 **virgulae** (Braun, 1927), n. comb.
47 **lindquisti** (Free., 1962)
48 **rubifoliella** (Clem., 1860)
49 **ulmella** (Braun, 1912)
50 **nyssaefoliella** (Cham., 1880), n.
  comb.
51 **quadrinotata** (Braun, 1917)
52 **populella** Bsk., 1907
53 **obrutella** (Zell., 1873)
  bosquella (Cham., 1878)
54 **chlorantis** Meyr., 1908
55 **heinrichi** Bsk., 1914
56 **castaneae** Bsk., 1913
57 **phleophaga** Bsk., 1914
58 **mesoloba** Davis, 1978

GLAUCOLEPIS Braun, 1917
59 **saccharella** (Braun, 1912)

FOMORIA Beirne, 1945
60 **pteliaeella** (Cham., 1880), n. comb.
61 **hypericella** (Braun, 1925), n. comb.

STIGMELLA Schr., 1802
  NEPTICULA Heyden, 1843
62 **crataegifoliella** (Clem., 1861)
63 **scintillans** (Braun, 1917)
64 **pomivorella** (Pack., 1870)
65 **chalybeia** (Braun, 1914)
66 **scinanella** Wilkinson & Scoble, 1979
67 **purpuratella** (Braun, 1917), n. comb.
68 **stigmaciella** Wilkinson & Scoble,
  1979
69 **taeniola** (Braun, 1925), n. comb.
70 **prunifoliella** (Clem., 1861), rev. stat.
  bifasciella (Clem., 1862), n. syn.
  serotinaeella (Cham., 1873)
71 **ceanothi** (Braun, 1910), n. comb.
72 **intermedia** (Braun, 1917)
73 **rhoifoliella** (Braun, 1912), n. comb.
74 **rhamnicola** (Braun, 1916), n. comb.
  rhamnella; (Braun, 1912), not H.-S.,
  1860

75 **diffasciae** (Braun, 1910), n. comb.
76 **gossypii** (Fbs. & Leonard, 1930), n. comb.
77 **cerea** (Braun, 1917), n. comb.
78 **rosaefoliella** (Clem., 1861)
  a. pectocatena Wilkinson & Scoble, 1979
79 **slingerlandella** (Kft., 1908)
80 **villosella** (Clem., 1861), n. comb.
  dallasiana (F. & B., 1876)
81 **apicialbella** (Cham., 1873), n. comb.
  leucostigma (Braun, 1912)
82 **fuscotibiella** (Clem., 1862)
  ciliaefuscella (Cham., 1873)
  furcotibiella Riley, 1891
  discolorella (Braun, 1912)
83 **canadensis** (Braun, 1914), ident. uncert.
84 **populetorum** (F. & B., 1878)
85 **aromella** Wilkinson & Scoble, 1979
86 **pallida** (Braun, 1912), n. comb.
87 **saginella** (Clem., 1861)
  quercicastanella (Cham., 1873)
  fuscocapitella (Cham., 1873)
88 **latifasciella** (Cham., 1878)
  macrocarpae (Free., 1967)
89 **nigriverticella** (Cham., 1875), n. comb.
  maculosella (Cham., 1880)
90 **castaneaefoliella** (Cham., 1875)
91 **flavipedella** (Braun, 1914), n. comb.
92 **corylifoliella** (Clem., 1861)
  virginiella (Clem., 1861)
  minimella (Cham., 1873)
  opulifoliella (Braun, 1914)
  paludicola (Braun, 1917)
  exasperata (Braun, 1930)
93 **ostryaefoliella** (Clem., 1861)
  caryaefoliella (Clem., 1861)
  obscurella (Braun, 1912)
94 **myricafoliella** (Bsk., 1900)
95 **juglandifoliella** (Clem., 1861)
96 **unifasciella** (Cham., 1875), n. comb.
97 **condaliafoliella** (Bsk., 1900), n. comb.
98 **tiliella** (Braun, 1912), n. comb.
99 **quercipulchella** (Cham., 1878)
  terminella (Braun, 1914)
100 **variella** (Braun, 1910)
101 **altella** (Braun, 1914), n. comb.
102 **procrastinella** (Braun, 1927)
103 **alba** Wilkinson & Scoble, 1979
104 **braunella** (W. Jones, 1933), n. comb.
105 **argentifasciella** (Braun, 1912), n. comb.
106 **amelanchierella** (Clem., 1861), n. comb., ident. uncert.
107 **anguinella** (Clem., 1861), n. comb., ident. uncert.
108 **belfragrella** (Cham., 1875), n. comb., ident. uncert.
  belfrageella (Braun, 1917), missp.
109 **grandisella** (Cham., 1880), n. comb., ident. uncert.
110 **platea** (Clem., 1861), n. comb., ident. uncert.
111 **resplendensella** (Cham., 1875), n. comb., ident. uncert.

MICROCALYPTRIS Braun, 1925
112 **thoracealbella** (Cham., 1873)
  badiocapitella (Cham., 1876)
113 **bircornutus** Davis, 1978
114 **punctulatus** (Braun, 1910), n. comb.

115 **tenuijuxtus** Davis, 1978
116 **scirpi** Braun, 1925

MANONEURA Davis, 1979
  OLIGONEURA Davis, 1978, preocc. by Bigot, 1878
117 **basidactyla** (Davis, 1978)

ARTAVERSALA Davis, 1978
118 **gilvafascia** Davis, 1978

OPOSTEGIDAE

by DONALD R. DAVIS

OPOSTEGA Zell., 1839
119 **cretea** Meyr., 1920
120 **bistrigulella** Braun, 1918
121 **albogaleriella** Clem., 1862
  napaeella Clem., 1872, n. syn.
122 **quadristrigella** Cham., 1875
  accessoriella F. & B., 1876
123 **nonstrigella** Cham., 1881, ident. uncert.
124 **kempella** Eyer, 1967
125 **scioterma** Meyr., 1920

TISCHERIIDAE

by DONALD R. DAVIS

TISCHERIA Zell., 1839
  EVEXIA Gistl, 1847
  PHILODOXA Gistl, 1848
  COPTOTRICHE Wlsm., 1890
  TISHERIA Bsk., 1903, missp.
126 **citrinipennella** Clem., 1859
  quercivorella Cham., 1877
127 **mediostriata** Braun, 1927
128 **consanguinea** Braun, 1972
129 **badiiella** Cham., 1875
  bodicella Cham., 1875, missp.
  nubila Braun, 1920
130 **lucida** Braun, 1972
131 **distincta** Braun, 1972
132 **subnubila** Braun, 1972
133 **concolor** Zell., 1875
134 **simulata** Braun, 1972
135 **purinosella** Cham., 1875
  pruinosella Cham., 1878, missp.
  albostraminea Wlsm., 1907
136 **discreta** Braun, 1972
137 **arizonica** Braun, 1972
138 **clemensella** Cham., 1878
  bicolor F. & B., 1878
139 **fuscomarginella** Cham., 1875
140 **castaneaeella** Cham., 1875
  castanella Wlsm., 1891, emend.
  cinerotunicella Braun, 1927
141 **perplexa** Braun, 1972
142 **sulphurea** F. & B., 1878
143 **zelleriella** Clem., 1859
  complanoides F. & B., 1873
  latipennella Cham., 1878
144 **quercitella** Clem., 1863
  tinctoriella Cham., 1875
145 **malifoliella** Clem., 1860
146 **crataegifoliae** Braun, 1972
147 **roseticola** F. & B., 1873
148 **agrimoniella** Braun, 1972
149 **aenea** F. & B., 1873

150 **splendida** Braun, 1972
151 **insolita** Braun, 1972
152 **confusa** Braun, 1972
153 **inexpectata** Braun, 1972
154 **amelanchieris** Braun, 1972
155 **admirabilis** Braun, 1925
156 **solidagonifoliella** Clem., 1859
  solidaginifoliella Wlsm., 1890, missp.
157 **astericola** Braun, 1972
158 **occidentalis** Braun, 1972
159 **heliopsisella** Cham., 1875
  nolckenii F. & B., 1876
160 **ambrosiaeella** Cham., 1875
  ambrosiella Wlsm., missp.
161 **helianthi** F. & B., 1878
162 **gregaria** Braun, 1972
163 **marginata** Braun, 1972
164 **heteroterae** F. & B., 1878
165 **longeciliata** F. & B., 1878
166 **pallidipennella** Braun, 1972
167 **ceanothi** Wlsm., 1890
168 **immaculata** Braun, 1915
169 **ambigua** Braun, 1915
170 **bifurcata** Braun, 1915
171 **omissa** Braun, 1927
172 **explosa** Braun, 1923
173 **pulvella** Cham., 1878

Incurvarioidea

INCURVARIIDAE

by DONALD R. DAVIS

INCURVARIINAE

INCURVARIA Haw., 1828
174 **vetulella** (Zett., 1840)
  pallidulella H.-S., 1851
  triangulifera Tr., 1869

LAMPRONIA Steph., 1829
175 **russatella** (Clem., 1860)
  tripunctella Wlsm., 1880
176 **oregonella** (Wlsm., 1880)
  gilletella (Bsk., 1915), n. syn.
177 **capitella** (Cl., 1759)
  flavimitrella (Hbn., 1814–17)
178 **taylorella** (Kft., 1907)
  quieta Braun, 1921, n. syn.
179 **trimaculella** (Dup., 1838)
  mesopilella H.-S., 1851
  quadrimaculella (Höfner, 1900)
180 **rubiella** (Bjerkander, 1781)
  variella (F., 1794)
  corticella (Haw., 1828)
  multipunctella Dup., 1838

PARACLEMENSIA Bsk., 1904
  BRACKENRIDGIA Bsk., 1903, preocc. by Eigenmann & Ulrich, 1902
181 **acerifoliella** (Fitch, 1854)
  luteiceps (Wlk., 1863)
  iridella (Cham., 1873)

VESPINA Davis, 1972
  CAREOSPINA Davis, 1972, preocc. by Peters, 1971
182 **quercivora** (Davis, 1972)

PHYLLOPORIA Heinemann, 1870
183 **bistrigella** (Haw., 1828)
    abalienella (Zett., 1840)
    subammanella (Staint., 1849)
    dilorella (H.-S., 1851)
    labradorella (Clem., 1864), n. syn.
    labradoriella (Wlsm., 1888), missp.
    aureovireus (Dietz, 1905), n. syn.
    aureovirens (B. & McD., 1917),
      missp.

TANYSACCUS Davis, 1978
184 **aenescens** (Wlsm., 1888)
185 **sublustris** (Braun, 1925)
186 **humulis** (Wlsm., 1888)

GREYA Bsk., 1903
    GRAYA Dietz, 1905, missp.
187 **obscuromaculata** (Braun, 1921), n.
    comb.
    augustella Blkmre., 1926, n. syn.
188 **sparsipunctella** (Wlsm., 1907), n.
    comb.
189 **punctiferella** (Wlsm., 1888)
    piperella (Bsk., 1904)
    speculella Blkmre., 1926, n. syn.
190 **variata** (Braun, 1921), n. comb.
191 **subalba** (Braun, 1921)
192 **solenobiella** (Wlsm., 1880)
193 **reticulata** (Riley, 1892), n. comb.
194 **politella** (Wlsm., 1888), n. comb.

TRIDENTAFORMA Davis, 1978
195 **fuscoleuca** (Braun, 1923)

PRODOXINAE

TEGETICULA Zell., 1873
    PRONUBA Riley, 1872, preocc. by
      Thomson, 1860
    PROMIBA Kby., 1874, missp.
    THIA Hy. Edw., 1888, preocc. by
      Leach, 1815
    THELETHIA Dyar, 1893, repl. name
    VALENTINIA Coolidge, 1909,
      preocc. by Wlsm., 1907
196 **synthetica** (Riley, 1892)
    paradoxa (Trelease, 1893)
197 **maculata** (Riley, 1881)
    a. **apicella** (Dyar, 1903)
    b. **extranea** (Hy. Edw., 1888)
    aterrima (Trelease, 1893)
198 **yuccasella** (Riley, 1872)
    alba Zell., 1873
    yuccaella (Boll, 1876), missp.
    intermedia (Riley, 1881)
    mexicana Bastida, 1962

PARATEGETICULA Davis, 1967
199 **pollenifera** Davis, 1967

PRODOXUS Riley, 1880
200 **quinquepunctella** (Cham., 1875)
    paradoxica Cham., 1878
    decipiens Riley, 1880
201 **y-inversum** Riley, 1892
202 **coloradensis** Riley, 1892
    lautus Ckll., 1897
    confluens Ckll., 1897
    rheumapterella (Dietz, 1905)

203 **ochrocarus** Davis, 1967
204 **sordidus** Riley, 1892
205 **marginatus** Riley, 1881
206 **pulverulentus** Riley, 1892
207 **cinereus** Riley, 1881
208 **aenescens** Riley, 1881

MESEPIOLA Davis, 1967
209 **specca** Davis, 1967

AGAVENEMA Davis, 1967
210 **barberella** (Bsk., 1915)
211 **pallida** Davis, 1967

ADELINAE

CHALCEOPLA Braun, 1921
    CYANAUGES Braun, 1919, preocc.
      by Philippi, 1865
212 **dietziella** (Kft., 1908)
213 **cyanella** (Bsk., 1915)
214 **sedella** (Bsk., 1915)
215 **cockerelli** (Bsk., 1915)
216 **simpliciella** (Wlsm., 1880)
    itoniella (Bsk., 1915), n. syn.
    ovata Braun, 1921, n. syn.
217 **discalis** Braun, 1925

NEMOPHORA Illiger &
    Hoffmannsegg, 1798
    ELASMION Hbn., 1822
    NEMOTOIS Hbn., 1825
    EPITYPHIA Hbn., 1825
    EUTYPHIA Hbn., 1825
    UCETIA Wlk., 1866
218 **bellela** (Wlk., 1863)

ADELA Latr., 1796
    CAPILLARIA Haw., 1828
    METALLITIS Sodoffsky, 1837
    CAUCHAS Zell., 1839
    AEDILIS Gistl, 1848
    DICTE Cham., 1873
219 **punctiferella** Wlsm., 1870
220 **singulella** Wlsm., 1880
221 **septentrionella** Wlsm., 1880
    septentrionelis McD., 1927, missp.
222 **oplerella** Powell, 1969
223 **thorpella** Powell, 1969
224 **flammeusella** Cham., 1876
    flamensella Cham., 1878, missp.
    lactimaculella Wlsm., 1880
    flammensella Kft., 1903, missp.
225 **trigrapha** Zell., 1876
    trifasciella Cham., 1876
    fasciella Cham., 1876
226 **eldorada** Powell, 1969
227 **caeruleella** Wlk., 1863
    bella Cham., 1873, n. syn.
    chalybeis Zell., 1873
    iochora Zell., 1877
    iochroa Wlsm., 1880, missp.
    aeruginosella Wlsm., 1890, n. syn.
228 **ridingsella** Clem., 1864
    corruscifasciella (Cham., 1873)
    schlaegeri Zell., 1873
    coruscifasciella Wlsm., 1880, missp.
229 **purpurea** Wlk., 1863
    biviella Zell., 1873
    purpura auth., missp.

HELIOZELIDAE
by DONALD R. DAVIS

HELIOZELA H.-S., 1853
230 **aesella** Cham., 1877
231 **gracilis** Zell., 1873

ANTISPILA Hbn., 1825
232 **cornifoliella** Clem., 1860
233 **freemani** Lafontaine, 1973
234 **nysaefoliella** Clem., 1860
    nyssaefoliella Cham., 1878, missp.
235 **eugeniella** Bsk., 1900
236 **isabella** Clem., 1860
    issabella Cham., 1878, missp.
237 **viticordifoliella** Clem., 1860
238 **major** Kft., 1907
239 **aurirubra** Braun, 1915
240 **ampelopsifoliella** Cham., 1874
    ampelopsiella Cham., 1874, missp.
    ampelopsisella Cham., 1874, missp.
241 **voraginella** Braun, 1927
242 **argentifera** Braun, 1927
243 **hydrangaeella** Cham., 1874
    hydrangiaeella Cham., 1878, missp.

COPTODISCA Wlsm., 1895
    ASPIDISCA Clem., 1860, preocc.
      by Ehrenberg, 1830
244 **diospyriella** (Cham., 1874)
245 **condaliae** Bsk., 1900
246 **ella** (Cham., 1871)
247 **lucifluella** (Clem., 1860)
248 **juglandiella** (Cham., 1874)
249 **magnella** Braun, 1916
250 **matheri** Lafontaine, 1974
251 **negligens** Braun, 1920
252 **ostryaefoliella** (Clem., 1861)
253 **saliciella** (Clem., 1861)
254 **splendoriferella** (Clem., 1860)
    pruniella (Clem., 1861)
    saccatella (Pack., 1889)
255 **arbutiella** Bsk., 1904
256 **kalmiella** Dietz, 1921
257 **ribesella** Braun, 1925
258 **cercocarpella** Braun, 1925
259 **quercicolella** Braun, 1927
    querciella Opler, 1971, missp.
260 **powellella** Opler, 1971

DITRYSIA

Tineoidea

TINEIDAE
by DONALD R. DAVIS

NEMAPOGON Schr., 1802
   DIAPHTHIRUSA Hbn., 1825
261 **acapnopennella** (Clem., 1863), n. comb.
   minutipulvella (Cham., 1875)
   minutipulnella (Dietz, 1905), missp.
262 **angulifasciella** (Dietz, 1905), n. comb.
263 **auropulvella** (Cham., 1873), n. comb.
   auripulvella (Dietz, 1905), missp.
264 **defectella** (Zell., 1873)
265 **geniculatella** (Dietz, 1905), n. comb.
266 **granella** (L., 1758)
   costotristrigella (Cham., 1873)
   fuscomaculella (Cham., 1873), n. syn.
   marmorella (Cham., 1875), n. syn.
   costistrigella (Dietz, 1905), missp.
   fascomaculella (Dietz, 1905), missp.
   nigroatomella (Dietz, 1905), n. syn.
267 **interstitiella** (Dietz, 1905), n. comb.
268 **molybdanella** (Dietz, 1905)
269 **multistriatella** (Dietz, 1905), n. comb.
270 **ophrionella** (Dietz, 1905), n. comb.
271 **oregonella** (Bsk., 1900)
272 **rileyi** (Dietz, 1905), n. comb.
   atriflua (Meyr., 1919), n. comb., n. syn.
273 **roburella** (Dietz, 1905), n. comb.
274 **tylodes** (Meyr., 1919), n. comb.
275 **variatella** (Clem., 1859), n. comb.
   apicisignatella (Dietz, 1905), n. syn.
   fulvisuffusella (Dietz, 1905), n. syn.
   personella (Pierce & Metcalfe, 1934), n. syn.

DOLEROMORPHA Braun, 1930
276 **porphyria** Braun, 1930

EUDARCIA Clem., 1860
   MEESSIA Hofmann, 1898, n. syn.
277 **eunitariaeella** Cham., 1873
   eunitaraeella Cham., 1873, missp.
   coemeteriiella Cham., 1873, emend.
   cunitariaeella (Rye, 1875), missp.
   eunitariella Cham., 1876, missp.
   coemitariella Cham., 1876, emend.
   imitatorella (Cham., 1876), n. syn.
   coemetarioeella Wlsm., 1882, missp.
   coemetariaeella Wlsm., 1882, missp.
   coemitariella Dietz, 1905, missp.
278 **simulatricella** Clem., 1860

DIACHORISIA Clem., 1860
279 **velatella** Clem., 1860
   fuscopulvella (Cham., 1873), n. syn.
   maculabella (Cham., 1873), n. syn.

BATHROXENA Meyr., 1919
   PELATES Dietz, 1905, preocc. by Cuvier & Valenciennes, 1829
280 **heteropalpella** (Dietz, 1905)

AUGOLYCHNA Meyr., 1922
281 **septemstrigella** (Cham., 1878)

LEUCOMELE Dietz, 1905
282 **miriamella** Dietz, 1905

OENOE Cham., 1874
283 **hybromella** Cham., 1874

HOMOSETIA Clem., 1863
   PITYS Cham., 1873
   PITYO Cham., 1876, missp.
   SEMELE Cham., 1875
   CALOSTINEA Dietz, 1905, n. syn.
284 **argentinotella** (Cham., 1876)
285 **argentistrigella** (Cham., 1873), ident. uncert.
286 **auricristatella** (Cham., 1873)
   aurocristatella Dietz, 1905, missp.
287 **bifasciella** (Cham., 1876)
   obscurella Dietz, 1905, n. syn.
288 **chrysoadspersella** Dietz, 1905
289 **costisignella** (Clem., 1863)
290 **cristatella** (Cham., 1875)
291 **fasciella** (Cham., 1873)
292 **fuscocristatella** (Cham., 1873), ident. uncert.
293 **marginimaculella** (Cham., 1875)
   maculimarginella (Dietz, 1905), missp.
   maculatella Dietz, 1905, n. syn.
294 **miscecristatella** (Cham., 1873)
295 **tricingulatella** (Clem., 1863)

STENOPTINEA Dietz, 1905, n. stat.
   CELESTICA Meyr., 1917, n. syn.
296 **auriferella** (Dietz, 1905), n. comb.
297 **ornatella** (Dietz, 1905), n. comb.

ISOCORYPHA Dietz, 1905
298 **chrysocomella** Dietz, 1905
299 **mediostriatella** (Clem., 1865)
   auristrigella (Cham., 1873)

HYBROMA Clem., 1862
300 **servulella** Clem., 1862
   aurosuffusella (Cham., 1873)

HOMOSTINEA Dietz, 1905
   HOMOTINEA Meyr., 1932, emend.
301 **curviliniella** Dietz, 1905

POMPOSTOLELLA Fletcher, 1940
   POMPOSTOLA Meyr., 1927, preocc. by Hbn., 1819
302 **charipepla** (Meyr., 1927)

ERECHTHIAS Meyr., 1880
   ERECHTIAS Ghesquière, 1940, missp.
   TINEXOTAXA Gozmány, 1968, n. syn.
303 **zebrina** (Butler, 1881)
   lanceolata (Wlsm., 1897)
   xenica (Meyr., 1911)
   travestita Gozmány, 1968, n. syn.

LEPIDOBREGMA Zimmerman, 1978
304 **minuscula** (Wlsm., 1897)

MEA Bsk., 1906
   PROGONA Dietz, 1905, preocc. by Berg, 1882
305 **bipunctella** (Dietz, 1905), mispl.
306 **skinnerella** (Dietz, 1905)
   floridella (Dietz, 1905), n. syn.

CHOROPLECA Durrant, 1914
   CYANE Cham., 1873, preocc. by Felder, 1861
   DIACHALASTIS Meyr., 1920
307 **vesaliella** (Cham., 1873)

SCARDIA Tr., 1830
   AGARICA Sodoffsky, 1837
   MOROPHAGA H.-S., 1854
   ATABYRIA Snell., 1884
   OSPHRETICA Meyr., 1910
   MICROSCARDIA Amsel, 1952
308 **approximatella** Dietz, 1905
309 **cryptophori** (Clarke, 1940), n. comb.
310 **fuscofasciella** (Cham., 1875)
   pravatella Bsk., 1908, n. syn.

FERNALDIA Grt., 1881
311 **anatomella** Grt., 1881
   fiskeella (Bsk., 1908), n. syn.
312 **coloradella** (Dietz, 1905), n. comb.

MOROPHAGOIDES Petersen, 1957
313 **berkeleyella** (Powell, 1967), n. comb.
314 **burkerella** (Bsk., 1903), n. comb.
   gracilis (Wlsm., 1907), n. comb., n. syn.
   caryophyllella (Bsk., 1908), n. comb., n. syn.
   errandella (Bsk., 1908), n. comb., n. syn.

DIATAGA Wlsm., 1914
315 **leptosceles** Wlsm., 1914

XYLESTHIA Clem., 1859
   XYLESTIA Dyar, 1903, missp.
316 **albicans** Braun, 1923
317 **pruniramiella** Clem., 1859
   clemensella Cham., 1873
   congeminatella Zell., 1873
   kearfottella Dietz, 1905

PHRYGANEOPSIS Wlsm., 1881
   SETONELLA McD., 1927
318 **brunnea** Wlsm., 1881
   buscki (McD., 1927)

KEARFOTTIA Fern., 1904
319 **albifasciella** Fern., 1904

HYPOPLESIA Bsk., 1906
   PARAPLESIA Dietz, 1905, preocc. by Felder, 1862
320 **busckiella** (Dietz, 1905)
   respersa (Meyr., 1919), n. comb., n. syn.
321 **dietziella** Bsk., 1913

DYOTOPASTA Bsk., 1907
   DYSTOPASTA McD., 1939, emend.
   PSEUDOXYLESTHIA Wlsm., 1907

322 **yumaella** (Kft., 1907)
   angustella (Wlsm., 1907)

TENAGA Clem., 1862
   LICHENOVORA Petersen, 1957, n.
   syn.
323 **pomiliella** Clem., 1862

CEPHITINEA Zagulajev, 1964
324 **obscurostrigella** (Cham., 1874), n.
   comb.
   longinella Zagulajev, 1964, n. syn.

EPISCARDIA Ragonot, 1895
   HAPLOTINEA Diakonoff &
   Hinton, 1956
325 **insectella** (F., 1794)
   misella (Zell., 1839)

APRETA Dietz, 1905
   EPICHAETA Dietz, 1905, n. syn.
326 **paradoxella** Dietz, 1905
   nepotella (Dietz, 1905), n. comb., n.
   syn.

AMYDRIA Clem., 1859
   CASAPE Wlk., 1864
   AMADRIA Cham., 1875, missp.
   AMADRYA Cham., 1875, missp.
   AMYDRYA Kft., 1903, missp.
327 **apachella** Dietz, 1905
328 **arizonella** Dietz, 1905
329 **brevipennella** Dietz, 1905
330 **clemensella** Cham., 1874, ident.
   uncert.
331 **confusella** (Dietz, 1905), mispl.
332 **curvistrigella** (Dietz, 1905), mispl.
   pandurella (Dietz, 1905), n. syn.
333 **dyarella** Dietz, 1905
334 **effrentella** Clem., 1859
   effrenatella Staint., 1872, emend.
   coloradella Dietz, 1905, n. syn.
335 **margoriella** Dietz, 1905
   marjorieella Dietz, 1905, missp.
   margorieella Dietz, 1905, missp.
   marjoriella Bsk., 1906, missp.
336 **obliquella** Dietz, 1905
337 **onagella** (Dietz, 1905), n. comb.
   a. **occidentella** (Dietz, 1905),
   ident. uncert.

ACROLOPHUS Poey, 1832
   ANAPHORA Clem., 1859
   DERCHIS Wlk., 1863
   HIBITA Wlk., 1863
   TIRASIA Wlk., 1863
   EDDARA Wlk., 1863, preocc. by
   Wlk., 1858
   BAZIRA Wlk., 1864
   PHLONGIA Wlk., 1864
   URBARA Wlk., 1864
   TACHASARA Wlk., 1865
   EUTHECA Grt., 1881, preocc. by
   Kiesenwetter, 1877
   EULEPISTE Wlsm., 1882
   PSEUDOCONCHYLIS Wlsm.,
   1884
   HOMONYMUS Wlsm., 1887
   NEOLOPHUS Wlsm., 1887
   HYPOCLOPUS Wlsm., 1887

   THYSANOSCELIS Wlsm., 1887
   THYSANOSKELIS Wlsm., 1887,
   missp.
   ANKISTROPHORUS Wlsm., 1887
   CAENOGENES Wlsm., 1887
   FELDERIA Wlsm., 1887
   ORTHOLOPHUS Wlsm., 1887
   PSEUDANAPHORA Wlsm., 1887
   SAPINELLA Kby., 1892
   ATOPOCERA Wlsm., 1897
   PILANOPHORA Wlsm., 1897
   THYSANOSEDES Druce, 1901,
   missp.
   HYPOCOLPUS Dyar, 1903, missp.
   ANOPHORA Kft., 1903, missp.
   APOCLISIS Wlsm., 1914
   ANAPHORINA Strand, 1932
338 **acanthogonus** Meyr., 1919
339 **acornus** Hasbrouck, 1964
340 **arcanella** (Clem., 1859)
341 **arizonellus** Wlsm., 1887
342 **baldufi** Hasbrouck, 1964
343 **bicornutus** Hasbrouck, 1964
344 **chiricahuae** Hasbrouck, 1964
345 **cockerelli** (Dyar, 1900)
346 **crescentella** (Kft., 1907)
347 **cressoni** (Wlsm., 1882)
348 **davisellus** Beutenmüller, 1887
349 **dorsimaculus** (Dyar, 1900)
350 **exaphristus** Meyr., 1919
351 **fervidus** Bsk., 1912
   antonellus (B. & McD., 1913)
352 **filicornus** (Wlsm., 1887)
   mexicanellus Beutenmüller, 1888
353 **forbesi** Hasbrouck, 1964
354 **furcatus** (Wlsm., 1887)
355 **griseus** (Wlsm., 1887)
   leucallactis Meyr., 1919
   a. **capitatus** Hasbrouck, 1964
356 **juxtatus** Hasbrouck, 1964
357 **kearfotti** (Dyar, 1903)
   diversus Bsk., 1912
358 **klotsi** Hasbrouck, 1964
359 **laticapitanus** (Wlsm., 1884)
   unistriganus (Dyar, 1903)
   a. **clarkei** Hasbrouck, 1964
   b. **heinrichi** Hasbrouck, 1964
   c. **leopardus** Bsk., 1910
   d. **occidens** Bsk., 1910
   flavicomus Bsk., 1912
360 **leucodocis** (Zell., 1877)
   medioliniella (Kft., 1907), n. syn.
361 **luriei** Hasbrouck, 1964
362 **macrogaster** (Wlsm., 1887)
   a. **bipectinicornus** Hasbrouck, 1964
   b. **laminicornus** Hasbrouck, 1964
   c. **unipectinicornus** Hasbrouck,
   1964
363 **macrophallus** Hasbrouck, 1964
364 **maculifer** (Wlsm., 1887)
365 **minor** (Dyar, 1903)
   coloradellus (Wlsm., 1907)
366 **mortipennella** (Grt., 1872)
   quadripunctellus (Dyar, 1900)
   carphologus Meyr., 1919
   zeela (Hasbrouck, 1964), unavail.,
   publ. in syn.
367 **morus** (Grt., 1881)
368 **panamae** Bsk., 1914
369 **parvipalpus** Hasbrouck, 1964

370 **persimplex** (Dyar, 1900)
371 **piger** (Dyar, 1900)
372 **plumifrontella** (Clem., 1859)
   bombycinus (Zell., 1873)
   cervinus Wlsm., 1887
   angustipennellus Beutenmüller, 1887
373 **popeanella** (Clem., 1859)
   agrotipennella (Grt., 1872)
   scardinus (Zell., 1873)
   morrisoni (Wlsm., 1887)
   confusellus (Dyar, 1900)
374 **propinquus** (Wlsm., 1887)
   tenuis (Wlsm., 1887)
   violaceellus Beutenmüller, 1887
   busckella (Haim., 1915)
375 **pseudohirsutus** Hasbrouck, 1964
   hirsutus Bsk., 1912, preocc. by
   Wlsm., 1887
376 **punctellus** (Bsk., 1907)
377 **pyramellus** (B. & McD., 1913)
378 **quadrellus** (B. & McD., 1913)
379 **seculatus** Hasbrouck, 1964
380 **serratus** Hasbrouck, 1964
381 **simulatus** Wlsm., 1882
382 **sinclairi** Hasbrouck, 1964
   a. **nelsoni** Hasbrouck, 1964
383 **texanella** (Cham., 1878)
   hulstellus Beutenmüller, 1887
   barnesii (Dyar, 1900)
384 **vanduzeei** Hasbrouck, 1964
385 **variabilis** (Wlsm., 1887)
386 **vauriei** Hasbrouck, 1964

DORATA Bsk., 1904
   DOROTA Kft., 1907, missp.
   PTEROLONCHE Wlsm., 1889,
   preocc. by Zell., 1847
387 **atomophora** Meyr., 1928
388 **inornatella** Bsk., 1904
   inorratella Kft., 1907, missp.
389 **lineata** (Wlsm., 1889)
   virgatella (Bsk., 1904)

PHEREOECA Hinton & Bradley,
   1956
390 **walsinghami** (Bsk., 1933)

TRYPTODEMA Dietz, 1905
391 **sepulchrella** Dietz, 1905

TINEA L., 1758
   AUTOSES Hbn., 1825
   ACEDES Hbn., 1825
   EDOSA Wlk., 1866
   CHRYSORYCTIS Meyr, 1886
   DYSTINEA Börner, 1925
392 **apicimaculella** Cham., 1875
393 **behrensella** Cham., 1875, ident. uncert.
394 **carnariella** Clem., 1859
395 **columbariella** Wocke, 1877
396 **croceoverticella** Cham., 1876
397 **dubiella** Staint., 1859
   turicencis Müller-Rutz, 1920
   bispinella Zagulajev, 1960
   tenerifi Zagulajev, 1966
398 **grumella** Zell., 1873
399 **irrepta** Braun, 1926
400 **mandarinella** Dietz, 1905
   bimaculella Cham., 1873, preocc. by
   Thunb., 1794, n. syn.

401 **misceella** Cham., 1873., ident. uncert.
402 **niveocapitella** Cham., 1875
   leucocapitella Bsk., 1903, n. syn.
403 **occidentella** Cham., 1880
   tuscanella Dietz, 1905, n. syn.
404 **pallescentella** Staint., 1851
   nigrifoldella Gregson, 1856
   horosema Meyr., 1931
   stimulatrix Meyr., 1931
405 **pellionella** L., 1758
   pelliomella Dietz, 1905, missp.
   flavifrontella; auth., larva
406 **straminella** Cham., 1873, ident. uncert.
407 **thoracestrigella** Cham., 1876, ident. uncert.
408 **translucens** Meyr., 1917
   metonella Pierce & Metcalf, 1934
   leonhardi Petersen, 1957
   margaritacea Gozmány, 1967
   fortificata Gozmány, 1968
409 **unomaculella** Cham., 1875
410 **xanthostictella** Dietz, 1905

NIDITINEA Petersen, 1957
   TINEIDIA Zagulajev, 1960
411 **spretella** (D. & S., 1775)
   fuscipunctella (Haw., 1828)
   nubilipennella (Clem., 1859), n. comb.
   ignotella (Wlk., 1864)
   frigidella (Pack., 1867)
   griseella (Cham., 1873), n. comb., n. syn.
412 **orleansella** (Cham., 1873), n. comb.

TRICHOPHAGA Rag., 1894
413 **tapetzella** (L., 1758)
   tapetiella (Zell., 1852), missp.
   palaestrica Butler, 1877

CERATOPHAGA Petersen, 1957
414 **vicinella** (Dietz, 1905), n. comb.

MONOPIS Hbn., 1825
   BLABOPHANES Zell., 1852
   HYALOSPILA H.-S., 1853
415 **crocicapitella** (Clem., 1859)
   hyalinella (Stgr., 1870)
   lombardica (Hering, 1889)
   heringi (Richardson, 1893)
416 **dorsistrigella** (Clem., 1859)
   subjunctella (Wlk., 1863)
417 **marginistrigella** (Cham., 1873)
   irrorella Dietz, 1905, n. syn.
418 **monachella** (Hbn., 1796)
   longella (Wlk., 1863)
419 **mycetophilella** Powell, 1967
420 **rusticella** (Hbn., 1796)
   saturella (Haw., 1828)
421 **spilotella** Tengström, 1848
   biflavimaculella (Clem., 1859)
   insignisella (Wlk., 1863)
   halospila Meyr., 1919

ECCRITOTHRIX Bradley & Hinton, 1956
422 **trimaculella** (Cham., 1873)

PRAEACEDES Amsel, 1954
   TITAENOSES Hinton & Bradley, 1956

423 **seminolella** (Beutenmüller, 1889), n. comb.
424 **thecophora** (Wlsm., 1908)
   deluccae Amsel, 1954

ELATOBIA H.-S., 1849
   TINEOMINA Stgr., 1892
   ABACOBIA Dietz, 1905
   DIETZIA Bsk., 1906
425 **carbonella** Dietz, 1905. rev. stat.

TINEOLA H.-S., 1853
426 **bisselliella** (Hummel, 1823)
   crinella (Tr., 1832)
   destructor (Steph., 1834)
   flavifrontella; auth., adult
   biselliella (H.-S., 1853), missp.
   lanariella (Clem., 1859)

TIQUADRA Wlk., 1863
   OSCELLA Wlk., 1864
   MANCHANA Wlk., 1866
   VENTIA Wlk., 1866
   ACUREUTA Zell., 1877
427 **inscitella** Wlk., 1863

SETOMORPHA Zell., 1852
   EPILEGIS Dietz, 1905
   APOTOMIA Dietz, 1905
   SEMIOTA Dietz, 1905
   TRISYNTOPA Lower, 1918
428 **rutella** Zell., 1852
   insectella; auth. not F., 1794
   rupicella Zell., 1852
   operosella Zell., 1873
   inamoenella Zell., 1873
   ruderella Zell., 1873
   multimaculella (Cham., 1878)
   dryas (Butler, 1881)
   corticinella Snell., 1885
   bogotatella; Alpheraki, 1889, not Wlk., 1864
   discipunctella Rebel, 1894
   majorella Dietz, 1905
   sigmoidella Dietz, 1905, n. syn.
   cariosella (Dietz, 1905)
   fractiliniella (Dietz, 1905)
   transversestrigella (Dietz, 1905)
   margalaestriata Keuchenius, 1917
   euryspoda (Lower, 1918)
   tineoides Dammerman, 1919
   nitella Voûte, 1935, missp.

LINDERA Blanchard, 1854
   PALPULA; Blanchard, 1854, not Tr., 1832
   SAFRA Wlk. 1864, preocc. by Wlk., 1863
   CHRESOTES Butler, 1881
   PARANEURA Dietz, 1905
   CERVITINEA Amsel, 1954
429 **tessellatella** (Blanchard, 1854)
   variegella (Blanchard, 1854)
   bogotatella Wlk., 1864
   simulella (Dietz, 1905)
   ehrhornella (Dietz, 1905), n. syn.
   cruciferella (Dietz, 1905), n. syn.

PHAEOSES Fbs., 1922
430 **sabinella** Fbs., 1922

OPOGONA Zell., 1853
   LOZOSTOMA Staint., 1859
   CONCHYLIOSPILA Wallgr., 1861
   CACHURA Wlk., 1864
   DENDRONEURA Wlsm., 1892
   HIEROXESTIS Meyr., 1892
   EXALA Meyr., 1912
431 **arizonensis** Davis, 1978
432 **floridensis** Davis, 1978
433 **omoscopa** (Meyr., 1892)
   apicalis Swezey, 1909
   praematura (Meyr., 1909)

OINOPHILA Steph., 1848
434 **v-flavum** (Haw., 1828)
   v-flava, missp.

PSYCHIDAE
by DONALD R. DAVIS

TALEPORIINAE

SOLENOBIA Dup., 1842
435 **walshella** Clem., 1862
436 **triquetrella** (Hbn., 1812)

PSYCHINAE

PSYCHE Schr., 1801
   FUMARIA Haw., 1811
   FUMEA Haw., 1812
437 **casta** (Pallas, 1767)

APTERONA Millière, 1857
438 **helix** (Siebold, 1850)

PROCHALIA B. & McD., 1913
439 **pygmaea** B. & McD., 1913

ZAMOPSYCHE Dyar, 1923
440 **commentella** Dyar, 1923

CRYPTOTHELEA Duncan, 1841
   PLATOECETICUS Pack., 1869
441 **nigrita** (B. & McD., 1913)
442 **gloverii** (Pack., 1869)
   jonesi (B. & Benj., 1922)
   pizote (Schaus, 1927)
   tristis; Vazquez, 1941

ASTALA Davis, 1964
443 **confederata** (G. & R., 1868)
   lepidopteris (Dyar, 1926)
444 **polingi** (B. & Benj., 1924)
445 **edwardsi** (Heylaerts, 1884)
   carbonaria (Pack., 1887)
   edwardsii (Dyar, 1903), missp.

HYALOSCOTES Butler, 1881
   HYALOSCOTUS Gaede, 1936, missp.
446 **coniferella** Hy. Edw., 1877, ident. uncert.
447 **fragmentella** Hy. Edw., 1877, ident. uncert.
448 **fumosa** Butler, 1881
449 **pithopoera** (Dyar, 1923)
   sheppardi Free., 1944

BASICLADUS Davis, 1964
450 tracyi (Jones, 1911)
   cacocnemos (Jones, 1922)
   traceyi (Dyar, 1923), missp.
451 celibatus (Jones, 1922)

COLONEURA Davis, 1964
451.1 fragilis (B. & McD., 1916)

OIKETICINAE

OIKETICUS Guilding, 1827
   OEKETICUS Lefebre, 1842,
   missp.
   OECETICUS Pack., 1864, missp.
452 toumeyi Jones, 1922
   mortonjonesi Vasquez, 1949
453 townsendi Townsend, 1894
   bonniwelli B. & Benj., 1924
   a. dendrokomos Jones, 1926
454 abbotii Grt., 1880
   abboti Holl., 1905, missp.

THYRIDOPTERYX Steph., 1835
   HYMENOPSYCHE Grt., 1865
455 meadii Hy. Edw., 1881
   meadi B. & McD., 1917, missp.
456 alcora Barnes, 1905
457 ephemeraeformis (Haw., 1803)
   coniferarum (Pack., 1864)
   pallidovenata Grossb., 1917
   vernalis Jones, 1923
   INCERTAE SEDIS
458 rileyi Heylaerts, 1884 (Chalia)
459 davidsoni Hy. Edw., 1877
   (Oiketicus)

OCHSENHEIMERIIDAE
by DONALD R. DAVIS

OCHSENHEIMERIA Hbn., 1825
   LEPIDOCERA Curt., 1831
   PHYGAS Tr., 1833
460 vacculella F. v. Röslerstamm, 1842

LYONETIIDAE
by DONALD R. DAVIS

EUPRORA Bsk., 1906
461 argentiliniella Bsk., 1906

PHILONOME Cham., 1874
   PHILLONOME Cham., 1875,
   missp.
   EURYNOME Cham., 1875, preocc.
   by Leach, 1814
   BUSCKIA Dyar, 1903
462 clemensella Cham., 1874
463 luteella (Cham., 1875)
464 albella (Cham., 1877)

BEDELLIA Staint., 1849
465 minor Bsk., 1900
466 somnulentella (Zell., 1847)
   staintoniella Clem., 1860

EULYONETIA Cham., 1880
467 inornatella Cham., 1880

LYONETIA Hbn., 1825
   ARGYROMIGES Curt., 1829
   ARGYROMIS Steph., 1829
468 alniella Cham., 1875
   aliella Kuroko, 1964, missp.
469 candida Braun, 1916
470 latistrigella Wlsm., 1882
471 saliciella Bsk., 1904
472 speculella Clem., 1862
   nidificansella (Pack., 1869)
   apicistrigella Cham., 1875
   gracilella Cham., 1876

ACANTHOCNEMES Cham., 1878
   CACONOME Dyar, 1903
473 fuscoscapulella Cham., 1878

PROLEUCOPTERA Bsk., 1902
474 smilaciella (Bsk., 1900)

PARALEUCOPTERA Heinr., 1918
475 albella (Cham., 1871)
475.1 heinrichi Jones, 1947

CORYTHOPHORA Braun, 1915
476 aurea Braun, 1915

LEUCOPTERA Hbn., 1825
   CEMIOSTOMA Zell., 1848
477 erythrinella Bsk., 1900
478 guettardella Bsk., 1900
479 laburnella (Staint., 1851)
   wailesella (Staint., 1858)
480 pachystimella Bsk., 1904
481 robinella Braun, 1925
482 spartifoliella (Hbn., 1810–13)
   punctaurella (Haw., 1828)

EXEGETIA Braun, 1918
483 crocea Braun, 1918

BUCCULATRIX Zell., 1839
   CEROCLASTIS Zell., 1848
484 fusicola Braun, 1920
485 solidaginiella Braun, 1963
486 montana Braun, 1920
487 magnella Cham., 1875
488 needhami Braun, 1956
489 longula Braun, 1963
490 simulans Braun, 1963
491 niveella Cham., 1875
492 parvinotata Braun, 1963
493 ochritincta Braun, 1963
494 viguierae Braun, 1963
495 micropunctata Braun, 1963
496 inusitata Braun, 1963
497 seneciensis Braun, 1963
498 bicristata Braun, 1963
499 cuneigera Meyr., 1919
   errans Braun, 1920
500 albaciliella Braun, 1910
501 ochristrigella Braun, 1910
502 eurotiella Wlsm., 1907
   chrysothamni Braun, 1925
503 tenebricosa Braun, 1925
504 ericameriae Braun, 1963
505 variabilis Braun, 1910
506 separabilis Braun, 1963
507 brunnescens Braun, 1963
508 evanescens Braun, 1963

509 benenotata Braun, 1963
510 floccosa Braun, 1923
511 flourensiae Braun, 1963
512 franseriae Braun, 1963
513 staintonella Cham., 1878
   pertenuis Braun, 1918
514 immaculatella Cham., 1875
515 agnella Clem., 1860
   capitealbella Cham., 1873
   capitialbella Cham., 1878, missp.
   albicapitella Cham., 1875
516 kimballi Braun, 1963
517 ivella Bsk., 1900
518 ambrosiaefoliella Cham., 1875
   rileyi F. & B., 1876
519 pallidula Braun, 1963
520 taeniola Braun, 1963
521 carolinae Braun, 1963
522 angustata F. & B., 1876
   crescentella Braun, 1916
523 adelpha Braun, 1963
524 plucheae Braun, 1963
525 eupatoriella Braun, 1918
526 polymniae Braun, 1963
527 speciosa Braun, 1963
528 subnitens Wlsm., 1914
529 sexnotata Braun, 1927
530 divisa Braun, 1925
531 illecebrosa Braun, 1963
532 insolita Braun, 1918
533 transversata Braun, 1910
534 koebelella Bsk., 1910
535 salutatoria Braun, 1925
536 leptalea Braun, 1963
537 arnicella Braun, 1925
538 tridenticola Braun, 1963
539 spectabilis Braun, 1963
540 seorsa Braun, 1963
541 angustisquamella Braun, 1925
542 columbiana Braun, 1963
543 sororcula Braun, 1963
544 nigripunctella Braun, 1923
545 atrosignata Braun, 1963
546 enceliae Braun, 1963
547 latella Braun, 1918
548 sporobolella Bsk., 1910
549 packardella Cham., 1873
550 albertiella Bsk., 1910
   tetrella Braun, 1910
551 coniforma Braun, 1963
552 platyphylla Braun, 1963
553 ochrisuffusa Braun, 1963
554 trifasciella Clem., 1866
   obscurofasciella Cham., 1873
555 quinquenotella Cham., 1875
556 domicola Braun, 1963
557 zophopasta Braun, 1963
558 litigiosella Zell., 1875
559 coronatella Clem., 1860
560 canadensisella Cham., 1875
561 improvisa Braun, 1963
562 polytita Braun, 1963
563 luteella Cham., 1873
564 recognita Braun, 1963
565 paroptila Braun, 1963
566 fugitans Braun, 1930
567 callistricha Braun, 1963
568 eugrapha Braun, 1963
569 cerina Braun, 1963
570 copeuta Meyr., 1919

571 locuples Meyr., 1919
572 ainsliella Murt., 1905
573 eclecta Braun, 1963
574 anaticula Braun, 1963
575 disjuncta Braun, 1963
576 ceanothiella Braun, 1918
577 pomifoliella Clem., 1860
   curvilineatella (Pack., 1869)
   pomonella Pack., 1880
578 ilecella Bsk., 1915
579 quadrigemina Braun, 1918
   althaeae Bsk., 1919
580 gossypiella Morrill, 1927
581 sphaeralceae Braun, 1963
582 thurberiella Bsk., 1914

GRACILLARIIDAE

by DONALD R. DAVIS

GRACILLARIINAE

CALOPTILIA Hbn., 1825
  CALOPTILIA Hbn., 1825,
    subgenus
  POECILOPTILIA Hbn., 1825
  ORNIX Tr., 1833
  CORISCIUM Zell., 1839
  ANTIOLOPHA Meyr., 1894
583 aceriella (Cham., 1881), n. comb.
584 acerifoliella (Cham., 1875), n. comb.
585 agrifoliella Opler, 1971
586 alnicolella (Cham., 1875), n. comb.
587 alnivorella (Cham., 1875), n. comb.
588 amphidelta (Meyr., 1918), n. comb.
589 anthobaphes (Meyr., 1921), n. comb.
590 asplenifoliatella (Darlington, 1944), n. comb.
591 atomosella (Zell., 1873), n. comb.
592 azaleella (Brants, 1913)
   azaleae (Bsk., 1914)
593 behrensella (Cham., 1876), n. comb.
594 belfrageella (Cham., 1875), n. comb.
   auriferella (F. & B., 1876)
594.1 betulivora McD., 1946
595 bimaculatella (Ely, 1915), n. comb.
596 blandella (Clem., 1864), n. comb.
   juglandivorella (Cham., 1873)
597 burgessiella (Zell., 1873), n. comb.
598 burserella (Bsk., 1900), n. comb.
599 canadensisella (McD., 1956), n. comb.
600 cornusella (Ely, 1915), n. comb.
601 coroniella (Clem., 1864), n. comb.
602 diversilobiella Opler, 1967
603 ferruginella (Braun, 1918), n. comb.
604 flavella (Ely, 1915), n. comb.
605 flavimaculella (Ely, 1915), n. comb.
606 fraxinella (Ely, 1915), n. comb.
   cuculipennella; auth.
607 glutinella (Ely, 1915), n. comb.
608 hypericella (Braun, 1918), n. comb.
609 invariabilis (Braun, 1927), n. comb.
610 juglandiella (Cham., 1872)
   juglandisnigraeella (Cham., 1878), nom. nud.
611 macranthes (Meyr., 1928), n. comb.
612 melanocarpae (Braun, 1925), n. comb.

613 minimella (Ely, 1915), n. comb.
614 murtfeldtella (Bsk., 1904), n. comb.
615 negundella (Cham., 1876), n. comb.
616 nondeterminata (Braun, 1939), n. comb.
617 obscuripennella (F. & B., 1876), n. comb.
618 ostryaeella (Cham., 1878), n. comb.
619 ovatiella Opler, 1969
620 packardella (Cham., 1872), n. comb.
   elegantella (F. & B., 1873)
   inornatella (Cham., 1876)
621 palustriella (Braun, 1910), n. comb.
622 paradoxa (F. & B., 1873), n. comb.
623 perseae (Bsk., 1920), n. comb.
624 populiella (Cham., 1875), n. comb.
625 porphyretica (Braun, 1923), n. comb.
626 pulchella (Cham., 1875), n. comb.
627 quercinigrella (Ely, 1915), n. comb.
628 reticulata (Braun, 1910)
629 rhodorella (McD., 1954), n. comb.
630 rhoifoliella (Cham., 1876)
631 ribesella (Cham., 1877), n. comb.
632 sanguinella (Beutenmüller, 1888)
   nigristrigella (Beutenmüller, 1888)
   ruptistrigella (Beutenmüller, 1888)
   shastella (Beutenmüller, 1888)
   fuscoocherella (Beutenmüller, 1889)
   ruptostrigella (Ely, 1917), missp.
   shastaella (Ely, 1917), missp.
633 sassafrasella (Cham., 1876), n. comb.
634 sauzalitoeella (Cham., 1876), n. comb.
   sauzalitella (Meyr., 1912), emend.
   sauzalitoella (Ely, 1917), missp.
635 scutellariella (Braun, 1923), n. comb.
636 sebastianiella (Bsk., 1900), n. comb.
637 serotinella (Ely, 1910), n. comb.
638 speciosella (Braun, 1939), n. comb.
639 stigmatella (F., 1781)
   ochracea (Haw., 1828)
   purpuriella (Cham., 1872)
   consimilella (F. & B., 1876)
640 strictella (Wlk., 1864), n. comb.
   adaptella (Wlk., 1864), n. comb.
641 superbifrontella (Clem., 1860), n. comb.
642 umbratella (Braun, 1927), n. comb.
643 vacciniella (Ely, 1915), n. comb.
644 violacella (Clem., 1860, n. comb.
   desmodifoliella (Clem., 1865), n. comb.

  GRACILLARIA Haw., 1828,
    subgenus
  GRACILARIA Zell., 1839, missp.
645 syringella (F., 1794)
   anastomosis (Haw., 1828)

NEUROSTROTA Ely, 1917
  NEUROSTRATA Ely, 1917, missp.
646 gunniella (Bsk., 1906)

MICRURAPTERYX Spuler, 1910
647 salicifoliella (Cham., 1872)

PARECTOPA Clem., 1860
  EUSPILOPTERYX Zell., 1847, preocc. by Steph., 1835
  EUSPILAPTERYX Spuler, 1910
648 albicostella Braun, 1925

649 bosquella (Cham., 1876)
   basquella (Cham., 1876), missp.
650 bumeliella Braun, 1939, mispl.
651 geraniella Braun, 1935
652 interpositella (F. & B., 1876)
653 lespedezaefoliella Clem., 1860
   lespedezifoliella Meyr., 1912, emend.
   mirabilis (F. & B., 1873)
654 occulta Braun, 1922
655 pennsylvaniella (Engel, 1907)
656 plantaginisella (Cham., 1872)
   geiella (Cham., 1874)
   erigeronella (Cham., 1877)
657 robiniella Clem., 1863
658 thermopsella (Cham., 1875)

NEUROLIPA Ely, 1917
659 randiella (Bsk., 1900)

APOPHTHISIS Braun, 1915
660 congregata Braun, 1923
661 pullata Braun, 1915

NEUROBATHRA Ely, 1917
662 bohartiella Opler, 1971
663 strigifinitella (Clem., 1860)
   duodecemlineella (Cham., 1872)
   quercifoliella (Cham., 1875)

CALLISTO Steph., 1834
  ORNIX; Zell., 1839, not Tr., 1833
664 denticulella (Thunb., 1794)
   guttea (Haw., 1828)

PARORNIX Spuler, 1910
665 alta (Braun, 1925)
666 arbitrella (Dietz, 1907)
667 arbutifoliella (Dietz, 1907)
668 boreasella (Clem., 1864)
669 conspicuella (Dietz, 1907)
670 crataegifoliella (Clem., 1860)
671 dubitella (Dietz, 1907)
672 festinella (Clem., 1860)
673 geminatella (Pack., 1869)
   prunivorella (Cham., 1873)
   pruinrosella? (Cham., 1875), missp.
   prunionella (Cham., 1876), missp.
674 innotata (Wlsm., 1907)
675 inusitatumella (Cham., 1873)
676 kalmiella (Dietz, 1907)
677 melanotella (Dietz, 1907)
678 obliterella (Dietz, 1907)
679 peregrinaella (Darlington, 1949), n. comb.
680 preciosella (Dietz, 1907)
681 quadripunctella (Clem., 1861)
   a. albifaciella (Dietz, 1907)
681.1 solitariella (Dietz, 1907), n. stat.
682 spiraeifoliella (Braun, 1918)
683 strobivorella (Dietz, 1907)
684 texanella (Bsk., 1906)
685 trepidella (Clem., 1860)
686 vicinella (Dietz, 1907)

CHILOCAMPYLA Bsk., 1900
687 dyariella Bsk., 1900

ACROCERCOPS Wallgr., 1881
688 affinis Braun, 1918
689 albinatella (Cham., 1872)

albanoteila (Cham., 1877), missp.
albinotella Meyr., 1912, emend.
690 **arbutella** Braun, 1925
691 **insulariella** Opler, 1971
692 **astericola** (F. & B., 1873)
693 **pnosmodiella** (Bsk., 1902)
onosmodiella (Ckll., 1905), emend.
694 **quinquistrigella** (Cham., 1875)
quinquestrigella (Cham., 1877), missp.
695 **rhombiderella** (F. & B., 1876)
696 **sideroxylonella** (Bsk., 1900)
sideroxylella Meyr., 1912, emend.
697 **strigosus** Braun, 1914

LEUCOSPILAPTERYX Spuler, 1910
698 **venustella** (Clem., 1860)
eupatoriella (Cham., 1872)

METRIOCHROA Bsk., 1900
OECOPHYLLEMBIUS Silvestri, 1908
699 **psychotriella** Bsk., 1900

LEUCANTHIZA Clem., 1859
700 **amphicarpeaefoliella** Clem., 1859
saundersella Cham., 1871
701 **dircella** Braun, 1914

MARMARA Clem., 1863
AESYLE Cham., 1875
702 **apocynella** Braun, 1915
703 **arbutiella** Bsk., 1903
704 **auratella** Braun, 1915
705 **basidendroca** Fitzgerald, 1973
706 **corticola** Fitzgerald, 1973
707 **elotella** (Bsk., 1909)
708 **fasciella** (Cham., 1875)
quinquenotella (Cham., 1877)
709 **fraxinicola** Braun, 1922
710 **fulgidella** (Clem., 1860)
711 **guilandinella** Bsk., 1900
712 **leptodesma** Meyr., 1928
713 **opuntiella** Bsk., 1907
714 **oregonensis** Fitzgerald, 1975
715 **pomonella** Bsk., 1915
716 **salictella** Clem., 1863
717 **serotinella** Bsk., 1915
718 **smilacisella** (Cham., 1875)

LITHOCOLLETINAE

PROTOLITHOCOLLETIS Braun, 1929
719 **lathyri** Braun, 1929

CREMASTOBOMBYCIA Braun, 1908
720 **ambrosiella** (Cham., 1871)
amoena (F. & B., 1878)
721 **grindeliella** (Wlsm., 1891)
722 **ignota** (F. & B., 1873)
bostonica (F. & B., 1873)
helianthisella (Cham., 1874), nom. nud.
helianthivorella (Cham., 1875)
acrinomeridis (F. & B., 1878)
elephantopodella (F. & B., 1878)
723 **solidaginis** (F. & B., 1876)
724 **verbesinella** (Bsk., 1900)

PHYLLONORYCTER Hbn., 1822
LITHOCOLLETIS Hbn., 1825
EUCESTIS Hbn., 1825
PHYLLORYCTER Wlsm., 1914, missp.
725 **aberrans** (Braun, 1930)
726 **aeriferella** (Clem., 1859)
727 **albanotella** (Cham., 1875)
subaureola (F. & B., 1878)
728 **alni** (Wlsm., 1891)
alnivorella (Cham., Oct. 1875), preocc. by Rag., April 1875
729 **alnicolella** (Wlsm., 1889)
730 **antiochella** (Opler, 1971), n. comb.
731 **apicinigrella** (Braun, 1908)
732 **arbutusella** (Braun, 1908)
arbutella (Meyr., 1912), emend.
733 **argentifimbriella** (Clem., 1859)
longestriata (F. & B., 1873)
fuscocostella (Cham., 1875)
longirostrata (Dyar, 1903), missp.
734 **argentinotella** (Clem., 1859)
735 **arizonella** (Braun, 1925)
736 **atomariella** (Zell., 1875)
737 **auronitens** (F. & B., 1873)
738 **basistrigella** (Clem., 1859)
intermedia (F. & B., 1873)
739 **bataviella** (Braun, 1908)
740 **blancardella** (F., 1781)
pyrariella (Tutt, 1898)
concomitella (Bankes, 1899)
741 **caryaealbella** (Cham., 1871)
caryalbella (Wlsm., 1891), missp.
742 **celtifoliella** (Cham., 1871)
743 **celtisella** (Cham., 1871)
nonfasciella (Cham., 1871)
pulsillifoliella (F. & B., 1876)
744 **clemensella** (Cham., 1871)
745 **comptoniella** (Darlington, 1949)
746 **crataegella** (Clem., 1859)
747 **cretaceella** (Braun, 1925), n. comb.
748 **deceptusella** (Cham., 1879)
749 **diaphanella** (F. & B., 1878)
750 **diversella** (Braun, 1916)
750.1 **emberizaepenella** (Bouché, 1834)
emberizipennella (Dup., 1843), emend.
751 **felinelle** Heinr., 1920
752 **fitchella** (Clem., 1860)
quercifoliella (Fitch, 1859), preocc. by Zell., 1839
quercetorum (F. & B., 1873)
quercitorum (Cham., 1875), missp.
753 **fragilella** (F. & B., 1878)
trifasciella (F. & B., 1873), preocc. by Haw., 1828
754 **gemmea** (F. & B., 1873)
755 **hagenii** (F. & B., 1873)
hageni (Cham., 1879), missp.
necopinusella (Cham., 1878)
756 **holodisci** (Braun, 1939)
757 **incanella** (Wlsm., 1889)
758 **insignis** (Wlsm., 1889)
759 **intermixtus** (Braun, 1930)
760 **inusitatella** (Braun, 1925)
761 **kearfottella** (Braun, 1908)
762 **kenora** (Free., 1970)
763 **ledella** (Wlsm., 1889)
764 **lucetiella** (Clem., 1859)
aenigmatella (F. & B., 1873)
765 **lucidicostella** (Clem., 1859)
ludicostella (Riley, 1891), missp.

766 **lysimachiaeella** (Cham., 1875)
767 **malimalifoliella** (Braun, 1908)
malifoliella (Meyr., 1912), emend.
768 **manzanita** (Braun, 1925)
769 **mariaeella** (Cham., 1875)
mariella (Riley, 1891), missp.
770 **martiella** (Braun, 1908)
771 **memorabilis** (Braun, 1839), n. comb.
772 **minutella** (F. & B., 1878)
773 **morrisella** (Fitch, 1859)
texanella (Zell., 1875)
texana (Cham., 1877), missp.
amphicarpeaeella (Cham., 1877)
amphicarpaeella (Riley, 1891), missp.
774 **nipigon** (Free., 1970)
775 **obscuricostella** (Clem., 1859)
virginiella (Cham., 1871)
776 **obsoletus** (F. & B., 1873)
777 **occitanicus** (F. & B., 1876)
778 **olivaeformis** (Braun, 1908)
oliviformis (Meyr., 1912), emend.
779 **ontario** (Free., 1970)
780 **oregonensis** (Wlsm., 1889)
781 **ostryaefoliella** (Clem., 1859)
mirifica (F. & B., 1873)
782 **pernivalis** (Braun, 1925)
783 **populiella** (Cham., 1878)
784 **propinquinella** (Braun, 1908)
785 **quercialbella** (Fitch, 1859)
quercibella (Cham., 1875)
quercipulchrella (Cham., 1878), nom. nud.
786 **restrictella** (Braun, 1939)
787 **rhododendrella** (Braun, 1935)
788 **ribefoliae** (Braun, 1939)
789 **rileyella** (Cham., 1875)
tenuistrigata (F. & B., 1876)
790 **robiniella** (Clem., 1859), mispl.
pseudacaciella (Fitch, 1859)
791 **salicifoliella** (Cham., 1875)
792 **salicivorella** (Braun, 1908)
793 **sandraella** (Opler, 1971), n. comb.
794 **scudderella** (F. & B., 1873)
795 **sexnotella** (Cham., 1880)
795 **symphoricarpaeella** (Cham., 1875)
symphoricarpella (F. & B., 1878)
symphoricarpella (Wlsm., 1890), emend.
boliella (Dyar, 1903)
797 **tiliacella** (Cham., 1871)
tiliaeella (Cham., 1874), missp.
tiliella (Wlsm., 1891), missp.
798 **tremuloidiella** (Braun, 1908)
799 **trinotella** (Braun, 1908)
800 **tritaenianella** (Cham., 1871)
tritenoeanella Cham., 1873, missp.
consimilella (F. & B., 1873)
tritaeniaella Cham., 1879, missp.
tritaeniella Wlsm., 1889, missp.
801 **uhlerella** (Fitch, 1859)
amorphaeella (Cham., 1877)
amorphae (F. & B., 1878)
802 **viburnella** (Braun, 1923)

CAMERARIA Chapman, 1902
803 **aceriella** (Clem., 1859), n. comb.
804 **aesculisella** (Cham., 1871)
805 **affinis** (F. & B., 1876)
806 **agrifoliella** (Braun, 1908)
807 **arcuella** (Braun, 1908)

808 **australisella** (Cham., 1878)
809 **bethunella** (Cham., 1871)
    lebertella (F. & B., 1878)
810 **betulivora** (Wlsm., 1891)
811 **caryaefoliella** (Clem., 1859)
    juglandiella (Clem., 1861)
812 **castaneaeella** (Cham., 1875)
    castanella (Wlsm., 1891), emend.
    castaneella (Meyr., 1912), emend.
813 **cervina** (Wlsm., 1907)
814 **chambersella** (Wlsm., 1889)
    quinquenotella (Cham., 1880), preocc.
      by Frey, 1855
815 **cincinnatiella** (Cham., 1871)
816 **conglomeratella** (Zell., 1875)
    bicolorella (Cham., 1878)
    obtusilobae (F. & B., 1878)
817 **corylisella** (Cham., 1871)
    coryliella (Cham., 1879), emend.
    bifasciella (Wlsm., 1907)
818 **eppelsheimii** (F. & B., 1878)
819 **fasciella** (Wlsm., 1891)
    unifasciella (Cham., 1875), preocc. by
      Tengström, 1865
820 **fletcherella** (Braun, 1908)
821 **gaultheriella** (Wlsm., 1889)
822 **guttifinitella** (Clem., 1859)
    toxicodendri (F. & B., 1878)
823 **hamadryadella** (Clem., 1859)
    alternatella (Zell., 1875)
    hamadryella (F. & B., 1878), missp.
    alternata (Cham., 1878), missp.
824 **hamameliella** (Bsk., 1903)
    hamamelis (Kft., 1903), missp.
825 **lentella** (Braun, 1908)
826 **leucothorax** (Wlsm., 1907), n. comb.
827 **macrocarpae** Free., 1970
828 **macrocarpella** (F. & B., 1878)
829 **mediodorsella** (Braun, 1908)
830 **nemoris** (Wlsm., 1889)
831 **obstrictella** (Clem., 1859)
    bifasciella (Cham., 1878)
    ceriferae (Wlsm., 1907)
832 **ostryarella** (Cham., 1871)
    ostryella (Meyr., 1912), emend.
833 **picturatella** (Braun, 1916)
834 **platanoidiella** (Braun, 1908)
835 **quercivorella** (Cham., 1879)
836 **saccharella** (Braun, 1908)
837 **superimposita** (Braun, 1925)
838 **tubiferella** (Clem., 1860)
839 **ulmella** (Cham., 1871)
    modesta (F. & B., 1876)
840 **umbellulariae** (Wlsm., 1889)
841 **wizlizeniella** Opler, 1971

    CHRYSASTER Kumata, 1961
842 **ostensackenella** (Fitch, 1859), n.
    comb.
    ornatella (Cham., 1871), n. comb.

    PORPHYROSELA Braun, 1908
843 **desmodiella** (Clem., 1859)
    gregariella (Murt., 1881)

PHYLLOCNISTINAE

PHYLLOCNISTIS Zell., 1848
    PHYLLOENISTIS Cham., 1875,
    missp.

    PHYLLOETIS Cham., 1876, missp.
    PHYLLOCNITIS Bsk., 1900,
    missp.
844 **ampelopsiella** Cham., 1871
845 **finitima** Braun, 1927
846 **insignis** F. & B., 1876
    erechtitisella Cham., 1878, nom. nud.
    erechtiisella McD., 1939, nom. nud.
847 **intermediella** Bsk., 1900
848 **liquidambarisella** Cham., 1875
849 **liriodendronella** Clem., 1863
    liriodendrella McD., 1939, missp.
850 **magnatella** Zell., 1873
851 **magnoliella** Fbs., 1923
    magnoliaeella Cham., 1880, nom.
    nud.
    magnoliaella Bsk., 1900, nom. nud.
    magnoliella Dyar, 1903, nom. nud.
852 **populiella** Cham., 1875
853 **vitegenella** Clem., 1859
854 **vitifoliella** Cham., 1871

Gelechioidea

OECOPHORIDAE

by RONALD W. HODGES

DEPRESSARIINAE

Depressariini

AGONOPTERIX Hbn., 1825
    EPELEUSTIA Hbn., 1825
    PINARIS Hbn., 1825
    TICHONIA Hbn., 1825
    HAEMYLIS Tr., 1832
    HAEMILIS Dup., 1838, missp.
    AGONOPTERYX auth., missp.
    SYLLOCHITIS Meyr., 1910
    CTENIOXENA Meyr., 1923
855 **gelidella** (Bsk., 1908)
856 **hyperella** Ely, 1910
    testifica (Meyr., 1920), repl. name
857 **lythrella** (Wlsm., 1889)
    arcuella Clarke, 1941
858 **nubiferella** (Wlsm., 1881)
859 **curvilineella** (Beutenmüller, 1889)
860 **muricolorella** (Bsk., 1902)
861 **oregonensis** Clarke, 1941
862 **clemensella** (Cham., 1876)
863 **clarkei** Keif., 1936
864 **atrodorsella** (Clem., 1863)
865 **pteleae** B. & Bsk., 1920
866 **eupatoriiella** (Cham., 1878)
    plummerella Bsk., 1908
867 **pulvipennella** (Clem., 1864)
    solidaginis (Wlsm., 1889)
868 **nigrinotella** (Bsk., 1908)
    costimacula Clarke, 1941
869 **walsinghamella** (Bsk., 1902)
    fernaldella (Wlsm., 1889), preocc. by
      Cham., 1878
870 **fusciterminella** Clarke, 1941
871 **chrautis** Hodges, 1974
872 **sabulella** (Wlsm., 1881)
    callosella B. & Bsk., 1920

873 **dammersi** Clarke, 1947
874 **cajonensis** Clarke, 1941
874.1 **alstroemeriana** (Cl., 1759)
    monilella (D. & S., 1775)
    puella (Hbn., 1796)
875 **toega** Hodges, 1974
876 **rosaciliella** (Bsk., 1904)
    rosiciliella (Meyr., 1922), emend.
    echinopanicis Clarke, 1941
877 **ciliella** (Staint., 1849)
    annexella (Zell., 1868)
878 **canadensis** (Bsk., 1902)
    pallidella (Bsk., 1904)
    terinella B. & Bsk., 1920
    sciadopa (Meyr., 1920)
    serrae Clarke, 1933
879 **arnicella** (Wlsm., 1881)
880 **flavicomella** (Engel, 1907)
881 **senicionella** (Bsk., 1902)
    seniciella (Bsk., 1902), missp.
882 **robiniella** (Pack., 1869)
    hilarella (Zell., 1873)
883 **cratia** Hodges, 1974
884 **thelmae** Clarke, 1941
885 **sanguinella** (Bsk., 1902)
886 **lecontella** (Clem., 1860)
887 **dimorphella** Clarke, 1941
888 **pergandeella** (Bsk., 1908)
889 **argillacea** (Wlsm., 1881)
    blacella B. & Bsk., 1920
890 **amissella** (Bsk., 1908)
891 **psoraliella** (Wlsm., 1881)
    murmurans (Meyr., 1927)
892 **hesphoea** Hodges, 1975
893 **antennariella** Clarke, 1941
    victori de Lesse & Viette, 1949
894 **nebulosa** (Zell., 1873)
895 **nervosa** (Haw., 1811)
    costosa (Haw., 1811)
    depunctella (Hbn., 1810–13)
    boicella (Freyer, 1835)
    dryadoxena (Meyr., 1920)
    blackmori Bsk., 1922
    venosata (Kautz, 1930)
896 **posticella** (Wlsm., 1881)
897 **latipalpella** B. & Bsk., 1920
898 **amyrisella** (Bsk., 1900)
    amyridella (Meyr., 1922), emend.

    DEPRESSARIODES Turati, 1925
    MARTYRHILDA Clarke, 1941
899 **canella** (Bsk., 1904), n. comb.
    cogitata (Braun, 1926), n. comb.
900 **umbraticostella** (Wlsm., 1881), n.
    comb.
901 **sordidella** (Clarke, 1941), n. comb.
902 **gracilis** (Wlsm., 1889), n. comb.
903 **thoracenigraeella** (Cham., 1875), n.
    comb.
    novi-mundi (Wlsm., 1881), n. comb.
    thoracinigrella (Meyr., 1922),
      emend., n. comb.
904 **thoracefasciella** (Cham., 1875), n.
    comb.
    thoracifasciella (Meyr., 1922),
      emend., n. comb.
    sphaeralceae (Clarke, 1941), n. comb.
905 **nechlys** (Hodges, 1974), n. comb.
906 **nivalis** (Braun, 1921), n. comb.
    jacobi (McD., 1944), n. comb.

907 **hildaella** (Clarke, 1941), n. comb.
908 **ciniflonella** (Lienig & Zell., 1846), n. comb.
    klamathianus (Wlsm., 1881), n. comb.
    kusnezovi (Krulikowsky, 1908)
    smolandiae (Palm, 1943), n. comb.
    isa (Clarke, 1947), n. comb.
909 **scabella** (Zell., 1873), n. comb.
    scabrella (Wlsm., 1881), missp., n. comb.
910 **fulvus** (Wlsm., 1882), n. comb.
    endryopa (Meyr., 1918), n. comb.

BIBARRAMBLA Clarke, 1941
911 **allenella** (Wlsm., 1882)

SEMIOSCOPIS Hbn., 1825
    EPIGRAPHIA Steph., 1829
912 **packardella** (Clem., 1863)
    eruditella (Grt., 1880)
913 **merricella** Dyar, 1902
    merrickella Meyr., 1922, emend.
914 **inornata** Wlsm., 1882
    inornatella Bsk., 1908, missp.
915 **megamicrella** Dyar, 1902
    braunae Clarke, 1941
916 **aurorella** Dyar, 1902
917 **mcdunnoughi** Clarke, 1941

DEPRESSARIA Haw., 1811
    VOLUCRUM Berthold, 1827
    SIGANOROSIS Wallgr., 1881
    SCHISTODEPRESSARIA Spuler, 1910
918 **atrostrigella** Clarke, 1941
919 **artemisiae** Nickerl, 1864
    dracunculi Clarke, 1933
920 **palousella** Clarke, 1941
921 **cinereocostella** Clem., 1864
    clausella Wlk., 1864
922 **pastinacella** (Dup., 1838)
    heraclei Haw., 1811, emend.
    ontariella Bethune, 1870
    heracliana; auth., not L., 1758
923 **juliella** Bsk., 1908
924 **daucella** (D. & S., 1775)
    rubricella (D. & S., 1775)
    apiella (Hbn., 1796)
    nervosa; auth., not Haw., 1811
925 **eleanorae** Clarke, 1941
926 **alienella** Bsk., 1904
    nymphidia Meyr., 1918
    corystopa Meyr., 1927
927 **artemisiella** McD., 1927
928 **constancei** Clarke, 1947
929 **betina** Clarke, 1947
930 **whitmani** Clarke, 1941
931 **schellbachi** Clarke, 1947
932 **angelicivora** Clarke, 1952
933 **leptotaeniae** Clarke, 1933
934 **yakimae** Clarke, 1941
935 **multifidae** Clarke, 1933
936 **moya** Clarke, 1947
937 **besma** Clarke, 1947
938 **pteryxiphaga** Clarke, 1952
939 **togata** Wlsm., 1889
    thustra Clarke, 1947
940 **armata** Clarke, 1952
941 **angustati** Clarke, 1941

NITES Hodges, 1974
942 **grotella** (Rob., 1869)
    symmochlota (Meyr., 1918)
943 **atrocapitella** (McD., 1944)
944 **betulella** (Bsk., 1902)
945 **maculatella** (Bsk., 1908)
946 **ostryella** (McD., 1943)

APACHEA Clarke, 1941
947 **barberella** (Bsk., 1902)

HIMMACIA Clarke, 1941
948 **huachucella** (Bsk., 1908)
949 **stratia** Hodges, 1974
950 **diligenda** (Meyr., 1928)

Amphisbatini

MACHIMIA Clem., 1860
951 **tentoriferella** Clem., 1860
    confertella (Wlk., 1864)
    fernaldella (Cham., 1878)
952 **trigama** (Meyr., 1928)

EUPRAGIA Wlsm., 1911
953 **hospita** Hodges, 1969
954 **banis** Hodges, 1974

PSILOCORSIS Clem., 1860
    PAEPIA Wlk., 1864
    HAGNO Cham., 1872
955 **quercicella** Clem., 1860
956 **cryptolechiella** (Cham., 1872)
    faginella (Cham., 1872)
    obsoletella (Zell., 1873)
    dubitatella (Zell., 1877)
    cressonella (Cham., 1878)
957 **reflexella** Clem., 1860
    ferruginosa (Zell., 1873)
    fletcherella Gibson, 1909
    caryae Clarke, 1941
958 **amydra** Hodges, 1961
959 **arguta** Hodges, 1961
959.1 **fatula** Hodges, 1975
960 **cirrhoptera** Hodges, 1961

ETHMIINAE

PYRAMIDOBELA Braun, 1923
    IDIOPTILA Meyr., 1927
961 **quinquecristata** (Braun, 1921)
961.1 **angelarum** Keif., 1936
961.2 **agyrtodes** (Meyr., 1927)

ETHMIA Hbn., 1819
    PSECADIA Hbn., 1825
    ANESYCHIA Hbn., 1825
    DISTHYMNIA Hbn., 1825
    MELANOLEUCA Steph., 1829
    AEDIA Dup., 1836
    CHALYBE Dup., 1836
    AZINIS Wlk., 1863
    TAMARRHA Wlk., 1864
    CERATOPHYSETIS Meyr., 1887
    THEOXENIA Wlsm., 1887
    BABAIAXA Bsk., 1902
    WILTSHIREIA Amsel, 1949
962 **umbrimarginella** Bsk., 1907
963 **lassenella** Bsk., 1908
964 **coquillettella** Bsk., 1967

965 **monachella** Bsk., 1910
966 **scylla** Powell, 1973
967 **brevistriga** Clarke, 1950
    a. **aridicola** Powell, 1973
968 **albitogata** Wlsm., 1907
969 **plagiobothrae** Powell, 1973
970 **minuta** Powell, 1973
971 **tricula** Powell, 1973
972 **charybdis** Powell, 1973
973 **albistrigella** (Wlsm., 1880)
    chrysurella (Dietz, 1905)
    chrysocomella; Bsk., 1905, not Dietz, 1905
    a. **icariella** Powell, 1973
974 **nadia** Clarke, 1950
975 **orestella** Powell, 1973
976 **semilugens** (Zell., 1872)
    multipunctella (Cham., 1874)
    semiopaca (Grt., 1881)
    plumbeella (Beutenmüller, 1889)
977 **epileuca** Powell, 1959
978 **apicipunctella** (Cham., 1875)
    zavalla Bsk., 1915
979 **arctostaphylella** (Wlsm., 1880)
    obscurella (Beutenmüller, 1888)
    mediella Bsk., 1913
980 **discostrigella** (Cham., 1877)
    a. **subcaerulea** (Wlsm., 1880)
981 **semitenebrella** Dyar, 1902
982 **macelhosiella** Bsk., 1907
983 **geranella** B. & Bsk., 1920
984 **timberlakei** Powell, 1973
985 **macneilli** Powell, 1973
986 **bipunctella** (F., 1775)
    echiella (D. & S., 1775)
    hochenwartiella (Rossi, 1790)
    bipunctelia (Uffeln, 1938), missp.
    griseicostella (Wiltshire, 1947)
987 **monticola** (Wlsm., 1880)
    a. **emmeli** Powell, 1973
    b. **fuscipedella** (Wlsm., 1888)
988 **caliginosella** Bsk., 1904
989 **hagenella** (Cham., 1878)
    a. **josephinella** Dyar, 1902
990 **mimihagenella** Powell, 1973
991 **burnsella** Powell, 1973
992 **zelleriella** (Cham., 1878)
    texanella (Cham., 1880)
993 **delliella** (Fern., 1891)
994 **bittenella** (Bsk., 1910)
995 **notatella** (Wlk., 1863)
    xanthorrhoa (Zell., 1877)
996 **confusella** (Wlk., 1863)
    strigosella (Wlk., 1864)
    ingricella (Mösch., 1890)
    strigosa (Ckll., 1891), missp.
997 **julia** Powell, 1973
998 **farrella** Powell, 1973
999 **longimaculella** (Cham., 1872)
    walsinghamella Beutenmüller, 1889
    a. **coranella** Dyar, 1902
1000 **semiombra** Dyar, 1902
1001 **albicostella** (Beutenmüller, 1889)
    mirusella; (Cham., 1877), not Cham., 1874
1002 **mirusella** (Cham., 1874)
    mirella Meyr., 1914, repl. name
1003 **trifurcella** (Cham., 1873)
1004 **marmorea** (Wlsm., 1888)
1005 **hodgesella** Powell, 1973

1006 sphenisca Powell, 1973
1007 prattiella Bsk., 1915

PSEUDETHMIA Clarke, 1950
1008 protuberans Clarke, 1950

PELEOPODINAE

Peleopodini

PSEUDEROTIS Clarke, 1956
1009 obiterella (Bsk., 1908)

DURRANTIA Bsk., 1908
DOLIDIRIA Bsk., 1912
1010 piperatella (Zell., 1873)
albella (Cham., 1874)
montivola Meyr., 1927

STENOMATINAE

ANTAEOTRICHA Zell., 1854
MESOPTYCHA Zell., 1854
BRACHILOMA Clem., 1863
HARPALYCE Cham., 1874
IDE Cham., 1880
AEDEMOSES Wlsm., 1912
1011 schlaegeri (Zell., 1854)
1012 lindseyi (B. & Bsk., 1920)
1013 unipunctella (Clem., 1863)
lithosina (Zell., 1873)
tortricella (Cham., 1874)
1014 leucillana (Zell., 1854)
algidella (Wlk., 1864)
1015 osseella (Wlsm., 1889)
querciella (Bsk., 1908)
1016 decorosella (Bsk., 1908)
decorasella (B. & Bsk., 1920), missp.
decorella (B. & Bsk., 1920), missp.
1017 furcata (Wlsm., 1889)
1018 irene (B. & Bsk., 1920)
1019 humilis (Zell., 1855)
nebeculosa (Zell., 1873)
canusella (Cham., 1874)
1020 agrioschista (Meyr., 1927)
1021 thomasi (B. & Bsk., 1920)
1022 haesitans (Wlsm., 1912)
hessitans (Heinr., 1921), missp.
hesitans Bsk., 1934, missp.
1023 fuscorectangulata Duckworth, 1964
1024 vestalis (Zell., 1873)
1025 manzanitae Keif., 1937

SETIOSTOMA Zell., 1875
1026 xanthobasis Zell., 1875
1027 fernaldella Riley, 1889

MENESTOMORPHA Wlsm., 1907
1028 oblongata Wlsm., 1907
1029 kimballi (Duckworth, 1964), n. comb.

MENESTA Clem., 1860
HYALE Cham., 1880
1030 tortriciformella Clem., 1860
liturella (Wlk., 1864)
coryliella (Cham., 1875)
albaciliaeella (Cham., 1875)
albiciliella Wlsm., 1911, missp.
albaciliella Braun, 1915, missp.
1031 melanella Murt., 1890

GONIOTERMA Wlsm., 1897
1032 mistrella (Bsk., 1907), n. comb.
1033 crambitella (Wlsm., 1889), n. comb.

OECOPHORINAE

Oecophorini

INGA Bsk., 1908
LYSIGRAPHA Meyr., 1914
PELOMIMAS Meyr., 1914
ORSIMACHA Meyr., 1914
SIDEROGRAPTIS Meyr., 1920
PHANERODOXA Meyr., 1921
EPIMORYCTIS Meyr., 1930
HOROMERISTIS Meyr., 1931
AGRIOTORNA Meyr., 1931
1034 sparsiciliella (Clem., 1864)
contrariella (Wlk., 1864)
inscitella (Wlk., 1864)
atropicta (Zell., 1875)
1035 cretacea (Zell., 1873)
1036 ciliella (Bsk., 1908)
humata (Meyr., 1914)
1037 obscuromaculella (Cham., 1878)
1038 canariella (Bsk., 1908)
1039 concolorella (Beutenmüller, 1888)
1040 proditrix Hodges, 1974
1041 rimatrix Hodges, 1974

DECANTHA Bsk., 1908
1042 boreasella (Cham., 1873)
1043 stecia Hodges, 1974
1044 tistra Hodges, 1974
1045 stonda Hodges, 1974

CALLIMA Clem., 1860
EPICALLIMA Dyar, 1903, repl. name
1046 argenticinctella Clem., 1860
1047 nathrax Hodges, 1974

DAFA Hodges, 1974
1048 formosella (D. & S., 1775)

BATIA Steph., 1834
DISCOLATA Spuler, 1910
CHIROCOMPA Meyr., 1914
CHIROCAMPA Morley & Rait-Smith, 1933, missp.
1049 lunaris (Haw., 1828)
metznerella (Tr., 1835)
begrandella (Dup., 1842)
clavella (H.-S., 1854)

FABIOLA Bsk., 1908
1050 shaleriella (Cham., 1875)
1051 tecta Braun, 1935
1052 lucidella (Bsk., 1912)
1053 edithella (Bsk., 1907)
amplicincta (Braun, 1923)
1054 quinqueferella (Wlsm., 1881), n. comb.

BRYMBLIA Hodges, 1974
1055 quadrimaculella (Cham., 1875)
dimidiella (Wlsm., 1888)

DENISIA Hbn., 1825
BLEPHAROCERA Cham., 1877,

preocc. by Agassiz, 1846
CHAMBERSIA Riley, 1891, n. syn.
1056 haydenella (Cham., 1877), n. comb.

ESPERIA Hbn., 1825
STENOPTERA Dup., 1838
1057 sulphurella (F., 1775)
orbonella (Hbn., 1810–13)
aucta (Krausse, 1915)

POLIX Hodges, 1974
1058 coloradella (Wlsm., 1888)
rostrigera (Meyr., 1919)

MATHILDANA Clarke, 1941
1059 newmanella (Clem., 1864)
1060 flipria Hodges, 1974

BORKHAUSENIA Hbn., 1825
AMAUROSETIA Steph., 1835
1061 nefrax Hodges, 1974

CAROLANA Clarke, 1941
1062 ascriptella (Bsk., 1908)
1063 golmeia Hodges, 1974

HOFMANNOPHILA Spuler, 1910
1064 pseudospretella (Staint., 1849)

MARTYRINGA Bsk., 1902
ANCHONOMA Meyr., 1910
SANTUZZA Heinr., 1920
1065 latipennis (Wlsm., 1882)
1066 ravicapitis Hodges, 1960

ENDROSIS Hbn., 1825
1067 sarcitrella (L., 1758)
fenestrella (Scop., 1763)
lactella (D. & S., 1775)
betulinella (Hbn., 1818–19)
kennecottella (Clem., 1860)
subditella (Wlk., 1864)
ferrestrella Cham., 1875, missp.
antarctica (Stgr., 1898)
kennicotella Dyar, 1903, missp.

EIDO Cham., 1873
VENILIA Cham., 1872, preocc. by Rafinesque, 1815
EUMEYRICKIA Bsk., 1902
1068 trimaculella (Fitch, 1856)
albapalpella (Cham., 1872)
albapulvella (Cham., 1875)
haustellata (Wlsm., 1882)

CARCINA Hbn., 1825
PHIBALOCERA Steph., 1829
1069 quercana (F., 1775)
fagana (D. & S., 1775)
cancella (Hbn., 1823–24)
cancrella Hbn., 1825, missp.
faganella (Tr., 1833), emend.
purpurana Millière, 1874

STATHMOPODA H.-S., 1853
BOOCARA Butler, 1880
PLACOSTOLA Meyr., 1887
ERINEDA Bsk., 1909
AGRIOSCELIS Meyr., 1913
KAKIVORIA Nagano, 1916

1070 **elyella** Bsk., 1909
1071 **pedella** (L., 1761)
    alucitella (D. & S., 1775)
    cylindrella (F., 1777)
    angustipennella (Hbn., 1796)
    cylindricus (F., 1798)
    fastuosella (Costa, 1836)

    IDIOGLOSSA Wlsm., 1881
    METAMORPHA F. & B., 1878,
      preocc. by Hbn., 1819
    IDIOSTOMA Wlsm., 1882
1072 **miraculosa** (Frey, 1878)
    americella (Wlsm., 1882)

    CYPHACMA Meyr., 1915
1073 **tragiae** Braun, 1942

Pleurotini

    PLEUROTA Hbn., 1825
    EUPLEURIS Hbn., 1825
    HOLOSCOLIA Zell., 1839
    PROTASIS H.-S., 1853
1074 **albastrigulella** (Kft., 1907)

CHIMABACHINAE

Chimabachini

    CHEIMOPHILA Hbn., 1825
    DASYSTOMA Curt., 1833
1075 **salicella** (Hbn., 1796)
    rufocrinitalis (Zett., 1840)

ELACHISTIDAE
by RONALD W. HODGES

COELOPOETINAE

    COELOPOETA Wlsm., 1907
1076 **glutinosi** Wlsm., 1907
    baldella B. & Bsk., 1920

ELACHISTINAE

    ONCEROPTILA Braun, 1948
1077 **cygnodiella** (Bsk., 1921)
1078 **eremonoma** Braun, 1948

    STEPHENSIA Staint., 1858
1079 **cunilae** Braun, 1930

    HEMIPROSOPA Braun, 1948
1080 **albella** (Cham., 1877)

    ELACHISTA Tr., 1833
    APHELOSETIA Steph., 1834
    CYCNODIA H.-S., 1853
    PHIGALIA Cham., 1875, preocc. by
      Dup., 1829
    ATACHIA Wocke, 1876
    NEAERA Cham., 1880, preocc. by
      Robineau-Desvoidy, 1830
    HECISTA Wallgr., 1881
    APHIGALIA Dyar, 1903
    IRENICODES Meyr., 1919
    EUPROTEODES Viette, 1954
1081 **epimicta** Braun., 1948

1082 **symmorpha** Braun, 1948
1083 **orestella** Bsk., 1908
1084 **synopla** Braun, 1948
1085 **spatiosa** Braun, 1948
1086 **aurocristata** Braun, 1921
1087 **controversa** Braun, 1923
1088 **albella** (Cham., 1875)
1089 **adempta** Braun, 1948
    albella; Cham., 1875, not Wood,
      1835
1090 **griseicornis** Meyr., 1932
1091 **acenteta** Braun, 1948
1092 **hololeuca** Braun, 1948
1093 **purissima** Braun, 1948
1094 **lamina** Braun, 1948
1095 **sincera** Braun, 1925
1096 **parvipulvella** Cham., 1875
1097 **coniophora** Braun, 1948
1098 **hiberna** Braun, 1948
1099 **patriodoxa** Meyr., 1932
1100 **irrorata** Braun, 1920
    philopatris Meyr., 1932
1101 **fuliginea** Braun, 1948
1102 **oxytypa** Braun, 1948
1103 **pusilla** F. & B., 1876
1104 **unifasciella** Cham., 1875
1105 **maculoscella** (Clem., 1860)
1106 **excelsicola** Braun, 1948
1107 **stramineola** Braun, 1921
1108 **leucofrons** Braun, 1920
1109 **albicapitella** Engel, 1907
1110 **sylvestris** Braun, 1920
1111 **nitidiuscula** Braun, 1948
1112 **texanica** F. & B., 1876
1113 **maritimella** McD., 1942
1114 **staintonella** Cham., 1878
1115 **cana** Braun, 1920
1116 **amideta** Braun, 1948
1117 **inaudita** Braun, 1927
1118 **praelineata** Braun, 1915
1119 **solitaria** Braun, 1922
1120 **radiantella** Braun, 1922
1121 **madarella** (Clem., 1860)
1122 **enitescens** Braun, 1921
1123 **argentosa** Braun, 1920

    BISELACHISTA Traugott-Olsen &
      Nielsen, 1977
1124 **cucullata** (Braun, 1921)
1125 **agilis** (Braun, 1921)
1126 **leucosticta** (Braun, 1948)
1127 **tanyopis** (Meyr., 1932)
1128 **salinaris** (Braun, 1925)

    COSMIOTES Clem., 1860
1129 **illectella** Clem., 1860
    praematurella (Clem., 1860)
    cristatella (Cham., 1860)
    albapalpella (Cham., 1876)
    illicitella Dyar., 1903, missp.
1130 **herbigrada** (Braun, 1925)
1131 **scopulicola** Braun, 1948

    DICRANOCTETES Braun, 1918
    DONACIVOLA Bsk., 1934
1132 **brachyelytrifoliella** (Clem., 1864)
    angularis Braun, 1918

BLASTOBASIDAE
by RONALD W. HODGES

SYMMOCINAE

    SYMMOCA Hbn., 1825
    SIMOCA Weiler, 1877, missp.
    PARASYMMOCA Rebel, 1903
    ASARISTA Meyr., 1935
    CONQUASSATA Gozmány, 1957
    SYMMOLETRIA Gozmány, 1963
1133 **signatella** (H.-S., 1854)

    OEGOCONIA Staint., 1854
    OECOGENIA auth., missp.
    OECOGONIA Harlmann, 1880,
      missp.
    OECOGONIA Reutti, 1898, emend.
    APATEMA Wlsm., 1900
    CLEROGENES Meyr., 1921
    MICROGONIA Popescu-Gorj &
      Capuse, 1965
1134 **quadripuncta** (Haw., 1828)
    novimundi (Bsk., 1915)

    SCEPTEA Wlsm., 1891
    SCEPTIA McD., 1939, missp.
1135 **aequepulvella** (Cham., 1872), n.
    comb.
    aberratella (Bsk., 1907), n. syn.

    GLYPHIDOCERA Wlsm., 1892
    HARPAGANDRA Meyr., 1918
1136 **barythyma** Meyr., 1929
1137 **democratica** Meyr., 1929
    aequepulvella; auth., not Cham., 1872
1138 **floridanella** Bsk., 1901
1139 **lactiflosella** (Cham., 1878)
1140 **lithodoxa** Meyr., 1929
1141 **meyrickella** Bsk., 1907
1142 **septentrionella** Bsk., 1904
    speratella Bsk., 1908, n. syn.
    isonephes Meyr., 1929, n. syn.
1143 **dimorphella** Bsk., 1907

    GERDANA Bsk., 1908
1144 **caritella** Bsk., 1908
    telemacha (Meyr., 1927), n. comb., n.
      syn.

BLASTOBASINAE

Blastobasini

    BLASTOBASIS Zell., 1855
    EPISTETUS Wlsm., 1894
    PROSTHESIS Wlsm., 1907
1145 **coenomorpha** Meyr., 1931
1146 **distinctella** Dietz, 1910
1147 **eriobotryae** Bsk., 1915
1148 **guilandinae** Bsk., 1900
1149 **hulstella** Dietz, 1910
1150 **maritimella** McD., 1961
1151 **plummerella** Dietz, 1910
    fuscopurpurella Dietz, 1910
    a. *simplicella* Dietz, 1910
1152 **sagitella** Dietz, 1910
1153 **segnella** Zell., 1873
1154 **yuccaecolella** Dietz, 1910

BLASTOBASOIDES McD., 1961
1155 **differtellus** McD., 1961

HYPATOPA Wlsm., 1907
1156 **titanella** McD., 1961

ZENODOCHIUM Wlsm., 1908
1157 **citricolella** (Cham., 1880)
1158 **coccivorella** (Cham., 1880)

VALENTINIA Wlsm., 1907
1159 **confectella** (Zell., 1873)
1160 **floridella** Dietz, 1910
1161 **fractilinea** (Zell., 1873)
1162 **glandulella** (Riley, 1871)
    nubilella (Zell., 1873)
1163 **nothrotes** Wlsm., 1907
1164 **quaintancella** Dietz, 1910
1165 **repartella** Dietz, 1910
1166 **retectella** (Zell., 1873)

EURESIA Dietz, 1910
1167 **pulchella** Dietz, 1910

CALOSIMA Dietz, 1910
1168 **argyrosplendella** Dietz, 1910
1169 **dianella** Dietz, 1910

ASAPHOCRITA Meyr., 1931
1170 **protypica** Meyr., 1931

HOLCOCERA Clem., 1863
    CATACRYPSIS Wlsm., 1907
    CYNOTES Wlsm., 1907
    PROSODICA Wlsm., 1907
1171 **aphidiella** Wlsm., 1907
1172 **augusti** Heinr., 1920
1173 **boreasella** Dietz, 1910
1174 **busckiella** Dietz, 1910
1175 **chalcofrontella** Clem., 1863
    a. **quisquiliella** (Zell., 1873)
    b. **fumerella** Dietz, 1910
    c. **minorella** Dietz, 1910
1176 **clemensella** Cham., 1874
1177 **confamulella** Heinr., 1921
1178 **crassicornella** Dietz, 1910
1179 **crescentella** Dietz, 1910
    a. **annulipes** Dietz, 1910
1180 **dives** Dietz, 1910
    basipallidella Dietz, 1910, form
1181 **elyella** Dietz, 1910
1182 **estriatella** Dietz, 1910
1183 **fluxella** (Zell., 1873)
1184 **funebra** Dietz, 1910
    a. **reductella** Dietz, 1910
1185 **gigantella** Cham., 1876
1186 **gibbociliella** Clem., 1863
1187 **iceryaeella** (Riley, 1887)
1188 **illibella** Dietz, 1910
1189 **inclusa** Dietz, 1910
1190 **inconspicua** (Wlsm., 1907)
1191 **insulatella** Dietz, 1910
1192 **interpunctella** Dietz, 1910
1193 **irenica** (Wlsm., 1907)
1194 **lepidophaga** Clarke, 1960
1195 **livorella** (Zell., 1873)
1196 **maligemmella** Murt., 1898
1197 **melanostriatella** Dietz, 1910
    melonostriatella, missp.

1198 **messelinella** Dietz, 1910
    spoliatella Dietz, 1910
1199 **modestella** Clem., 1863
1200 **morrisoni** (Wlsm., 1907)
1201 **nana** Dietz, 1910
1202 **nigrostriata** Wlsm., 1907
1203 **nucella** (Wlsm., 1907)
1204 **ochrocephala** Dietz, 1910
1205 **panurgella** Heinr., 1920
1206 **paradoxa** Powell, 1976
1207 **plagiatella** Dietz, 1910
1208 **punctiferella** (Clem., 1863)
    a. **subsenella** (Zell., 1873)
    b. **texanella** (Wlsm., 1907)
1209 **purpurocomella** Clem., 1863
1210 **pusilla** Dietz, 1910
1211 **rufopunctella** Dietz, 1910
1212 **sciaphilella** (Zell., 1873)
    triangularisella Cham., 1875
1213 **simulella** Dietz, 1910
1214 **spretella** Dietz, 1910
1215 **stygna** (Wlsm., 1907)
1216 **tartarella** Dietz, 1910
1217 **ursella** (Wlsm., 1907)
1218 **vestaliella** Dietz, 1910
1219 **zelleriella** Dietz, 1910
    a. **annectella** Dietz, 1910

HOLCOCERINA McD., 1961
1220 **confluentella** (Dietz, 1910)
1221 **immaculella** (McD., 1930)
1222 **simplicis** McD., 1961
1223 **simuloides** McD., 1961

EUBOLEPIA Dietz, 1910
1224 **anomalella** Dietz, 1910
1225 **gargantuella** Heinr., 1920

Pigritiini

PLOIOPHORA Dietz, 1900
1226 **ampla** Dietz, 1900
1227 **fidella** Dietz, 1900

PIGRITIA Clem., 1860
1228 **angustipennella** Dietz, 1900
1229 **arizonella** Dietz, 1900
1230 **basilarella** Dietz, 1900
1231 **confusella** Dietz, 1900
1232 **laticapitella** Clem., 1860
    aufugella (Zell., 1873)
    luteopulvella (Cham., 1875)
1233 **mediofasciella** Dietz, 1900
1234 **obscurella** Dietz, 1900
1235 **ornatella** Dietz, 1900
1236 **purpurella** Dietz, 1900
1237 **spoliatella** Dietz, 1910
1238 **tristella** Dietz, 1900

EPIGRITIA Dietz, 1900
1239 **ochrocomella** (Clem., 1863)
    pallidotinctella Dietz, 1900
    heidemannella Dietz, 1900

DRYOPERIA Coolidge, 1909
    DRYOPE Cham., 1874, preocc. by
      Robineau-Desvoidy, 1830
1240 **canariella** (Dietz, 1900)
1241 **discopunctella** (Dietz, 1900)
1242 **fenyesella** (Dietz, 1900)

1243 **fuscosuffusella** (Dietz, 1900)
1244 **grisella** (Dietz, 1900)
1245 **minnicella** (Dietz, 1900)
1246 **murtfeldtella** (Cham., 1874)
    erratella (Dietz, 1910)
1247 **occidentella** (Dietz, 1900)
1248 **ochreella** (Clem., 1863)
1249 **tenebrella** (Dietz, 1900)

PSEUDOPIGRITIA Dietz, 1900
1250 **argyreella** Dietz, 1900
1251 **dorsomaculella** Dietz, 1900
1252 **equitella** Dietz, 1900
1253 **fraternella** Dietz, 1900

COLEOPHORIDAE

COLEOPHORINAE

by BARRY WRIGHT

COLEOPHORA Hbn., 1822
    EUPISTA Hbn., 1825
    APISTA Hbn., 1825
    HAPLOPTILIA Hbn., 1825
    PORRECTARIA Haw., 1828
    DAMOPHILA Curt., 1832
    ASTYAGES Steph., 1834
    METALLOSETIA Steph., 1834
    CASAS Wallgr., 1881
    CASIGNETA Wallgr., 1881
    CORYTHANGELA Meyr., 1897
    CASIGNETELLA Strand, 1928
    CALARITANIA Mariani, 1943
    HERINGIELLA Börner, 1944
    TOLLEOPHORA Capuse, 1971
    IONESCUMIA Capuse, 1971
    STOLLIA Capuse, 1971
    RAZOWSKIA Capuse, 1971
    ORGHIDANIA Capuse, 1971
    FREDERICKOENIGIA Capuse,
      1971
    SUIREIA Capuse, 1971
    ZAGULAJEVIA Capuse, 1971
    AMSELIPHORA Capuse, 1971
    NEMESIA Capuse, 1971
    ZANGHERIPHORA Capuse, 1971
    BOURGOGNEJA Capuse, 1971
    AURELIANIA Capuse, 1971
    BACESCUIA Capuse, 1971
    KLINZIGEDIA Capuse, 1971
    VLADDELIA Capuse, 1971
    KLIMESCHJA Capuse, 1971
    GLASERIA Capuse, 1971
    VALVULONGIA Capuse, 1971
    FALKOVITSHIA Capuse, 1972
    HELOPHAREA Falkovitsh, 1972
    CRICOTECHNA Falkovitsh, 1972
    PLEGMIDIA Falkovitsh, 1972
    AGAPALSA Falkovitsh, 1972
    PHYLLOSCHEMA Falkovitsh,
      1972
    BIMA Falkovitsh, 1972
    SYSTROPHOECA Falkovitsh,
      1972
    APORIPTURA Falkovitsh, 1972
    SYMPHYPODA Falkovitsh, 1972
    OEDICAULA Falkovitsh, 1972
    ARGYRACTINIA Falkovitsh, 1972
    CHNOOCERA Falkovitsh, 1972

ORTHOGRAPHIS Falkovitsh, 1972
PHAGOLAMIA Falkovitsh, 1972
MONOTEMACHIA Falkovitsh, 1972
CORETHROPOEA Falkovitsh, 1972
CHARACIA Falkovitsh, 1972
PERYGRA Falkovitsh, 1972
PERYGRIDIA Falkovitsh, 1972
LUZULINA Falkovitsh, 1972
CARPOCHENA Falkovitsh, 1972
ABARASCHIA Capuse, 1973
AMSELGHIA Capuse, 1973
ARDANIA Capuse, 1973
ASCLERIDUCTIA Capuse, 1973
BARASCHIA Capuse, 1973
BENANDERPIA Capuse, 1973
CALCOMARGINIA Capuse, 1973
CALEOPHORA Capuse, 1973, missp.
COROTHROPOEA Capuse, 1973, missp.
CORNULIVALVULIA Capuse, 1973, missp.
DUMITRESCUMIA Capuse, 1973
ECEBALIA Capuse, 1973
GLOBULIA Capuse, 1973
HAMULIELLA Capuse, 1973
HELVALBIA Capuse, 1973
KASYFIA Capuse, 1973
KUZNETZOVVLIA Capuse, 1973
LATISACCULIA Capuse, 1973
LONGIBACILLIA Capuse, 1973
LUCIDAESIA Capuse, 1973
LVARIA Capuse, 1973
MEMBRANIA Capuse, 1973
METAPISTA Capuse, 1973
MULTICOLORIA Capuse, 1973
NEUGENVIA Capuse, 1973
NOSYRISLIA Capuse, 1973
ORTOHGRAPHIS Capuse, 1973, missp.
OUDEJANSIA Capuse, 1973
PARAVALVULIA Capuse, 1973
PATZAKIA Capuse, 1973
POSTVINCULIA Capuse, 1973
PROGLASERIA Capuse, 1973
QUADRATIA Capuse, 1973
RHAMNIA Capuse, 1973
SACCULIA Capuse, 1973
SCLERIDUCTIA Capuse, 1973
TOLLSIA Capuse, 1973
TUBERCULIA Capuse, 1973
ULNA Capuse, 1973
DUCTISPIRA Capuse, 1973
KLIMESCHJOSEFIA Capuse, 1975
IONNEMESIA Capuse, 1973
1254 **malivorella** Riley, 1878
    cinerella Cham., 1878
    multipulvella Cham., 1878
    castipennella Wlsm., 1882
    atlantica Heinr., 1920
1255 **sacramenta** Heinr., 1914
    anatipennella; auth.
1256 **tiliaefoliella** Clem., 1861
    tilliaefoliella Cham., 1878, missp.
    tiliafoliella Heinr., 1920, missp.
1257 **atromarginata** Braun, 1914
1258 **albovanescens** Heinr., 1926
    currucipennella; Wlsm., not Zell., 1882

1259 **discostriata** Wlsm., 1882
1260 **elaeagnisella** Kft., 1908
    elaegnisella (McD., 1933), missp.
1261 **querciella** Clem., 1861
1262 **rosaefoliella** Clem., 1864
    ciliaeochrella Cham., 1874
1263 **laurentella** McD., 1944
1264 **vancouverensis** McD., 1944
1265 **annulicola** Braun, 1925
1266 **asterophagella** McD., 1944
1267 **wyethiae** Wlsm., 1882
1268 **monardella** (McD., 1933)
1269 **vernoniaeella** Cham., 1878
    veroniaeella Heinr., 1920, missp.
    vernoniaella Heinr., 1923, missp.
1270 **argentella** Cham., 1875
1271 **pruniella** Clem., 1861
    nigralineella Cham., 1876
    nigerlineella Cham., 1878
    ochrella Cham., 1878
    volkei Heinr., 1917
    piperata Braun, 1925
    innotabilis Braun, 1927
1272 **leucochrysella** Clem., 1863
1273 **kalmiella** (McD., 1936)
1274 **canadensisella** McD., 1955
1275 **salicivorella** McD., 1945
1276 **gaylussaciella** Heinr., 1915
    peregrinaevorella McD., 1954
1277 **cornivorella** McD., 1945
1278 **viburniella** Clem., 1861
    viburnella auth., missp.
1279 **affiliatella** McD., 1945
1280 **multicristatella** McD., 1954
1281 **dissociella** McD., 1955
1282 **vacciniivorella** McD., 1955
1283 **cretaticostella** Clem., 1860
1284 **murinella** Tengström, 1847
1285 **rupestrella** McD., 1955
1286 **ledi** Staint., 1860
1287 **persimplexella** McD., 1955
1288 **manitoba** Bsk., 1915
1289 **accordella** Wlsm., 1882
1290 **kearfottella** B. & Bsk., 1920
1291 **laticornella** Clem., 1860
    caryaefoliella Clem., 1861
1292 **corylifoliella** Clem., 1861
    coryliella Clem., 1861, missp.
1293 **juglandella** McD., 1946
1294 **lentella** Heinr., 1915
1295 **ostryae** Clem., 1861
    rufoluteella Cham., 1874
    carpinella Heinr., 1923
1296 **alniella** Heinr., 1914
1297 **cornella** Wlsm., 1882
    albiantennaella Wild, 1915
1298 **alnifoliae** Barasch, 1934
    alnivorella McD., 1946
1299 **umbratica** Braun, 1914
1300 **comptoniella** (McD., 1926)
    betulivora McD., 1946
    limosipennella; auth., part
1301 **ulmifoliella** McD., 1946
    limosipennella; auth., part
1302 **granifera** Braun, 1919
1303 **astericola** Heinr., 1920
1304 **paludoides** McD., 1957
    paludicola McD., 1945, preocc. by Staint., 1887
1305 **glaucella** Wlsm., 1882

1306 **polemoniella** Braun, 1919
1307 **cerasivorella** Pack., 1870
    nigrella Cham., 1878
    occidentalis Cham., 1878
    fletcherella Fern., 1892
    occidentis; auth.
    serratella; auth, not. L., 1761
1308 **serratella** (L., 1761)
    fuscedinella Zell., 1849
    metallicella Hodgkinson, 1892
    salmani Heinr., 1929
    insulicola McD., 1945, preocc. by Toll, 1942
    parasalmani Oudejans, 1971
1309 **irroratella** Wlsm., 1882
1310 **demissella** Braun, 1914
    pruniella; Wlsm., 1908, not Clem., 1861
1311 **laricella** (Hbn., 1814–17)
1312 **asterosella** McD., 1944
1313 **rosaevorella** McD., 1946
1314 **acutipennella** Wlsm., 1882
1315 **bistrigella** Cham., 1875
1316 **rosacella** Clem., 1864
1317 **viscidiflorella** Wlsm., 1882
1318 **heinrichella** (McD., 1933)
1319 **monardae** McD., 1945
1320 **lynosyridella** Wlsm., 1882
1321 **mcdunnoughiella** Oudejans, 1971
    dubiella McD., 1946; preocc. by Baker, 1888
1322 **entoloma** Bsk., 1913
1323 **sparsipuncta** Heinr., 1929
1324 **crinita** Braun, 1921
1325 **seminella** McD., 1946
1326 **atriplicivora** Ckll., 1898
1327 **suaedae** Bsk., 1915
1328 **acamtopappi** Bsk., 1915
1329 **quadristrigella** Bsk., 1913
1330 **simulans** McD., 1961
1331 **versurella** Zell., 1849
    thalassella McD., 1940
1332 **ericoides** Braun, 1919
    ericodes Heinr., 1923, missp.
1333 **subapicis** Braun, 1940
1334 **triplicis** McD., 1940
1335 **puberuloides** McD., 1956
1336 **texanella** Cham., 1878
1337 **duplicis** Braun, 1921
1338 **rugosae** McD., 1956
1339 **acuminatoides** McD., 1958
1340 **bidens** Braun, 1940
1341 **nemorella** McD., 1956
1342 **intermediella** McD., 1940
1343 **dextrella** Braun, 1940
1344 **detractella** McD., 1961
1345 **prepostera** Braun, 1923
1346 **trilineella** Cham., 1875
1347 **littorella** McD., 1940
1348 **salinoidella** McD., 1945
1349 **lineapulvella** Cham., 1874
    lapidicornis Wlsm., 1907
    amaranthella Braun, 1919
1350 **quadruplex** McD., 1940
1351 **chambersella** Dyar, 1903
    artemisicolella Cham., 1877, preocc.
1352 **sparsipulvella** Cham., 1877
    sparispulrella Cham., 1875, missp.
1353 **ochrostriata** Wlsm., 1882
1354 **basistrigella** Cham., 1877

1355 **nigrostriata** Wlsm., 1882
1356 **bella** Wlsm., 1882
1357 **angentialbella** Cham., 1874
1358 **quadrilineella** Cham., 1878
1359 **cervinella** McD., 1946
1360 **tenuis** Wlsm., 1882
1361 **borea** Braun, 1921
1362 **sparsiatomella** McD., 1941
1363 **pulchricornis** Wlsm., 1897
1364 **vagans** Wlsm., 1908
1365 **cratipennella** Clem., 1864
    gigantella Cham., 1874
    shaleriella Cham., 1874
    tamesis Waters, 1929
1366 **benestrigatella** McD., 1941
1367 **brunneipennis** Braun, 1921
1368 **bidentella** McD., 1941
1369 **suaedicola** Ckll., 1898
1370 **biforis** Braun, 1921
1371 **infuscatella** Clem., 1860
1372 **coenosipennella** Clem., 1860
    caenosipennella auth., missp.
1373 **contrariella** McD., 1955
1374 **sexdentatella** McD., 1958
1375 **caespititiella** Zell., 1839
1376 **latronella** McD., 1940
1377 **glissandella** McD., 1942
1378 **glaucicolella** Wood, 1892
1379 **alticolella** Zell., 1849
1380 **fagicorticella** Cham., 1874
    fagicosticella Cham., 1874, incorr.
      orig. spell.
1381 **bispinatella** McD., 1954
1382 **dentiferoides** McD., 1958
1383 **luteocostella** Cham., 1875
1384 **concolorella** Clem., 1863
    unicolorella Cham., 1874
1385 **maritella** McD., 1941
1386 **viridicuprella** Wlsm., 1882
1387 **spissicornis** (Haw., 1828)
    coruscipennella Clem., 1860
    auropurpuriella Cham., 1874
    corruscipennella Cham., 1878, missp.
    auropurpurella McD., 1944, missp.
1388 **trifolii** (Curt., 1832)
    frischella; auth.
1389 **apicialbella** Braun, 1920
    apicella; Braun, 1919, not Staint., 1858
1390 **portulacae** Ckll., 1898
1391 **aeneusella** Cham., 1874, nom. nud.
1392 **aenusella** Cham., 1878, ident. uncert.
1393 **albacostella** Cham., 1875, mispl.
1394 **biminimmaculella** Cham., 1878,
    mispl.
1395 **fuscostrigella** Cham., 1878, mispl.
1396 **indefinitella** Oudejans, 1971, repl.
    name
    zelleriella Cham., 1874, preocc. by
      Heinemann, 1854
1397 **inornatella** Cham., 1880, mispl.
1398 **octagonella** Wlsm., 1882, mispl.

BATRACHEDRINAE
by RONALD W. HODGES

BATRACHEDRA H.-S., 1853
    EUSTAINTONIA Spuler, 1910
1399 **praeangusta** (Haw., 1828)
    turdipennella (Koll., 1832)
    clemensella Cham., 1877

1400 **striolata** Zell., 1875
    pulvella (Cham., 1876)
1401 **curvilineella** (Cham., 1872)
1402 **folia** Hodges, 1966
1403 **salicipomonella** Clem., 1865
1404 **illusor** Hodges, 1966
1405 **testor** Hodges, 1966
1406 **busiris** Hodges, 1966
1407 **calator** Hodges, 1966
1408 **concitata** Meyr., 1928
1409 **elucus** Hodges, 1966
1410 **garritor** Hodges, 1966
1411 **scitator** Hodges, 1966
1412 **hageter** Hodges, 1966
1413 **enormis** Meyr., 1928
1414 **mathesoni** Bsk., 1916
1415 **libator** Hodges, 1966
1416 **decoctor** Hodges, 1966

CHEDRA Hodges, 1966
1417 **inquisitor** Hodges, 1966
1418 **pensor** Hodges, 1966

DUOSPINA Hodges, 1966
1419 **abolitor** Hodges, 1966
1420 **trichella** (Bsk., 1908)
    concors (Meyr., 1917)

HOMALEDRA Bsk., 1900
1421 **heptathalama** Bsk., 1900
1422 **sabalella** (Cham., 1880)

MOMPHIDAE
by RONALD W. HODGES

MOMPHA Hbn., 1825
    LAVERNA Curt., 1839
    LOPHOPTILUS Sircom, 1848
    CYPHOPHORA H.-S., 1853
    PSACAPHORA H.-S., 1853
    ANYBIA Staint., 1854
    WILSONIA Clem., 1864
    LEUCOPHRYNE Cham., 1875
1423 **albapalpella** (Cham., 1875)
1424 **albella** (Cham., 1875)
1425 **annulata** (Braun, 1923)
1426 **argentimaculella** (Murt., 1900)
1427 **bicristatella** (Cham., 1879)
1428 **bifasciella** (Cham., 1876)
1429 **bottimeri** Bsk., 1940
1430 **brevivittella** (Clem., 1864)
    oenotheraesemenella (Cham., 1876)
    oenotheraevorella (Cham., 1880)
1431 **canicinctella** (Clem., 1863), n. comb.
1432 **capella** Bsk., 1940
1433 **cephalonthiella** (Cham., 1871)
    cephalanthiella (Cham., 1875), missp.
1434 **circumscriptella** (Zell., 1873)
1435 **claudiella** Kft., 1907
1436 **coloradella** (Cham., 1877)
1437 **communis** (Braun, 1925)
1438 **conturbatella** (Hbn., 1818–19)
1439 **deceptella** (Braun, 1921)
1440 **definitella** (Zell., 1873)
    unicristatella (Cham., 1875)
1441 **difficilis** (Braun, 1923)
1442 **edithella** (B. & Bsk., 1920)
1443 **eloisella** (Clem., 1860)
    magnatella (Zell., 1873)

    oenotheraeella (Cham., 1875)
    lyonetiella (Cham., 1875)
1444 **ignobilisella** (Cham., 1875)
    ignotilisella (Cham., 1875), missp.
1445 **luciferella** (Clem., 1860)
1446 **metallifera** (Wlsm., 1882)
1447 **minimella** (Cham., 1880)
1448 **murtfeldtella** (Cham., 1875)
    albocapitella (Cham., 1875)
    grissella (Cham., 1875)
    obscurusella (Cham., 1875)
    parvicristatella (Cham., 1875)
1449 **nuptialis** Meyr., 1922
1450 **passerella** (Bsk., 1909)
1451 **pecosella** Bsk., 1907
1452 **purpuriella** (Bsk., 1909)
1453 **rufocristatella** (Cham., 1875)
1454 **sexstrigella** (Braun, 1921)
1455 **stellella** Bsk., 1906
1456 **terminella** (Westwood, 1851)
    engelella Bsk., 1906
1457 **tricristatella** (Cham., 1875)
    grandisella (Cham., 1875)
    subiridescens (Wlsm., 1882)
1458 **unifasciella** (Cham., 1876)

SYNALLAGMA Bsk., 1907
1459 **busckiella** Engel, 1907

AGONOXENIDAE
by RONALD W. HODGES

BLASTODACNINAE

Blastodacnini

BLASTODACNA Wocke, 1876
1460 **bicristatella** (Cham., 1875)
    placendiella (Bsk., 1908)
    sublustris (Meyr., 1922)
1461 **curvilineella** (Cham., 1872)

GLYPHIPTERYX Curt., 1827
    CHRYSOCLISTA Staint., 1854
1462 **cambiella** (Bsk., 1915)
1463 **linneella** (Cl., 1759)
1464 **villella** (Bsk., 1904)

Parametriotini

AETIA Cham., 1880
    TETANOCENTRIA Rebel, 1902
    PLATYBATHRA Meyr., 1912
    CHAETOCAMPA Bsk., 1926
1465 **bipunctella** Cham., 1880
    crotonella (Bott., 1926)

COSMOPTERIGIDAE
by RONALD W. HODGES

ANTEQUERINAE

ANTEQUERA Clarke, 1941
1466 **acertella** (Bsk., 1913)

EUCLEMENSIA Grt., 1878
    HAMADRYAS Clem., 1864,
      preocc. by Hbn., 1806

1467 **bassettella** (Clem., 1864)
1468 **schwarziella** Bsk., 1901

COSMOPTERIGINAE

COSMOPTERIX Hbn., 1825
   COSMOPTERYX Zeller, 1839,
     missp.
1469 **nitens** Wlsm., 1889
1470 **sinelinea** Hodges, 1978
1471 **molybdina** Hodges, 1962
1472 **pulchrimella** Cham., 1875
1473 **bendidia** Hodges, 1962
1474 **attenuatella** (Wlk., 1864)
   flavofasciata Woll., 1879
   mimetis Meyr., 1897
   antillia Fbs., 1931
   superba Gozmány, 1960
1475 **clandestinella** Bsk., 1906
1476 **montisella** Cham., 1875
   unicolorella Wlsm., 1889
1477 **magophila** Meyr., 1919
1478 **gracilens** Hodges, 1962
1479 **dapifera** Hodges, 1962
1480 **delicatella** Wlsm., 1889
1481 **dicacula** Hodges, 1962
1482 **lespedezae** Wlsm., 1882
1483 **opulenta** Braun, 1919
1484 **chisosensis** Hodges, 1978
1485 **quadrilineella** Cham., 1878
1486 **minutella** Beutenmüller, 1889
1487 **abdita** Hodges, 1962
1488 **inopis** Hodges, 1962
1489 **chalybaeella** Wlsm., 1889
1490 **gemmiferella** Clem., 1860
1491 **bacata** Hodges, 1962
1492 **damnosa** Hodges, 1962
1493 **clemensella** Staint., 1860
   hermodora Meyr., 1919
1494 **scirpicola** Hodges, 1962
1495 **ebriola** Hodges, 1962
1496 **fernaldella** Wlsm., 1882
1497 **floridanella** Beutenmüller, 1889
   nigrapunctella Bsk., 1900
1498 **facunda** Hodges, 1962

PEBOPS Hodges, 1978
1499 **ipomoeae** (Bsk., 1900)

TANYGONA Braun, 1923
1500 **lignicolorella** Braun, 1923

ERALEA Hodges, 1962
1501 **albalineella** (Cham., 1878)
   striata Hodges, 1962
1502 **abludo** Hodges, 1978

MELANOCINCLIS Hodges, 1962
1503 **lineigera** Hodges, 1962
1504 **nigrilineella** (Cham., 1878)
1505 **sparsa** Hodges, 1978
1506 **gnoma** Hodges, 1978
1507 **vibex** Hodges, 1978

STAGMATOPHORA H.-S., 1853
   ETEOBALEA Hodges, 1962
   PARASTAGMATOPHORA Riedl,
     1965
1508 **sexnotella** (Cham., 1878)
1509 **wyattella** B. & Bsk., 1920

1510 **iridella** Bsk., 1907
   niphochrysa Meyr., 1930
1511 **enchrysa** (Hodges, 1962)

PYRODERCES H.-S., 1853
   ANATRACHYNTIS Meyr., 1915
   SATHROBROTA Hodges, 1962
1512 **rileyi** (Wlsm., 1882)
   stigmatophora (Wlsm., 1897)
1513 **badia** (Hodges, 1962)
1514 **albistrigella** (Mösch., 1890)

LIMNAECIA Staint., 1851
   LIMNOECIA auth., missp.
1515 **phragmitella** Staint., 1851

TELADOMA Bsk., 1932
1516 **helianthi** Bsk., 1932
1517 **astigmatica** (Meyr., 1928)
1518 **tonia** Hodges, 1978
1519 **incana** Hodges, 1962
1520 **murina** Hodges, 1962
1521 **habra** Hodges, 1978
1522 **nebula** Hodges, 1978
1523 **exigua** Hodges, 1978

TRICLONELLA Bsk., 1901
   ANORCOTA Meyr., 1920
   PHARMACOPTIS Meyr., 1932
1524 **pergandeella** Bsk., 1901
1525 **xuthocelis** Hodges, 1962
1526 **antidectis** (Meyr., 1914)
1527 **determinatella** (Zell., 1873)
   australisella (Cham., 1875)
1528 **bicoloripennis** Hodges, 1962

ANONCIA Clarke, 1941
1529 **conia** (Wlsm., 1907)
   marinensis (Keif., 1935)
1530 **brunneipes** Hodges, 1962
1531 **sphacelina** (Keif., 1935)
1532 **fasciata** (Wlsm., 1907)
1533 **alboligula** Hodges, 1962
1534 **slales** Hodges, 1978
1535 **glacialis** Hodges, 1962
1536 **bitoqua** Hodges, 1978
1537 **porriginosa** Hodges, 1962
1538 **flegax** Hodges, 1978
1539 **aciculata** (Meyr., 1928)
1540 **mones** Hodges, 1978
1541 **naclia** Hodges, 1978
1542 **callida** Hodges, 1962
1543 **nocticola** Hodges, 1962
1544 **nebritis** Hodges, 1962
1545 **psepsa** Hodges, 1978
1546 **leucoritis** (Meyr., 1927)
   mentzeliae Clarke, 1942
1547 **longa** (Meyr., 1927)
1548 **furvicosta** Hodges, 1962
1549 **diveni** (Heinr., 1921)
1550 **orites** (Wlsm., 1907)
1551 **fregeis** Hodges, 1978
1552 **loexya** Hodges, 1978
1553 **noscres** Hodges, 1978
1554 **episcia** (Wlsm., 1907)
1555 **venis** Hodges, 1978
1556 **piperata** Hodges, 1962
1557 **mosa** Hodges, 1978
1558 **smogops** Hodges, 1978
1559 **psentia** Hodges, 1978

CHRYSOPELEIINAE

PERIPLOCA Braun, 1919
1560 **orichalcella** (Clem., 1864)
   concolorella (Cham., 1875)
   purpuriella Braun, 1919
1561 **arsa** Hodges, 1978
1562 **ceanothiella** (Cosens, 1908)
1563 **gleditschiaeella** (Cham., 1876)
   gleditschiaeela Hodges, 1962, missp.
1564 **intermedia** Hodges, 1978
1565 **repanda** Hodges, 1978
1566 **hostiata** Hodges, 1969
1567 **teres** Hodges, 1978
1568 **tridens** Hodges, 1978
1569 **hortatrix** Hodges, 1969
1570 **opinatrix** Hodges, 1969
1571 **atrata** Hodges, 1962
1572 **mimula** Hodges, 1962
1573 **laeta** Hodges, 1962
1574 **cata** Hodges, 1962
1575 **devia** Hodges, 1969
1576 **soror** Hodges, 1978
1577 **nigra** Hodges, 1962
1578 **fessa** Hodges, 1962
1579 **gulosa** Hodges, 1962
1580 **facula** Hodges, 1962
1581 **juniperae** Hodges, 1978
1582 **funebris** Hodges, 1962
1583 **serrulata** Hodges, 1978
1584 **dentella** Hodges, 1978
1585 **dipapha** Hodges, 1969
1586 **labes** Hodges, 1969

SISKIWITIA Hodges, 1969
1587 **alticolans** Hodges, 1969
1588 **latebra** Hodges, 1978
1589 **falcata** Hodges, 1978

PRISTEN Hodges, 1978
   NEOPLOCA Hodges, 1964, preocc.
     by Matsumura, 1927
1590 **corusca** (Hodges, 1964)

SYNPLOCA Hodges, 1964
1591 **gumia** Hodges, 1964

STILBOSIS Clem., 1860
   AEAEA Cham., 1874
   AMAUROGRAMMA Braun, 1919
1592 **juvantis** (Hodges, 1964)
1593 **dulcedo** (Hodges, 1964)
1594 **venifica** (Hodges, 1964)
1595 **venatrix** (Hodges, 1964)
1596 **ostryaeella** (Cham., 1874)
1597 **quadricustatella** (Cham., 1880)
   quadricristatella auth., missp.
1598 **stipator** (Hodges, 1964)
1599 **risor** (Hodges, 1964)
1600 **victor** (Hodges, 1964)
1601 **placatrix** (Hodges, 1969)
1602 **sagana** (Hodges, 1964)
1603 **rhynchosiae** (Hodges, 1964)
1604 **extensa** (Braun, 1919)
1605 **pagina** Hodges, 1978
1606 **ornatrix** Hodges, 1978
1607 **scleroma** Hodges, 1978
1608 **rotunda** Hodges, 1978
1609 **tesquella** Clem, 1860
   quinquicristatella (Cham., 1881)

1610 **nubila** Hodges, 1964
1611 **lonchocarpella** Bsk., 1934

CHRYSOPELEIA Cham., 1874
1612 **purpuriella** Cham., 1874

WALSHIA Clem., 1864
1613 **particornella** (Bsk., 1919)
1614 **elegans** Hodges, 1978
1615 **miscecolorella** (Cham., 1875)
    miscecalonella (Cham., 1875), missp.
1616 **amorphella** Clem., 1864
1617 **floridensis** Hodges, 1978
1618 **exemplata** Hodges, 1961
1619 **similis** Hodges, 1961
1620 **dispar** Hodges, 1961

NEPOTULA Hodges, 1964
1621 **secura** Hodges, 1964

AFEDA Hodges, 1978
1622 **biloba** Hodges, 1978

PERIMEDE Cham., 1874
1623 **erransella** Cham., 1874
1624 **battis** Hodges, 1962
1625 **latris** Hodges, 1962
1626 **grandis** Hodges, 1978
1627 **parilis** Hodges, 1969
1628 **circitor** Hodges, 1969
1629 **erema** Hodges, 1969
1630 **maniola** Hodges, 1969
1631 **ricina** Hodges, 1962
1632 **falcata** Braun, 1919

SORHAGENIA Spuler, 1910
    CYSTIOECETES Braun, 1915
1633 **nimbosa** (Braun, 1915)
1634 **cracens** Hodges, 1978
1635 **daedala** Hodges, 1964
1636 **baucidis** Hodges, 1969
1637 **pexa** Hodges, 1969

ITHOME Cham., 1875
    ERIPHIA Cham., 1875, preocc. by
    Latr., 1817
1638 **concolorella** (Cham., 1875)
    unimaculella Cham., 1875
    unomaculella auth., missp.
1639 **curvipunctella** (Wlsm., 1892)
    quinquepunctata (Fbs., 1931)
1640 **ferax** Hodges, 1962
1641 **edax** Hodges, 1962
1642 **aquila** Hodges, 1978
1643 **lassula** Hodges, 1962
1644 **simulatrix** Hodges, 1978

OBITHOME Hodges, 1964
1645 **punctiferella** (Bsk., 1906)

SCYTHRIDIDAE
by RONALD W. HODGES

SCYTHRIS Hbn., 1825
    BUTALIS Tr., 1833, preocc. by
    Boie, 1826
    AROTRURA Wlsm., 1888
    COLINITA Bsk., 1907
1646 **albacostella** (Cham., 1875), n. comb.

1647 **albapenella** (Cham., 1875)
1648 **albilineata** (Wlsm., 1888)
1649 **altisierrae** Keif., 1937
1650 **anthracina** Braun, 1923
1651 **aterrimella** (Wlk., 1864)
1652 **basilaris** (Zell., 1855)
    flavifrontella (Clem., 1860)
1653 **charon** Meyr., 1918
1654 **confinis** Braun, 1920
1655 **eboracensis** (Zell., 1855)
1656 **epilobiella** McD., 1942
1657 **eburnea** (Wlsm., 1888)
    arizoniella (Kft., 1907)
1658 **fissirostris** Meyr., 1928
1659 **fuscicomella** (Clem., 1860)
1660 **graminivorella** Braun, 1920
1661 **hemidictyas** Meyr., 1928
1662 **impositella** (Zell., 1855)
    matutella (Clem., 1860)
    monstratella (Wlk., 1864)
    dorsipallidella (Cham., 1875)
    buristriga (Cham., 1875)
    brevistrigella (Cham., 1875)
    immaculatella (Cham., 1875)
    brevistriga (Cham., 1878), missp.
1663 **interrupta** Braun, 1920
1664 **magnatella** Bsk., 1904
1665 **mixaula** Meyr., 1916
1666 **ochristriata** (Wlsm., 1888)
1667 **oxyplecta** Meyr., 1916
1668 **pacifica** McD., 1927
1669 **perspicellella** (Wlsm., 1888)
1670 **pilosella** (Zell., 1873)
1671 **piratica** Meyr., 1928
1672 **plausipennella** (Cham., 1875)
    planipennella (Cham., 1878), emend.
1673 **quadriguttella** (Thunb., 1794)
    tristella (Hbn., 1796)
    chenopodiella (Hbn., 1810–13)
1674 **reducta** Braun, 1923
1675 **scintillifera** Braun, 1927
1676 **sponsella** (Bsk., 1907)
1677 **suffusa** (Wlsm., 1888)
1678 **trivinctella** (Zell., 1873)
1679 **ypsilon** Braun, 1920

ARENISCYTHRIS Powell, 1976
1680 **brachypteris** Powell, 1976

GELECHIIDAE
by RONALD W. HODGES

ANOMOLOGINAE

NEALYDA Dietz, 1900
1681 **bifidella** Dietz, 1900
1682 **kinzelella** Bsk., 1900
1683 **phytolaccae** Clarke, 1946
1684 **pisoniae** Bsk., 1900

METZNERIA Zell., 1839
    CLEODORA Steph., 1834, preocc.
    by Péron & Lesueur, 1810
    PARASIA Dup., 1846
    ARCHIMETZNERIA Amsel, 1936
1685 **lappella** (L., 1758)
1686 **paucipunctella** Zell., 1839
    zimmermanni Hering, 1941
    confusalis Lucas, 1956

MEGACRASPEDUS Zell., 1839
    NEDA Cham., 1874, preocc. by
    Mulsant, 1850
    PYCNOBATHRA Lower, 1901
    AUTONEDA Bsk., 1903, repl.
    name
    TOXOCERAS Chrétien, 1915
1687 **plutella** (Cham., 1874)

ISOPHRICTIS Meyr., 1917
1688 **actiella** B. & Bsk., 1920
1689 **actinopa** Meyr., 1929
1690 **anteliella** (Bsk., 1903)
1691 **canicostella** (Wlsm., 1888)
1692 **cilialineella** (Cham., 1874)
1693 **dietziella** (Bsk., 1903)
1694 **magnella** (Bsk., 1903)
1695 **modesta** (Wlsm., 1888)
1696 **occidentalis** Braun, 1925
1697 **pallidastrigella** (Cham., 1874)
1698 **pallidella** (Cham., 1874)
1699 **pennella** (Bsk., 1907)
1700 **rudbeckiella** Bott., 1926
    denotata Bott., 1926, form
1701 **sabulella** (Wlsm., 1888)
1702 **similiella** (Cham., 1872)
    solaniiella (Cham., 1873), repl. name
    piscipellis (Zell., 1873)
    solaniella (Cham., 1875), missp.
    piscipalis (Cham., 1878), missp.
    similella Meyr., 1926, missp.
1703 **striatella** (D. & S., 1775)
    tanacetella (Schr., 1802)
1704 **tophella** (Wlsm., 1888)
1705 **trimaculella** (Cham., 1874), n. comb.
    touceyella (Bsk., 1903), n. comb.

MONOCHROA Heinemann, 1870
    CATABRACHMIA Rebel, 1909
1706 **absconditella** (Wlk., 1864), n. comb.
    palpiannulella (Cham., 1872), n.
    comb.
1707 **angustipennella** (Clem., 1863), n.
    comb.
    kearfottella (Bsk., 1903), n. comb.
1708 **disconotella** (Cham., 1878), n. comb.
1709 **discriminata** (Meyr., 1923), n. comb.
1710 **gilvolinella** (Clem., 1863), n. comb.
1711 **fragariae** (Bsk., 1919), n. comb.
1712 **harrisonella** (Bsk., 1904), n. comb.
1713 **monactis** (Meyr., 1923), n. comb.
1714 **perterrita** (Meyr., 1923), n. comb.
1715 **placidella** (Zell., 1874), n. comb.
    natalella (Bsk., 1904), n. comb.
1716 **quinquepunctella** (Bsk., 1903), n. comb.

CHRYSOESTHIA Hbn., 1825
    MICROSETIA Steph., 1829
    CHRYSIA Bruand, 1850
    NOMIA Clem., 1860, preocc. by
    Latr., 1804
    CHRYSOPORA Clem., 1860
    NANNODIA Heinemann, 1870
1717 **drurella** (F., 1775)
    myllerella (F., 1794)
    zinckeella (Hbn., 1810–13)
    hermannella; auth.
1718 **lingulacella** (Clem., 1860)
    hermannella; auth.
    arminiella (F. & B., 1878)

1719 **sexguttella** (Thunb., 1794)
    aurofasciella (Steph., 1834)
    naeviferella (Dup., 1843)
1720 **versicolorella** (Kft., 1908)

    ENCHRYSA Zell., 1873
1721 **dissectella** Zell., 1873
    youngella (Kft., 1905)

    THEISOA Cham., 1874
    HELICE Cham., 1873, preocc. by
      de Haan, 1835
    CACELICE (Bsk., 1902)
1722 **constrictella** (Zell., 1873)
    bifasciella (Cham., 1874)
1723 **multifasciella** (Cham., 1875)
1724 **pallidochrella** (Cham., 1873)
    gleditschiaeella (Cham., 1877)
    permolestella (Bsk., 1902)

GELECHIINAE

    STEREOMITA Braun, 1922
1725 **andropogonis** Braun, 1922

    ARISTOTELIA Hbn., 1825
    ERGATIS Heinemann, 1870,
      preocc. by Blackwall, 1841
    EUCATOPTUS Wlsm., 1897
1726 **adceanotha** Keif., 1935
1727 **adenostomae** Keif., 1933
1728 **amelanchierella** Braun, 1925
1729 **aquosa** Meyr., 1925
    suffusella (Cham., 1872), preocc. by
      Douglas, 1850
1730 **argentifera** Bsk., 1903
1731 **bifasciella** Bsk., 1903
1732 **callens** Meyr., 1923
1733 **callirhoda** Meyr., 1923
1734 **devexella** Braun, 1925
1735 **eldorada** Keif., 1936
    eldorado McD., 1939, missp.
1736 **elegantella** (Cham., 1874)
    superbella (Cham., 1875), nom.
      nud.
1737 **eumeris** Meyr., 1923
1738 **fungivorella** (Clem., 1864)
1739 **hexacopa** Meyr., 1929
1740 **intermediella** (Cham., 1878)
1741 **iospora** Meyr., 1929
1742 **isopelta** Meyr., 1929
1743 **ivae** Bsk., 1900
1744 **lespedezae** Braun, 1930
1745 **lindanella** B. & Bsk., 1920
1746 **melanaphra** Meyr., 1923
1747 **molestella** (Zell., 1873)
1748 **monilella** B. & Bsk., 1920
1749 **nigrobasiella** Clarke, 1932
1750 **ochroxysta** Meyr., 1929
1751 [see 1957]
1752 **physaliella** (Cham., 1872)
1753 **planitia** Braun, 1925
1754 **primipilana** Meyr., 1923
1755 **psoraleae** Braun, 1930
1756 **pudibundella** (Zell., 1873)
1757 **pullusella** (Cham., 1874)
    minimella (Cham., 1874)
1758 **rhamnina** Keif., 1933
1759 **rhoisella** Bsk., 1934
1760 **robusta** Braun, 1921

1761 **roseosuffusella** (Clem., 1860)
    bellela (Wlk., 1865)
1762 **rubidella** (Clem., 1860)
    rubensella (Cham., 1872), n. syn.
1763 **salicifungiella** (Clem., 1864)
1764 **urbaurea** Keif., 1933

    NUMATA Bsk., 1906
1765 **bipunctella** Bsk., 1906

    GLAUCE Cham., 1875
1766 **pectenalaeella** Cham., 1875

    NAERA Cham., 1875
    LEUCE Cham., 1875, repl. name
1767 **fuscocristatella** Cham., 1875
    belfragesella (Cham., 1879)

    EVIPPE Cham., 1873
    PHAETUSA Cham., 1875, preocc.
      by Wagler, 1832
    THOLEROSTOLA Meyr., 1917
1768 **abdita** Braun, 1925
1769 **laudatella** (Wlsm., 1907)
1770 **leuconota** (Zell., 1873)
    plutella (Cham., 1875)
1771 **prunifoliella** Cham., 1873

    AGNIPPE Cham., 1872
1772 **biscolorella** Cham., 1872
1773 **crinella** Keif., 1927
1774 **evippeella** Bsk., 1906
1775 **fuscopulvella** Cham., 1872

    TOSCA Heinr., 1920
1776 **elachistella** (Bsk., 1906)
1777 **plutonella** Heinr., 1920
1778 **pollostella** (Bsk., 1906)

    ARGYROLACIA Keif., 1936
1779 **bifida** Keif., 1936

    RECURVARIA Haw., 1828
    TELEA Steph., 1834, preocc. by
      Hbn., 1819
    APHANAULA Meyr., 1895
    HINNEBERGIA Spuler, 1910
    MICROLECHIA Turati, 1924
1780 **ceanothiella** Braun, 1921
1781 **consimilis** Braun, 1930, mispl.
1782 **francisca** Keif., 1928
1783 **nanella** (D. & S., 1775)
    crataegella Bsk., 1903
1784 **stibomorpha** Meyr., 1929
1785 **taphiopis** Meyr., 1929
1786 **vestigata** Meyr., 1929, mispl.

    COLEOTECHNITES Cham., 1880
    EVAGORA Clem., 1860, preocc. by
      Péron & Lesueur, 1810
    EIDOTHEA Cham., 1873, preocc.
      by Risso, 1826
    EIDOTHOA Cham., 1873, missp.
    EUCORDYLEA Dietz, 1900
    PULICALVARIA Free., 1963
1787 **albicostatus** (Free., 1965)
1788 **alnifructella** (Bsk., 1915)
1789 **apicitripunctella** (Clem., 1860)
    attritella (Wlk., 1864)
    abietisella (Pack., 1883)

1790 **ardas** (Free, 1960)
1791 **argentiabella** (Cham., 1874)
1792 **atrupictella** (Dietz, 1900)
1793 **australis** (Free., 1963)
1794 **bacchariella** (Keif., 1927)
1795 **biopes** (Free., 1960)
1796 **blastovora** (McLeod, 1962)
1797 **canusella** (Free., 1957)
1798 **carbonarius** (Free., 1965)
1799 **chilcotti** (Free., 1963)
1800 **citriella** (Cham., 1880)
1801 **colubrinae** (Bsk., 1903)
1802 **condignella** (Bsk., 1929)
1803 **coniferella** (Kft., 1907)
1804 **cristatella** (Cham., 1875)
1805 **vagatioella** (Cham., Oct. 1873)
    dorsivittella (Zell., Dec. 1873)
1806 **ducharmei** (Free., 1962)
1807 **elucidella** (B. & Bsk., 1920)
1808 **eryngiella** (Bott., 1926)
1809 **florae** (Free., 1960)
1810 **gallicola** (Bsk., 1915)
1811 **gibsonella** (Kft., 1907)
1812 **granti** (Free., 1965)
1813 **huntella** (Keif., 1936)
1814 **invictella** (Bsk., 1908)
1815 **juniperella** (Kft., 1903)
1816 **laricis** (Free., 1965)
1817 **lewisi** (Free., 1960)
1818 **mackiei** (Keif., 1932)
1819 **macleodi** (Free., 1965)
1820 **martini** (Free., 1965)
1821 **milleri** (Bsk., 1914)
1822 **moreonella** (Heinr., 1920)
1823 **nigritus** Hodges, n. name
    niger (Bsk., 1903), preocc. by (Haw., 1828)
1824 **obliquistrigella** (Cham., 1872)
1825 **occidentis** (Free., 1965)
1826 **piceaella** (Kft., 1903)
    niger (Kft., 1903), preocc. by (Haw.,
      1828)
    obscurella (Kft., 1903), repl. name
1827 **pinella** (Bsk., 1906)
1828 **quercivorella** (Cham., 1872)
    gilviscopella (Zell., 1873)
1829 **resinosae** (Free., 1960)
1830 **stanfordia** (Keif., 1933)
1831 **starki** (Free., 1957)
1832 **thujaella** (Kft., 1903)
1833 **variella** (Cham., 1872)

    SINOE Cham., 1873
1834 **robiniella** (Fitch, 1859)
    fuscopalidella Cham., 1873

    EXOTELEIA Wallgr., 1881
    PARALECHIA Bsk., 1903
    HERINGIA Spuler, 1910, preocc.
      by Rondani, 1856
    HERINGIOLA Strand, 1917, repl. name
1835 **burkei** Keif., 1932
1836 **californica** (Bsk., 1907)
1837 **dodecella** (L., 1758)
    annulicornis (Steph., 1834)
1838 **graphicella** (Bsk., 1903), n. comb.
1839 **nepheos** Free., 1967
1840 **pinifoliella** (Cham., 1880)

    TRYPANISMA Clem., 1860
1841 **prudens** Clem., 1860

quinqueannulella (Cham., 1872)
fagella (Bsk., 1903), n. syn.

TAYGETE Cham., 1873
1842 **attributella** (Wlk., 1864)
    difficilisella (Cham., 1872)
1843 **citrinella** (B. & Bsk., 1920), mispl.
1844 **decemmaculella** (Cham., 1872), n.
    comb., mispl.
    bicostomaculella (Cham., 1877), n.
      syn.
    thoracella (Wlsm., 1888), n. syn.
    osteosema (Meyr., 1929), n. syn.
1845 **gallaegenitella** (Clem., 1864), mispl.
    geminella (Riley, 1871)
1846 **saundersella** (Cham., 1876), mispl.
1847 **sylvicolella** (Bsk., 1903), mispl.

LEUCOGONIELLA Fletcher, 1940
    LEUCOGONIA Meyr., 1929,
      preocc. by Hampson, 1910
1848 **californica** (Keif., 1930)
1849 **distincta** (Keif., 1935)
1850 **subsimella** (Clem., 1860)

AROGALEA Wlsm., 1910
1851 **cristifasciella** (Cham., 1878)
    inscripta (Wlsm., 1882)

ATHRIPS Billberg, 1820
    RHYNCHOPACHA Stgr., 1871
    EPITHECTIS Meyr., 1895
    LEOBATUS Wlsm., 1904
    ZIMINIOLA Gerasimov, 1930
    CREMONA Bsk., 1934
1852 **mouffetella** (L., 1758)
    pedisequella (Hbn., 1796)
    punctifera (Haw., 1828)
1853 **pruinosella** (Lienig & Zell., 1846)
1854 **rancidella** (H.-S., 1854)
    triatomaea (Muhlig, 1864)
    vepretella (Zell., 1871)
    superfetella (Peyerimhoff, 1877)
    cotoneastri (Bsk., 1934)
    cerasivorella (Kuznetzov, 1960)

TELPHUSA Cham., 1872
    ADRASTEIA Cham., 1872
    ADRASTIA Kby., 1874, missp.
1855 **alexandriacella** (Cham., 1872), ident.
    uncert.
1856 **fasciella** (Cham., 1872), ident. uncert.
1857 **latifasciella** (Cham., 1875), mispl.
1858 **longifasciella** (Clem., 1863)
    curvistrigella Cham., 1872
    obliquifasciella (Cham., 1879)
    lutraula (Meyr., 1923)
1859 **sedulitella** (Bsk., 1910)
    agrifolia Braun, 1921

PSEUDOCHELARIA Dietz, 1900
1860 **arbutina** (Keif., 1930)
1861 **manzanitae** (Keif., 1930)
1862 **pennsylvanica** Dietz, 1900
1863 **scabrella** (Bsk., 1913)
1864 **walsinghami** Dietz, 1900

NEOTELPHUSA Janse, 1958
1865 **praefixa** (Braun, 1921), n. comb.
1866 **querciella** (Cham., 1872), n. comb.

PSEUDOTELPHUSA Janse, 1958
1867 **amelanchierella** (Braun, 1930), n.
    comb.
1868 **basifasciella** (Zell., 1873), n. comb.
1869 **belangerella** (Cham., 1875), n. comb.
    oronella (Wlsm., 1882), n. comb.
1870 **betulella** (Bsk., 1903), n. comb.
1871 **fuscopunctella** (Clem., 1863)
1872 **incana** Hodges, 1969
1873 **palliderosacella** (Cham., 1878), n.
    comb.
1874 **quercinigracella** (Cham., 1872)
    fragmentella (Zell., 1873), n. comb.

XENOLECHIA Meyr., 1895
1875 **aethiops** (Humphreys & Westwood,
    1845)
1876 **quinquecristatella** (Cham., 1878), n.
    comb.
1877 **basistrigella** (Zell., 1873), n. comb.
1878 **ontariensis** Keif., 1933
1879 **querciphaga** Keif., 1933
1880 **velatella** (Bsk., 1907)

TELEIODES Sattler, 1960
    TELEIA Heinemann, 1870, preocc.
     by Hbn., 1825
1881 **sequax** (Haw., 1828)

TELEIOPSIS Sattler, 1960
1882 **baldiana** (B. & Bsk., 1920), n. comb.

LITA Tr., 1833
1883 **barnesiella** (Bsk., 1903)
1884 **deoia** Hodges, 1966
1885 **dialis** Hodges, 1966
1886 **geniata** Hodges, 1966
1887 **incicur** Hodges, 1966
1888 **invariabilis** (Kft., 1908)
1889 **jubata** Hodges, 1966
1890 **maenadis** Hodges, 1966
1891 **nefrens** Hodges, 1966
1892 **obnubila** Hodges, 1966
1893 **pagella** Hodges, 1966
1894 **princeps** (Bsk., 1910)
1895 **puertella** (Bsk., 1916)
1896 **recens** Hodges, 1966
1897 **rectistrigella** (B. & Bsk., 1920)
1898 **sexpunctella** (F., 1974)
    virgella (Thunb., 1794), n. syn.
    longicornis (Curtis, 1827), n. syn.
    histrionella (Geyer, 1832), n. syn.
    zebrella Tr., 1833, n. syn.
    alpicola (Frey, 1867), n. syn.
    alternatella (Kft., 1908), n. syn.
    petulans (Braun, 1925), n. syn.
1899 **sironae** Hodges, 1966
1900 **texanella** (Cham., 1880)
    chambersella Dyar, 1903
1901 **thaliae** Hodges, 1966
1902 **variabilis** (Bsk., 1903)
1903 **veledae** Hodges, 1966

ARLA Clarke, 1942
1904 **diversella** (Bsk., 1916)
1905 **tenuicornis** Clarke, 1942

NEODACTYLOTA Bsk., 1903
1906 **basilica** Hodges, 1966
1907 **egena** Hodges, 1966

1908 **liguritrix** Hodges, 1966
1909 **snellenella** (Wlsm., 1888)

EUDACTYLOTA Wlsm., 1911
1910 **abstemia** Hodges, 1966
1911 **barberella** (Bsk., 1903)
1912 **diadota** Hodges, 1966
1913 **iobapta** (Meyr., 1927)

FRISERIA Bsk., 1939
1914 **acaciella** (Bsk., 1906)
1915 **caieta** Hodges, 1966
1916 **cockerelli** (Bsk., 1903)
    lindenella (Bsk., 1903)
    malindella (Bsk., 1910)
    sarcochlora (Meyr., 1929)
1917 **nona** Hodges, 1966

RIFSERIA Hodges, 1966
1918 **fuscotaeniaella** (Cham., 1878)

SRIFERIA Hodges, 1966
1919 **cockerella** (Bsk., 1903), n. comb.
1920 **fulmenella** (Bsk., 1910)
    prorepta (Meyr., 1923), repl. name
1921 **oxymeris** (Meyr., 1929), n. comb.

BRYOTROPHA Heinemann, 1870
    MNIOPHAGA Pierce & Daltry,
     1938
    ADELPHOTROPHA Gozmány,
     1955
1922 **branella** (Bsk., 1908)
1923 **clandestina** (Meyr., 1923), n. comb.
1924 **pullifimbriella** (Clem., 1863), n.
    comb.
1925 **tahavusella** (Fbs., 1922), n. comb.

DELTOPHORA Janse, 1950
1926 **duplicata** Sattler, 1979
1927 **glandiferella** (Zell., 1873)
1928 **sella** (Cham., 1874)
    a. **atacta** (Meyr., 1927)
    b. **californica** Sattler, 1979

GELECHIA Hbn., 1825
    GUENEA Bruand, 1850
    CIRRHA Cham., 1872
    OESEIS Cham., 1875
1929 **albisparsella** (Cham., 1872)
    fuscoluteella (Cham., 1872), n. syn.
    platanella (Cham., 1872), repl. name
1930 **anarsiella** Cham., 1877
1931 **badiomaculella** Cham., 1872, ident.
    uncert.
1932 **benitella** B. & Bsk., 1920
1933 **bianulella** (Cham., 1875)
    ocellella Cham., 1877
    melanchlora (Meyr., 1929)
1934 **bistrigella** (Cham., 1872), ident.
    uncert.
1935 **brumella** Clem., 1864, ident. uncert.
1936 **capiteochrella** Cham., 1875, ident.
    uncert.
1937 **caudatae** Clarke, 1934
1938 **desiliens** Meyr., 1923
1939 **discostrigella** Cham., 1875, ident.
    uncert.
1940 **dromicella** Bsk., 1910
1941 **dyariella** Bsk., 1903

1942 **flexurella** Clem., 1860, ident. uncert.
1943 **gracula** (Meyr., 1929)
    diaconalis (Meyr., 1929)
1944 **grisaeella** (Cham., 1872), ident. uncert.
1945 **griseochrella** Cham., 1875, ident. uncert.
1946 **lynceella** Zell., 1873
    trilineella Cham., 1877
1947 **maculatusella** Cham., 1875, ident. uncert.
1948 **mandella** Bsk., 1904
1949 **mimella** Clem., 1860, ident. uncert.
1950 **monella** Bsk., 1904
1951 **mundata** (Meyr., 1929)
1952 **obscurella** Cham., 1872, ident. uncert.
    perobscurella Wlsm., 1903, repl. name
1953 **obscurosuffusella** Cham., 1878
    canopulvella Cham., 1878
1954 **ocherfuscella** Cham., 1875, ident. uncert.
    ochreofuscella Meyr., 1925, repl. name
1955 **packardella** Cham., 1877, ident. uncert.
1956 **pallidagriseella** Cham., 1874, ident. uncert.
1957 **palpialbella** Cham., 1875, ident. uncert.
1958 **panella** Bsk., 1903
1959 **parvipulvella** Cham., 1874, ident. uncert.
1960 **ribesella** Cham., 1875
1961 **rileyella** (Cham., 1872)
1962 **sabinella** Zell., 1839
1963 **thoracestrigella** Cham., 1875, ident. uncert.
1964 **thymiata** (Meyr., 1929)
1965 **unistrigella** Cham., 1873, ident. uncert.
1966 **versutella** Zell., 1873
1967 **wacoella** Cham., 1874, ident. uncert.

GNORIMOSCHEMA Bsk., 1900
    LERUPSIA Riedl, 1965
    NEOSCHEMA Povolný, 1967
1968 **alaricella** Bsk., 1908
1969 **albangulatum** Braun, 1926
1970 **albimarginella** (Cham., 1875)
1971 **ambrosiella** (Cham., 1875)
1972 **baccharisella** Bsk., 1903
1973 **banksiella** Bsk., 1903
1974 **batanella** Bsk., 1903
1975 **busckiella** Kft., 1903
1976 **collinusella** (Cham., 1877)
1977 **compsomorpha** Meyr., 1929
1978 **contrarium** Braun, 1921
1979 **coquillettella** Bsk., 1902
1980 **dudiella** Bsk., 1903
1981 **marmorella** (Cham., 1875), rev. stat.
    emancipatum (Meyr., 1925), repl. name
1982 **ericameriae** Keif., 1933
1983 **faustella** Bsk., 1910
1984 **florella** Bsk., 1903
1985 **gallaeasterella** (Kellicott, 1878)
    gallaediplopappi (Fyles, 1890)
    ceasiella (Brodie, 1909)
    gallaeasteris Meyr., 1925, emend.

1986 **gallaesolidaginis** (Riley, 1869)
1987 **gibsoniella** Bsk., 1915
1988 **grisella** (Cham., 1872)
    discomaculella (Cham., 1872)
1989 **inexpertum** (Meyr., 1925)
    simpliciella (Cham., 1875), preocc. by Staint., 1858
1990 **klotsi** Povolný, 1967
1991 **lipatiella** (Bsk., 1909)
1992 **milleriella** (Cham., 1875)
1993 **minor** (Bsk., 1906)
1994 **nordlandicolella** (Strand, 1902)
    eucaustum Meyr., 1929
1995 **octomaculella** (Cham., 1875)
1996 **pedmontella** (Cham., 1877)
1997 **salinaris** Bsk., 1911
1998 **saphirinella** (Cham., 1875)
1999 **semicyclionella** Bsk., 1903
2000 **septentrionella** Fyles, 1911
2001 **serratipalpella** (Cham., 1877)
2002 **splendoriferella** Bsk., 1904
2003 **sporomochla** Meyr., 1929
2004 **subterraneum** Bsk., 1911
2005 **terracottella** Bsk., 1900
2006 **triocellella** (Cham., 1877)
2007 **valesiella** (Stgr., 1877)
    charcoti (Meyr., 1934)
    alaskense Povolný, 1967
2008 **vastificum** Braun, 1929
2009 **versicolorella** (Cham., 1872)
2010 **washingtoniella** Bsk., 1904

PHTHORIMAEA Meyr., 1902
2011 **operculella** (Zell., 1873)
    terrella (Wlk., 1864), preocc. by D. & S. 1775
    solanella (Bdv., 1874)
    tabacella (Rag., 1879)
    sedata (Butler, 1880)

SCROBIPALPULA Povolný, 1964
2012 **artemisiella** (Kft., 1903)
    axenopis (Meyr., 1919)
2013 **chiquitella** (Bsk., 1910)
2014 **erigeronella** (Braun, 1921), n. comb.
2015 **henshawiella** (Bsk., 1903)
    ochreostrigella (Cham., 1877), preocc. by Cham., 1875
2016 **hodgesi** (Povolný, 1967), n. comb.
2017 **lutescella** (Clarke, 1934)
2018 **ochroschista** (Meyr., 1929)
2019 **polemoniella** (Braun, 1925), n. comb.
2020 **potentella** (Keif., 1936)
2021 **radiatella** (Bsk., 1904)
2022 **sacculicola** (Braun, 1925), n. comb.
2023 **semirosea** (Meyr., 1929)

SCROBIPALPA Janse, 1951
2024 **atriplex** (Bsk., 1910), n. comb.
2025 **atriplicella** (F. v. Röslerstamm, 1839)
    detersella (Clem., 1860), preocc. by Zell., 1847
    brackenridgella (Bsk., 1903), repl. name
    chenopodiella (Bsk., 1916)
2026 **consueta** (Braun, 1925), n. comb.
2027 **macromaculata** (Braun, 1925), n. comb.
2028 **monumentella** (Cham., 1877), n. comb.

2029 **obsoletella** (F. v. Röslerstamm, 1841)
    miscitatella (Clarke, 1932)
2030 **scutellariaeela** (Cham., 1873), n. comb.

EXCEPTIA Povolný, 1967
2031 **neopetrella** (Keif., 1936)

SYMMETRISCHEMA Povolný, 1967
2032 **capsicum** (Bradley & Povolný, 1965)
2033 **fercularium** (Meyr., 1929)
2034 **inexpectatum** Povolný, 1967
2035 **lavernella** (Cham., 1874)
    physalivorella (Cham., 1875)
2036 **lectuliferum** (Meyr., 1929)
2037 **pallidochrella** (Cham., 1872), n. comb.
2038 **plaesiosema** (Turner, 1919)
    melanoplintha (Meyr., 1926)
    tuberosella (Bsk., 1931)
2039 **striatella** (Murt., 1900)

CARYOCOLUM Gregor & Povolný, 1954
2040 **cassella** (Wlk., 1864), n. comb.
2041 **protecta** (Braun, 1965), n. comb.
2042 **subtractella** (Wlk., 1864), n. comb.

PTYCERATA Ely, 1910
    SCROBIPALPOPSIS Povolný, 1967, n. syn.
2043 **arnicella** (Clarke, 1942), n. comb.
2044 **busckella** Ely, 1910
2045 **petrella** (Bsk., 1915), n. comb.
2046 **tetradymiella** (Bsk., 1903), n. comb.

KEIFERIA Bsk., 1939
2047 **lycopersicella** (Wlsm., 1897)
    lenta (Meyr., 1917)
    lycopersicella (Bsk., 1928), preocc. by Wlsm., 1897
    elmorei (Keif., 1936)

TILDENIA Povolný, 1967
2048 **altisolani** (Keif., 1937)
2049 **glochinella** (Zell., 1873)
    peniculo (Heinr., 1946)
2050 **inconspicuella** (Murt., 1883)
    cinerella (Murt., 1881), preocc. by Cl., 1759

FRUMENTA Bsk., 1939
2051 **nephelomicta** (Meyr., 1930), n. comb.
2052 **nundinella** (Zell., 1873)
    beneficentella (Murt., 1881)

AGONOCHAETIA Povolný, 1965
    SAUTEREOPSIS Povolný, 1965
2053 **conspersa** (Braun, 1921), n. comb.

CHIONODES Hbn., 1825
2054 **abdominella** (Bsk., 1903)
2055 **abella** (Bsk., 1903)
2056 **abradescens** (Braun, 1921), n. comb.
2057 **acerella** Sattler, 1967
2058 **acrina** (Keif., 1933)
2059 **agriodes** (Meyr., 1927)
2060 **arenella** (Fbs., 1922)
2061 **argentipunctella** (Ely, 1910)
2062 **aristella** (Bsk., 1903)

2063 **bicolor** Clarke, 1947
2064 **bicostomaculella** (Cham., 1872)
    quercifoliella (Cham., 1872)
2065 **braunella** (Keif., 1931)
    a. *arborei* (Keif., 1931)
2066 **canofusella** Clarke, 1947
2067 **ceanothiella** (Bsk., 1904)
    marinensis (Keif., 1935)
2068 **chrysopyla** (Keif., 1935)
2069 **continuella** (Zell., 1839)
    trimaculella (Pack., 1867)
    albomaculella (Cham., 1875)
2070 **dammersi** (Keif., 1936)
2071 **dentella** (Bsk., 1903)
2072 **discoocellella** (Cham., 1872)
    violaceofusca (Zell., 1873)
2073 **figurella** (Bsk., 1912)
2074 **flavicorporella** (Wlsm., 1882), n. comb.
2075 **fluvialella** (Bsk., 1908)
2076 **fondella** (Bsk., 1906)
2077 **formosella** (Murt., 1881), rev. stat.
    vernella (Murt., 1883)
2078 **fructuarius** (Braun, 1925)
2079 **fuscomaculella** (Cham., 1872), n. comb.
    maculimarginella (Cham., 1874), n. syn.
    caryaevorella (Pack., 1886), n. syn.
2080 **gilvomaculella** (Clem., 1863), n. comb.
    johnstoni Clarke, 1947, n. syn.
2081 **grandis** Clarke, 1947
2082 **halycopa** (Meyr., 1927)
2083 **helicostictus** (Meyr., 1929)
2084 **hibiscella** (Bsk., 1903)
2085 **iridescens** Clarke, 1947
2086 **kincaidella** (Bsk., 1907)
    laguna (Bsk., 1912), n. comb., n. syn.
    coticola (Bsk., 1913), n. syn.
    chloroschema (Meyr., 1923), n. syn.
    notochlora (Meyr., 1929), n. syn.
    acanthocarpae Clarke, 1947, n. syn.
2087 **labradoricus** (Mösch., 1864), n. comb.
2088 **loetae** Clarke, 1942
2089 **lophosella** (Bsk., 1910)
    lophella (Meyr., 1925), missp.
2090 **lugubrella** (F., 1794)
    luctificella (Hbn., 1813)
    lunatella (Zett., 1839)
2091 **luteogeminatus** (Clarke, 1935)
2092 **mariona** (Heinr., 1921), n. comb.
2093 **mediofuscella** (Clem., 1863)
    vagella (Wlk., 1864)
    fuscoochrella (Cham., 1872)
    liturosella (Zell., 1873)
    rhedaria (Meyr., 1923)
2094 **metallicus** (Braun, 1921)
2095 **nanodella** (Bsk., 1910)
2096 **negundella** (Heinr., 1920)
2097 **nigrobarbatus** (Braun, 1925)
2098 **notandella** (Bsk., 1916)
2099 **obscurusella** (Cham., 1872)
    fuscopulvella (Cham., 1872)
    obscurella (Wlsm., 1903)
    asema Clarke, 1947
2100 **occidentella** (Cham., 1875)
2101 **occlusus** (Braun, 1925)
2102 **ochreostrigella** (Cham., 1875)
2103 **paralogella** (Bsk., 1916)

2104 **pereyra** Clarke, 1947
2105 **periculella** (Bsk., 1910)
2106 **permactus** (Braun, 1925)
2107 **petalumensis** Clarke, 1947
2108 **phalacrus** (Wlsm., 1911), n. comb.
2109 **pinguicula** (Meyr., 1929)
2110 **pseudofondella** (Bsk., 1908)
2111 **psilopterus** (B. & Bsk., 1920)
2112 **retiniella** (B. & Bsk., 1920)
    langei (Keif., 1936)
2113 **sabinianus** Powell, 1959
2114 **salicella** Sattler, 1967
2115 **seculaella** (Clarke, 1932)
2116 **sistrella** (Bsk., 1903)
2117 **terminimaculella** (Kft., 1908)
2118 **tessa** Clarke, 1947
2119 **thoraceochrella** (Cham., 1872)
    nigrimaculella (Bsk., 1903)
2120 **trichostola** (Meyr., 1923)
2121 **trophella** (Bsk., 1903)
2122 **vanduzeei** (Keif., 1935)
2123 **viduella** (F., 1794)
    leucomella (Quensel, 1802)
    luctiferella (H.-S., 1856)
    labradoriella (Clem., 1863)
2124 **whitmanella** Clarke, 1942
2125 **xanthophilella** (B. & Bsk., 1920)

NEOFACULTA Gozmány, 1955
2126 **infernella** (H.-S., 1854)

FILATIMA Bsk., 1939
2127 **abactella** (Clarke, 1932)
2128 **albicostella** Clarke, 1942
    angustipennis Sattler, 1961
2129 **albilorella** (Zell., 1873)
    trifasciella (Cham., 1875)
2130 **albipectus** (Wlsm., 1911), n. comb.
2131 **arizonella** (Bsk., 1903)
2132 **aulaea** (Clarke, 1932)
2133 **betulae** Clarke, 1947
2134 **biforella** (Bsk., 1909)
    promonitrix (Meyr., 1927), n. syn.
2135 **bigella** (Bsk., 1913)
    spilosella (B. & Bsk., 1920)
2136 **biminimaculella** (Cham., 1880), rev. stat.
2137 **catacrossa** (Meyr., 1927)
2138 **collinearis** (Meyr., 1927)
2139 **confusatella** (Darlington, 1949), n. comb.
2140 **cushmani** Clarke, 1942
2141 **dimissae** (Keif., 1931)
2142 **depuratella** (Bsk., 1910)
2143 **epulatrix** Hodges, 1969
2144 **frugalis** (Braun, 1925)
2145 **fuliginea** (Meyr., 1929)
2146 **glycyrhizaeella** (Cham., 1877), n. comb.
    lepidotae (Clarke, 1934), n. syn.
2147 **golovina** Clarke, 1947
2148 **gomphopis** (Meyr., 1927)
2149 **hemicrossa** (Meyr., 1927)
2150 **inquilinella** (Bsk., 1910)
2151 **isocrossa** (Meyr., 1927)
    virgea Clarke, 1947
2152 **monotaeniella** (Bott., 1926)
2153 **natalis** (Heinr., 1920)
2154 **neotrophella** (Heinr., 1921)
2155 **nigripectus** (Wlsm., 1911), n. comb.

2156 **normifera** (Meyr., 1927)
2157 **nucifer** (Wlsm., 1911)
2158 **obidenna** Clarke, 1947
2159 **obscuroocellella** (Cham., 1875)
2160 **obscurosuffusella** (Cham., 1878)
    canopulvella (Cham., 1878)
    monopa (Meyr., 1927), n. syn.
    epigypsa (Meyr., 1927), n. syn.
2161 **ochreosuffusella** (Cham., 1874)
    depressostrigella (Cham., 1874)
    asbolodes (Meyr., 1927), n. syn.
2162 **ornatifimbriella** (Clem., 1864)
    unctulella (Zell., 1873)
    amorphaeella (Cham., 1877), n. syn.
2163 **perpensa** Clarke, 1947
2164 **persicaeella** (Murt., 1899)
    confusella (Cham., 1875), preocc. by Heinemann, 1870
2165 **platyochra** Clarke, 1947
2166 **pravinominella** (Cham., 1878)
    quadrimaculella (Cham., 1875), preocc. by Cham., 1874
2167 **procedes** Clarke, 1947
2168 **prognosticata** (Braun, 1925)
2169 **pseudacaciella** (Cham., 1872)
    caecella (Zell., 1873)
    pseudoacaciella (Bsk., 1903), missp.
2170 **roceliella** Clarke, 1942
2171 **saliciphaga** (Keif., 1937)
2172 **serotinella** (Bsk., 1903)
2173 **shastaella** (Gaede, 1937)
    albifemorella (Clarke, 1932), preocc. by Hofmann, 1867
    clarkella Bsk., 1939, rep. name
2174 **sperryi** Clarke, 1947
2175 **spinigera** Clarke, 1947
2176 **striatella** (Bsk., 1903)
    rivulata (Meyr., 1927), n. syn.
2177 **tephrinopa** (Meyr., 1929)
2178 **tridentata** Clarke, 1947
2179 **vaccinii** Clarke, 1947
2180 **vaniae** Clarke, 1947
2181 **xanthuris** (Meyr., 1927), rev. stat.

AROGA Bsk., 1914
2182 **acharnaea** (Meyr., 1927)
2183 **alleriella** Bsk., 1940
2184 **argutiola** Hodges, 1974
2185 **camptogramma** (Meyr., 1931)
2186 **chlorocrana** (Meyr., 1927)
2187 **coloradensis** (Bsk., 1903)
    speculifera (Meyr., 1931), n. syn.
2188 **eldorada** (Keif., 1936)
    eldorado Bsk., 1939, missp.
2189 **epigaeella** (Cham., 1881). rev. stat.
2190 **eriogonella** (Clarke, 1935)
2191 **leucanieella** (Bsk., 1910)
    leucaniella (Meyr., 1925), missp.
2192 **morenella** (Bsk., 1908)
    moreonella (Keif., 1936), missp.
2193 **paraplutella** (Bsk., 1910)
2194 **paulella** (Bsk., 1903)
2195 **rigidae** (Clarke, 1935)
2196 **thoracealbella** (Cham., 1874), n. comb.
    minimaculella (Cham., 1874), n. comb.
2197 **trachycosma** (Meyr., 1923)
2198 **trialbamaculella** (Cham., 1875)
2199 **trilineella** (Cham., 1877)
    bispiculata (Meyr., 1923), n. syn.
    hipposaris (Meyr., 1927), n. syn.

2200 **unifasciella** (Bsk., 1903)
2201 **websteri** Clarke, 1942
2202 **xyloglypta** (Meyr., 1923)

    FASCISTA Bsk., 1939
2203 **bimaculella** (Cham., 1872)
    ternariella (Zell., 1873)
    sylvaecolella (Cham., 1878)
2204 **cercerisella** (Cham., 1872)
    olympiadella (Zell., 1873)
    cercerella (Meyr., 1925), missp.
2205 **quinella** (Zell., 1873)

    FACULTA Bsk., 1939
2206 **inaequalis** (Bsk., 1910), n. comb.
    inaequalis (Wlsm., 1911), n. comb.,
      preocc. by Bsk., 1910
    anisectis (Meyr., 1923), n. comb.
    clistrodoma (Meyr., 1923), n. comb., n.
      syn.
2207 **triangulella** (Bsk., 1907)

    MIRIFICARMA Gozmány, 1955
    HELINA Guenée, 1849, preocc. by
      Robineau-Desvoidy, 1830
2208 **flamella** (Hbn., 1825)
    formosella (Hbn., 1796), preocc. by
      D. & S., 1775

    STEGASTA Meyr., 1904
2209 **bosqueella** (Cham., 1875)
    basqueella (Cham., 1875), missp.
    costipunctella (Mösch., 1890)

    POLYHYMNO Cham., 1874
    COPOCERCIA Zell., 1877
    OEGOCONIODES Matsumura,
      1931
2210 **acaciella** Bsk., 1900
2211 **luteostrigella** Cham., 1874
    fuscostrigella Cham., 1876

    CALLIPRORA Meyr., 1914
2212 **sexstrigella** (Cham., 1874), n. comb.
    thermogramma Meyr., 1929, n. syn.

    SOPHRONIA Hbn., 1825
2213 **primella** Bsk., 1907
2214 **roseicrinella** Bsk., 1909
2215 **teretracma** Meyr., 1927

    YMELDIA Hodges, 1965
2216 **janae** Hodges, 1965

ANACAMPSINAE

    APROAEREMA Durrant, 1897
    SCHUETZEIA Spuler, 1910
2217 **anthyllidella** (Hbn., 1810–13)
    sparsiciliella (Barrett, 1891)

    SYNCOPACMA Meyr., 1925
    HARPAGUS Steph., 1834, preocc.
      by Vigors, 1824
    STOMOPTERYX; auth.
2218 **adversa** (Braun, 1930), n. comb.
2219 **crotalariella** (Bsk., 1900), n. comb.
2220 **metadesma** (Meyr., 1927), n. comb.
2221 **nigrella** (Cham., 1875), n. comb.
2222 **palpilineella** (Cham., 1875), n. comb.

    UNTOMIA Bsk., 1906
2223 **albistrigella** (Cham., 1872)
2224 **untomiella** Bsk., 1906

    BATTARISTIS Meyr., 1914
    DUVITA Bsk., 1916
2225 **concinusella** (Cham., 1877), rev. stat.,
      n. comb.
2226 **cyclella** (Bsk., 1903), n. comb.
2227 **nigratomella** (Clem., 1863), n. comb.
    apicilineella (Clem., 1863), n. comb.
    apicistrigella (Cham., 1872), n. comb.
2228 **pasadenae** (Keif., 1935), n. comb.
2229 **vittella** (Bsk., 1926), n. comb.

    ANACAMPSIS Curt., 1827
    TACHYPTILIA Heinemann, 1870
    AGRIASTIS Meyr., 1914
2230 **agrimoniella** (Clem., 1860)
    aduncella (Zell., 1868)
    aderusella missp.
2231 **argyrothamniella** Bsk., 1900
2232 **comparanda** (Meyr., 1929)
2233 **conclusella** (Wlk., 1864)
    grissefasciella (Cham., 1875)
2234 **coverdalella** Kft., 1903
2235 **fragariella** Bsk., 1904
2236 **fullonella** (Zell., 1873)
    rufusella (Cham., 1873)
    subruberella (Cham., 1874)
    rubescens (Wlsm., 1881)
2237 **innocuella** (Zell., 1873)
2238 **kearfottella** (Bsk., 1903)
2239 **lacteusochrella** (Cham., 1875)
2240 **lagunculariella** Bsk., 1900
2241 **levipedella** (Clem., 1863)
2242 **lupinella** Bsk., 1901
2243 **niveopulvella** (Cham., 1875)
2244 **nonstrigella** Bsk., 1906
2245 **paltodoriella** Bsk., 1903
2246 **populella** (Cl., 1759)
    boeberana (F., 1787)
    laticinctella Steph., 1834
    tremulella (Dup., 1839)
2247 **psoraliella** B. & Bsk., 1920
2248 **rhoifructella** (Clem., 1860)
    consonella (Zell., 1873)
    quadrimaculella (Cham., 1874)
    ochreocostella (Cham., 1878)
2249 **sacramenta** Keif., 1933
2250 **tephriasella** (Cham., 1872), rev. stat.
2251 **tristrigella** (Wlsm., 1882)

    COMPSOLECHIA Meyr., 1918
2252 **crescentifasciella** (Cham., 1874)

    STROBISIA Clem., 1860
    SYSTASIOTA Wlsm., 1910
2253 **iridipennella** Clem., 1860
    aphroditella Cham., 1872
2254 **proserpinella** F. & B., 1878

    HOLOPHYSIS Wlsm., 1910
    HOPLOPHYSIS McD., 1939, missp.
2255 **emblemella** (Clem., 1860)

CHELARIINAE

    PROSTOMEUS Bsk., 1903
2256 **brunneus** Bsk., 1903

    ANARSIA Zell., 1839
    ANANARSIA Amsel, 1959
2257 **lineatella** Zell., 1839
    pruniella Clem., 1860

    HYPATIMA Hbn., 1825
    CHELARIA Haw., 1828
    ALLOCOTA Meyr., 1904
    CYNESTOMORPHA Meyr., 1904
    DEUTEROPTILA Meyr., 1904
    SEMODICTIS Meyr., 1909
    ALLOCOTANIANA Strand, 1913
    EPISACTA Turner, 1919
2258 **zesticopa** (Meyr., 1929)

    EPILECHIA Bsk., 1939
2259 **catalinella** (Bsk., 1907)
    tehuacana (Bsk., 1913)

    SITOTROGA Heinemann, 1870
    NESOLECHIA Meyr., 1921
    SYNGENOMICTIS Meyr., 1927
2260 **cerealella** (Olivier, 1789)
    hordei (Kby., 1815)
    arctella (Wlk., 1864)
    melanarthra (Lower, 1900)
    palearis (Meyr., 1913)

    PECTINOPHORA Bsk., 1917
2261 **gossypiella** (Saund., 1844)

    PLATYEDRA Meyr., 1895
2262 **subcinerea** (Haw., 1828)
    vilella (Zell., 1847)
    bathrosticta (Meyr., 1936)

DICHOMERIDINAE

    BRACHMIA Hbn., 1825
    CLADODES Heinemann, 1870,
      preocc. by Solier, 1849
    CERATOPHORA Heinemann,
      1870, preocc. by Gray, 1832–35
    EUDODACLES Snell., 1889
    AULACOMIMA Meyr., 1904
    APETHISTIS Meyr., 1908
2263 **badia** Braun, 1921
2264 **casca** Braun, 1925
2265 **chambersella** (Murt., 1874)
    subalbusella (Cham., 1874)
    parvipulvella (Cham., 1874)
    inaequepulvella (Cham., 1875)
2266 **discoannulella** (Cham., 1875)
2267 **fernaldella** (Bsk., 1903), n. comb.
2268 **hystricella** Braun, 1921
2269 **melanocarpa** Meyr., 1929
2270 **melantherella** (Bsk., 1900), n. comb.
2271 **trimaculella** (Cham., 1874), n. comb.

    BRACHYACMA Meyr., 1886
    LATHONTOGENES Wlsm.,
      1897
    PARASPISTES Meyr., 1905
    LIPATIA Bsk., 1910
2272 **palpigera** (Wlsm., 1891)
    crotalariella (Bsk., 1910)

    DICHOMERIS Hbn., 1818
2273 **bidiscomaculella** (Cham., 1874),
      mispl.

2274 **bipunctella** (Wlsm., 1882), mispl.
2275 **condaliavorella** (Bsk., 1900), mispl.
2276 **deflecta** Bsk., 1909, mispl.
2277 **georgiella** (Wlk., 1866), mispl.
    bimaculella (Cham., 1877), n. syn.
    mollis B. & Bsk., 1920, n. syn.
2278 **glenni** Clarke, 1947, mispl.
2279 **hirculella** Bsk., 1909, mispl.
2280 **hypochloa** Wlsm., 1911, mispl.
2281 **ligulella** Hbn., 1818
    contubernatella (Fitch, 1853)
    pometella (Harr., 1853)
    malifoliella (Fitch, 1854)
    flavivittella (Clem., 1864)
    quercipomonella (Cham., 1872)
    reedella (Cham., 1872)
    ruderella (Cham., 1878), missp.
2282 **marginella** (F., 1781), mispl.
2283 **punctidiscella** (Clem., 1863), mispl.
    straminiella (Cham., 1872)
2284 **rustica** (Wlsm., 1892), mispl.
2285 **suffusella** (Cham., 1874), ident.
    uncert.
2286 **vacciniella** Bsk., 1915, mispl.
    nephanthes (Meyr., 1929), n. syn.
2287 **ventrella** (Fitch, 1854), mispl.
    unicipunctella (Clem., 1860)
    caryaefoliella (Cham., 1872), n. syn.
    querciella (Cham., 1872)
    roseocostella (Wlsm., 1882), n. syn.
    trinotella (Coquillett, 1883), n. syn.

ANORTHOSIA Clem., 1860
    SAGARITIS Cham., 1872, preocc.
    by Billberg, 1820
    ANORTHODISCA Gaede, 1937,
    missp.
2288 **punctipennella** Clem., 1860
    gracilella (Cham., 1872)

TRICHOTAPHE Clem., 1860
    BEGOE Cham., 1872
    MALACOTRICHA Zell., 1873
2289 **alacella** Clem., 1862, mispl.
    goodellella (Cham., 1881)
    ochripalpella (Zell., 1873)
2290 **barnesiella** Bsk., 1907
2291 **bilobella** (Zell., 1873), rev. stat.
2292 **citrifoliella** (Cham., 1880), mispl.
2293 **costarufoella** (Cham., 1874), mispl.
2294 **delotella** (Bsk., 1909), mispl.
    mexicana (Wlsm., 1911), n. syn.
2295 **flavocostella** (Clem., 1860), mispl.
2296 **griseella** (Cham., 1874), ident. uncert.
2297 **inserrata** (Wlsm., 1882)
2298 **juncidella** Clem., 1860
    pallipalpis (Wlk., 1864)
    dubitella (Cham., 1872)
    hallipalpis (Riley, 1891), missp.
2299 **leuconotella** Bsk., 1904
2300 **levisella** Fyles, 1904
2301 **serrativittella** (Zell., 1873)
    plutella (Cham., 1874)
2302 **setosella** Clem., 1860
    costolutella (Cham., 1872)
    eupatoriella (Cham., 1872), n. syn.
    dolabella (Zell., 1873), n. syn.
2303 **simpliciella** Bsk., 1904
    hemiclina Meyr., 1929, n. syn.
2304 **stipendaria** Braun, 1925

2305 **trinotella** Bsk., 1906, mispl.
2306 **washingtoniella** Bsk., 1906

THELYASCETA Meyr., 1923
2307 **nonstrigella** (Cham., 1878)
2308 **purpureofusca** (Wlsm., 1882), rev.
    stat.

CARBATINA Meyr., 1913
2309 **picrocarpa** Meyr., 1913
    iothalles (Fbs., 1939), n. comb., n.
    syn.

EPICORTHYLIS Zell., 1873
2310 **inversella** Zell., 1873

LECITHOCERINAE

DEOCLONA Bsk., 1903
    PROCLESIS Wlsm., 1911
    LIOCLEPTA Meyr., 1922, n. syn.
2311 **yuccasella** Bsk., 1903

Copromorphoidea

COPROMORPHIDAE
by JOHN B. HEPPNER

LOTISMA Bsk., 1909
2312 **trigonana** (Wlsm., 1879)
    kincaidiella (Bsk., 1904)

ALUCITIDAE
by JOHN B. HEPPNER

ALUCITA L., 1758
    ORNEODES Latr., 1796
    EUCHIRADIA Hbn., 1825
2313 **hexadactyla** L., 1758
    polydactyla Hbn., 1813
    poecilodactyla Steph., 1835
    montana Cockerell, 1889, nom. nud.

CARPOSINIDAE
by JOHN B. HEPPNER

CARPOSINA H.-S., 1855
2314 **niponensis** Wlsm., 1900, extralim.
    a. ottawana Kft., 1907
    nicholsana Fbs., 1923
2315 **fernaldana** Bsk., 1907
2316 **simulator** Davis, 1969
2317 **biloba** Davis, 1969

BONDIA Newman, 1856
2318 **comonana** (Kft., 1907)
    euryleuca (Meyr., 1912)
2319 **crescentella** (Wlsm., 1882)
2320 **shastana** Davis, 1969
2321 **spicata** Davis, 1969
2322 **fidelis** Meyr., 1913
2323 **fuscata** Davis, 1969

TESUQUEA Klots, 1936
2324 **hawleyana** Klots, 1936

EPERMENIIDAE
by JOHN B. HEPPNER

EPERMENIINAE

Epermeniini

EPERMENIA Hbn., 1825
    CALOTRIPIS Hbn., 1825
    TICHOTRIPIS Hbn., 1825
    CHAULIODUS Tr., 1833, preocc.
    by Schneider, 1810
    LOPHONOTUS Steph., 1834
    CHAULIOMORPHA Blanchard,
    1840
    CALOTRYPIS H.-S., 1854, missp.
    HEYDENIA Hofmann, 1868,
    preocc. by Foerster, 1856
    CATAPLECTICA Wlsm., 1894
    EPIMENIA Kft., 1903, missp.
    ACANTHEDRA Meyr., 1917
    EPERMENIOLA Gaedike, 1968
2325 **imperialella** Bsk., 1906
2326 **stolidota** (Meyr., 1917)
2327 **californica** Gaedike, 1977
2328 **albapunctella** Bsk., 1908
2329 **cicutaella** Kft., 1903
    alameda Braun, 1923
2330 **pimpinella** Murt., 1900
2331 **lomatii** Gaedike, 1977
2332 **infracta** Braun, 1926
2333 **strictelloides** Gaedike, 1977

OCHROMOLOPINAE

OCHROMOLOPIS Hbn., 1825
2334 **ramapoella** (Kft., 1903)
    metrothetis (Meyr., 1921)
    bidentata (Braun, 1926)

PAROCHROMOLOPIS Gaedike, 1977
2335 **floridana** Gaedike, 1977

GLYPHIPTERIGIDAE
by JOHN B. HEPPNER

ABRENTHIA Bsk., 1915
2336 **cuprea** Bsk., 1915

GLYPHIPTERIX Hbn., 1825
    HERIBEIA Steph., 1829
    AECHMIA Tr., 1833
    AECIMIA Bdv., 1836, missp.
    GLYPHIPTERYX Zell., 1839,
    emend.
    GLYPHITERYX F. v. Rösler-
    stamm, 1841, missp.
    ANACAMPSOIDES Bruand, 1850,
    nom. nud.
    GLYPIPTERYX Staint., 1854,
    missp.
    GLYPHOPTERYX H.-S., 1854,
    missp.
    GLYPHIPTORYX Mann &
    Rogenhofer, 1878, missp.
    GLYPHPTIERYX Turati, 1879,
    missp.
    GLYPHYTERYX Hamp., 1918,
    missp.

2337 **circumscriptella** Cham., 1881
    circumscripta Dyar, 1900, missp.
2338 **quadragintapunctata** Dyar, 1900
2339 **bifasciata** Wlsm., 1881
2340 **unifasciata** Wlsm., 1881
2341 **haworthana** (Steph., 1834)
    haworthella (Steph., 1829), nom.
      nud.
    zonella (Zett., 1839)
    howarthana Jordan, 1886, missp.
2342 **californiae** Wlsm., 1881
2343 **saurodonta** Meyr., 1913
2344 **montisella** Cham., 1875
    montinella Cham., 1877, missp.
    montella Meyr., 1913, emend.
2345 **lanista** Meyr., 1918, mispl.
2346 **impigritella** Clem., 1863, mispl.
    exoptatella Cham., 1875

Yponomeutoidea

PLUTELLIDAE

by JOHN B. HEPPNER & W. DONALD DUCKWORTH

ARAEOLEPIA Wlsm., 1881
2347 **subfasciella** Wlsm., 1881

ELLABELLA Bsk., 1925
    PROBOLACMA Meyr., 1927
    SPILOGENES Meyr., 1938
2348 **editha** Bsk., 1925
2349 **melanoclista** (Meyr., 1927)

EUCALANTICA Bsk., 1904
2350 **polita** (Wlsm., 1881)

EUCERATIA Wlsm., 1881
2351 **castella** Wlsm., 1881
2352 **securella** Wlsm., 1881

HOMADAULA Lower, 1899
    PARAPRAYS Rebel, 1910
    STICHOTACTIS Meyr., 1930
    HAMADAULA Moriuti, 1977, missp.
2353 **anisocentra** Meyr., 1922
    usuguronis Matsumura, 1931
    albizziae Clarke, 1943

MELITONYMPHA Meyr., 1927
2354 **heteraula** Meyr., 1927

PLINIACA Bsk., 1907
2355 **bakerella** Bsk., 1907
2356 **sparsisquamella** Bsk., 1907

PLUTELLA Schr., 1802
    ANADETIA Hbn., 1825
    EUOTA Hbn., 1825
    CEROSTOMA; auth., not Latr., 1802
    CREAGRIA Sodoffsky, 1837
2357 **albidorsella** Wlsm., 1881
2358 **armoraciae** Bsk., 1912
    amoraciae Marsh, 1913, missp.
    monochlora Meyr., 1914
2359 **dammersi** Bsk., 1934

2360 **interrupta** Wlsm., 1881
2361 **notabilis** Bsk., 1904
2362 **omissa** Wlsm., 1889
2363 **porrectella** (L., 1758)
    hesperidella (Hbn., 1796)
    vigilaciella Clem., 1860
2364 **poulella** Bsk., 1904
2365 **vanella** Wlsm., 1881
2366 **xylostella** (L., 1758)
    maculipennis (Curt., 1832)
    annulatellus; Wood., 1839, not Curt.,
      1832
    cruciferarum Zell., 1843
    brassicella Fitch, 1856
    limbipennella Clem., 1860
    mollipedella Clem., 1860
    cicerella (Rondani, 1876)
    galeatella (Mabille, 1888)
    dubiosella Beutenmüller, 1889
    dudiosalla (Moriuti, 1977), missp.

YPSOLOPHA Latr., 1796
    YPSOLOPHUS F., 1798
    CEROSTOMA Latr., 1802
    HYPSOLOPHA Billberg, 1820
    THERISTIS Hbn., 1825
    HARPIPTERIX Hbn., 1825
    ABEBAEA Hbn., 1825
    HARPIPTERYX Tr., 1833, emend.
    CHAETOCHILUS Steph., 1834
    HARPEPTERYX Sodoffsky, 1837, emend.
    PTEROXIA Gn., 1834, nom. nud.
    HYPOLEPIA Gn., 1845, nom. nud.
    HYPSILOPHUS Agassiz, 1846, emend.
    HARPOPTERYX Agassiz, 1846, emend.
    CREDEMNON Wallgr., 1880
    PERICLYMENOBIUS Wallgr., 1880
    TRACHOMA Wallgr., 1880
    PLUTELOPTERA Cham., 1880
    PLUTELLOPTERA Wlsm., 1881, missp.
    MAPA Strand, 1911
    PYCNOPOGON Chrétien, 1922
    CREDEMNA Fbs., 1923, missp.
    CHALCONYMPHA Meyr., 1931
2367 **aleutianella** (Beutenmüller, 1889)
2368 **angelicella** (Bsk., 1903)
2369 **arizonella** (Bsk., 1903)
2370 **barberella** (Bsk., 1903)
2371 **canariella** (Wlsm., 1881)
2372 **cervella** (Wlsm., 1881)
    subsylvella (Wlsm., 1889)
2373 **cockerella** (Bsk., 1903)
2374 **delicatella** (Bsk., 1903)
2375 **dentella** (F., 1775)
    xylostella; auth., not L., 1758
    harpella (D. & S., 1775)
2376 **dentiferella** (Wlsm., 1881)
2377 **dorsimaculella** (Kft., 1907)
2378 **electropa** (Meyr., 1914)
2379 **elongata** (Braun, 1925)
2380 **falciferella** (Wlsm., 1881)
    ordinalis (Meyr., 1914)
2381 **flavistrigella** (Bsk., 1906)
2382 **frustella** (Wlsm., 1881)
2382 **gerdanella** (Bsk., 1903)

2383 **leptaula** (Meyr., 1927)
2385 **lyonothamnae** (Powell, 1967)
2386 **maculatella** (Bsk., 1906)
2387 **manella** (Bsk., 1903)
2388 **nella** (Bsk., 1903)
2389 **oliviella** (Bsk., 1903)
2390 **querciella** (Bsk., 1903)
2391 **rubrella** (Dyar., 1902)
2392 **senex** (Wlsm., 1889)
    koebelella (Dyar, 1900)
2393 **schwarziella** (Bsk., 1903)
2394 **striatella** (Bsk., 1903)
2395 **sublucella** (Wlsm., 1881)
2396 **undulatella** (Bsk., 1906)
2397 **unicipunctella** (Bsk., 1903)
2398 **ustella** (Cl., 1759)
    radiatella (Don., 1794)
    variella (Hbn., 1796)
    fissella (Hbn., 1796)
    lutarella (Hbn., 1796)
    byssinella (Hbn., 1810–13)
    quinquepunctatus Haw., 1828
    lutosus Haw., 1828
    flaviciliatus Haw., 1828
    unitella (Tr., 1833)
    rufimitrella Steph., 1834
    fulvella (Dup., 1838)
    ochrella (Cham., 1880)
2399 **vintrella** (Bsk., 1906)
2400 **walsinghamiella** (Bsk., 1903)
    instabilella; auth., not Mann, 1866

YPONOMEUTIDAE

by JOHN B. HEPPNER & W. DONALD DUCKWORTH

ATTEVA Wlk., 1854
    POECILOPTERA Clem., 1860, preocc. by Latr., 1829
    AMBLOTHRIDIA Wallgr., 1861
    CORINEA Wlk., 1863
    OETA Grt., 1865
    CARTHARA Wlk., 1866, preocc. by Wlk., 1865
    SYNADIA Wlk., 1866
    SCINTILLA Gn., 1879, preocc. by Deshayes, 1856
    SYBLIS Gn., 1879
2401 **punctella** (Cram., 1781)
    pastulella (F., 1787)
    subtilis (Hbn., 1819)
    aurea (Fitch, 1856)
    compta (Clem., 1860)
    gemmata (Grt., 1873)
    fastuosa (Zell., 1877)
    floridana (Neum., 1891), n. syn.
    edithella Bsk., 1908, n. syn.
    exquisita Bsk., 1912., n. syn.

EUCATAGMA Bsk., 1901
2402 **amyrisella** Bsk., 1901
    amyridella Meyr., 1914, emend.

LACTURA Wlk., 1854
    DIANASA Wlk., 1854
    MIEZA Wlk., 1854
    SARBENA Wlk., 1865, preocc. by Wlk., 1862
    THEMISCYRA Wlk., 1865

CYPTASIA Wlk., 1866
BUXETA Wlk., 1866
ENAEMIA Zell., 1872
PSEUDOTALARA Druce, 1885
PSEUDOCAPRIMA Wlsm., 1900
EPIDICTICA Turner, 1903
HEDYCHARIS Turner, 1903
ERIOPYRRHA Meyr., 1913
2403 atrolinea (B. & McD., 1913)
2404 basistriga (B. & McD., 1913)
2405 pupula (Hbn., 1827–31)
   laeta (Geyer, 1832)
   igninix (Wlk., 1854)
   crassivenella (Zell., 1872)
   crassinervella (Slosson, 1896)
2406 psammitis (Zell., 1872)
2407 subfervens (Wlk., 1854)
2408 rhodocentra (Meyr., 1913)

OCNEROSTOMA Zell., 1847
2409 piniariella Zell., 1847
2410 strobivorum Free., 1960

ORINYMPHA Meyr., 1927
2411 aetherias Meyr., 1927

PODIASA Bsk., 1900
2412 chiococcella Bsk., 1900

SWAMMERDAMIA Hbn., 1825
2413 caesiella (Hbn., 1796)
   heroldella Hbn., 1825
   nubeculella (Tengström, 1848)
   griseocapitella (Staint., 1851)
   nanivora Staint., 1871
   castaneae Bsk., 1914
   cuprescens Braun, 1918
2414 pyrella (Villers, 1789)
   cerasiella (Hbn., 1810–13)
   passerella (Zett., 1839)
   variegata Tengström, 1869

URODUS H.-S., 1854
TRICHOSTIBAS Zell., 1863
PARATIQUADRA Wlsm., 1897
2415 parvula (Hy. Edw., 1881)
   calligera; auth., not Zell., 1877

YPONOMEUTA Latr., 1796
HYPHANTES Hbn., 1806, suppr.
   (ICZN Op. 97)
ERMINEA Haw., 1811
COENYPHANTES Hbn., 1822
NYGMIA Hbn., 1825, preocc. by
   Hbn., 1820
HYPONOMEUTA Sodoffsky, 1837,
   emend.
TEINOPTILA Sauber, 1902
2416 atomocella Dyar, 1902
   diaphora Wlsm., 1907
   atomosella Meyr., 1914, emend.
2417 euonymella Cham., 1872
   orbimaculella Cham., 1873
2418 leucothorax Meyr., 1913
2419 martinella (Wlk., 1863), n. comb.
   afflictella (Wlk., 1864), n. comb.
2420 multipunctella Clem., 1860
   ordinatella Wlk., 1863
2421 padella (L., 1758)
   helicella Freyer, 1842

   mahalebella Gn., 1845
   diffluella Heinemann, 1870
2422 plumbella (D. & S., 1775)
2423 semialbus Meyr., 1913

ZELLERIA Staint., 1849
2424 arizonica Braun, 1940
2425 celastrusella Kft., 1903
   celastrella (Meyr., 1914), emend.
2426 gracilariella Bsk., 1904
   ribesella Bsk., 1904
2427 haimbachi Bsk., 1915
2428 ochroplagiata (Braun, 1918)
2429 parnassiae Braun, 1940
2430 pyri Clarke, 1942
2431 retiniella Fbs., 1923
2432 semitincta Meyr., 1930

ARGYRESTHIIDAE
by JOHN B. HEPPNER & W. DONALD
DUCKWORTH

ARGYRESTHIA Hbn., 1825
ARGYROSETIA Steph., 1829
OLIGOS Tr., 1830
EDERESA Curt., 1833
ISMENE Steph., 1834
BLASTOTERE Ratzeburg, 1840
2433 abies Free., 1972
2434 affinis Braun, 1940
2435 alternatella Kft., 1908
2436 altissimella Cham., 1877
2437 annettella Bsk., 1907
2438 apicimaculella Cham., 1874
   visaliella Cham., 1875
2439 arceuthobiella Bsk., 1916
2440 aureoargentella Brower, 1953
2441 austerella Zell., 1873
2442 belangerella Cham., 1875
2443 bolliella Bsk., 1907
2444 calliphanes Meyr., 1913
2445 canadensis Free., 1972
2446 castaneella Bsk., 1915
2447 chalcochrysa Meyr., 1913
2448 columbia Free., 1972
2449 conjugella Zell., 1839
   maculosa Tengström, 1847
   aerariella Staint., 1871
2450 cupressella Wlsm., 1890
2451 deletella Zell., 1873
2452 eugeniella Bsk., 1916
2453 franciscella Bsk., 1915
2454 flexilis Free., 1960
2455 freyella Wlsm., 1890
   abdominalis; auth., not Zell., 1839
2456 furcatella Bsk., 1916
2457 goedartella (L., 1758)
   semiargentella (Don., 1793)
   literella (Haw., 1828)
   splendida Reutti, 1898
   aurescentella Uffeln, 1930
2458 inscriptella Bsk., 1907
2459 laricella Kft., 1908
2460 libocedrella Bsk., 1916
2461 mariana Free., 1972
2462 media Braun, 1914
2463 mesocausta Meyr., 1913
2464 monochromella Bsk., 1921
2465 montella Cham., 1877

2466 nymphocoma Meyr., 1919
2467 oreasella Clem., 1860
   andereggiella Zell., 1873
   oreadella Meyr., 1914, emend.
2468 pallidella Braun, 1918
2469 pedmontella Cham., 1877
   pedemontella Meyr., 1914, emend.
2470 picea Free., 1972
2471 pilatella Braun, 1910
2472 plicipunctella Wlsm., 1890
2473 pseudotsuga Free., 1972
2474 pygmaeella (Hbn., 1810–13)
   pygmaella auth., missp.
   semifasciella Steph., 1834
   capilella Strand, 1901
   alpina Müller-Rutz, 1920
   hyperboreella Strand, 1920
2475 quadristrigella Zell., 1873
2476 quercicolella Cham., 1877
2477 rileiella Bsk., 1907
2478 ruidosa Braun, 1940
2479 subreticulata Wlsm., 1882
2480 thoracella Bsk., 1907
2481 thuiella (Pack., 1871)
2482 trifasciae Braun, 1910
   triplicata Meyr., 1914
2483 tsuga Free., 1972
2484 undulatella Cham., 1874

DOUGLASIIDAE
by JOHN B. HEPPNER & W. DONALD
DUCKWORTH

TINAGMA Zell., 1839
DOUGLASIA Staint., 1854
2485 giganteum Braun., 1921
2486 leucaspis Braun, 1926
2487 obscurofasciella (Cham., 1881)
   crenulellum Engel, 1907
2488 ochreomaculella (Cham., 1875)
2489 pulverilinea Braun, 1921

ACROLEPIIDAE
by JOHN B. HEPPNER & W. DONALD
DUCKWORTH

ACROLEPIOPSIS Gaedike., 1970
ARGIOPE Cham., 1873, preocc. by
   Audouin, 1827
2490 incertella (Cham., 1872), n. comb.
   dorsimaculella (Cham., 1873), n. comb.
2491 leucoscia (Meyr., 1927), n. comb.
2492 reticulosa (Braun, 1927), n. comb.

HELIODINIDAE
by JOHN B. HEPPNER & W. DONALD
DUCKWORTH

CYCLOPLASIS Clem., 1864
2493 panicifoliella Clem., 1864

SCELORTHUS Bsk., 1900
2494 pisoniella Bsk., 1900

LAMPROLOPHUS Bsk., 1900.
EMBOLA Wlsm., 1909
2495 lithella Bsk., 1900

HELIODINES Staint., 1854
  AETOLE Cham., 1875
  AETOLA Frey, 1884, missp.
  HELIODINIDES Turner, 1941,
    missp.
2496 **albaciliella** Bsk., 1910
  albiciliella Meyr., 1913, emend.
2497 **bella** (Cham., 1875)
2498 **ciccella** B. & Bsk., 1920
2499 **extraneella** Wlsm., 1881
2500 **ionis** Clarke, 1952
2501 **metallicella** Bsk., 1909
2502 **nyctaginella** Gibson, 1914
2503 **perichalca** Meyr., 1912
2504 **sexpunctella** Wlsm., 1892
2505 **tripunctella** Wlsm., 1892
2506 **unipunctella** Wlsm., 1892

SCHRECKENSTEINIA Hbn., 1825
  CHRYSOCORYS Curt., 1833
2507 **erythriella** (Clem., 1860)
2508 **felicella** (Wlsm., 1880)
2509 **festaliella** (Hbn., 1818–19)
  scisscella (Haw., 1828)
  montandonella (Dup., 1838)

LITHARIAPTERYX Cham., 1876
2510 **abroniaeella** Cham., 1876
  abroniella Meyr., 1913, emend.
2511 **jubarella** Comst., 1940
2512 **mirabilinella** Comst., 1940

Sesioidea

SESIIDAE

by JOHN B. HEPPNER & W. DONALD
DUCKWORTH

TINTHIINAE

Pennisetiini

PENNISETIA Dehne, 1850
  BEMBECIA auth., not Hbn., 1819
  ANTHRENOPTERA Swinhoe,
    1892
  LOPHOCNEMA Turner, 1917
  DIAPYRA Turner, 1917
  GLOSSECIA Hamp., 1919
2513 **marginata** (Harr., 1839)
  pleciaeformis (Wlk., 1856)
  odyneripennis (Wlk., 1856)
  rubi (Riley, 1874)
  flavipes (Hulst, 1881)
  albicoma (Hulst, 1883), var.

Tinthiini

ZENODOXUS G. & R., 1868
2514 **canescens** Hy. Edw., 1881
  sidae Engelh., 1946, race
2515 **heucherae** Hy. Edw., 1881
  potentillae Hy. Edw., 1881
2516 **maculipes** G. & R., 1868
2517 **mexicanus** Beutenmüller, 1897
2518 **palmii** (Neum., 1891)

  palmiana (Dalla Torre, 1925), repl.
    name
  wissadulae Engelh., 1946
  sphaeralceae Engelh., 1946, race
  incanae Engelh., 1946, race
2519 **rubens** Engelh., 1946
  bexari Engelh., 1946, race
2520 **sidalceae** Engelh., 1946

PARANTHRENINAE

Cissuvorini

CISSUVORA Engelh., 1946
2521 **ampelopsis** Engelh., 1946

Paranthrenini

PARANTHRENE Hbn., 1819
  MEMYTHRUS Newman, 1832
  PARANTHRENA H.-S., 1846,
    emend.
  SCIAPTERON Stgr., 1854
  TARSA Wlk., 1856
  PSEUDOSESIA Felder, 1861
  PSEUDOSETIA Bdv., 1875, missp.
  PRAMILA Moore, 1879
  FATUA Hy. Edw., 1882, preocc. by
    Dejean, 1833
  SCIAPTERUM Bartel, 1912,
    emend.
  PARANTHRENELLA Strand,
    1916
  PSEUDOSECIA Dalla Torre &
    Strand, 1925, missp.
  NOKONO Matsumura, 1931
  LEPTOCIMBICINA Bryk, 1947
2522 **asilipennis** (Bdv., 1829)
  denudatum (Harr., 1839)
  vespipenne (H.-S., 1854)
  bombyciformis (Wlk., 1856)
  vespipennis (Bdv., 1875)
  championi (Druce, 1883)
2523 **dollii** (Neum., 1894)
  castaneum (Beutenmüller, 1897), var.
  fasciventris Engelh., 1946, form
2524 **tabaniformis** (Rottemburg, 1775)
  asiliformis; D. & S., 1775, not
    Rottemburg, 1775
  vespiformis; Newman, 1832, not L.,
    1761
  tricincta (Harr., 1839)
  serratiformis (Freyer, 1842)
  denotata (Hy. Edw., 1882)
  oslari Engelh., 1946, form
2525 **fenestrata** B. & L., 1922
2526 **robiniae** (Hy. Edw., 1880)
  perlucida (Busck, 1915)
  palescens Engelh., 1946, form
2527 **simulans** (Grt., 1881)
  palmii (Hy. Edw., 1887)
  luggeri (Hy. Edw., 1891)

VITACEA Engelh., 1946
2528 **admiranda** (Hy. Edw., 1882)
2529 **cupressi** (Hy. Edw., 1881)
2530 **polistiformis** (Harr., 1854)
  seminole (Neum., 1894)
  huron Engelh., 1946, form
2531 **scepsiformis** (Hy. Edw., 1881)

ALBUNA Hy. Edw., 1881
  HARMONIA Hy. Edw., 1882,
    preocc. by Mulsant, 1846
  PARHARMONIA Beutenmüller,
    1894, repl. name
2532 **fraxini** (Hy. Edw., 1881)
  morrisoni (Hy. Edw., 1882)
  vitriosa Engelh., 1946, ♂ form
2533 **pyramidalis** (Wlk., 1856)
  hylotomiformis (Wlk., 1856)
  nomadaepennis (Bdv., 1869)
  rubescens (Hulst, 1881)
  montana Hy. Edw., 1881
  tanaceti Hy. Edw., 1881
  vancouverensis Hy. Edw., 1881
  coloradensis Hy. Edw., 1881
  torva Hy. Edw., 1881
  beutenmuelleri Skin., 1903

EUHAGENA Hy. Edw., 1881
  LARUNDA Hy. Edw., 1881,
    preocc. by Leach, 1815
  GAEA Beutenmüller, 1896, repl.
    name
2534 **emphytiformis** (Wlk., 1856)
  solituda (Hy. Edw., 1881)
2535 **nebraskae** Hy. Edw., 1881
  coloradensis (Beutenmüller, 1893)
  mormoni Engelh., 1946, form
  intensa Engelh., 1946, form

SESIINAE

Melittiini

MELITTIA Hbn., 1819
  EUMALLOPODA Wallgr., 1858
  PARASA Wallgr., 1863, preocc. by
    Moore, 1858
  PANSA Wallgr., 1865, repl. name
  PODERIS Bdv., 1875, nom. nud.
  MELITHA Kby., 1879, missp.
  MELITTA Druce, 1892, missp.
  PREMELITTIA LeCerf, 1916
  NEOSPHECIA LeCerf, 1916
  MELITTINA LeCerf, 1917
2536 **cucurbitae** (Harr., 1828)
  satyriniformis Hbn., 1827–31
  ceto (Westwood, 1848)
  amoena Hy. Edw., 1882
2537 **calabaza** Duckworth & Eichlin, 1973
2538 **snowii** Hy. Edw., 1882
2539 **grandis** (Stkr., 1881)
  beckeri Druce, 1892
  hermosa Engelh., 1946, var.
2540 **gloriosa** Hy. Edw., 1880
  superba; B. & L., 1922, not Roths.,
    1909
  lindseyi B. & Benj., 1925
  barnesi Dalla Torre, 1925
2541 **magnifica** Beutenmüller, 1900

Sesiini

SESIA F., 1775
  TROCHILIUM Scop., 1777
  AEGERIA F., 1807
  SETIA Oken, 1815, suppr. (ICZN
    Op. 417)
  SPHECIA Hbn., 1819

SETIA Meigen, 1830, emend.
SOMETIA Meigen, 1830, missp.
TROCHILUM Wlk., 1854, missp.
TROCHILIA Speyer & Speyer,
    1858, emend.; preocc. by
    Dujardin, 1840
SPHECODOPTERA Hamp., 1893
GLOSSOSPHECIA Hamp., 1919
SPHECOPTERA Dalla Torre &
    Strand, 1925, missp.
EUSPHECIA LeCerf, 1937
2542 apiformis (Cl., 1759)
    crabroniformis (D. & S., 1775)
2543 tibialis (Harr., 1839)
    flavitibia (Wlk., 1856)
    pacificum (Hy. Edw., 1881)
    californicum (Neum., 1891)
    minimum (Neum., 1891)
    dyari (Ckll., 1908), var.
    anonyma (Strand, 1925), var.
    melanoformis (Engelh., 1946), var.

Osminiini

CALASESIA Beutenmüller, 1899
2544 coccinea (Beutenmüller, 1898)

OSMINIA LeCerf, 1917
    SIGNAPHORA Engelh., 1946
2545 ruficornis (Hy. Edw., 1881)
    minuta (Hy. Edw., 1881)
    candescens (Hy. Edw., 1882)
    marcia (Druce, 1889)

Synanthedonini

SYNANTHEDON Hbn., 1819
    CONOPIA Hbn., 1819
    AUSTROSETIA Felder, 1874
    TEINOTARSINA Felder, 1874
    PYRRHOTAENIA Grt., 1875
    ICHNEUMENOPTERA Hamp.,
        1893, n. syn.
    VESPAMIMA Beutenmüller, 1894
    SANNINOIDEA Beutenmüller, 1896
    THAMNOSPHECIA Spuler, 1910
    CANOPIA Wileman & South, 1918,
        missp.
    SCABISA Matsumura, 1931
    RAMOSIA Engelh., 1946
    SYLVORA Engelh., 1946
    SYNATHEDON Wolfsberger,
        1961, missp.
    TIPULIA Králiček & Povolný,
        1977, n. syn.
2546 acerrubri Engelh., 1925
2547 geliformis (Wlk., 1856)
2548 richardsi (Engelh., 1946)
2549 scitula (Harr., 1839)
    gallivorum (Westwood, 1854)
    hospes (Walsh, 1867)
    corusca (Hy. Edw., 1881)
    aemula (Hy. Edw., 1883)
2550 pictipes (G. & R., 1868)
    inusitata (Hy. Edw., 1881)
2551 rhododendri (Beutenmüller, 1909)
2552 rileyana (Hy. Edw., 1881)
    brunneipennis (Hy. Edw., 1881)
    hyperici (Hy. Edw., 1881)
    austini (Engelh., 1946)

2553 tipuliformis (Cl., 1759)
    salmachus (L., 1758), unused sr. syn.
    tipula (Retzius, 1783)
2554 acerni (Clem., 1860)
    acericolum (Germadius, 1874)
    tepperi (Hy. Edw., 1881)
    buscki (Engelh., 1946), race
2555 fatifera Hodges, 1962
2556 viburni Engelh., 1925
2557 alleri (Engelh., 1946)
2558 arctica (Beutenmüller, 1900)
2559 bolteri (Hy. Edw., 1883)
2560 canadensis Duckworth & Eichlin,
    1973
2561 culiciformis (L., 1758)
    culex (Retzius, 1783)
    thynniformis (Laspeyres, 1801)
    americana (Beutenmüller, 1896), var.
    biannulata Bartel, 1902, var.
    flavocingulata (Spuler, 1910), ab.
    triannulata (Spuler, 1910), ab.
2562 dominicki Duckworth & Eichlin, 1973
2563 fulvipes (Harr., 1839)
2564 helenis (Engelh., 1946)
2565 pyri (Harr., 1830)
    koebelei (Hy. Edw., 1881)
2566 refulgens (Hy. Edw., 1881)
    marica (Beutenmüller, 1899)
    seminole (Beutenmüller, 1899)
    marcia auth., missp.
2567 rubrofascia (Hy. Edw., 1881)
2568 saxifragae (Hy. Edw., 1881)
    henshawii (Hy. Edw., 1882)
2569 sigmoidea (Beutenmüller, 1897)
2570 albicornis (Hy. Edw., 1881)
2571 decipiens (Hy. Edw., 1881)
    imperfecta (Hy. Edw., 1881)
    nicotianae (Hy. Edw., 1881)
    rubristigma (Kellicott, 1892)
2572 proxima (Hy. Edw., 1881)
    modesta (Kellicott, 1892)
2573 sapygaeformis (Wlk., 1856)
    floridensis (Grt., 1875)
2574 arizonensis (Beutenmüller, 1916)
2575 arkansasensis Duckworth & Eichlin,
    1973
2576 bibionipennis (Bdv., 1869)
    rutilans (Hy. Edw., 1881)
    lupini (Hy. Edw., 1881)
    perplexa (Hy. Edw., 1881)
    impropria (Hy. Edw., 1881)
    aureola (Hy. Edw., 1881)
    neglecta (Hy. Edw., 1881)
    washingtonia (Hy. Edw., 1881)
    hemizoniae (Hy. Edw., 1881)
    madariae (Hy. Edw., 1881)
    hemizonae Sm., 1891, missp.
2577 castaneae (Bsk., 1913)
2578 chrysidipennis (Bdv., 1869)
    tacoma (Beutenmüller, 1898)
    wallowa (Engelh., 1946), race
2579 kathyae Duckworth & Eichlin, 1977
2580 mellinipennis (Bdv., 1836)
    artemisiae (Hy. Edw., 1881)
    seneciodes (Hy. Edw., 1881)
    senecioides (Sm., 1891), missp.
2581 polygoni (Hy. Edw., 1881)
    fragariae (Hy. Edw., 1881)
    helianthi (Hy. Edw., 1881)
    achillae (Hy. Edw., 1881)

    eremocarpi (Hy. Edw., 1881)
    meadii (Hy. Edw., 1881)
    orthocarpi (Hy. Edw., 1881)
    praestans (Hy. Edw., 1882)
    behrensii (Hy. Edw., 1882)
    animosa (Hy. Edw., 1883)
    elda (Hy. Edw., 1885)
    semipraestans (Ckll., 1908), var.
    achillaeae Dalla Torre & Strand,
        1925, missp.
    behrensi Dalla Torre & Strand, 1925,
        missp.
2582 resplendens (Hy. Edw., 1881)
2583 exitiosa (Say, 1823)
    persica (Thomas, 1824)
    pepsidiformis (Hbn., 1827–31)
    xiphiaeformis (Bdv., 1875)
    graefi (Hy. Edw., 1881)
    opalescens (Hy. Edw., 1881)
    fitchii (Hy. Edw., 1882), var.
    pacifica (Riley, 1891)
    luminosa (Neum., 1894), var.
    edwardsii (Beutenmüller, 1900), var.
    graefii (Beutenmüller, 1900), missp.
    barnesii (Beutenmüller, 1900), var.
    fitchi (Dalla Torre & Strand, 1925),
        missp.
2584 novaroensis (Hy. Edw., 1881)
    piceae (Dyar, 1904)
    brunneri (Bsk., 1914)
2585 pini (Kellicott, 1881)
2586 sequoiae (Hy. Edw., 1881)
    superba (Hy. Edw., 1881)
    pinorum (Behrens, 1889)

PALMIA Beutenmüller, 1896
2587 praecedens (Hy. Edw., 1883)

PODOSESIA Mösch., 1879, repl.
    name
    GROTEA Mösch., 1876, preocc. by
        Cresson, 1864
2588 aureocincta Purrington & Nielsen,
    1977
2589 syringae (Harr., 1839)
    longipes (Mösch., 1876)
    fraxini (Lugger, 1891)

SANNINA Wlk., 1856
    SAUNINA Bdv., 1875, missp.
    SOSPITA Hy. Edw., 1882, preocc.
        by Rafinesque, 1815
    PHEMONOE Hy. Edw., 1882,
        repl. name
2590 uroceriformis Wlk., 1856
    quinquecaudata (Ridings, 1862)
    uroceripennis Bdv., 1875

CARMENTA Hy. Edw., 1881
2591 albociliata (Engelh., 1925)
2592 anthracipennis (Bdv., 1875)
    sanborni Hy. Edw., 1881
    morula (Hy. Edw., 1881)
2593 apache Engelh., 1946
2594 arizonae (Beutenmüller, 1898)
2594.1 armasata (Druce, 1892)
2595 auritincta (Engelh., 1925)
2596 bassiformis (Wlk., 1856)
    lustrans (Grt., 1880)
    aureopurpura (Hy. Edw., 1880)

bolli (Hy. Edw., 1881)
sexfasciata (Hy. Edw., 1881)
consimilis (Hy. Edw., 1881)
eupatorii (Hy. Edw., 1881)
imitata (Hy. Edw., 1881)
aureopurpurea (Sm., 1891), missp.
bollii (Sm., 1891), missp.
2597 **corni** (Hy. Edw., 1881)
infirma (Hy. Edw., 1881)
2598 **engelhardti** Duckworth & Eichlin, 1973
2599 **giliae** (Hy. Edw., 1881)
vitrina (Neum., 1891)
deceptiva (Beutenmüller, 1894)
woodgatei Engelh., 1946, race
2600 **ithacae** (Beutenmüller, 1897)
2601 **mariona** (Beutenmüller, 1900)
2602 **mimuli** (Hy. Edw., 1881)
torrancia Engelh., 1946
2603 **odda** Duckworth & Eichlin, 1977
2604 **ogalala** Engelh., 1946
2605 **pallene** (Druce, 1889)
2606 **phoradendri** Engelh., 1946
2607 **prosopis** (Hy. Edw., 1882)
2608 **pyralidiformis** (Wlk., 1856)
nigella (Hulst, 1881)
aurantis Engelh., 1946, var.
2609 **querci** (Hy. Edw., 1882)
comes (Heinr., 1920)
2610 **rubricincta** (Beutenmüller, 1909)
2611 **subaerea** (Hy. Edw., 1883)
2612 **suffusata** Engelh., 1946
2613 **tecta** (Hy. Edw., 1882)
2614 **texana** (Hy. Edw., 1881)
wittfeldii (Hy. Edw., 1883)
2615 **verecunda** (Hy. Edw., 1881)
nigra Beutenmüller, 1894
florissantella (Ckll., 1908)
florisantella (Dalla Torre & Strand, 1925), missp.
hirsulta (Engelh., 1946)
2616 **welchelorum** Duckworth & Eichlin, 1977
2617 **wellerae** Duckworth & Eichlin, 1976

PENSTEMONIA Engelh., 1946
2618 **clarkei** Engelh., 1946
2619 **dammersi** Engelh., 1946
brevifolia Engelh., 1946
2620 **edwardsii** (Beutenmüller, 1894)
utahensis (Beutenmüller, 1909)
2621 **hennei** Engelh., 1946

ALCATHOE Hy. Edw., 1882
ALCALTHOE Riley & Howard, 1891, missp.
ALCOTHOE Patch, 1908, missp.
2622 **carolinensis** Engelh., 1925
autumnalis Engelh., 1946
2623 **caudata** (Harr., 1839)
walkeri Neum., 1894, var.
annettella Engelh., 1946, race
2624 **pepsioides** Engelh., 1925
atra Engelh., 1925, var.
ferrugata Engelh., 1946, race
2625 **verrugo** (Druce, 1884)
corvinus Engelh., 1946, var.

HYMENOCLEA Engelh., 1946
2626 **palmii** (Beutenmüller, 1902)

CHOREUTIDAE
by JOHN B. HEPPNER

BRENTHIINAE

BRENTHIA Clem., 1860
MICROAETHIA Cham., 1878, nom. nud.
2627 **pavonacella** Clem., 1860
amphicarpeoeana (Cham., 1878), nom. nud.
pavonicella Cham., 1878, missp.

CHOREUTINAE

ANTHOPHILA Haw., 1811
SIMAETHIS Leach, 1815
XYLOPODA Berthold, 1827
XYLOPODA Latr., 1829, preocc. by Berthold, 1827
SIMOETHIS Desmarest, 1848, missp.
SYMAETHIS Bruand, 1850, missp.
SIMAETIS Kautz, 1931, missp.
SIMETHIS Bleszynski, Razowski, & Zukowski, 1965, missp.
ANTOPHILA Bleszynski, Razowski, & Zukowski, 1965, missp., not Hbn., 1806
SIAMETHIS Klimesch, 1968, missp.
2628 **alpinella** (Bsk., 1904)

PROCHOREUTIS Diakonoff & Heppner, 1980
2629 **inflatella** (Clem., 1863)
virginiella (Clem., 1864)
2630 **sororculella** Dyar, 1900
2631 **dyarella** Kft., 1902
2632 **extrincicella** Dyar, 1900
extrinsecella Meyr., 1913, emend.
2633 **pernivalis** Braun, 1921

CALOREAS Heppner, 1977
2634 **apocynoglossa** (Heppner, 1976)
2635 **occidentella** (Dyar, 1900)
2636 **coloradella** (Dyar, 1900)
2637 **caliginosa** (Braun, 1921)
2638 **schausiella** (Bsk., 1907)
2639 **augustella** (Clarke, 1933)
2640 **multimarginata** (Braun, 1925)
melanifera (Keif., 1937)
2641 **leucobasis** (Dyar, 1900)

TEBENNA Billberg, 1820
PORPE Hbn., 1825
TEBEUNA Danilevsky, 1969, missp.
2642 **silphiella** (Grt., 1881)
2643 **balsamorrhizella** (Bsk., 1904)
2644 **gemmalis** (Hulst, 1886)
2645 **immutabilis** (Braun, 1927)
2646 **piperella** (Bsk., 1904)
2647 **gnaphaliella** (Kft., 1902)
gnaphiella (Mosher, 1916), missp.
2648 **onustana** (Wlk., 1864)
ohiensis (Zell., 1875)
ohioensis (Kft., 1902), missp.
2649 **carduiella** (Kft., 1902)
busckiella (Kft., 1902)
carduella (Meyr., 1913), emend.

CHOREUTIS Hbn., 1825
HEMEROPHILA Hbn., 1806, suppr. (ICZN Op. 97)
EUTROMULA Frölich, 1828
CHOREUTES Tr., 1835, emend.
MACROPIA O. Costa, 1836
CHORENTES Morris, 1872, missp.
ENTOMOLOMA Rag., 1875, nom. nud.
CHORENTIS Turner, 1898, missp.
ORCHEMIA Fern., 1900, preocc. by Gn., 1845
HEMEROPHILA Fern., 1900, preocc. by Hbn., 1817
CHOREUTIDIA Sauber, 1902
ALLONONYMA Bsk., 1904
ALLONYMA Fracker, 1915, missp.
CLOREUTIS Kautz, 1931, missp.
CHOREUTHIS Hackman, 1947, missp.
CHLOREUTIS Viette, 1947, missp.
ALLONOMYIA Fgn., 1975, missp.
2650 **pariana** (Cl., 1759)
lutosa (Haw., 1811)
parialis (Tr., 1829), emend.
pariava (Desmarest, 1849), missp.
parina (Procter, 1946), missp.
2651 **diana** (Hbn., 1819–22)
dianalis (Tr., 1835), emend.
decorana (Zett., 1839)
luridana (Wlk., 1863)
vicarialis (Zell., 1875)
vicarilis (Dyar, 1900), missp.
2652 **betuliperda** (Dyar, 1902)

TORTYRA Wlk., 1863
CHOREGIA Felder & Rogenhofer, 1875
CHOREGIA Zell., 1877, preocc. by Felder & Rogenhofer, 1875
2653 **slossonia** (Fern., 1900)

HEMEROPHILA Hbn., 1817
GAURIS Hbn., 1821
WALSINGHAMIA Riley, 1889
GUARIS Fern., 1900, missp.
2654 **dyari** Bsk., 1900
2655 **diva** (Riley, 1889)

Cossoidea

COSSIDAE
by RONALD W. HODGES

HYPOPTINAE

HYPOPTA Hbn., 1820
2656 **palmata** B. & McD., 1910

INGUROMORPHA Hy. Edw., 1888
RAVIGIA Dyar, 1905
POMERIA B. & McD., 1911
INGURIMORPHA Turner, 1918, missp.
JNGUROMORPHA Dalla Torre, 1923, missp.
2657 **itzalana** (Stkr., 1900)

2658 **arcifera** (Dyar, 1906)
    gabriel (Dyar, 1913)
2659 **basalis** (Wlk., 1856)
    slossonae Hy. Edw., 1889, emend.
    slossoni Hy. Edw., 1889, incorr. orig.
      spell.

    GIVIRA Wlk., 1856
    EUGIVIRA Schaus, 1901
    LENTAGENA Dyar, 1905
2660 **mucida** (Hy. Edw., 1882)
2661 **arbeloides** (Dyar, 1899)
    felicoma Dyar, 1913
    a. **rufescens** B. & McD., 1911
2662 **theodori** (Dyar, 1893)
    kunzei Dyar, 1923
2663 **durangona** (Schaus, 1901)
2664 **carla** Dyar, 1923
2665 **cornelia** (N. & D., 1893)
    caerulea (Dalla Torre, 1923), missp.
2666 **lucretia** B. & McD., 1913
2667 **ethela** (N. & D., 1893)
2668 **anna** (Dyar, 1898)
2669 **marga** B. & McD., 1910
2670 **lotta** B. & McD., 1910
2671 **francesca** (Dyar, 1909)
2672 **minuta** B. & McD., 1910
2673 **cleopatra** B. & McD., 1912

COSSULINAE

    COSSULA Bailey, 1882
    COSTRIA Schaus, 1901
    HEMIPECTEN Dyar, 1905, preocc.
      by Adams & Reeve, 1848
    SCHAUSIANA Strand, 1910
    HEMIPECTRONA Schaus, 1921
2674 **magnifica** (Stkr., 1876)
    magnifica Bailey, 1882, preocc. by
      Stkr., 1876
    norax (Druce, 1898)

COSSINAE

    ACOSSUS Dyar, 1905
2675 **centerensis** (Lint., 1877)
2676 **populi** (Wlk., 1856)
    a. **angrezi** (Bailey, 1882)
    b. **orc** (Stkr., 1893)
    generosus Dyar, 1925
2677 **undosus** (Lint., 1878)
    brucei (French, 1890)
    nodosus (Kby., 1892), missp.
2678 **connectus** B. & McD., 1916

    COMADIA B. & McD., 1911
    HETEROCOMA B. & McD., 1918
2679 **henrici** (Grt., 1882)
2680 **suaedivora** Brown & Allen, 1973
2681 **dolli** B. & Benj., 1923
2682 **intrusa** B. & Benj., 1923
2683 **arenae** Brown, 1976
2684 **subterminata** B. & Benj., 1923
    fusca B. & Benj., 1923
2685 **sperata** Brown, 1976
2686 **bertholdi** (Grt., 1880)
    edwardi (N. & D., 1893)
    edwardsi (Dalla Torre, 1923), missp.
    engelhardti B. & Benj., 1923
    stabilis B. & Benj., 1923

    a. **indistincta** Brown, 1976
    b. **polingi** B. & Benj., 1927
2687 **alleni** Brown, 1976
2688 **manfredi** (Neum., 1884)
2689 **redtenbacheri** (Hammerschmidt, 1848)
    agavis Blasquez, 1870
    redtenbachi (Druce, 1887), missp.
    chilodora (Dyar, 1910)
2890 **albistriga** (B. & McD., 1918)

    FANIA B. & McD., 1911
2691 **nana** (Stkr., 1876)

    PRIONOXYSTUS Grt., 1882
    XYSTUS Grt., 1874, preocc. by
      Schoenherr, 1826
    PRIONYXSTUS B. & McD., 1911,
      missp.
2692 **piger** (Grt., 1866)
    baccharidis Clarke, 1952, n. syn.
2693 **robiniae** (Peck, 1818)
    crepera (Hanr., 1835)
    plagiatus (Wlk., 1856)
    querciperda (Pack., 1864), preocc. by
      Fitch, 1859, ab.
    reticulatus (Lint., 1878), ab.
    quercus Ehrmann, 1893, ab.
    a. **zabolicus** (Stkr., 1898)
    b. **mixtus** B. & Benj., 1923
    c. **subnigrus** B. & Benj., 1923
    d. **flavotinctus** B. & Benj., 1923
2694 **macmurtrei** (Guér., 1829)
    querciperda (Fitch, 1859)

    MIACORA Dyar, 1905
    TORONIA B. & McD., 1911
2695 **perplexa** (N. & D., 1893)
2696 **luzena** (Barnes, 1905)

ZEUZERINAE

    HAMILCARA B. & McD., 1910
2697 **ramosa** (Schaus, 1892)
    ramuscula (Dyar, 1906)
2698 **atra** B. & McD., 1910
2699 **gilensis** B. & McD., 1910

    ZEUZERA Latr., 1804
    LATAGIA Hbn., 1820
2700 **pyrina** (L., 1761)
    aesculi (L., 1767)
    decipiens Kby., 1892

Tortricoidea

TORTRICIDAE

by JERRY A. POWELL

OLETHREUTINAE

Olethreutini

    EPISIMUS Wlsm., 1892
2701 **argutanus** (Clem., 1860)
    hamameliella (Clem., 1861)
    allutana (Zell., 1879)
2702 **augmentanus** (Zell., 1877)

2703 **tyrius** Heinr., 1923

    CACOCHARIS Wlsm., 1892
2704 **cymotoma** (Meyr., 1917)

    BACTRA Steph., 1834
2705 **lancealana** (Hbn., 1796–99)
    lanceolana (Hbn., 1822)
2706 **furfurana** (Haw., 1811)
2707 **verutana** Zell., 1875
    a. **albipuncta** Heinr., 1926
    b. **chrysea** Heinr., 1926
2708 **maiorina** Heinr., 1923
2709 **priapeia** Heinr., 1923
2710 **sinistra** Heinr., 1926

    ENDOPIZA Clem., 1860
    PARALOBESIA Obr., 1953
    LOBESIA; auth., part, not Gn., 1845
2711 **liriodendrana** (Kft., 1904), n. comb.
    magnoliana (Kft., 1907)
2712 **viteana** Clem., 1860
    vitivorana (Pack., 1869)
    botrana; auth.
2713 **monotropana** (Heinr., 1926), n. comb.
2714 **sambuci** (Clarke, 1953), n. comb.
2715 **cypripediana** (Fbs., 1924), n. comb.
2716 **rhoifructana** (Kft., 1904), n. comb.
2717 **yaracana** (Kft., 1907), n. comb.
    signifera (Meyr., 1912), repl. name,
      n. comb.
2718 **spiraeifoliana** (Heinr., 1923), n.
      comb.
2719 **exasperana** (McD., 1938), n. comb.
2720 **palliolana** (McD., 1938), n. comb.
2721 **piceana** (Free., 1941), n. comb.
2722 **aemulana** (Heinr., 1926), n. comb.
2723 **vernoniana** (Kft., 1907), n. comb.
    ambrosiana (Kft., 1907), n. comb.
2724 **aruncana** (Kft., 1907), n. comb.
2725 **slingerlandana** (Kft., 1904), n. comb.
2726 **blandula** (Heinr., 1926), n. comb.
2727 **cyclopiana** (Heinr., 1926), n. comb.

    LOBESIA Gn., 1845
    POLYCHROSIS Rag., 1894
    LOMASCHIZA Lower, 1901
    BYRSOPTERA Lower, 1901
    STERIPHOTIS Meyr., 1911
2728 **carduana** (Bsk., 1907)
2729 **spiraeae** (McD., 1938)

    AHMOSIA Heinr., 1926
2730 **galbinea** Heinr., 1926
2731 **aspasiana** (McD., 1922)

    ENDOTHENIA Steph., 1852
    ORTHOTAENIA; Staint., 1859,
      not Steph., 1829
    TANIVA Heinr., 1926
    TIA Heinr., 1926
    HULDA Heinr., 1926
2732 **montanana** (Kft., 1907)
    kingi McD., 1927
2733 **heinrichi** McD., 1929
2734 **rubipunctana** (Kft., 1907)
2735 **sordulenta** Heinr., 1926
2736 **melanosticta** (Wlsm., 1895)
    flavillana (Dyar, 1903)
2737 **affiliana** McD., 1942

2738 **hebesana** (Wlk., 1863)
   inexpertana (Wlk., 1863)
   fullerea (Riley, 1870)
2739 **daeckeana** (Kft., 1907)
2740 **conditana** (Wlsm., 1879)
2741 **microptera** Clarke, 1953
2742 **infuscata** Heinr., 1923
2743 **quadrimaculana** (Haw., 1811)
   antiguana (Hbn., 1811–13)
   antiquana (Hbn., 1822)
   a. **nubilana** (Clem., 1865)
   ventulana (Wlsm., 1879)
2744 **gentianaeana** (Hbn., 1796–99)
   gentiana (Hbn., 1809–10)
   gentianana (Hbn., 1825)
2745 **albolineana** (Kft., 1907)
   abietana (Fern., 1908)
   piceae (Bsk., 1916)
2746 **vulgana** (McD., 1922)
2747 **impudens** (Wlsm., 1884)

ATERPIA Gn., 1845
   ESIA Heinr., 1926
2748 **approximana** (Heinr., 1919)

EUMAROZIA Heinr., 1926
2749 **malachitana** (Zell., 1875)

ZOMARIA Heinr., 1926
2750 **interruptolineana** (Fern., 1882)
2751 **rosaochreana** (Kft., 1907)
2752 **andromedana** (B. & McD., 1917)

APOTOMIS Hbn., 1825
   APHANIA Hbn., 1825
   LIMMA Hbn., 1825
   ANTITHESIA Steph., 1829
   BRACHYTAENIA Steph., 1852
2753 **capreana** (Hbn., 1814–17)
2754 **paludicolana** (Brower, 1953), n. comb.
2755 **funerea** (Meyr., 1920)
   youngana (McD., 1922), n. comb.
2756 **frigidana** (Pack., 1867), n. comb.
   moeschleri (Kennel, 1900), n. comb.
2757 **spinulana** (McD., 1938), n. comb.
2758 **brevicornutana** (McD., 1938), n. comb.
2759 **tertiana** (McD., 1922), n. comb.
2760 **bifida** (McD., 1938), n. comb.
2761 **afficticia** (Heinr., 1926), n. comb.
2762 **strigosa** (Heinr., 1926), n. comb.
2763 **albeolana** (Zell., 1875), n. comb.
2764 **apateticana** (McD., 1922), n. comb.
   deceptana (McD., 1922), preocc. by Kft., 1905
2765 **deceptana** (Kft., 1905)
   salicaceana (Free., 1957)
2766 **dextrana** (McD., 1923)
   deceptana; auth., part, not Kft., 1905
2767 **infida** (Heinr., 1926)
2768 **removana** (Kft., 1907), n. comb.

PSEUDOSCIAPHILA Obr., 1966
   SCIAPHILA; auth., not Tr., 1829
2769 **duplex** (Wlsm., 1905)
   thallasana (McD., 1922)

ORTHOTAENIA Steph., 1829
   BADEBECIA Heinr., 1926

2770 **undulana** (D. & S., 1775)
   urticana (Hbn., 1796–99)
   campestrana (Zell., 1875)
   dilutifuscana (Wlsm., 1879)

PHAECASIOPHORA Grt., 1873
2771 **confixana** (Wlk., 1863)
   perductana (Wlk., 1863)
   mutabilana (Clem., 1865)
2772 **niveiguttana** Grt., 1873
2773 **inspersa** Heinr., 1931

OLETHREUTES Hbn., 1822
   ARGYROPLOCE Hbn., 1825
   ROXANA Steph., 1834
   MIXODIA Gn., 1845
   EXARTEMA Clem., 1860
   LOXOTERMA Bsk., 1906
   CELYPHOIDES Obr., 1960
2774 **monetiferana** (Riley, 1881)
2775 **nitidana** (Clem., 1860)
2776 **furfurana** (McD., 1922)
   foedana; auth., not Clem., 1865
2777 **comandrana** (Clarke, 1953)
2778 **olivaceana** (Fern., 1882)
   bolandana (McD., 1922)
2779 **fraternana** (McD., 1924)
2780 **subnubilus** (Heinr., 1923)
2781 **electrofuscus** (Heinr., 1923)
2782 **rusticana** (McD., 1922)
2783 **zelleriana** (Fern., 1875)
   nitidana; Zell., 1875, not Clem., 1860
   trepidula (Heinr., 1926)
2784 **footiana** (Fern., 1882)
2785 **atrodentana** (Fern., 1882)
2786 **punctana** (Wlsm., 1903)
   connana (Heinr., 1923)
   cornana (Heinr., 1926)
2787 **connectus** (McD., 1935)
2788 **inornatana** (Clem., 1860)
2789 *number omitted*
2790 **mediopartitus** (Heinr., 1923)
2791 **exoletus** (Zell., 1875)
2792 **bicolorana** (McD., 1922)
2793 **tenebricus** (Heinr., 1926)
2794 **quadrifidus** (Zell., 1875)
2795 **tiliana** (Heinr., 1923)
2796 **sciotana** (Heinr., 1923)
2797 **appalachiana** (Braun, 1951)
2798 **betulana** (Heinr., 1926)
   zelleriana; auth., not Fern., 1875
2799 **clavana** (Wlk., 1863)
2800 **nigrana** (Heinr., 1923)
2801 **viburnana** (McD., 1935)
2802 **hippocastana** (Kft., 1907)
2803 **merrickana** (Kft., 1907)
2804 **hamameliana** (McD., 1944)
2805 **corylana** (Fern., 1882)
2806 **ochrosuffusana** (Heinr., 1923)
   ochrisuffusana Fbs., 1923, missp.
2807 **brunneopurpuratus** (Heinr., 1923)
2808 **ferrugineana** (Riley, 1881)
2809 **fagigemmeana** (Cham., 1878)
2810 **sericorana** (Wlsm., 1879)
2811 **melanomesa** (Heinr., 1923)
2812 **valdana** (McD., 1922)
   micantana (Fbs., 1923)
2813 **baccatana** (McD., 1942)
2814 **versicolorana** (Clem., 1860)

2815 **brevirostratus** (Heinr., 1926)
2816 **galevora** (McD., 1956)
2817 **permundana** (Clem., 1860)
   meanderana (Wlk., 1863)
   gaylussaciana (Heinr., 1926)
2818 **submissana** (McD., 1922)
2819 **nanana** (McD., 1922)
   quebecense (Heinr., 1923)
   quebecensis Fbs., 1923, missp.
2820 **malana** (Fern., 1882)
2821 **appendicea** (Zell., 1875)
   versicolorana; auth., not Clem., 1860
2822 **concinnana** (Clem., 1865)
   foedana (Clem., 1865)
   a. **terminana** (McD., 1922)
2823 **fasciatana** (Clem., 1860)
   decisana (Wlk., 1862)
   albofasciata (Zell., 1875)
2824 **troglodana** (McD., 1922)
2825 **exaeresima** (Heinr., 1926)
2826 **lacunana** (Free., 1941)
2827 **ferriferana** (Wlk., 1863)
   gratiosana (Clem., 1865)
   usticana (Zell., 1875)
2828 **griseoalbana** (Wlsm., 1879)
2829 **osmundana** (Fern., 1879)
   ochromediana Kft., 1907
2830 **auricapitana** (Wlsm., 1879)
2831 **agilana** (Clem., 1860)
   albiciliana; Bsk., 1909 not Fern., 1882
2832 **albiciliana** (Fern., 1882)
2833 **siderana** Tr., 1834, extralim.
   a. **chalybeana** (Wlsm., 1879)
2834 **sordidana** (McD., 1922)
2835 **galaxana** Kft., 1907
   a. **glitranana** Kft., 1907
2836 **constellatana** (Zell., 1875)
2837 **astrologana** (Zell., 1875)
   coronana Kft., 1907
2838 **coruscana** (Clem., 1860)
   ferrolineana (Wlk., 1863)
   argyroelana (Zell., 1875)
2839 **puncticostana** (Wlk., 1863)
   murina (Pack., 1867)
   a. **major** (Wlsm., 1895)
2840 **nordeggana** (McD., 1922)
2841 **kennethana** McD., 1941
2842 **heinrichana** (McD., 1927)
2843 **minaki** (McD., 1929)
2844 **deprecatorius** Heinr., 1926
2845 **carolana** (McD., 1922)
2846 **polluxana** (McD., 1922)
2847 **glaciana** (Mösch., 1860)
   dealbana (Wlk., 1863)
   fuscalbana (Zell., 1875)
   castorana (McD., 1922)
2848 **bipartitana** (Clem., 1860)
   similisana (Wlk., 1863)
   caesialbana (Zell., 1875)
2849 **trinitana** (McD., 1931)
2850 **schulziana** (F., 1777)
   bentleyana (Curt., 1835)
2851 **intermistana** (Clem., 1865)
   turfosana; Mösch., 1864, not H.-S., 1848
   tessellana (Pack., 1867)
   septentrionana (Mösch., 1883), preocc. by Curt., 1831
   boreana Rebel, 1901, repl. name.
   intermedia Dufrane, 1957, missp.

2852 **septentrionana** (Curt., 1831)
    ?primariana (Wlk., 1863)
    ?fulvifrontana (Pack., 1867)
2853 **inquietana** (Wlk., 1863)
2854 **bowmanana** (McD., 1923)
2855 **mengelana** (Fern., 1894)
    groenlandicana (B.-H., 1896)
2856 **costimaculana** (Fern., 1882)
2857 **devotana** Kft., 1907
2858 **buckellana** (McD., 1922)
    a. **albidula** Heinr., 1926
2859 **cespitana** (Hbn., 1814–17)
    flavofasciana (Humphreys &
        Westwood, 1844)
    instrutana (Clem., 1865)
    poana (Zell., 1875)

HEDYA Hbn., 1825
    EPISAGMA Hbn., 1825
    PENDINA Tr., 1829
    PENTHINA Tr., 1830
    HEDIA emend.
2860 **separatana** (Kft., 1907)
    dimidiana; (Fern., 1882), not
        Sodoffsky, 1830
2861 **ochroleucana** (Frölich, 1828)
    nimbatana (Clem., 1860)
    contrariana (Wlk., 1863)
    consanguinana (Wlsm., 1879)
2862 **nubiferana** (Haw., 1811)
    variegana (Hbn., 1796–99)
2863 **chionosema** (Zell., 1875)
2864 **cyanana** (Murt., 1880)

TSINILLA Heinr., 1931
2865 **lineana** (Fern., 1901)

EVORA Heinr., 1926
2866 **hemidesma** (Zell., 1875)

Eucosmini

RHYACIONIA Hbn., 1825
    ORTHOTAENIA Steph., 1834
    RETINIA Gn., 1845
    EVETRIA; auth., not Hbn., 1825
    EVERTIA Matsumura, 1931, missp.
2867 **buoliana** (D. & S., 1775)
    gemmana (Hbn., 1818–19)
    bouliana Frölich, 1828, missp.
    pallasana Sodoffsky, 1830
    pinicolana; Steph., 1852, var., not
        Doubleday, 1849
    buolina Neugebauer, 1950, missp.
2868 **rigidana** (Fern., 1880)
2869 **subtropica** Miller, 1961
2870 **multilineata** Powell, 1978
2871 **pasadenana** (Kft., 1907)
2872 **zozana** (Kft., 1907)
    matutina (Meyr., 1912), repl. name
    montana (Bsk., 1912)
2873 **neomexicana** (Dyar, 1903)
    offectalis; Ckll., 1901, not Hulst,
        1886
2874 **salmonicolor** Powell, 1978
2875 **monophylliana** (Kft., 1907)
2876 **martinana** Powell, 1978
2877 **adana** Heinr., 1923
2878 **jenningsi** Powell, 1978
2879 **busckana** Heinr., 1923

2880 **blanchardi** Miller, 1978
2881 **fumosana** Powell, 1978
2882 **frustrana** (Comstock, 1880)
2883 **bushnelli** (Bsk., 1914)
2884 **sonia** Miller, 1967
2885 **aktita** Miller, 1978
2886 **subcervinana** (Wlsm., 1879)
2887 **pallifasciata** Powell, 1978
2888 **versicolor** Powell, 1978

PETROVA Heinr., 1923
2889 **comstockiana** (Fern., 1879)
2890 **taedana** Miller, 1978
2891 **wenzeli** (Kft., 1910)
    virginiana (Bsk., 1914)
2892 **albicapitana** (Bsk., 1914)
2893 **arizonensis** (Heinr., 1920)
2894 **metallica** (Bsk, 1914)
2895 **luculentana** (Heinr., 1920)
2896 **sabiniana** (Kft., 1907)
2897 **edemoidana** (Dyar, 1903)
2898 **gemistrigulana** (Kft., 1905)
2899 **pallipennis** McD., 1938
2900 **burkeana** (Kft., 1907)
2901 **picicolana** (Dyar, 1906)
2902 **houseri** Miller, 1959

BARBARA Heinr., 1923
2903 **colfaxiana** (Kft., 1907)
    a. **coloradensis** (Heinr., 1920)
    b. **taxifoliella** (Bsk., 1914)
2904 **ulteriorana** (Heinr., 1920)
2905 **mappana** Free., 1941

SPILONOTA Steph., 1829
    TMETOCERA Led., 1859
2906 **ocellana** (D. & S., 1775)
    pyrifoliana (Clem., 1860)
    oculana (Harr., 1862)

STREPSICRATES Meyr., 1888
    PHTHINOLOPHUS Dyar, 1903
2907 **smithiana** (Wlsm., 1891)
    indentana (Dyar, 1903), n. syn.
    imminens (Meyr., 1917)

PHANETA Steph., 1852
    THIODIA; auth., not Hbn., 1825
    CALOSETIA Staint., 1859
    IOPLOCAMA Clem., 1860
2908 **radiatana** (Wlsm., 1879), n. comb.
2909 **albertana** (McD., 1925), n. comb.
2910 **essexana** (Kft., 1907), n. comb.
2911 **awemeana** (Kft., 1907), n. comb.
2912 **indeterminana** (McD., 1925), n.
    comb.
2913 **umbrastriana** (Kft., 1907), n. comb.
2914 **roseoterminana** (Kft., 1907), n.
    comb.
2915 **ferruginana** (Fern., 1882)
2916 **formosana** (Clem., 1860)
    sagittana (Wlk., 1863)
    steroreana (Zell., 1875)
    a. **subcandida** (Heinr., 1929)
2917 **altana** (McD., 1927), n. comb.
2918 **corculana** (Zell., 1874), n. comb.
    aspidiscana; Wlsm., 1879, not Hbn.,
        1822
2919 **annetteana** (Kft., 1907), n. comb.
2920 **scotiana** (McD., 1958), n. comb.

2921 **citricolorana** (McD., 1942), n. comb.
2922 **amphorana** (Wlsm., 1879), n. comb.
2923 **decempunctana** (Wlsm., 1879), n.
    comb.
2924 **refusana** (Wlk., 1863)
2925 **autumnana** (McD., 1942)
2926 **verna** Miller, 1971
2927 **ochrocephala** (Wlsm., 1895), n. comb.
2928 **raracana** (Kft., 1907), n. comb.
    fastidiosa (Meyr., 1912), repl. name
2929 **ochroterminana** (Kft., 1907), n.
    comb.
2930 **perfuscana** (Heinr., 1923), n. comb.
2931 **crispana** (Clem., 1865)
2932 **alterana** (Heinr., 1923), n. comb.
2933 **marmontana** (Kft., 1907), n. comb.
2934 **sinestrigana** (McD., 1938), n. comb.
2935 **oregonensis** (Heinr., 1923), n. comb.
2936 **tomonana** (Kft., 1907), n. comb.
    limigena (Meyr., 1912), repl. name
2937 **parmatana** (Clem., 1860)
2938 **modernana** (McD., 1925), n. comb.
2939 **fasciculatana** (McD., 1938), n. comb.
2940 **convergana** (McD., 1925), n. comb.
2941 **mormonensis** (Heinr., 1923), n.
    comb.
2942 **delphinus** (Heinr., 1923), n. comb.
2943 **latens** (Heinr., 1929), n. comb.
2944 **columbiana** (Wlsm., 1879), n. comb.
2945 **insignata** (Heinr., 1924), n. comb.
2946 **apacheana** (Wlsm., 1884), n. comb.
2947 **influana** (Heinr., 1923), n. comb.
2948 **sublapidana** (Wlsm., 1879), n. comb.
2949 **lapidana** (Wlsm., 1879), n. comb.
    lepidana (Heinr., 1929), missp., n.
        comb.
2950 **kokana** (Kft., 1907), n. comb.
    chortaea (Meyr., 1912), repl. name,
        n. comb.
    sororiana (Heinr., 1923), n. comb.
2951 **ornatula** (Heinr., 1924), n. comb.
2952 **elongana** (Wlsm., 1879), n. comb.
2953 **rupestrana** (McD., 1925), n. comb.
2954 **vernalana** (McD., 1942), n. comb.
2955 **transversa** (Wlsm., 1895), n. comb.
2956 **tarandana** (Mösch., 1874), n. comb.
2957 **nepotinana** (Heinr., 1923), n. comb.
2958 **complicana** (McD., 1925), n. comb.
2959 **spectana** (McD., 1938), n. comb.
2960 **tenuiana** (Wlsm., 1879), n. comb.
2961 **migratana** (Heinr., 1923), n. comb.
2962 **cinereolineana** (Heinr., 1923), n.
    comb.
2963 **misturana** (Heinr., 1923), n. comb.
2964 **parvana** (Wlsm., 1879), n. comb.
2965 **fertoriana** (Heinr., 1923), n. comb.
2966 **crassana** (McD., 1938), n. comb.
2967 **alatana** (McD., 1938), n. comb.
2968 **clavana** (Fern., 1882)
2969 **indagatricana** (Heinr., 1923), n.
    comb.
2970 **argenticostana** (Wlsm., 1879), n.
    comb.
2971 **spiculana** (Zell., 1875), n. comb.
2972 **dorsiatomana** (Kft., 1905), n. comb.
2973 **striatana** (Clem., 1860)
    albicepsana (Wlk., 1863), n. comb.
    trivittana (Zell., 1875), n. comb.
    a. **occidentalis** (Heinr., 1923), n.
        comb.

2974 **implicata** (Heinr., 1931), n. comb.
2975 **delphinoides** (Heinr., 1923), n. comb.
2976 **pallidarcis** (Heinr., 1923), n. comb.
2977 **modicellana** (Heinr., 1923), n. comb.
2978 **minimana** (Wlsm., 1879), n. comb.
2979 **subminimana** (Heinr., 1923), n. comb.
2980 **pallidicostana** (Wlsm., 1879), n. comb.
2981 **perangustana** (Wlsm., 1879), n. comb.
2982 **kiscana** (Kft., 1907), n. comb.
speculigera (Meyr., 1912), repl. name, n. comb.
2983 **salmicolorana** (Heinr., 1923), n. comb.
2984 **artemisiana** (Wlsm., 1879), n. comb.
2985 **infimbriana** (Dyar, 1904), n. comb.
a. **candidula** (Heinr., 1924), n. comb.
2986 **octopunctana** (Wlsm., 1895), n. comb.
2987 **youngi** (McD., 1925), n. comb.
2988 **setonana** (McD., 1927), n. comb.
2989 **scalana** (Wlsm., 1879), n. comb.
2990 **festivana** (Heinr., 1923), n. comb.
2991 **segregata** (Heinr., 1924), n. comb.
2992 **castrensis** (McD., 1929), n. comb.
2993 **camdenana** (McD., 1925), n. comb.
2994 **montanana** (Wlsm., 1884), n. comb.
triangulana (Kft., 1905), n. comb.
2995 **benjamini** (Heinr., 1923), n. comb.
2996 **griseocapitana** (Wlsm., 1879), n. comb.
2997 **pastigiata** (Heinr., 1929), n. comb.
2998 **olivaceana** (Riley, 1881), n. comb.
2999 **verniochreana** (Heinr., 1923), n. comb.
3000 **imbridana** (Fern., 1905)
3001 **granulatana** (Kft., 1908), n. comb.
3002 **grindeliana** (Bsk., 1906), n. comb.
3003 **stramineana** (Wlsm., 1879), n. comb.
3004 **umbraticana** (Heinr., 1923), n. comb.
3005 **offectalis** (Hulst, 1886), n. comb.
obliterana (Wlsm., 1895), n. comb.
3006 **bucephaloides** (Wlsm., 1891), n. comb.
3007 **southamptonensis** (Heinr., 1935), n. comb.

EUCOSMA Hbn., 1823
CATOPTRIA Gn., 1845 preocc. by Hbn., 1825
PYGOLOPHA Led., 1852
AFFA Wlk., 1863
EXENTERA Grt., 1877
EXENTERELLA Grt., 1883
PALPOCRINIA Kennel, 1919
EUCOSMOIDES Obr., 1946
3008 **quinquemaculana** (Rob., 1869)
3009 **robinsonana** (Grt., 1872)
quintana (Zell., 1875)
tryonana (Kft., 1905)
3010 **hazelana** Klots, 1936
3011 **crambitana** (Wlsm., 1879)
3012 **fandana** Kft., 1907
argyraula Meyr., 1912, repl. name
3013 **canariana** Kft., 1907
3014 **ridingsana** (Rob., 1869)
argentifurcatana (Grt., 1876)
hipeana (Grt., 1876)

3015 **fernaldana** (Grt., 1880)
3016 **betana** McD., 1942
3017 **magnidicana** Heinr., 1923
3018 **caniceps** (Wlsm., 1884)
3019 **gandana** Kft., 1907
chloroleuca Meyr., 1912, repl. name
3020 **avalona** McD., 1938
3021 **adamantana** (Gn., 1845)
3022 **spaldingana** Kft., 1907
3023 **sandiego** Kft., 1908
sandiegana Meyr., 1912, repl. name
3024 **gilletteana** Dyar, 1903
3025 **optimana** Dyar, 1893
3026 **agassizii** (Rob., 1869)
3027 **laticurva** Heinr., 1929
3028 **dapsilis** Heinr., 1929
3029 **bolanderana** (Wlsm., 1879)
3030 **ragonoti** (Wlsm., 1895)
barnesiana Dyar, 1903
3031 **serpentana** (Wlsm., 1895)
3032 **ophionana** McD., 1925
3033 **heathiana** Kft., 1907
3034 **langstoni** Powell, 1963
3035 **morrisoni** (Wlsm., 1884)
3036 **lathami** Fbs., 1937
3037 **agricolana** (Wlsm., 1879)
a. **argentialbana** (Wlsm., 1879)
ochreana (Clem., 1865), unused sr. syn.
pergandeana (Fern., 1905)
flavana (Fern., 1905)
3038 **smithiana** (Wlsm., 1895)
argentialbana; Heinr., 1923, part, not Wlsm., 1879
britana McD., 1927
3039 **barbara** Miller, 1974
3040 **costastrigulana** Kft., 1907
3041 **comatulana** (Zell., 1875)
3042 **vagana** McD., 1925
3043 **albiguttana** (Zell., 1875)
3044 **graciliana** Kft., 1905
3045 **galenapunctana** Kft., 1908
3046 **monogrammana** (Zell., 1875)
3047 **atomosana** (Wlsm., 1879)
3048 **uta** Clarke, 1953
3049 **serapicana** Heinr., 1923
3050 **watertonana** McD., 1925
3051 **glomerana** (Wlsm., 1879)
sandana Kft., 1907, n. syn.
griphodes Meyr., 1912, repl. name
3052 **circulana** Hbn., 1823
a. **gemellana** Heinr., 1923
3053 **fraudabilis** Heinr., 1923
3054 **perdricana** (Wlsm., 1879)
kandana Kft., 1907
argillacea Meyr., 1912, repl. name
3055 **louisana** McD., 1944
3056 **russeola** Heinr., 1929
3057 **luridana** (Wlsm., 1879)
3058 **consociana** Heinr., 1923
3059 **irroratana** (Wlsm., 1879)
3060 **subflavana** (Wlsm., 1879)
3061 **handana** Kft., 1907
ceramitis Meyr., 1912, repl. name
caramitis Heinr., 1923, missp.
3062 **sepulchrana** Meyr., 1927
3063 **immaculana** Kft., 1907
3064 **maculatana** (Wlsm., 1879)
3065 **sonomana** Kft., 1907
3066 **gloriola** Heinr., 1931

3067 **bobana** Kft., 1907
antichroma Meyr., 1912, repl. name
3068 **ponderosa** Powell, 1968
3069 **monoensis** Powell, 1968
3070 **franclemonti** Powell, 1968
3071 **recissoriana** Heinr., 1920
3072 **cocana** Kft., 1907
rhodophaea Meyr., 1912, repl. name.
3073 **monitorana** Heinr., 1920
3074 **tocullionana** Heinr., 1920
3075 **siskiyouana** (Kft., 1907)
3076 **crymalana** Powell, 1968
3077 **momana** Kft., 1907
metaschista Meyr., 1912, repl. name.
3078 **grotiana** Kft., 1908
3079 **palabundana** Heinr., 1923
3080 **lolana** Kft., 1907
leucomalla Meyr., 1912, repl. name
3081 **dodana** Kft., 1907
spilophora Meyr., 1912, repl. name
3082 **fofana** Kft., 1907
annulata Meyr., 1912, repl. name
3083 **invicta** (Wlsm., 1895)
3084 **sperryana** McD., 1942
3085 **eburata** Heinr., 1929
3086 **subinvicta** Kft., 1907
3087 **snyderana** Kft., 1907
3088 **emaciatana** (Wlsm., 1884)
3089 **totana** Kft., 1907
spodias Meyr., 1912, repl. name
3090 **persolita** Heinr., 1929
3091 **matutina** (Grt., 1873)
3092 **larana** (Wlsm., 1879)
3093 **exclusoriana** Heinr., 1923
3094 **heinrichi** McD., 1925
3095 **shastana** (Wlsm., 1879)
3096 **grandiflavana** (Wlsm., 1879)
3097 **hyponomeutana** (Wlsm., 1895)
3098 **giganteana** (Riley, 1881)
a. **minorata** Heinr., 1924
3099 **bipunctella** (Wlk., 1863)
worthingtoniana (Fern., 1878)
3100 **bilineana** Kft., 1907
3101 **denverana** Kft., 1907
3102 **williamsi** Powell, 1963
3103 **graziella** Blanchard, 1968
3104 **mediostriata** (Wlsm., 1895)
3105 **excerptionana** Heinr., 1923
3106 **abstemia** Meyr., 1932
bactrana Heinr., 1923, preocc. by Kennel, 1901
3107 **biplagata** (Wlsm., 1895)
3108 **primulana** (Wlsm., 1879)
3109 **hasseanthi** Clarke, 1952
3110 **gomonana** Kft., 1907
discipula Meyr., 1912, repl. name
3111 **dilatana** (Wlsm., 1895)
3112 **nandana** Kft., 1907
chersaea Meyr., 1912, repl. name
3113 **aprilana** (Grt., 1877)
3114 **landana** Kft., 1907
isospora Meyr., 1912, repl. name
3115 **simplex** McD., 1925
3116 **dorsisignatana** (Clem., 1860)
distigmana (Wlk., 1863)
clavana (Zell., 1875)
a. **diffusana** Kft., 1905
b. **similana** (Clem., 1860)
confluana Kft., 1905
c. **engelana** Kft., 1908

3117 **hennei** Clarke, 1947
3118 **graduatana** (Wlsm., 1879)
3119 **juncticiliana** (Wlsm., 1879)
3120 **derelecta** Heinr., 1929
    juncticiliana; Heinr., 1923, not
      Wlsm., 1879
3121 **excusabilis** Heinr., 1923
3122 **wandana** Kft., 1907
    eumaea Meyr., 1912, repl. name
3123 **mandana** Kft., 1907
    amanda Meyr., 1912, repl. name
3124 **fulminana** (Wlsm., 1879)
3125 **rusticana** Kft., 1905
3126 **mobilensis** Heinr., 1923
3127 **sombreana** Kft., 1905
    phaeodes Meyr., 1920
3128 **pandana** Kft., 1907
    sardiopa Meyr., 1912, repl. name
3129 **fiskeana** Kft., 1905
3130 **nuntia** Heinr., 1929
3131 **inquadrana** (Wlsm., 1884)
3132 **pulveratana** (Wlsm., 1884)
3133 **consobrinana** Heinr., 1923
3134 **aspidana** (Wlsm., 1884)
3135 **hohana** Kft., 1907
    syrtodes Meyr., 1912, repl. name
3136 **jejunana** McD., 1942
3137 **biquadrana** (Wlsm., 1879)
3138 **mirosignata** Heinr., 1929
3139 **suadana** Heinr., 1923
3140 **aeana** McD., 1942
3141 **canana** (Wlsm., 1879)
3142 **cataclystiana** (Wlk., 1863)
3143 **conspiciendana** Heinr., 1923
3144 **floridana** Kft., 1907
3145 **fuscana** Kft., 1907
3146 **liturana** (Wlsm., 1879)
3147 **petalonota** Meyr., 1937
3148 **urnigera** Meyr., 1937

PELOCHRISTA Led., 1859
    CALLIMOSEMA Clem., 1865
    PSEUDEUCOSMA Obr., 1946
3149 **argenteana** (Wlsm., 1895), n. comb.
3150 **idahoana** (Kft., 1907), n. comb.
3151 **scintillana** (Clem., 1865)
    dodecana (Zell., 1875)
    a. *randana* (Kft., 1907)
      paraglypta (Meyr., 1912), repl.
        name
3152 **fratruelis** (Heinr., 1923), n. comb.
3153 **pallidipalpana** (Kft., 1905), n. comb.
3154 **popana** (Kft., 1907), n. comb.
    carcharias (Meyr., 1912), repl. name,
      n. comb.
3155 **daemonicana** (Heinr., 1923), n. comb.
3156 **occipitana** (Zell., 1875), n. comb.
3157 **reversana** (Kft., 1907), n. comb.
3158 **perpropinqua** (Heinr., 1929), n. comb.
3159 **tahoensis** (Heinr., 1923), n. comb.
    shastana; Heinr., 1923, part, not
      Wlsm., 1879
    subdivita (Heinr., 1929)
3160 **palpana** (Wlsm., 1879), n. comb.
3161 **fuscosparsa** (Wlsm., 1895), n. comb.
3162 **corosana** (Wlsm., 1884), n. comb.
3163 **palousana** (Kft., 1907), n. comb.
3164 **expolitana** (Heinr., 1923), n. comb.
3165 **rorana** (Kft., 1907)
    sceletopa (Meyr., 1912), repl. name

3166 **metariana** (Heinr., 1923), n. comb.
3167 **passerana** (Wlsm., 1879), n. comb.
3168 **zomonana** (Kft., 1907), n. comb.
    explosa (Meyr., 1912), repl. name, n.
      comb.
3169 **womonana** (Kft., 1907), n. comb.
    semnitis (Meyr., 1912), repl. name, n.
      comb.
3170 **vandana** (Kft., 1907)
    pholas (Meyr., 1912), repl. name

EPIBLEMA Hbn., 1825
    CACOCHROEA Led., 1859
    MONOSPHRAGIS Clem., 1860
    EURYPTYCHIA Clem., 1865
3171 **boxcana** (Kft., 1907)
    aspista (Meyr., 1912), repl. name
3172 **strenuana** (Wlk., 1863)
    exvagana (Wlk., 1863)
    flavocellana (Clem., 1865)
    subversana (Zell., 1875)
    minutana (Kft., 1905)
    antaxia (Meyr., 1920), repl. name
3173 **abruptana** (Wlsm., 1879)
3174 **numerosana** (Zell., 1875)
3175 **grossbecki** (Heinr., 1923)
3176 **praesumptiosum** Heinr., 1923
3177 **separationis** Heinr., 1923
3178 **deflexana** Heinr., 1923
3179 **ochraceana** Fern., 1901
3180 **sosana** (Kft., 1907)
    pelina (Meyr., 1912), repl. name
3181 **insidiosana** Heinr., 1923
3182 **symbolaspis** Meyr., 1927
3183 **exacerbatricana** Heinr., 1923
3184 **tripartitana** (Zell., 1875)
3185 **benignatum** McD., 1925
3186 **scudderiana** (Clem., 1860)
    saligneana (Clem., 1865)
    affusana (Zell., 1875)
3187 **kennebecana** (Kft., 1907)
3188 **discretivana** (Heinr., 1921)
3189 **obfuscana** (Dyar, 1888)
3190 **desertana** (Zell., 1875)
3191 **rudei** Powell, 1975
3192 **carolinana** (Wlsm., 1895)
3193 **arizonana** Powell, 1975
3194 **hirsutana** (Wlsm., 1879)
3195 **radicana** (Wlsm., 1879)
    vomonana (Kft., 1907)
    serangias (Meyr., 1912), repl. name
    gratuitana Heinr., 1923
3196 **walsinghami** (Kft., 1907)
3197 **periculosana** Heinr., 1923
3198 **iowana** McD., 1935
3199 **naoma** Clarke, 1953
3200 **lyallana** McD., 1935
3201 **infelix** Heinr., 1923
3202 **otiosana** (Clem., 1860)
    inclinana (Zell., 1875)
3203 **brightonana** (Kft., 1907)
3204 **tandana** (Kft., 1907)
    trapezitis (Meyr., 1912), repl. name
3205 **resumptana** (Wlk., 1863)
    abbreviatana (Wlsm., 1879)
3206 **dorsisuffusana** (Kft., 1908)
3207 **macneilli** Powell, 1975

NOTOCELIA Hbn., 1825
    ASPIS Tr., 1829, preocc. by

      Laurenti, 1768
    ASPIDIA Dup., 1834, repl. name
    PARDIA Gn., 1845
3208 **trimaculana** (Haw., 1811)
    suffusana (Dup., 1843)
3209 **purpurissatana** Heinr., 1923
3210 **illotana** (Wlsm., 1879), n. comb.
3211 **culminana** (Wlsm., 1879), n. comb.

SULEIMA Heinr., 1923
3212 **helianthana** (Riley 1881)
3213 **daracana** (Kft., 1907)
    profana (Meyr., 1912), repl. name
3214 **skinnerana** Heinr., 1923
3215 **lagopana** (Wlsm., 1884)
3216 **baracana** (Kft., 1907)
    caracana (Kft., 1907)
    oxyleuca (Meyr., 1912), repl. name
    famosa (Meyr., 1912), repl. name.
3217 **cinerodorsana** Heinr., 1923

SONIA Heinr., 1923
3218 **constrictana** (Zell., 1875)
3219 **canadana** McD., 1925
3220 **vovana** (Kft., 1907)
    fraternana (Bsk., 1907)
    typicodes (Meyr., 1912), repl. name.
3221 **comstocki** Clarke, 1952
3222 **filiana** (Bsk., 1907)

GYPSONOMA Meyr., 1895
3223 **fasciolana** (Clem., 1864)
    blakeana (Grt., 1873)
3224 **nebulosana** (Pack., 1866)
3225 **parryana** (Curt., 1835)
3226 **haimbachiana** (Kft., 1907)
3227 **substitutionis** Heinr., 1923
3228 **salicicolana** (Clem., 1864)
    saliciana (Clem., 1864)
3229 **adjuncta** Heinr., 1924

PROTEOTERAS Riley, 1881
3230 **aesculana** Riley, 1881
3231 **implicatum** Heinr., 1924
3232 **willingana** (Kft., 1904)
3233 **crescentana** Kft., 1907
3234 **naracana** Kft., 1907
    praesinospila Meyr., 1912, repl. name
3235 **moffatiana** Fern., 1905
3236 **arizonae** Kft., 1907
3237 **obnigrana** Heinr., 1923

ZEIRAPHERA Tr., 1829
3238 **claypoleana** (Riley, 1882)
    instrutana; Claypole, 1881, not
      Clem., 1865
3239 **pacifica** Free., 1966
3240 **canadensis** Mutuura & Free., 1967
    ratzeburgiana; auth., not Ratzeburg, 1840
3241 **improbana** (Wlk., 1863)
    diniana; auth., not Gn., 1845
    griseana; auth., not Steph., 1835
    diffiniana (Wlk., 1863)
    destitutana (Wlk., 1863)
    indivisana (Wlk., 1863)
    pseudotsugana (Kft., 1904)
3242 **fortunana** (Kft., 1907)
3243 **unfortunana** Powell, n. name
    destitutana; Mutuura & Free.,
      1967, not Wlk., 1863

3244 **hesperiana** Mutuura & Free., 1967
3245 **vancouverana** McD., 1925

PSEUDEXENTERA Heinr., 1940
   EXENTERA; auth., not Grt., 1877
3246 **cressoniana** (Clem., 1864)
3247 **mali** Free., 1942
3248 **oregonana** (Wlsm., 1879)
3249 **caryana** McD., 1940
3250 **senatrix** (Heinr., 1924), n. comb.
3251 **spoliana** (Clem., 1864)
3252 **haracana** (Kft., 1907), n. comb.
   resoluta (Meyr., 1912), repl. name, n.
     comb.
3253 **faracana** (Kft., 1907), n. comb.
   ultrix (Meyr., 1912), repl. name, n.
     comb.
3254 **maracana** (Kft., 1907), n. comb.
   praescripta (Meyr., 1912), repl.
     name, n. comb.
3255 **kalmiana** McD., 1959
3256 **habrosana** (Heinr., 1923)
3257 **costomaculana** (Clem., 1860)
   bipustulana (Wlk., 1863)
3258 **virginiana** (Clem., 1864)

GRETCHENA Heinr., 1923
   GWENDOLINA Heinr., 1923
   GRETCHINA Fbs., missp.
3259 **deludana** (Clem., 1864)
3260 **concubitana** Heinr., 1923
3261 **watchungana** (Kft., 1907)
3262 **dulciana** Heinr., 1923
3263 **bolliana** (Slingerland, 1896)
3264 **amatana** Heinr., 1923
3265 **delicatana** Heinr., 1923
3266 **biangulana**; auth., not Wlsm., 1879
3267 **semialba** McD., 1925
3268 **concitatricana** (Heinr., 1923)

GRISELDA Heinr., 1923
3269 **radicana** Heinr., 1923
   radicana; Heinr., 1923, not Wlsm.,
     1879

CHIMOPTESIS Powell, 1964
3270 **chrysopyla** Powell, 1964
3271 **matheri** Powell, 1964
3272 **gerulae** (Heinr., 1923)
3273 **pennsylvaniana** (Kft., 1907)

CROCIDOSEMA Zell., 1847
3274 **plebejana** Zell., 1847
   ptiladelpha Meyr., 1917

HENDECANEURA Wlsm., 1900
3275 **shawiana** (Kft., 1907)

RHOPOBOTA Led., 1859
   NORMA Heinr., 1923
   KUNDRYA Heinr., 1923
3276 **unipunctana** (Haw., 1811)
   naevana (Hbn., 1814–17)
   ilicifoliana (Kft., 1907)
   a. **geminana** (Steph., 1852)
     luctiferana (Wlk., 1863)
     vacciniana (Pack., 1869)
3277 **dietziana** (Kft., 1907)
3278 **finitimana** (Heinr., 1923)

EPINOTIA Hbn., 1825
   ASTATIA Hbn., 1825
   ASTHENIA Hbn., 1825
   EVETRIA Hbn., 1825
   PANOPLIA Hbn., 1825
   POECILOCHROMA Steph., 1829
   PAEDISCA Tr., 1830
   STEGANOPTYCHA Steph., 1834
   HYPERMECIA Gn., 1845
   PHLAEODES Gn., 1845
   PAMPLUSIA Gn., 1845
   CARTELLA Steph., 1852
   HALONOTA Steph., 1852
   LITHOGRAPHIA Steph., 1852
   CURTELLA Staint., 1859, missp.
   CATASTEGA Clem., 1861
   PROTEOPTERYX Wlsm., 1879
   EPINOTIS Kft., 1907, missp.
   NEURASTHENIA Pierce &
     Metcalf, 1922
   HAMULIGERA Obr., 1946
3279 **lantana** (Bsk., 1910)
   polyphaea (Turner, 1926)
   turnocosma (Turner, 1946)
   phaedropa (Turner, 1946)
3280 **stroemiana** (F., 1781)
   similana (Hbn., 1793)
   bimaculana (Don., 1808)
3281 **sperana** McD., 1935
3282 **myricana** McD., 1933
3283 **solandriana** (L., 1758)
3284 **ethnica** Heinr., 1923
3285 **pulsatillana** (Dyar, 1903)
   a. **siskiyouensis** Heinr., 1923
3286 **medioviridana** (Kft., 1908)
3287 **perplexana** (Fern., 1901)
3288 **castaneana** (Wlsm., 1895)
3289 **johnsonana** (Kft., 1907)
3290 **madderana** (Kft., 1907)
3291 **laracana** (Kft., 1907)
   navalis (Meyr., 1912), repl. name
3292 **vertumnana** (Zell., 1875)
   celtisana (Riley, 1881)
3293 **atristriga** Clarke, 1953
   atistriga auth., missp.
3294 **zandana** (Kft., 1907)
   peristicta (Meyr., 1912), repl. name
3295 **xandana** (Kft., 1907)
   yandana (Kft., 1907)
   atacta (Meyr., 1912), repl. name
   nothrodes (Meyr., 1912), repl. name
3296 **albicapitana** (Kft., 1907)
3297 **hopkinsana** (Kft., 1907)
   a. **cupressi** Heinr., 1923
3298 **subviridis** Heinr., 1929
3299 **fumoviridana** Heinr., 1923
3300 **subplicana** (Wlsm., 1879)
   basipunctana (Wlsm., 1879), n. syn.
3301 **improvisana** Heinr., 1923
3302 **rectiplicana** (Wlsm., 1879)
3303 **corylana** McD., 1925
3304 **solicitana** (Wlk., 1863)
   packardiana (Clem., 1864)
   tephrinana (Zell., 1875)
3305 **hamptonana** (Kft., 1907)
3306 **nisella** (Cl., 1759)
3307 **criddleana** (Kft., 1907)
3308 **albangulana** (Wlsm., 1879)
3309 **walkerana** (Kft., 1907)
3310 **transmissana** (Wlk., 1863)

3311 **removana** McD., 1935
3312 **momonana** (Kft., 1907)
   sanifica (Meyr., 1912), repl. name
3313 **terracoctana** (Wlsm., 1879)
3314 **miscana** (Kft., 1907)
   semalea (Meyr., 1912), repl. name
3315 **silvertoniensis** Heinr., 1923
3316 **marmoreana** Heinr., 1923
3317 **digitana** Heinr., 1923
3318 **nigralbanoidana** McD., 1929
3319 **nigralbana** (Wlsm., 1879)
3320 **ruidosana** Heinr., 1923
3321 **heucherana** Heinr., 1923
3322 **sagittana** McD., 1925
3323 **emarginana** (Wlsm., 1879)
3324 **crenana** (Hbn., 1814–17)
   columbia (Kft., 1904)
   albidorsana (Kft., 1904)
   mediostriana (Kft., 1904)
3325 **cercocarpana** (Dyar, 1903)
3326 **bigemina** Heinr., 1923
3327 **bicordana** Heinr., 1923
3328 **arctostaphylana** (Kft., 1904)
3329 **keiferana** Lange, 1937
3330 **unica** Heinr., 1923
3331 **infuscana** (Wlsm., 1879)
3332 **aporema** (Wlsm., 1914)
   opposita Heinr., 1931
3333 **timidella** (Clem., 1861)
3334 **aceriella** (Clem., 1861)
   signatana (Clem., 1864)
   variana (Clem., 1864)
   subnisana (Zell., 1875)
3335 **nonana** (Kft., 1907)
   carphologa (Meyr., 1912), repl. name
3336 **normanana** Kft., 1907
3337 **balsameae** Free., 1966
3338 **nanana** (Tr., 1835)
   domonana (Kft., 1907)
   piceafoliana (Kft., 1908)
   efficax (Meyr., 1912), repl. name
3339 **tsugana** Free., 1967
3340 **meritana** Heinr., 1923
3341 **aridos** Free., 1961
3342 **lomonana** (Kft., 1907)
   veneratrix (Meyr., 1912), repl. name
3343 **purpuriciliana** (Wlsm., 1879)
3344 **medioplagata** (Wlsm., 1895)
3345 **cruciana** (L., 1761)
   augustana (Hbn., 1811–13)
   direptana (Wlk., 1863)
   vilisana (Wlk., 1863)
   cockleana (Kft., 1904)
   alaskae Heinr., 1923, n. syn.
   lepida Heinr., 1924, n. syn.
3346 **plumbolineana** Kft., 1907
   russata Heinr., 1924, n. syn.
3347 **septemberana** Kft., 1907
3348 **vagana** Heinr., 1923
3349 **seorsa** Heinr., 1924
3350 **kasloana** McD., 1925
3351 **lindana** (Fern., 1892)
3352 **trossulana** (Wlsm., 1879)
3353 **biangulana** (Wlsm., 1879)
   signiferana Heinr., 1923

ANCYLIS Hbn., 1825
   EPICHARIS Hbn., 1825, preocc.
     by Klug, 1807
   ANCHYLOPERA Steph., 1829

PHOXOPTERIS Tr., 1830
ANTICLEA Steph., 1834, preocc.
  by Steph., 1831
PHILALCEA Steph., 1835
PHOXOPTERYX Sodoffsky, 1837,
  emend.
SIDERIA Gn., 1845
SIDEREA auth., missp.
3354 nubeculana (Clem., 1860)
3355 subaequana (Zell., 1875)
  a. kincaidiana (Fern., 1900)
3356 galeamatana (McD., 1956)
3357 sheppardana (McD., 1956)
3358 discigerana (Wlk., 1863)
  spiraeifoliana; Heinr., 1923, not
    Clem., 1860
  metamelana; Heinr., 1923, not Wlk.,
    1863
3359 metamelana (Wlk., 1863)
  discoferana (Wlk., 1863)
  angulifasciana (Zell., 1875)
  intermediana (Kft., 1907)
3360 tenebrica (Heinr., 1929)
3361 semiovana (Zell., 1875)
3362 columbiana (McD., 1955)
  shastensis (McD., 1955), n. syn.
3363 simuloides (McD., 1955)
  sierrae (McD., 1955), n. syn.
  litoris (McD., 1955), n. syn.
3364 maritima Dyar, 1904
3365 spiraeifoliana (Clem., 1860)
  spireaefoliana (Clem., 1860), incorr.
    orig. spell.
  spiraeifoliana Wlsm., 1879, emend.
3366 laciniana (Zell., 1875)
3367 burgessiana (Zell., 1875)
  murtfeldtiana (Riley, 1881)
  pruni (Heinr., 1923), n. syn.
3368 mira Heinr., 1929
  furvescens (Heinr., 1929), n. syn.
3369 fuscociliana (Clem., 1864)
  dubiana Clem., 1864
3370 platanana (Clem., 1860)
  marcidana (Zell., 1875)
3371 rhoderana (McD., 1954)
3372 brauni (Heinr., 1931)
3373 definitivana (Heinr., 1923)
3374 comptana (Frölich, 1828)
  conflexana (Wlk., 1863)
  pulchellana (Clem., 1864)
  lamiana (Clem., 1864)
  fragariae (Walsh & Riley, 1869)
  amblygona (Zell., 1875)
  floridana (Zell., 1875), n. syn.
  cometana (Wlsm., 1879), n. syn.
3375 divisana (Wlk., 1863)
3376 apicana (Wlk., 1866)
3377 muricana (Wlsm., 1879)
  a. cornifoliana (Riley, 1881)
3378 carbonana Heinr., 1923
  uncana; auth., not Hbn., 1799
3379 diminutana (Haw., 1811)
  diminuatana Kft., 1905, missp.
3380 goodelliana (Fern., 1882)
3381 albafascia Heinr., 1929
3382 unguicella (L. 1758)
  plagosana (Clem., 1864)
3383 pacificana (Wlsm., 1879)
3384 mediofasciana (Clem., 1864)
3385 torontana (Kft., 1907)

3386 tineana (Hbn., 1796–99)
  ocellana (Clem., 1864)
  leucophaleratana (Pack., 1866)
3387 albacostana Kft., 1905
3388 cordiae Bsk., 1933

HYSTRICHOPHORA Wlsm., 1879
  HYSTRICOPHORA auth., missp.
3389 leonana Wlsm., 1879
  a. aurantiana Wlsm., 1879
3390 paradisiae Heinr., 1923
3391 stygiana (Dyar, 1903)
  a. californiae Heinr., 1923
3392 roessleri (Zell., 1875)
3393 ostentatrix Heinr., 1923
3394 asphodelana (Kft., 1907)
  a. seraphicana Heinr., 1923
3395 taleana (Grt., 1878)
3396 ochreicostana (Wlsm., 1884)
3397 loricana (Grt., 1880)
3398 decorosa Heinr., 1929
3399 vestaliana (Zell., 1875)

Laspeyresiini

GODITHA Heinr., 1926
3400 bumeliana Heinr., 1926
  rumeliana Heinr., 1926, missp.

DICHRORAMPHA Gn., 1845
  DICRORAMPHA Westwood, 1854,
    missp.
  LIPOPTYCHA Led., 1859
  LIPOPTYCHODES Obr., 1953
  DICHRORAMPHODES Obr., 1953
  PARALIPOPTYCHA Obr., 1958
3401 kana (Bsk., 1906)
  planiloqua (Meyr., 1912), repl. name
3402 capitana (Bsk., 1906)
3403 britana (Bsk., 1906)
  alpinana; Fern., 1903, not Tr., 1830
3404 simulana (Clem., 1860)
  aurisignana (Zell., 1875)
3405 immaculata McD., 1946
3406 bittana (Bsk., 1906)
3407 incanana (Clem., 1860)
  nigromaculana (Kft., 1907)
3408 vancouverana McD., 1935
3409 radicicolana Wlsm., 1879
3410 banana (Bsk., 1906)
  sordescens (Meyr., 1912), repl. name
3411 piperana (Bsk., 1900)
3412 sedatana (Bsk., 1906)
  plumbana; Fern., 1903, not Scop., 1763
3413 dana (Kft., 1907)
  aequorea (Meyr., 1912), repl. name
  a. bradorensis McD., 1930
3414 leopardana (Bsk., 1906)

SATRONIA Heinr., 1926
3415 tantilla Heinr., 1926

RICULA Heinr., 1926
3416 maculana (Fern., 1901)

TALPONIA Heinr., 1926
3417 plummeriana (Bsk., 1906)

PAMMENE Hbn., 1825
  HEMIMENE Hbn., 1825

PSEUDOTOMIA Steph., 1829
PALLA Billberg, 1820, preocc. by
  Hbn., 1819
HEUSIMENE Steph., 1834
PYRODES Gn., 1845, preocc. by
  Audinet-Serville, 1832
HEMEROSIA Steph., 1852
PHTHOROBLASTIS Led., 1859
SPHAEROECA Meyr., 1895,
  preocc. by Lauterborn, 1894
PAMENE Rebel, 1901, missp.
METASPHAEROECA Fern., 1908,
  repl. name
3418 ocliferia (Heinr., 1926), n. comb.
3419 felicitana Heinr., 1923
3420 signifera (Heinr., 1926), n. comb.
3421 paula (Heinr., 1926), n. comb.
3422 bowmanana (McD., 1927), n. comb.

LARISA Miller, 1978
3423 subsolana Miller, 1978

ETHELGODA Heinr., 1926
3424 texanana (Wlsm., 1879)

SEREDA Heinr., 1923
3425 tautana (Clem., 1865)
  perfluana (Zell., 1875)
  lautana (Fern., 1882), emend.

GRAPHOLITA Tr., 1829
  GRAPHOLITHA Tr., 1830, emend.
  EUSPILA Steph., 1834
  ASPILA Steph., 1834
  EPHIPPIPHORA Dup., 1834
  OPADIA Gn., 1845
  ENDOPISA Gn., 1845
  STIGMONOTA Gn., 1845
  EUDOPISA Desmarest, 1857,
    missp.
  COPTOLOMA Led., 1859
  ENDOPSIA auth., missp.
3426 molesta (Bsk., 1916)
3427 libertina Heinr., 1926
3428 packardi Zell., 1875
  pyricolana (Murt., 1891)
3429 prunivora (Walsh, 1868)
3430 angeleseana (Kft., 1907)
3431 caeruleana Wlsm., 1879
  zana (Kft., 1907)
  vana (Kft., 1907)
  xanthospora (Meyr., 1912), repl.
    name
  eoleuca (Meyr., 1912), repl. name
3432 boulderana McD., 1942
3433 vitrana Wlsm., 1879
3434 fana (Kft., 1907)
  oenochroa (Meyr., 1912). repl. name
3435 conversana Wlsm., 1879
  wana (Kft., 1907)
  cupida (Meyr., 1912), repl. name
3436 imitativa Heinr., 1926
3437 lunatana Wlsm., 1879
3438 eclipsana Zell., 1875
3439 interstinctana (Clem., 1860)
  scitana (Wlk., 1863)
  distema Grt., 1873
3440 edwardsiana (Kft., 1907)
3441 lana (Kft., 1907)
  placerana (Kft., 1907)

vancouverana (Kft., 1907)
  chrysotypa (Meyr., 1912), repl. name
3442 **dyarana** (Kft., 1907)
3443 **tristrigana** (Clem., 1865)
  saundersana (Kft., 1907)

OFATULENA Heinr., 1926
3444 **duodecemstriata** (Wlsm., 1884)
3445 **luminosa** Heinr., 1926

CORTICIVORA Clarke, 1951
3446 **clarki** Clarke, 1951

CYDIA Hbn., 1825
  LASPEYRESIA Hbn., 1825,
    preocc. by R. L., 1817
  EUCELIS Hbn., 1825
  ERMINEA Kby. & Spence, 1826
  SEMASIA Steph., 1829
  CARPOCAPSA Tr., 1829
  COCCYX Tr., 1829
  ENCELIS Steph., 1834, missp.
  CARPOCAMPA Harr., 1841,
    emend.
  TRYCHERIS Gn., 1845
  ORCHEMIA Gn, 1845
  CERATA Steph., 1852
  CROBYLOPHORA Kennel, 1910
  EUCELLS Caradja, 1916, missp.
  HEDULIA Heinr., 1926, n. syn.
  LASPERESIA Wu, 1938, missp.
  KENNELIOLA Paclt, 1951
  LESPEYRESIA Gozmány, 1957,
    missp.
  PSEUDOTOMOIDES Obr., 1959
  ERMINIA Obr., 1959., missp.
  CROBILOPHORA Obr., 1959,
    missp.
3447 **coniferana** (Ratzeburg, 1840)
  separatana (H.-S., 1851)
3448 **bracteatana** (Fern., 1880), n. comb.
  pallidibasalis (Heinr., 1920), n. comb.
  a. *cornutana* (Dyar, 1903), n. comb.
3449 **laricana** (Bsk., 1916), n. comb.
3450 **rana** (Fbs., 1924), n. comb.
3451 **resinosae** (Free., 1962), n. comb.
3452 **inopiosa** (Heinr., 1926), n. comb.
3453 **confusana** (McD., 1935), n. comb.
3454 **obnisa** (Heinr., 1926), n. comb.
3455 **strobilella** (L., 1758)
3456 **larimana** (Wlsm., 1895), n. comb.
3457 **garacana** (Kft., 1907), n. comb.
  septicola (Meyr., 1912), repl. name
3458 **membrosa** (Heinr., 1926), n. comb.
3459 **multilineana** (Kft., 1908), n. comb.
3460 **ingrata** (Heinr., 1926), n. comb.
3461 **albimaculana** (Fern., 1879), n. comb.
  articulatana (Kft., 1908), n. comb.
3462 **palmetum** (Heinr., 1928), n. comb.
3463 **populana** (Bsk., 1916), n. comb.
3464 **lacustrina** (Miller, 1976), n. comb.
3465 **flexiloqua** (Heinr., 1926), n. comb.
3466 **youngana** (Kft., 1907), n. comb.
3467 **nigricana** (F., 1794)
  proximana (Haw., 1811)
  dandana (Kft., 1907)
  ratifera (Meyr., 1912), repl. name
  novimundi (Heinr., 1920)
3468 **perstructana** (Wlk., 1863), n. comb.
3469 **candana** (Fbs., 1923), n. comb.

3470 **grandicula** (Heinr., 1926), n. comb.
3471 **caryana** (Fitch, 1856), n. comb.
  caryae (Shimer, 1869), n. comb.,
    emend.
3472 **fletcherana** (Kft., 1907), n. comb.
3473 **pseudotsugae** (Evans, 1969), n. comb.
3474 **tana** (Kft., 1907), n. comb.
  cirrhas (Meyr., 1912), repl. name
3475 **cupressana** Kft., 1907
3476 **prosperana** (Kft., 1907)
  succedana; (Wlsm., 1879), not D. &
    S., 1775
3477 **costastrigulana** (McD., 1935), n. comb.
3478 **leucobasis** (Bsk., 1916), n. comb.
3479 **gallaesaliciana** (Riley, 1881), n.
    comb.
3480 **lautiuscula** (Heinr., 1926), n. comb.
3481 **americana** (Wlsm., 1879), n. comb.
3482 **flavicollis** Wlsm., 1897
3483 **ninana** Riley, 1883
3484 **colorana** Kft., 1907
3485 **erotella** (Heinr., 1923), n. comb.
3486 **toreuta** (Grt., 1873), n. comb.
3487 **ingens** (Heinr., 1926), n. comb.
3488 **anaranjada** (Miller, 1959), n. comb.
3489 **piperana** Kft., 1907
3490 **miscitata** (Heinr., 1926), n. comb.
3491 **injectiva** (Heinr., 1926), n. comb.
3492 **pomonella** (L., 1758)
  pomonana (Tr., 1830)
  a. *simpsoni* (Bsk., 1903)
3493 **deshaisiana** (Luc., 1858), n. comb.
  saltitans (Westwood, 1858), n. comb.
  sebastianiae (Riley, 1892), n. comb., n.
    syn.

MELISSOPUS Riley, 1881
  MELLISOPUS Fern., 1882, missp.
  MELLIOPUS Pack., 1890, missp.
  MELLISSOPUS Fern., 1908,
    missp.
3494 **latiferreanus** (Wlsm., 1879)
  aurichalceana Riley, 1881
  inquilina (Kft., 1907)

ECDYTOLOPHA Zell., 1875
  GYMNANDROSOMA Dyar, 1904
3495 **punctidiscana** (Dyar, 1904)
3496 **desotana** (Heinr., 1926)
3497 **insiticiana** Zell., 1875
3498 **mana** (Kft., 1907)
  thaliastis (Meyr., 1912), repl. name
3499 **islandana** (Kft., 1907)
  insulicola (Meyr., 1912), repl. name

PSEUDOGALLERIA Rag., 1884
3500 **inimicella** (Zell., 1872)

TORTRICINAE

Tortricini

CROESIA Hbn., 1825
  ARGYROTOZA Steph., 1829
  ARGYROTOSA Curt., 1831,
    emend.
  ARGYROTOXA Agassiz, 1846,
    emend.
  CROEESIA Fern., 1908, missp.
  ARGROTOXA Pierce & Metcalf,
    1922, emend.

3501 **forskaleana** (L., 1758)
  forskähleana (F., 1775), missp.
  forskolana auth., missp.
  forskahliana (F., 1781), missp.
  forskäliana (Haw., 1811), missp.
  forskaeleana (Zinck., 1821), missp.
  forscaleana (Hbn., 1825), missp.
  forskaliana (Westwood, 1845), missp.
  forscaeleana (Werneburg, 1864),
    missp.
  folskaleana (Kennel, 1910), missp.
  forsskaleana (Kloet & Hincks, 1945),
    missp.
3502 **albicomana** (Clem., 1865)
  bergmanniana; auth., not L., 1758
3503 **semipurpurana** (Kft., 1905)
  dorsipurpurana (Kft., 1907)
3504 **curvalana** (Kft., 1907)
3505 **holmiana** (L., 1758)

ACLERIS Hbn., 1825
  PERONEA Curt., 1824, preocc. by
    Rafinesque, 1815
  LOPAS Hbn., 1825
  RHACODIA Hbn., 1825
  ECLECTIS Hbn., 1825
  TELEIA Hbn., 1825
  OXIGRAPHA Hbn., 1825
  AMELIA Hbn., 1825
  RHOCODIA Hbn., 1826, missp.
  LEPTOGRAMMA Steph., 1829
  GLYPHISIA Steph., 1829
  CHEIMATOPHILA Steph., 1829
  TERAS Tr., 1829
  PHLOIOPHILA Dup., 1834
  GLYPHIPTERA Dup., 1834
  OXYGRAPHA Staint., 1859,
    emend.
  ALCERIS Fern., 1903, missp.
  PHYLACOPHORA Filippi, 1931
3506 **macdunnoughi** Obr., 1963
  latifasciana; auth., not. Haw., 1811
3507 **comariana** (Zell., 1846)
  meincki (Amsel, 1930)
  comparana; McD., 1934, not Hbn., 1823
3508 **caliginosana** (Wlk., 1863)
3509 **ptychogrammos** (Zell., 1875)
3510 **nivisellana** (Wlsm., 1879)
3511 **rhombana** (D. & S., 1775)
  reticulata (Ström, 1783)
  obscurana (Don., 1804)
  ciliana (Haw., 1811)
  contaminana (Haw., 1811)
  dimidiana (Frölich, 1828)
  unicolorana (Strand, 1902)
  reticulana (Kennel, 1908), missp.
3512 **tripunctana** (Hbn., 1796–99)
  bifidana (Haw., 1811)
  ferrugana; auth., not D. & S., 1775
  proteana (Gn., 1845)
  sabulana (Gn., 1845)
  obliterana (H.-S., 1851)
  virgulana (Reutti, 1898)
  galacteana (Krulikowsky, 1903)
3513 **caryosphena** (Meyr., 1937)
3514 **cervinana** (Fern., 1882)
  americana Fern., 1903
3515 **santacrucis** Obr., 1963
  sanctacrucis Powell, 1964, missp.
3516 **comandrana** (Fern., 1892)

3517 **subnivana** (Wlk., 1863)
    deflectana (Rob., 1859)
    peculiana (Zell., 1875)
3518 **braunana** (McD., 1934)
3519 **kearfottana** (McD., 1934)
3520 **fuscana** (B. & Bsk., 1920)
3521 **semiannula** (Rob., 1869)
    stadiana (B. & Bsk., 1920)
3522 **implexana** (Wlk., 1863)
    gallicolana (Clem., 1864)
    heindelana Fern., 1905
3523 **cornana** (McD., 1933)
3524 **simpliciana** (Wlsm., 1879)
3525 **forbesana** (McD., 1934)
3526 **negundana** (Bsk., 1940)
3527 **schalleriana** (L., 1761), extralim.
    a. **viburnana** (Clem., 1860)
        schalleriana; auth.
3528 **okanagana** (McD., 1940)
3529 **oxycoccana** (Pack., 1869)
3530 **variegana** (D. & S., 1775)
    asperana (F., 1777)
    osbeckiana (Thunb., 1784)
    albildgaardana (F., 1794)
    cristana; auth., not D. & S., 1775
    nyctemerana (Hbn., 1814–17)
    blandiana (Charpentier, 1821)
    blandana (Hbn., 1825)
    aspersana (Wood, 1839), missp., not
        Hbn., 1814–17
    cirrana (Curt., 1835)
    albana (Humphreys & Westwood,
        1845)
    insignana (H.-S., 1851)
    caeruleoatrana (Strand, 1916)
3531 **hastiana** (L., 1758)
    coronana (Thunb., 1784)
    scabrana; auth., not D. & S., 1775
    pulverosana (Wlk., 1863)
    pastiana (Murt., 1893), missp.
3532 **fragariana** Kft., 1904
    permutana; auth., not Dup., 1836
3533 **celiana** (Rob., 1869)
    albilineana Kft., 1907
3534 **walkerana** (McD., 1934)
    caryosphena; de Lesse and Viette,
        1949, not Meyr., 1937
    boreana Wolff, 1964
3535 **keiferi** Powell, 1964
3536 **robinsoniana** (Fbs., 1923)
    flavivittana; auth., not Clem., 1864
    clemensiana (Fbs., 1923)
3537 **britannia** Kft., 1904
    brittania auth., missp.
3538 **klotsi** Obr., 1963
3539 **chalybeana** (Fern., 1882)
3540 **logiana** (Cl., 1759), extralim.
    a. **placidana** (Rob., 1869)
        parisiana; auth.
        scabrana; auth.
        niveana; auth., part
        boscana; auth., part
        trisignana (Rob., 1869)
3541 **senescens** (Zell., 1874)
3542 **flavivittana** (Clem., 1864)
    hastiana; auth.
    perspicuana (Rob., 1869)
3543 **maculidorsana** (Clem., 1864)
    hypericana (Ely, 1910)
3544 **clarkei** Obr., 1963

3545 **minuta** (Rob., 1869)
    vaccinivorana (Pack., 1870)
    malivorana (LeBaron, 1871)
    cinderella (Riley, 1872)
    variolana (Zell., 1875)
3546 **paracinderella** Powell, 1964
3547 **gloverana** (Wlsm., 1879)
3548 **variana** (Fern., 1886)
    angusana (Fern., 1892)
3549 **maccana** (Tr., 1835)
    ?marmorana (Curt., 1835)
    abietana; auth., part
    torquana (Zett., 1840)
    leporinana (Zett., 1840)
    effractana; auth., part
    fishiana (Fern., 1882)
3550 **youngana** (McD., 1934)
3551 **inana** (Rob., 1869)
3552 **scabrana** (D. & S., 1775)
    elevana (F., 1787)
    psorana (Frölich, 1828)
    insulana (Krulikowsky, 1903)
    nigrobasis (Hauder, 1912)
    griseus (Hauder, 1912)
    pernix (Müller-Rutz, 1924)
3553 **bowmanana** (McD., 1934)
3554 **aenigmana** Powell, 1964
3555 **lipsiana** (D. & S., 1775)
    sudoriana (Hbn., 1822)
    ?strigulana (Frölich, 1828)
    sudorana (Frölich, 1828), missp.
3556 **nigrolinea** (Rob., 1869)
    ferruginiguttana (Fern., 1882), ident.
        uncert.
    disputabilis Obr., 1963
3557 **maximana** (B. & Bsk., 1920)
    maxima Frost, 1926, missp.
3558 **busckana** (McD., 1934)
3559 **emargana** (F., 1775), extralim.
    a. **blackmorei** Obr., 1963
        effractana; auth., not F.
        caudana; auth., not F.
        emargana; auth., not F., 1775
3560 **foliana** (Wlsm., 1879)
3561 **hudsoniana** (Wlk., 1863)
    brewsteriana (Rob., 1869)
3562 **incognita** Obr., 1963
3563 **capizziana** Obr., 1963

APOTOFORMA Bsk., 1934
    EMERALDA Diakonoff, 1960
3564 **rotundipennis** (Wlsm., 1897)

Cnephasiini

EULIA Hbn., 1825
    LOPHODERUS Steph., 1829
3565 **ministrana** (L., 1758)
    ferrugana (Hbn., 1822)
    subfasciana Steph., 1834
    livoniana Dup., 1845
    infuscanus (Strand, 1901)
    dilutana (Strand, 1901)

CNEPHASIA Curt., 1826
    SPHALEROPTERA Gn., 1845
    ANOPLOCNEPHASIA Real, 1953
    BRACHYCNEPHASIA Real, 1953
3566 **longana** (Haw., 1811)
    ictericana (Haw., 1811)

    eganana (Haw., 1811)
    expallidana (Haw., 1811)
    lutosana (Hbn., 1822)
    capillana (Gn., 1845)
    loewiana (Zell., 1847)
    stratana (Zell., 1847)
    insolitana (H.-S., 1851)
    luridalbana (H.-S., 1851)
    gratana (Laharpe, 1860)
    icterana Hodgkinson, 1874., missp.
    ongana Meyr., 1912, missp.
    minor Real, 1953
3567 **interjectana** (Haw., 1811)
    asseclana (D. & S., 1775), ident.
        uncert.
    virgaureana (Tr., 1835)
    oleraceana (Gibson, 1916)
    virgaurenana Bentinck, 1936, missp.

EANA Billberg, 1820
    EUTRACHIA Hbn., 1822
    ABLABIA Hbn., 1825
    NEPHODESME Hbn., 1825
    ARGYROPTERA Dup., 1834
    NEPHODESMA Steph., 1834,
        emend.
    HYPOSTEPHANUNCIA Real,
        1951
    OPOROPSAMMA Gozmány, 1954
3568 **argentana** (Cl., 1759)
    goiiana (L., 1761)
    govana (L., 1767)
    margaritalis (Hbn., 1796)
    magnana (Hbn., 1811–13)
3569 **georgiella** (Hulst, 1887)
    subargentana Obr., 1963, n. syn.
    argentana; auth., not Cl., 1759
3570 **osseana** (Scop., 1763)
    quadripunctana (Haw., 1811)
    pratana (Hbn., 1811–13)
    cantiana (Curt., 1826)
    boreana (Zett., 1840)
    stelviana (Millière, 1874)
    biformana (Hauder, 1913)
3571 **niveosana** (Pack., 1866)
    pratana; auth., not Hbn., 1811–13
    osseana; auth., not Scop., 1763
3572 **idahoensis** Obr., 1963

DECODES Obr., 1961
3573 **basiplaganus** (Wlsm., 1879)
    fragariana; Fbs., 1923, not Bsk.,
        1919
3574 **fragarianus** (Bsk., 1919)
    elapsa (Meyr., 1922)
3575 **montanus** Powell, 1961
3576 **lundgreni** Powell, 1965
3577 **bicolor** Powell, 1961
3578 **johnstoni** Powell, 1961
3579 **aneuretus** Powell, 1961
3580 **horarianus** (Wlsm., 1879)

DORITHIA Powell, 1964
3581 **semicirculana** (Fern., 1882)
3582 **peroneana** (B. & Bsk., 1920), n. comb.

ANOPINA Obr., 1962
3583 **triangulana** (Kft., 1908)
3584 **ednana** (Kft., 1907)
3585 **eleonora** Obr., 1962

3586 **arizonana** (Wlsm., 1884)
3587 **silvertonana** Obr., 1962
3588 **ainslieana** Obr., 1962
3589 **wellingtoniana** (Kft., 1907), n. comb.

ACROPLECTIS Meyr., 1927
3590 **haemanthes** Meyr., 1927
INCERTAE SEDIS
3591 **dorsistriatana** Wlsm., 1884 (Phalonia)

Archipini

PANDEMIS Hbn., 1825
3592 **cerasana** (Hbn., 1786)
   ribeana (Hbn., 1796–99)
   grossulariana (Steph., 1829)
   obscura (Schöyen, 1882)
   balticola Strand, 1917
3593 **lamprosana** (Rob., 1869)
   albaniana; auth., not Wlk., 1863
3594 **limitata** (Rob., 1869)
   limitana Fbs., 1923, missp.
3595 **canadana** Kft., 1905
3596 **pyrusana** Kft., 1907
   albaniana; Wlsm., 1879, not Wlk.,
     1863
   pyrana Meyr., 1912, repl. name

ARGYROTAENIA Steph., 1852
3597 **velutinana** (Wlk., 1863)
   triferana (Wlk., 1863)
   lutosana (Clem., 1865)
   incertana (Clem., 1865)
3598 **montezumae** (Wlsm., 1914), extralim.
   impositana (Wlsm., 1914)
   a. **huachucensis** Obr., 1961
3599 **floridana** Obr., 1961
3600 **kimballi** Obr., 1961
3601 **repertana** Free., 1944
   gloverana; Fbs., 1923, not Wlsm.,
     1879
3602 **pinatubana** (Kft., 1905)
   politana; auth., not Haw., 1811
   pinitubana (Meyr., 1913), emend.
3603 **tabulana** Free, 1944
3604 **spaldingiana** Obr., 1961
3605 **gogana** (Kft., 1907)
   crepuscularis (Meyr., 1912), repl.
     name
3606 **amatana** (Dyar, 1901)
   chioccana (Kft., 1907)
   chiococcana (Meyr., 1912), emend.
3607 **occultana** Free., 1942
   ?lutosana; Rob., 1869, not Clem.,
     1865
3608 **coloradana** (Fern., 1882)
3609 **provana** (Kft., 1907)
   invidana (B. & Bsk., 1920)
3610 **niscana** (Kft., 1907)
   camerata (Meyr., 1912), repl. name
3611 **lignitaenia** Powell, 1965
3612 **franciscana** (Wlsm., 1879)
   kearfotti Obr., 1961
   a. **insulana** Powell, 1964
3613 **isolatissima** Powell, 1964
3614 **citrana** (Fern., 1889)
   purata Meyr., 1930, part
3615 **cupressae** Powell, 1960
   a. **beyeria** Powell, 1960
3616 **paiuteana** Powell, 1960

3617 **burroughsi** Obr., 1961
3618 **dorsalana** (Dyar, 1903)
   dimorphana (B. & Bsk., 1920)
3619 **lautana** Powell, 1960
3620 **klotsi** Obr., 1961
3621 **quadrifasciana** (Fern., 1882)
3622 **juglandana** (Fern., 1879)
3623 **quercifoliana** (Fitch, 1858)
   trifurculana (Zell., 1875)
3624 **alisellana** (Rob., 1869)
3625 **mariana** (Fern., 1882)
3626 **burnsorum** Powell, 1960
3627 **ivana** (Fern., 1901)
3628 **martini** Powell, 1960
3629 **graceana** Powell, 1960
3630 **cockerellana** (Kft., 1907)

CHORISTONEURA Led., 1859
3631 **obsoletana** (Wlk., 1863)
   transiturana (Wlk., 1863)
   vesperana (Clem., 1865)
   sanbornana (Rob., 1869)
   seminolana (Kft., 1907), n. syn.
3632 **fractivittana** (Clem., 1865)
   fumosa (Rob., 1869)
3633 **parallela** (Rob., 1869)
3634 **zapulata** (Rob., 1869)
   symphoricarpana (Kft., 1905), n. syn.
3635 **rosaceana** (Harr., 1841)
   vicariana (Wlk., 1863)
   gossypiana (Pack., 1869)
3636 **albaniana** (Wlk., 1863)
   arcticana (Mösch., 1874)
   kukukana (Kft., 1907)
   albariana (Barrett, 1887), missp.
3637 **conflictana** (Wlk., 1863)
3638 **fumiferana** (Clem., 1865)
   nigridia (Rob., 1869)
3639 **retiniana** (Wlsm., 1879)
   lindseyana Obr., 1962
   viridis Free., 1967
3640 **occidentalis** Free., 1967
3641 **biennis** Free., 1967
3642 **orae** Free., 1967
3643 **pinus** Free., 1953
   a. **maritima** Free., 1967
3644 **lambertiana** (Bsk., 1915)
   a. **ponderosana** Obr., 1962
   b. **subretiniana** Obr., 1962
3645 **carnana** (B. & Bsk., 1920)
   a. **californica** Powell, 1964
3646 **spaldingiana** Obr., 1962

CUDONIGERA Obr. & Powell, 1977
3647 **houstonana** (Grt., 1873)
   retana (Wlsm., 1879)

ARCHIPS Hbn., 1822
   CACOECIA Hbn., 1825
   ARCHIPPUS Free., 1958
3648 **argyrospila** (Wlk., 1863)
   furvana (Rob., 1869)
   v-signatana (Pack., 1875)
   a. **columbiana** (McD., 1923)
   b. **vividana** (Dyar, 1902)
3649 **mortuana** Kft., 1907
3650 **rosana** (L., 1758)
   ameriana (L., 1758)
   amerina (L., 1761)
   laevigana (D. & S., 1775)

   variana (F., 1787)
   americana (Gmel., 1788)
   levigana (Illiger, 1801)
   oxyacanthana (Haw., 1811)
   acerana Hbn., 1822
   hewittana (Bsk., 1920)
3651 **eleagnana** (McD., 1923)
3652 **myricana** (McD., 1923)
3653 **semiferana** (Wlk., 1863)
   flaccidana (Rob., 1869)
3654 **negundana** (Dyar, 1902)
3655 **fervidana** (Clem., 1860)
   paludana (Rob., 1869)
   palludana (Beutenmüller, 1892), missp.
3656 **georgiana** (Wlk., 1863)
3657 **magnoliana** (Fern., 1892)
3658 **purpurana** (Clem., 1865)
   gurgitana (Rob., 1869)
   lintneriana (Grt., 1873)
   guritana (Darlington, 1947), missp.
3659 **infumatana** (Zell., 1875)
3660 **grisea** (Rob., 1869)
   brauniana (Kft., 1907)
3661 **cerasivorana** (Fitch, 1856)
3662 **rileyana** (Grt., 1868)
   fervidana (Wlk., 1863), preocc. by
     Clem., 1860
3663 **oporana** (L., 1758)
   piceana (L., 1758)
   operana (Wilkes, 1773), missp.
   fulvana (D. & S., 1775)
   pyrastrana (Hbn., 1822)
   congenerana (Tr., 1829)
   amerinana (Dup., 1834)
   sauberiana (Sorhagen, 1882)
3664 **striana** (Fern., 1905)
3665 **alberta** (McD., 1923)
3666 **dissitana** (Grt., 1879)
3667 **packardiana** (Fern., 1886)
3668 **tsuganus** (Powell, 1962), n. comb.

ARCHEPANDEMIS Mutuura, 1978
3669 **borealis** (Free., 1965)
3670 **coniferana** Mutuura, 1978
3671 **morrisana** Mutuura, 1978

SYNDEMIS Hbn., 1825
3672 **afflictana** (Wlk., 1863)
   fuscolineana (Clem., 1865)
   musculana; Wlsm., 1879, not Hbn.,
     1796–99

LOZOTAENIA Steph., 1829
3673 **hesperia** Powell, 1962
3674 **rindgei** Obr., 1962

APHELIA Hbn., 1825
3675 **alleniana** (Fern., 1882)
   trentonana (McD., 1923)
   a. **rindgeorum** Obr., 1959
3676 **koebelei** Obr., 1959
3677 **septentrionalis** Obr., 1959

CACOECIMORPHA Obr., 1954
3678 **pronubana** (Hbn., 1822)
   ambustana (Frölich, 1830)
   hermineana (Dup., 1835)
   insolatana (Luc., 1848)
   perochreana (H.-S., 1856)
   pronuba (Stenton, 1928), missp.

CLEPSIS Gn., 1845
    SICLOBOLA Diakonoff, 1947
3679 **listerana** (Kft., 1907), n. comb.
3680 **fucana** (Wlsm., 1879)
    victoriana (Bsk., 1922)
    busckana Keif., 1933
3681 **kearfotti** Obr., 1962
3682 **persicana** (Fitch, 1856)
    blandana (Clem., 1864)
    fragariana (Pack., 1869)
    conigerana (Zell., 1875)
    a. **forbesi** Obr., 1962
3683 **unifasciana** (Dup., 1843)
    ?consimilana (Hbn., 1822)
    ?externana (Evers., 1844)
    productana (Zell., 1847)
    obliterana (H.-S., 1851)
    xylotoma (Meyr., 1891)
3684 **clemensiana** (Fern., 1879)
    nervosana (Kft., 1907)
3685 **moeschleriana** (Wocke, 1862)
    algidana (Mösch., 1862)
    gelidana (Mösch., 1862)
3685.1 **danilevski** Kostiuk, 1973
3686 **melaleucana** (Wlk., 1863)
    invexana (Wlk., 1863)
    semifuscana (Clem., 1864)
3687 **flavidana** (McD., 1923)

PTYCHOLOMA Steph., 1829
    SMICROTES Clem., 1860
3688 **peritana** (Clem., 1860)
    inconclusana (Wlk., 1863)
3689 **virescana** (Clem., 1865)
    sescuplana (Zell., 1875)
    glaucana (Wlsm., 1879)

ADOXOPHYES Meyr., 1881
3690 **furcatana** (Wlk., 1863)
3691 **negundana** (McD., 1923)

DITULA Steph., 1829
    BATODES Gn., 1845
3692 **angustiorana** (Haw., 1811)
    rotundana (Haw., 1811)
    dumeriliana (Dup., 1836)

XENOTEMNA Powell, 1964
3693 **pallorana** (Rob., 1869)
    lata (Rob., 1869), n. syn.

Niasomini

NIASOMA Bsk., 1940
3694 **metallicana** (Wlsm., 1895)

Sparganothidini

SPARGANOTHIS Hbn., 1825
    SPARGONOTHIS Hbn., 1826,
      missp.
    SPARGANOTHRIS Steph., 1834,
      missp.
    OENOPHTHIRA Dup., 1845
    OENECTRA Gn., 1845
    AENECTRA Doubleday, 1850,
      emend.
    ONECTRA Wocke, 1861, missp.
    BEGUNNA Wlk., 1863
    LEPTORIS Clem., 1865

CENOPIS Zell., 1875
    SPARGANYTHIS Matsumura,
      1931, missp.
3695 **sulfureana** (Clem., 1860)
    gratana (Wlk., 1863)
    gallivorana (Clem., 1864)
    virginiana (Clem., 1864)
    fulvoroseana (Clem., 1864)
    sulphureana B. & Bsk., 1920, missp.
3696 **belfrageana** (Zell., 1875)
    euphronopa Meyr., 1927, n. syn.
3697 **lycopodiana** (Kft., 1907)
3698 **bistriata** Kft., 1907
3699 **tristriata** Kft., 1907
3700 **caryae** (Rob., 1869)
3701 **pulcherrimana** (Wlsm., 1879)
3702 **taracana** Kft., 1907
    procax Meyr., 1912, repl. name
3703 **demissana** (Wlsm., 1879)
3704 **distincta** (Wlsm., 1884)
    solidana Free., 1941, n. syn.
3705 **rubicundana** (H.-S., 1856)
    hudsoniana Free., 1940, n. syn.
3706 **xanthoides** (Wlk., 1863)
    breviornatana (Clem., 1865)
    irrorea (Rob., 1869), n. syn.
    a. **inconditana** (Wlsm., 1879), rev.
      stat.
3707 **daphnana** McD., 1961
3708 **salinana** McD., 1961
3709 **striata** (Wlsm., 1884)
3710 **violaceana** (Rob., 1869)
3711 **unifasciana** (Clem., 1864)
    puritana (Rob., 1869)
3712 **vocaridorsana** Kft., 1905
3713 **tunicana** (Wlsm., 1879)
    californiana (Wlsm., 1879)
3714 **senecionana** (Wlsm., 1879), rev. stat.
    rudana (Wlsm., 1879)
3715 **umbrana** B. & Bsk., 1920
    putmanana Free., 1940, n. syn.
3716 **diluticostana** (Wlsm., 1879)
    quercana (Fern., 1882)
3717 **flavibasana** (Fern., 1882)
3718 **karacana** (Kft., 1907)
    tempestiva Meyr., 1912, repl. name
3719 **pilleriana** (D. & S., 1775)
    vitis (Dantic, 1786)
    pillerana (F., 1794)
    vitana (F., 1794)
    luteolana (Hbn., 1822)
    danticana (Walcken, 1836)
    obscurana (Preissecker, 1936)
3720 **reticulatana** (Clem., 1860)
    fulgidipennana (Blanchard, 1840),
      unused sr. syn.
    subauratana (Wlk., 1863)
    mesospila (Zell., 1875)
    gracilana (Wlsm., 1879), n. syn.
    ferreana Bsk., 1915, n. syn.
3721 **albicaudana** Bsk., 1915
3722 **directana** (Wlk., 1863)
    testulana (Zell., 1875), n. syn.
3723 **chambersana** (Kft., 1907)
3724 **saracana** (Kft., 1907)
    austera (Meyr., 1912), repl. name
3725 **pettitana** (Rob., 1869)
3726 **acerivorana** MacKay, 1952
3727 **niveana** (Wlsm., 1879)
    groteana (Fern., 1882), n. syn.

3728 **cana** (Rob., 1869)
3729 **machimiana** B. & Bsk., 1920
3730 **hydeana** Klots, 1936
3731 **lentiginosana** (Wlsm., 1879), n. comb.

PLATYNOTA Clem., 1860
    CERORRHINETA Zell., 1877
    CERATORRHINETA Kirby, 1878,
      emend.
    PHYLACTERITIS Meyr., 1922
3732 **flavedana** Clem., 1860
    tinctana (Wlk., 1863), n. syn.
    concursana (Wlk., 1863)
    laterana (Rob., 1869)
    iridana B. & Bsk., 1920, n. syn.
3733 **viridana** B. & Bsk., 1920
3734 **yumana** (Kft., 1907)
3735 **larreana** (Comst., 1939)
3736 **stultana** Wlsm., 1884
    chiquitana B. & Bsk., 1920
3737 **nigrocervina** Wlsm., 1895
3738 **labiosana** (Zell., 1875)
    rubiginis Wlsm., 1913, n. syn.
3739 **calidana** (Zell., 1877)
3740 **idaeusalis** (Wlk., 1859)
    sentana Clem., 1860
    dioptrica (Meyr., 1922)
3741 **semiustana** Wlsm., 1884
3742 **scotiana** (McD., 1961), n. comb.
3743 **exasperatana** (Zell., 1875)
3744 **wenzelana** (Haim., 1915)
3745 **rostrana** (Wlk., 1863)
    restitutana (Wlk., 1863)
    connexana (Wlk., 1863)
    repandana (Wlk., 1863)
    saturatana (Wlk., 1863)
    egana (Wlk., 1866)

SYNNOMA Wlsm., 1879
3746 **lynosyrana** Wlsm., 1879

COELOSTATHMA Clem., 1860
3747 **discopunctana** Clem., 1860

AMORBIA Clem., 1860
    HENDECASTEMA Wlsm., 1879
    PTYCHAMORBIA Wlsm., 1892
3748 **humerosana** Clem., 1860
3749 **cuneana** (Wlsm., 1879)
    adumbrana (Wlsm., 1879)
    essigana Bsk., 1929, n. syn.
3750 **synneurana** B. & Bsk., 1920

CHLIDANOTINAE

Hilarographini

HILAROGRAPHA Zell., 1877
3751 **jonesi** Brower, 1953
3752 **youngiella** Bsk., 1922
    olympica Braun, 1923
    youngella Brower, 1953, missp.
3753 **regalis** (Wlsm., 1881)

COCHYLIDAE
by JERRY A. POWELL

AETHES Billberg, 1820
  PHALONIA Hbn., 1825
  CHLIDONIA Hbn., 1825
  DAPSILIA Hbn., 1825
  LOZOPERA Steph., 1829
  ARGYROPTERA Dup., 1834
  CHROSIS Gn., 1845
  ARGYRIDIA Steph., 1852
3754 **fernaldana** (Wlsm., 1879)
3755 **smeathmanniana** (F., 1781)
3756 **kindermannana** (Tr., 1830)
3757 **deutschiana** (Zett., 1840)
  fucostrigana (Clem., 1864)
  chalcana (Pack., 1866)
3758 **rutilana** (Hbn., 1818)
3759 **intactana** (Wlsm., 1879)

COCHYLIS Tr., 1830
  CONCHYLIS Sodoffsky, 1837
  PONTOTURANIA Obr., 1943
  ACORNUTIA Obr., 1944
  COCHYLICHROA Obr. &
    Swatschek, 1958
  PARACOCHYLIS Razowski, 1960
  BREVICORNUTIA Razowski, 1960
  NEOCOCHYLIS Razowski, 1960
  LONGICORNUTIA Razowski,
    1960
3760 **felix** (Wlsm., 1895)
3761 **parallelana** Wlsm., 1879
3762 **transversana** Wlsm., 1879
3763 **pimana** (Bsk., 1907)
3764 **nana** (Haw., 1811)
  carneana Gn., 1845
  ochreoalbana Wlk., 1863
3765 **campicolana** Wlsm., 1879
3766 **parvimaculana** Wlsm., 1879
3767 **formonana** (Kft., 1907), n. comb.
  myrinitis Meyr., 1912, repl. name
3768 **dilutana** Wlsm., 1879

THYRAYLIA Wlsm., 1897
3769 **bunteana** (Rob., 1869)

HENRICUS Bsk., 1943
  HEINRICHIA Bsk., 1940, preocc.
    by Streesemann, 1931
3770 **macrocarpanus** (Wlsm., 1895)
3771 **infernalis** (Heinr., 1920)
3772 **brevipalpatus** McD., 1944
3773 **fuscodorsanus** (Kft., 1904), n. comb.
3774 **contrastanus** (Kft., 1907), n. comb.
3775 **banus** (Kft., 1907), n. comb.
  rhodites (Meyr., 1912), repl. name
3776 **huachucanus** (Kft., 1907), n. comb.
3777 **umbrabasanus** (Kft., 1908)

IRAZONA Razowski, 1964
3778 **comes** (Wlsm., 1884)

PHALONIDIA LeMarch, 1933
  PIERCEA Filipjev, 1940
3779 **latipunctana** (Wlsm., 1879)
3780 **saxicolana** (Wlsm., 1879)

LORITA Bsk., 1939
3781 **abornana** Bsk., 1939

CAROLELLA Bsk., 1939
  PHARMACIS Hbn., 1823, preocc.
    by Hbn., 1820
3782 **sartana** (Hbn., 1823), n. comb.
3783 **bimaculana** (Rob., 1869), n. comb.
3784 **erigeronana** (Riley, 1881), n. comb.
3785 **deceptana** (Bsk., 1907), n. comb.
3786 **mexicana** (Bsk., 1907)
3787 **vitellinana** (Zell., 1875), n. comb.
3788 **argyoplaca** (Meyr., 1931), n. comb.
3789 **beevorana** Comst., 1940
3790 **busckana** Comst., 1939
3791 **willettana** Comst., 1939

HYSTEROSIA Steph., 1852
  IDIOGRAPHIS Led., 1859
  PROPIRA Durrant, 1914
  HYSTEROPHORA Obr., 1943
3792 **fulviplicana** (Wlsm., 1879)
  homanana Kft., 1907
  refuga Meyr., 1912, repl. name
  komonana Kft., 1907
  fermentata Meyr., 1912, repl. name
3793 **aegrana** (Wlsm., 1879)
3794 **aureoalbida** Wlsm., 1895
3795 **canariana** B. & Bsk., 1920
3796 **waracana** Kft., 1907
  dicax Meyr., 1912, repl. name
3797 **villana** Bsk., 1907
3798 **cartwrightana** Kft., 1907
3799 **terminana** Bsk., 1907
  merrickana Kft., 1907
3800 **perspicuana** B. & Bsk., 1920
3801 **birdana** Bsk., 1907
3802 **riscana** Kft., 1907
  vincta Meyr., 1912, repl. name
3803 **pecosana** Kft., 1907
3804 **baracana** Bsk., 1907
  tiscana Kft., 1907
  vigilans Meyr., 1912, repl. name
3805 **modestana** Bsk., 1907
  INCERTAE SEDIS
3806 **albidana** Wlk., 1866 (Simaethis)
  winniana Kft., 1905 (Phalonia)
3807 **angulatana** Rob., 1869 (Conchylis)
3808 **augustana** Clem., 1860 (Conchylis)
  dorsimaculana Rob., 1869, repl. name
3809 **argentilimitana** Rob., 1869
    (Conchylis)
3810 **atomosana** Bsk., 1907 (Phalonia)
3811 **aureana** Bsk., 1907 (Phalonia)
3812 **aurorana** Kft., 1907 (Phalonia)
3813 **baboquivariana** Kft., 1907 (Tortrix)
3814 **basiochreana** Kft., 1907 (Phalonia)
3815 **biscana** Kft., 1907 (Phalonia)
  giscana Kft., 1907 (Phalonia), n. syn.
  ixeuta Meyr., 1912, repl. name
3816 **bomonana** Kft., 1907 (Conchylis)
  cyamitis Meyr., 1912, repl. name
3817 **carmelana** Kft., 1907 (Phalonia)
  obispoana Kft., 1907 (Phalonia), n. syn.
3818 **cephalanthana** Heinr., 1921
    (Phalonia)
3819 **discana** Kft., 1907 (Phalonia)
  cricota Meyr., 1912, repl. name
3820 **edwardsiana** Wlsm., 1884 (Conchylis)
3821 **elderana** Kft., 1907 (Phalonia)
  helonoma Meyr., 1912, repl. name
3822 **flaccosana** Wlk., 1863 (Conchylis)
  confusana Rob., 1869 (Tortrix)

3823 **foxcana** Kft., 1907 (Phalonia)
  liquida Meyr., 1912, repl. name
3824 **fulvotinctana** Wlsm., 1884
    (Conchylis)
3825 **glaucofuscana** Zell., 1875 (Conchylis)
3826 **grandis** Bsk., 1907 (Phalonia)
3827 **gunniana** Bsk., 1907 (Phalonia)
3828 **hoffmanana** Kft., 1907 (Phalonia)
3829 **hollandana** Kft., 1907 (Phalonia)
3830 **hospes** Wlsm., 1884 (Conchylis)
3831 **hubbardana** Bsk., 1907 (Phalonia)
3832 **interruptofasciana** Rob., 1869
    (Conchylis)
3833 **labeculana** Rob., 1869 (Conchylis)
3834 **lavana** Bsk., 1907 (Phalonia)
3835 **leguminana** Bsk., 1907 (Phalonia)
3836 **lepidana** Clem., 1860 (Argyrolepia)
3837 **louisiana** Bsk., 1907 (Phalonia)
3838 **maiana** Kft., 1907 (Phalonia)
3839 **marloffiana** Bsk., 1907 (Phalonia)
  nonlavana Kft., 1907 (Phalonia)
3840 **nomonana** Kft., 1907 (Phalonia)
  voluntaria Meyr., 1912, repl. name
3841 **obliquana** Kft., 1907 (Phalonia)
3842 **oenotherana** Riley, 1881 (Conchylis)
3843 **plummeriana** Bsk., 1907 (Phalonia)
3844 **promptana** Rob., 1869 (Conchylis)
3845 **punctadiscana** Kft., 1908 (Phalonia)
3846 **rana** Bsk., 1907 (Phalonia)
  funesta Meyr., 1912, repl. name
3847 **romonana** Kft., 1907 (Phalonia)
  officiosa Meyr., 1912, repl. name
3848 **schwarziana** Bsk., 1907 (Phalonia)
3849 **scissana** Wlk., 1863 (Conchylis)
3850 **seriatana** Zell., 1875 (Conchylis)
3851 **spartinana** B. & McD., 1916 (Phalonia)
3852 **straminoides** Grt., 1873 (Conchylis)
3853 **sublepidana** Kft., 1907 (Phalonia)
3854 **temerana** Bsk., 1907 (Phalonia)
  cincinnatana Kft., 1907 (Phalonia)
3855 **toxcana** Kft., 1907 (Phalonia)
  baryzela Meyr., 1912, repl. name
3856 **vachelliana** Kft., 1907 (Phalonia)
3857 **viscana** Kft., 1907 (Phalonia)
  peganitis Meyr., 1912, repl. name
3858 **voxcana** Kft., 1907 (Phalonia)
  omphacitis Meyr., 1912, repl. name
3859 **wiscana** Kft., 1907 (Phalonia)
  acropeda Meyr., 1912, repl. name
3860 **yuccatana** Bsk., 1907 (Phalonia)
3861 **zaracana** Kft., 1907 (Phalonia)
3862 **ziscana** Kft., 1907 (Phalonia)
  fabicola Meyr., 1912, repl. name
3863 **zoxcana** Kft., 1907 (Phalonia)
  telifera Meyr., 1912, repl. name

Hesperioidea

HESPERIIDAE
by LEE D. MILLER & F. MARTIN BROWN

PYRRHOPYGINAE

PYRRHOPYGE Hbn., 1819
  TAMYRIS Swainson, 1820
  PYRRHOPYGA Westwood, 1852,
    missp.

PACHYRHOPALA Wallgr., 1858
YANGUNA Watson, 1893
APYRROTHRIX Linds., 1921
APYRRHOTHRIX Evans, 1951,
  missp.
3864 araxes (Hewitson, 1867), extralim.
  cyrillus (Plötz, 1879)
  a. arizonae Godm. & Salvin, 1893

PYRGINAE

PHOCIDES Hbn., 1819
ERYCIDES Hbn., 1819
DYSENIUS Scudder, 1872
3865 pigmalion (Cram., 1779), extralim.
  tenuistriga Mabille & Boullet, 1912
  vulcanides Röber, 1925
  iocularis Röber, 1925
  disparilis Röber, 1925
  a. okeechobee (Worthington, 1881)
    batabano; auth., not Luc., 1857
3866 palemon (Cram., 1777), extralim.
  polybius (F., 1793)
  cruentus Hbn., 1819
  gunderi R. C. Williams & Bell, 1931
  a. lilea (Reak., 1867)
    albicilla (H.-S., 1869)
    socius (Butler & H. Druce, 1872)
    cruentus (Scudder, 1872) , preocc.
      by Hbn., 1819
    sanguinea (Scudder, 1872)
    decolor Mabille, 1880
    albiciliata Röber, 1925
3867 urania (Westwood, 1852)
  texana (Scudder, 1872)

PROTEIDES Hbn., 1819
DICRANASPIS Mabille, 1878
3868 mercurius (F., 1787)
  idas (Cram., 1779), preocc. by L.,
    1758
  a. sanantonio (Luc., 1857)

EPARGYREUS Hbn., 1819
ERIDAMUS Burmeister, 1875
3869 zestos (Gey., 1832)
  oberon (Worthington, 1881)
  arsaces Mabille, 1903
3870 clarus (Cram., 1779)
  tityrus (F., 1775), preocc. by Poda, 1761
  obliteratus Scudder, 1889
  smythi R. C. Williams, 1927
  argentosus Hayward, 1933
  argenteola (Matsumura, 1940)
  a. huachuca Dixon, 1955
3871 exadeus (Cram., 1779), extralim.
  a. cruza Evans, 1952

POLYGONUS Hbn., 1825
ACOLASTUS Scudder, 1872
NENNIUS Kby., 1902
3872 leo (Gmel., 1790), extralim.
  amyntas (F., 1775), preocc. by Poda,
    1761
  a. savigny (Latr., 1824)
  b. arizonensis (Skin., 1911)
3873 manueli Bell & W. P. Comstock, 1948

CHIOIDES Linds., 1921
3874 catillus (Cram., 1780), extralim.

tarchon (Hbn., 1825)
longicauda (Sepp, 1848)
  a. albofasciatus (Hewitson, 1867)
3875 zilpa (Butler, 1874)
  a. namba Evans, 1952

AGUNA R. C. Williams, 1927
TMETOCERUS Poujade, 1895,
  preocc. by Hartert, 1891
3876 asander (Hewitson, 1867)
  panthius (H.-S., 1869)
  scheba (Plötz, 1882)
  euthemides (Mabille & Boullet, 1917)
3877 claxon Evans, 1952
  metophis; auth., not (Latr., 1824)

TYPHEDANUS Butler, 1870
3878 undulatus (Hewitson, 1867)
  sumichrasti (Scudder, 1872)
  nicasius (Plötz, 1880)
  elongatus (Plötz, 1880)

POLYTHRIX Watson, 1893
3879 mexicana H. A. Freeman, 1969
3880 octomaculata (Sepp, 1848)
  decurtata (H.-S., 1869)
  calenus (Mabille, 1888)
  maculata (Seitz, 1925), missp.
  elegans (Hayward, 1933)
3881 procera (Plötz, 1880)
  aelius (Plötz, 1880)
  auginulus (Godm. & Salvin, 1893)
  callicina (Schaus, 1902)

ZESTUSA Linds., 1925
PLESTIA Mabille, 1888, preocc. by
  Stål, 1870
3882 dorus (Edw., 1882)

CODATRACTUS Linds., 1921
HETEROPIA Mabille, 1889,
  preocc. by Carter, 1886
3883 alcaeus (Hewitson, 1867)
  montezuma (Scudder, 1872)
3884 melon (Godm. & Salvin, 1893)
3885 arizonensis (Skin., 1905)

URBANUS Hbn., 1807
GONIURUS Hbn., 1819
EUDAMUS Swainson, 1821
LYROPTERA Plötz, 1881
3886 proteus (L., 1758)
3887 pronta Evans, 1952
3888 esmeraldus (Butler, 1877)
  platowii (Plötz, 1880)
3889 dorantes (Stoll, 1790)
  proteus; Hbn., 1807, not L., 1758
  torones (Hbn., 1821)
  atletes (C. & R. Felder, 1862)
  amisus (Hewitson, 1867)
  protillus (H.-S., 1869)
  retractus (Plötz, 1880)
  velinus (Plötz, 1880)
  kefersteinii (Plötz, 1880)
  rauterbergi (Skin., 1895)
  a. santiago (Luc., 1856)
    cariosa (H.-S., 1862)
    corydon (Butler, 1870)
    larius (Plötz, 1880)
3890 teleus (Hbn., 1821)

dorantes; Hbn., 1807, not Stoll, 1790
eurycles (Latr., 1824)
zalanthus (Plötz, 1880)
latipennis (Mabille & Boullet, 1891)
3891 tanna Evans, 1952
3892 simplicius (Stoll, 1790)
  dorantes; Hbn., 1807, not Stoll, 1790
  pilatus (Plötz, 1880)
  gracilicauda (Plötz, 1880)
  theimei (Ehrmann, 1907)
  borja (Ehrmann, 1907)
3893 procne (Plötz, 1880)
3894 doryssus (Swainson, 1821)
  orion (Drury, 1782), preocc. by
    Pallas, 1771
  brachius (Gey., 1832)
  leucites (Mabille, 1888)
  cleopatra (Ehrmann, 1907)
  interruptus (R. C. Williams, 1926)
3895 albimargo (Mabille, 1875)
  dominicus (Plötz, 1886)
  triptolemus (Ehrmann, 1907)

ASTRAPTES Hbn., 1819
TELEGONUS Hbn., 1819
EUTHYMELE Mabille, 1878
THYMELE; auth., not F., 1807
3896 fulgerator (Walch, 1775), extralim.
  mercatus (F., 1793)
  fulminator (Sepp, 1848)
  a. azul (Reak., 1867)
    misitra (Plötz, 1881)
    albifasciatus (Röber, 1925)
3897 egregius (Butler, 1870)
  brevicauda (Plötz, 1884)
3898 alardus (Stoll, 1790), extralim.
  grullus (Mabille, 1888)
  fabrici (Ehrmann, 1918)
  a. latia Evans, 1952
3899 gilberti H. A. Freeman, 1969
  hopfferi; auth., not Plötz, 1882
3900 galesus (Mabille, 1888), extralim.
  subflavus (R. C. Williams, 1927)
  a. cassius Evans, 1952
  hahneli; auth., not. Stgr., 1888
3901 anaphus (Cram., 1777), extralim.
  leucogramma (Sepp, 1848)
  anaphides (Mabille & Boullet, 1912)
  a. annetta Evans, 1952

AUTOCHTON Hbn., 1823
CECROPS Hbn., 1818, preocc. by
  Leach, 1816
CECROPTERUS H.-S., 1869
RHABDOIDES Scudder, 1889
3902 cellus (Bdv. & Leconte, 1834)
  festus (Gey., 1837)
  aereofuscus (Gunder, 1925)
  leilae (A. H. Clark, 1934)
3903 pseudocellus (Coolidge & Clemence,
  1911)

ACHALARUS Scudder, 1872
MURGARIA Watson, 1893
3904 lyciades (Gey., 1832)
  lycidas (J. E. Smith, 1797), preocc.
    by Cram., 1777
  hedysarum Scudder, 1889
3905 casica (H.-S., 1869)
  epigena (Butler, 1870)

epigona (Godm., & Salvin, 1894),
missp.
3906 **albociliatus** (Mabille, 1877)
3907 **toxeus** (Plötz, 1882)
coyote (Skin., 1892)
3908 **jalapus** (Plötz, 1882)

THORYBES Scudder, 1872
LINTNERIA Edw., 1877, preocc.
by Butler, 1877
COCCEIUS Godm. & Salvin, 1894
3909 **bathyllus** (J. E. Smith, 1797)
?daunus (Cram., 1777)
?syloson (Mabille, 1903
3910 **pylades** (Scudder, 1870)
immaculata (Skin., 1911)
albosuffusa H. A. Freeman, 1943
integra Lanktree, 1968
3911 **diversus** Bell, 1927
3912 **mexicanus** (H.-S., 1869)
ananius (Plötz, 1882)
a. **nevada** Scudder, 1872
aemilea (Skin., 1893)
b. **dobra** Evans, 1952
3913 **confusis** Bell, 1922
corusis Evans, 1952, missp.
3914 **drusius** (Edw., 1883)
paucipuncta Dyar, 1917
3915 **valerianus** (Plötz, 1882)
?mysie Dyar, 1904

CABARES Godm. & Salvin, 1894
3916 **potrillo** (Luc., 1857)
paterculus (H.-S., 1863)

CELAENORRHINUS Hbn., 1819
ANCISTROCAMPA C. & R.
Felder, 1862
HANTANA Moore, 1881
GEHLOTA Doherty, 1889
NARGA Mabille, 1891
APALLAGA Strand, 1911
3917 **fritzgaertneri** (Bailey, 1880)
variegatus Godm. & Salvin, 1894
3918 **stallingsi** H. A. Freeman, 1946

DYSCOPHELLUS Godm. & Salvin, 1893
DYSCOPHUS Burmeister, 1878,
preocc. by Grandidier, 1872
NASCUS; auth., not Watson, 1893
3919 **euribates** (Stoll, 1782)
nicias (F., 1787)
hesus (Westwood, 1852)
etias (Hewitson, 1867)
gaurus (Plötz, 1882)
tychios (Plötz, 1882)

SPATHILEPIA Butler, 1870
3920 **clonius** (Cram., 1776)

COGIA Butler, 1870
PHOEDINUS Godm. & Salvin,
1894, preocc. by
Guér.-Méneville, 1838
ANAPERUS Mabille & Boullet,
1919, preocc. by Troschel, 1844
CAICELLA Hemming, 1934
3921 **calchas** (H.-S., 1869)
terranea (Butler, 1872)
anacreon (Plötz, 1882)

3922 **hippalus** (Edw., 1882)
gila (Plötz, 1886)
3923 **outis** (Skin., 1894)
3924 **caicus** (H.-S., 1869)
schaefferi (Plötz, 1882)
a. **moschus** (Edw., 1882)

NISONIADES Hbn., 1819
3925 **rubescens** (Mösch., 1876)
bromias (Godm. & Salvin, 1894)
triangularus (Mabille, 1897)
clara (Mabille & Boullet, 1916)
nigra (Mabille & Boullet, 1916)

PELLICIA H.-S., 1870
HEMIPTERIS Mabille, 1889
3926 **angra** Evans, 1953
3927 **arina** Evans, 1953

BOLLA Mabille, 1903
3928 **clytius** (Godm. & Salvin, 1897)
3929 **brennus** (Godm. & Salvin, 1896)

STAPHYLUS Godm. & Salvin, 1896
SCANTILLA Godm. & Salvin,
1896
3930 **ceos** (Edw., 1870)
3931 **mazans** (Reak., 1866)
3932 **hayhurstii** (Edw., 1870)

GORGYTHION Godm. & Salvin,
1896
3933 **begga** (Kby., 1870), extralim.
tucumana (Burmeister, 1878)
a. **pyralinus** (Mösch., 1876)
marginata Schaus, 1902

SOSTRATA Godm. & Salvin, 1895
3934 **bifasciata** (Mén., 1829), extralim.
a. **nordica** Evans, 1953

CARRHENES Godm. & Salvin, 1895
3935 **canescens** (R. Felder, 1869)

XENOPHANES Godm. & Salvin,
1895
3936 **trixus** (Stoll, 1784)
tryxus auth., missp.
salvianus (F., 1793)
ruatanensis Godm. & Salvin, 1895
euphemie (Ehrmann, 1907)
perplexus Bell, 1942
ruatensis Evans, 1953, missp.

SYSTASEA Edw., 1877
LINTNERIA Edw., 1877, preocc.
by Butler, 1877
PLESIOCERA Mabille, 1891,
preocc. by Macquart, 1841
3937 **pulverulenta** (R. Felder, 1869)
taeniatus (Plötz, 1884)
filipalpis (Mabille, 1891)
zampa; auth., not Edw., 1876
3938 **zampa** (Edw., 1876)
evansi (Bell, 1941)

ACHLYODES Hbn., 1819
EANTIS Bdv., 1836
SEBALDIA Mabille, 1903
ACHYLODES auth., missp.

3939 **thraso** (Hbn., 1807), extralim.
peruvianus (Mabille & Boullet, 1917)
a. **tamenund** (Edw., 1871)

GRAIS Godm. & Salvin, 1894
3940 **stigmatica** (Mabille, 1883)
fumosus (Plötz, 1884)

TIMOCHARES Godm. & Salvin, 1896
3941 **ruptifasciatus** (Plötz, 1884)

CHIOMARA Godm. & Salvin, 1899
3942 **asychis** (Stoll, 1780), extralim.
dilucida (Plötz, 1884)
palica (Mabille, 1888)
a. **georgina** (Reak., 1868)
pelagica; Evans, 1953, not Weeks,
1891

GESTA Evans, 1953
3943 **gesta** (H.-S., 1863), extralim.
bigutta (Prittwitz, 1868)
brusus (Burmeister, 1878)
blanda (Plötz, 1884)
a. **invisus** (Butler & H. Druce, 1872)
gorgona (Plötz, 1884)
llano (Dodge, 1903)

EPHYRIADES Hbn., 1819
OILEIDES Hbn., 1825
MELANTHES Mabille, 1903
BRACHYCORYNE Mabille, 1904
3944 **brunneus** (H.-S., 1864), extralim.
electra (Lint., 1881)
a. **floridensis** Bell &
W. P. Comstock, 1948

ERYNNIS Schrank, 1801
THYMELE F., 1807
ASTYCUS Hbn., 1822
THANAOS Bdv., 1834
HALLIA Tutt, 1906
ERYNNIDES Burns, 1964
3945 **icelus** (Scudder & Burgess, 1870)
bautista (Plötz, 1884)
hamamaelidis (Scudder, 1889)
3946 **brizo** (Bdv. & Leconte, 1834)
a. **somnus** (Lint., 1881)
b. **burgessi** (Skin., 1914)
c. **lacustra** (Wgt., 1905)
3947 **juvenalis** (F., 1793)
costalis (Westwood, 1852)
ennius (Scudder & Burgess, 1870)
plautus (Scudder & Burgess, 1870)
a. **clitus** (Edw., 1883)
maestus (Godm. & Salvin, 1899)
3948 **telemachus** Burns, 1960
plautus; auth., not Scudder &
Burgess, 1870
3949 **propertius** (Scudder & Burgess, 1870)
tibullus (Scudder & Burgess, 1870)
3950 **meridianus** Bell, 1927
3951 **scudderi** (Skin., 1914)
3952 **horatius** (Scudder & Burgess, 1870)
virgilius (Scudder & Burgess, 1870)
petronius (Lint., 1881)
3953 **tristis** (Bdv., 1852)
a. **tatius** (Edw., 1882)
albomarginatus (Godm. & Salvin,
1899)

3954 **martialis** (Scudder, 1869)
    quercus (Butler, 1870)
    ausonius (Lint., 1872)
3955 **pacuvius** (Lint., 1878)
    a. **lilius** (Dyar, 1904)
    b. **perniger** (Grinnell, 1905)
    c. **callidus** (Grinnell, 1905)
3956 **zarucco** (Luc., 1857)
    terentius (Scudder & Burgess, 1870)
    ovidius (Scudder & Burgess, 1870)
    naevius (Lint., 1881)
    diogenes (Plötz, 1884)
3957 **funeralis** (Scudder & Burgess, 1870)
    clericalis (Burmeister, 1878)
    australis (Mabille, 1883)
3958 **lucilius** (Scudder & Burgess, 1870)
3959 **baptisiae** (Fbs., 1936)
3960 **afranius** (Lint., 1878)
3961 **persius** (Scudder, 1863)
    a. **borealis** (Cary, 1907)
    b. **avinoffi** (Holl., 1930)
    c. **fredericki** H. A. Freeman, 1943

PYRGUS Hbn., 1819
    SCELOTRIX Rambur, 1858
    BREMERIA Tutt, 1906, preocc. by
      Alpheraky, 1892
    TELEOMORPHA Warr., 1926
    HEMITELEOMORAPHA Warr.,
      1926
    ATELEOMORPHA Warr., 1926
    HELIOPYRGUS Herrera, 1957
3962 **centaureae** (Rambur, 1840), extralim.
    a. **freija** (Warr., 1924)
    fasciata (Warr., 1926)
    b. **wyandot** (Edw., 1863)
    c. **loki** Evans, 1953
3963 **ruralis** (Bdv., 1852)
    caespitatis (Bdv., 1852)
    ricara (Edw., 1865)
    petreius (Edw., 1870)
3964 **xanthus** Edw., 1878
    mcdunnoughi (Oberth., 1913)
    macdunnoughi dos Passos, 1964,
      emend.
3965 **scriptura** (Bdv., 1852)
3966 **communis** (Grt., 1872)
    tessellata (Scudder, 1872), preocc. by
      Hewitson, 1866
    albovittata (Grt., 1873)
    insolatrix Plötz, 1884
    varus Plötz, 1884
    skinneri (Gunder, 1927)
3967 **albescens** Plötz, 1884
    occidentalis Skin., 1906
3968 **oileus** (L., 1767)
    syrichtus (F., 1775)
    montivagus Reak., 1866
    ajutrix Plötz, 1884
    fumosa (Reverdin, 1919)
3969 **philetas** Edw., 1881

HELIOPETES Billberg, 1820
    LEUCOSCIRTES Scudder, 1872
3970 **domicella** (Erichson, 1848)
    nearchus (Edw., 1882)
    aconita (Plötz, 1884)
    willi; dos Passos, 1964, not Plötz, 1884
3971 **ericetorum** (Bdv., 1852)
    alba (Edw., 1867)

3972 **lavianus** (Hewitson, 1868)
    pastor (R. Felder, 1869)
    oceanus (Edw., 1871)
3973 **macaira** (Reak., 1866)
    nivea (Scudder, 1872)
    locutia (Hewitson, 1875)
    eulalia (Plötz, 1885)
    orbigera (Mabille, 1888)
    cnemus Godm. & Salvin, 1897
3974 **arsalte** (L., 1758)
    niveus (Cram., 1775)
    menalcus (F., 1775)
    crameri Billberg, 1820
    figara (Butler, 1870)

CELOTES Godm. & Salvin, 1899
3975 **nessus** (Edw., 1877)
    notabilis (Stkr., 1878)
    radiatus (Plötz, 1884)
3976 **limpia** Burns, 1974

PHOLISORA Scudder, 1872
    HESPEROPSIS Dyar, 1905
3977 **catullus** (F., 1793)
3978 **mejicana** (Reak., 1866)
    mexicana Godm. & Salvin, 1897,
      missp.
3979 **libya** (Scudder, 1878)
    a. **lena** (Edw., 1882)
3980 **alpheus** (Edw., 1876)
    a. **oricus** Edw., 1879
    arizonensis (Mabille & Boullet, 1917)
3981 **gracielae** MacNeill, 1970

HETEROPTERINAE

CARTEROCEPHALUS Led., 1852
    AUBERTIA Oberth., 1896
    PAMPHILIDIA Lind., 1925
    STEROPTES; auth., not Bdv., 1832
3982 **palaemon** (Pallas, 1771), extralim.
    paniscus (F., 1775)
    brontes (D. & S., 1775)
    a. **mandan** (Edw., 1863)
    mesapano (Scudder, 1868)
    skada (Edw., 1870)
    skada (Edw., 1871), preocc. by
      Edw., 1870

PIRUNA Evans, 1955
    BUTLERIA; auth., not Kby., 1871
3983 **pirus** (Edw., 1878)
    semicaeca (Mabille & Boullet, 1917)
3984 **polingii** (Barnes, 1900)
3985 **microstictus** (Godm., 1900)
3986 **haferniki** H. A. Freeman, 1970

HESPERIINAE

SYNAPTE Mabille, 1904
    GODMANIA Skin. & Ramsden,
      1923, preocc. by Horváth, 1919
    CYMAENES; auth., not Scudder,
      1872
3987 **malitiosa** (H.-S., 1865), extralim.
    a. **pecta** Evans, 1955
3988 **salenus** (Mabille, 1883)

CORTICEA Evans, 1955
3989 **corticea** (Plötz, 1883)

CALLIMORMUS Scudder, 1872
3990 **saturnus** (H.-S., 1869)
    tenera (Plötz, 1883)

VIDIUS Evans, 1955
3991 **perigenes** (Godm., 1900)

MONCA Evans, 1955
3992 **tyrtaeus** (Plötz, 1883)
    telata; auth., not H.-S., 1869

NASTRA Evans, 1955
3993 **lherminier** (Latr., 1824)
    lherminieri auth., emend.
    fusca (G. & R., 1867)
3994 **julia** (H. A. Freeman, 1945)
3995 **neamathla** (Skin. & R. C. Williams,
      1923)

CYMAENES Scudder, 1872
    MEGISTIAS Godm., 1900
3996 **tripunctus** (H.-S., 1865)
    jamaca (Schaus, 1902)
    sinepuncta (Avinoff & Shoumatoff,
      1946)
3997 **odilia** (Burmeister, 1878), extralim.
    corescene (Schaus, 1902)
    a. **trebius** (Mabille, 1891)
    isus (Godm., 1900)

LEREMA Scudder, 1872
    SAREGA Mabille, 1904
3998 **accius** (J. E. Smith, 1797)
    monoco (Scudder, 1864)
    punctella (G. & R., 1867)
    nortonii (Edw., 1867)
    parumpunctata (H.-S., 1869)
    pattenii Scudder, 1872
    phocylides (Plötz, 1883)
3999 **liris** Evans, 1955

PERICHARES Scudder, 1872
4000 **philetes** (Gmel., 1790), extralim.
    coridon (F., 1775), preocc. by Poda,
      1761
    phocion (F., 1793)
    julianus (Turton, 1806)
    trinidad (Luc., 1857)
    a. **adela** (Hewitson, 1867)
    dolores (Reak., 1868)
    marmorata Scudder, 1872

RHINTHON Godm., 1900
4001 **osca** (Plötz, 1883)
    cabella (Plötz, 1886)
    chiriquensis (Mabille, 1889)
    biserta (Schaus, 1902)

DECINEA Evans, 1955
4002 **percosius** (Godm., 1900)

CONGA Evans, 1955
4003 **chydaea** (Butler, 1877)
    valo (Mabille, 1891)
    vala (Mabille, 1891)
    actor (Mabille, 1891)
    orope; Seitz, 1925, not Plötz, 1886

ANCYLOXYPHA C. Felder, 1862
    ANCYLOXIPHA auth., missp.

4004 **numitor** (F., 1793)
   bion (F., 1798)
   puer (Hbn., 1823)
   marginatus (Harr., 1862)
   longleyi French, 1897
4005 **arene** (Edw., 1871)
   myrtis (Edw., 1882)
   euphrasia (Plötz, 1884)
   leporina (Plötz, 1884)
   isidorus (Plötz, 1884)

OARISMA Scudder, 1872
   PARADOPAEA Godm., 1900
4006 **poweshiek** (Parker, 1870)
   garita; (Plötz, 1884), not Reak., 1866
4007 **garita** (Reak., 1866)
   hylax (Edw., 1871)
   poweshiek; (Godm., 1900), not
     Parker, 1870
   lena; (Wgt., 1905), not Edw., 1882
4008 **edwardsii** (Barnes, 1897)
   garita; (Godm., 1900), not Reak., 1866

COPAEODES Speyer, 1877
4009 **aurantiacus** (Hewitson, 1868)
   waco (Edw., 1868)
   simplex (R. Felder, 1869)
   procris (Edw., 1871), ♂
   macra (Plötz, 1884)
   candida Wgt., 1890
   nanus; Watson, 1893, not H.-S., 1869
4010 **minimus** (Edw., 1870)
   procris (Edw., 1871), ♀
   singularis; (Plötz, 1884), not H.-S.,
     1865
   aurantiaca; Godm., 1900, not
     Hewitson, 1868
   rayata B. & McD., 1913

ADOPAEOIDES Godm., 1900
   NEADOPAEA Hayward, 1941
4011 **prittwitzi** (Plötz, 1884)
   simplex; auth., not R. Felder, 1869

THYMELICUS Hbn., 1819
   ADOPOEA Billberg, 1820
   THYMELINUS Steph., 1835,
     missp.
   PELION Kby., 1858
   THYMETICUS Edw., 1871. missp.
   ADOPAEA McD., 1938, missp.
4012 **lineola** (Ochs., 1808)
   virgula (Hbn., 1808–1813)
   ludoviciae (Mabille, 1883)
   pallida Tutt, 1896
NOTE: many other palearctic synonyms not
included

HYLEPHILA Billberg, 1820
   EUTHYMUS Scudder, 1872
   ANDINUS Hayward, 1940
   CORDILLANA Hayward, 1941
4013 **phyleus** (Drury, 1773)
   phylaeus auth., missp.
   phareus (Panzer, 1785)
   carin (Hbn., 1823)
   bucephalus (Steph., 1828)
   hala (Butler, 1870)
   brettus; (Holl., 1898), not Bdv. &
     Leconte, 1834

   brettoides; Wgt., 1905, not Edw., 1883
   pallida Hayward, 1944

YVRETTA Hemming, 1935
   CHAEREPHON Godm., 1900,
     preocc. by Dobson, 1878
4014 **rhesus** (Edw., 1878)
   axius (Plötz, 1883)
4015 **carus** (Edw., 1883), extralim.
   a. **subreticulata** (Plötz, 1883)
     citrus (Mabille, 1889)

PSEUDOCOPAEODES Skin. &
   R. C. Williams, 1923
4016 **eunus** (Edw., 1881)
   wrightii (Edw., 1882)

STINGA Evans, 1955
4017 **morrisoni** (Edw., 1878)

HESPERIA F., 1793
   PAMPHILA F., 1807
   SYMMACHIA Sodoffsky, 1837,
     preocc. by Hbn., 1819
   OCYTES Scudder, 1872
   ANTHOMASTER Scudder, 1872
   URBICOLA Tutt, 1905
4018 **uncas** Edw., 1863
   ridingsii Reak., 1866
   unkas Mabille, 1904, missp.
   uncus F. M. Brown, Eff & Rotger,
     1956, missp.
   a. **lasus** (Edw., 1884)
   b. **macswaini** MacNeill, 1964
4019 **juba** (Scudder, 1872)
   comma; Bdv., 1852, not L., 1758
   viridis; Wgt., 1905, not Edw., 1883
   colorado; Wgt., 1905, not Scudder, 1874
   ogdenensis (Holl., 1931)
   nevada; Garth, 1934, not Scudder,
     1874
4020 **comma** (L., 1758), extralim.
NOTE: many palearctic synonyms not included
   a. **manitoba** (Scudder, 1874)
     colorado; Dyar, 1905, not Scudder,
       1874
   b. **assiniboia** (Lyman, 1892)
   c. **laurentina** (Lyman, 1892)
   d. **borealis** Linds., 1942
   e. **harpalus** (Edw., 1881)
     cabelus (Edw., 1881)
     idaho (Edw., 1883)
     manitoba; (Wgt., 1905), not
       Scudder, 1874
     colorado; (Wgt., 1905), not
       Scudder, 1874
     oregonia; (Wgt., 1905), not Edw.,
       1883
     viridis; Comst., 1927, not Edw.,
       1883
     leussleri; Linds., 1942, part, not
       Linds., 1940
     ruricola; dos Passos, 1960, not
       Bdv., 1852
   f. **yosemite** Leussler, 1933
     colorado; Garth, 1935, not
       Scudder, 1874
     harpalus; Linds., 1940, not Edw.,
       1881
   g. **leussleri** Linds., 1940

   h. **tildeni** H. A. Freeman, 1956
   i. **dodgei** (Bell, 1927)
   j. **oregonia** (Edw., 1883)
     ruricola; (Wgt., 1905), not Bdv.,
       1852
     juba; Comst., 1927, not Scudder,
       1872
     nevada; Comst., 1927, not
       Scudder, 1874
     manitoba; Blkmre., 1927, not
       Scudder, 1874
     colorado; Blkmre., 1927, not
       Scudder, 1874
     idaho; Leighton, 1946, not Edw.,
       1883
   k. **hulbirti** Linds., 1939
   l. **ochracea** Linds., 1941
   m. **colorado** (Scudder, 1874)
   n. **susanae** L. Miller, 1962
4021 **woodgatei** (R. C. Williams, 1914)
   liberia; Evans, 1955, not Plötz, 1883
4022 **ottoe** Edw., 1866
4023 **leonardus** Harr., 1862
   stallingsi H. A. Freeman, 1943
4024 **pawnee** Dodge, 1874
   montana (Skin., 1911)
   ogallala (Leussler, 1921)
4025 **pahaska** (Leussler, 1938)
   a. **williamsi** Linds., 1940
     columbia; Evans, 1955, not
       Scudder, 1872
   b. **martini** MacNeill, 1964
4026 **columbia** (Scudder, 1872)
   sylvanoides; (Scudder, 1874), not
     Bdv., 1852
   zabulon; (Kby., 1877), not Bdv. &
     Leconte, 1834
   colorado; (Edw., 1877), not Scudder,
     1874
   california (Wgt., 1905)
   erynnioides (Dyar, 1907)
   californica; (Skin., 1920), not
     Mabille, 1883
   idaho; (Seitz, 1924), not Edw., 1883
4027 **metea** Scudder, 1864
   a. **licinus** (Edw., 1871)
     horus Edw., 1871
     belfragei H. A. Freeman, 1944
4028 **viridis** (Edw., 1883)
4029 **attalus** (Edw., 1871)
   seminole (Scudder, 1872)
   quaiapen (Scudder, 1889)
   a. **slossonae** (Skin., 1890)
4030 **meskei** (Edw., 1877)
   a. **straton** (Edw., 1881)
     stratton Bell, 1938, missp.
4031 **dacotae** (Skin., 1911)
4032 **lindseyi** (Holl., 1930)
4033 **sassacus** Harr., 1862
   a. **manitoboides** (Fletcher, 1889)
4034 **miriamae** MacNeill, 1959
4035 **nevada** (Scudder, 1874)
   colorado; (Scudder, 1874), part, not
     Scudder, 1874

POLITES Scudder, 1872
   HEDONE Scudder, 1872
   LIMOCHORES Scudder, 1872
   PYRRHOSIDIA Scudder, 1874
   PYRRHOSYDIA auth., missp.

4036 **coras** (Cram., 1775)
peckius (Kby., 1837)
wamsutta (Harr., 1862)
4037 **sabuleti** (Bdv., 1852)
genoa (Plötz, 1883)
a. **tecumseh** (Grinnell, 1903)
chispa (Wgt., 1905)
b. **chusca** (Edw., 1873)
comstocki (Gunder, 1925)
4038 **mardon** (Edw., 1881)
4039 **draco** (Edw., 1871)
4040 **baracoa** (Luc., 1857)
amadis (H.-S., 1863)
myus (French, 1885)
4041 **themistocles** (Latr., 1824)
taumas (F., 1787), preocc. by Hufn.,
1766
thaumas (F., 1793), preocc. by
Hufn., 1766
phocion (F., 1798), preocc. by F.,
1781
cernes (Bdv. & Leconte, 1834)
ahaton (Harr., 1862)
origines; auth.
4042 **origenes** (F., 1793)
manataaqua (Scudder, 1864)
cernes; (Harr., 1862), not Bdv. &
Leconte, 1834
origines Linds., Bell &
R. C. Williams, 1931, missp.
a. **rhena** (Edw., 1878)
alcina (Skin., 1893)
4043 **mystic** (Edw., 1863)
weetamoo (Scudder, 1889)
nubs (Scudder, 1889)
a. **dacotah** (Edw., 1871)
pallida (Skin., 1911)
4044 **sonora** (Scudder, 1872)
sylvanoides; (Wgt., 1905), not Bdv.,
1852
columbia; (Wgt., 1905), not Scudder,
1872
a. **siris** (Edw., 1881)
b. **utahensis** (Skin., 1911)
4045 **vibex** (Gey., 1832)
brettus (Bdv. & Leconte, 1834)
wingina (Scudder, 1863)
osyka (Edw., 1867), ♀
morganta (Plötz, 1883)
unna (Plötz, 1883)
margarita Draudt, 1924, missp.
a. **praeceps** (Scudder, 1872)
lumida (Mösch., 1878)
golenia (Mösch., 1878)
?zenckei (Plötz, 1883)
combinata (Plötz, 1883)
hypozona (Dyar, 1919)
b. **brettoides** (Edw., 1883)
clara (Plötz, 1883)
stigma (Skin., 1896)

WALLENGRENIA Berg, 1897
CATIA Godm., 1900
4046 **otho** (J. E. Smith, 1797)
pustula (Gey., 1832)
4047 **egeremet** (Scudder, 1864)
aetna (Scudder, 1872)
ursa (Worthington, 1880)
cinna (Plötz, 1883)

POMPEIUS Evans, 1955
4048 **verna** (Edw., 1862)
pottawattomie (Worthington, 1880)
vetulina (Plötz, 1883)
sigida (Mabille, 1891)
sequoyah (H. A. Freeman, 1942)

ATALOPEDES Scudder, 1872
PANSYDIA Scudder, 1872
4049 **campestris** (Bdv., 1852)
sylvanoides (Bdv., 1852), ♀
kedema (Butler, 1870)
augustus (Plötz, 1883)
amphissa (Plötz, 1883)
flaveola (Mabille, 1891)
a. **huron** (Edw., 1863)

ATRYTONE Scudder, 1872
ANATRYTONE Dyar, 1905
4050 **arogos** (Bdv. & Leconte, 1834)
vitellius; (J. E. Smith, 1797), not F.,
1793
mutius (Plötz, 1883)
a. **iowa** (Scudder, 1869)
4051 **delaware** (Edw., 1863)
logan (Edw., 1863)
jowa; (Plötz, 1884), not *iowa*
Scudder, 1869
a. **lagus** (Edw., 1881)

PROBLEMA Skin. & R. C. Williams,
1924
4052 **byssus** (Edw., 1880)
a. **kumskaka** (Scudder, 1887)
4053 **bulenta** (Bdv. & Leconte, 1834)

OCHLODES Scudder, 1872
AUGIADES; auth., not Hbn., 1819
4054 **sylvanoides** (Bdv., 1852)
agricola; (Plötz, 1883), not Bdv.,
1852
francisca (Plötz, 1883)
a. **pratincola** (Bdv., 1852), ♀
nemorum; (Wgt., 1905), not Bdv.,
1852
b. **napa** (Edw., 1865)
amanda (Plötz, 1883)
milo; (Wgt., 1905), not Edw., 1883
4055 **agricola** (Bdv., 1852)
yreka (Edw., 1866)
milo (Edw., 1883)
a. **verus** (Edw., 1881)
nemorum; (Skin., 1900), not Bdv.,
1852
pratincola; (Wgt., 1905), not Bdv.,
1852
b. **nemorum** (Bdv., 1852)
pratincola (Bdv., 1852), ♂
milo; Seitz, 1924, not Edw., 1883
4056 **snowi** (Edw., 1877)
4057 **yuma** (Edw., 1873)
scudderi (Skin., 1899)

POANES Scudder, 1872
PHYCANASSA Scudder, 1872
PARATRYTONE; Dyar, 1905, not
Godm., 1900
4058 **massasoit** (Scudder, 1864)
suffusa (Laurent, 1892)
hughi A. H. Clark, 1931

a. **chermocki** Andersen &
Simmons, 1976
4059 **hobomok** (Harr., 1862)
pocahontas (Scudder, 1864), ♀
quadaquina (Scudder, 1868)
quadraquina (Kby., 1871), missp.
zabulon; (Scudder, 1889), not (Bdv.
& Leconte, 1834)
freidlei Watson, 1920
pallida Watson, 1921
alfaratta Holl., 1930, ♀
ridingsii F. H. & R. L. Chermock,
1940
4060 **zabulon** (Bdv. & Leconte, 1834)
erratica (Plötz, 1883)
ogeechensis (Scudder, 1889), nom.
nud.?
pocahontas; (Holl., 1898), not
Scudder, 1864
4061 **taxiles** (Edw., 1881)
4062 **aaroni** (Skin., 1890)
a. **howardi** (Skin., 1896)
4063 **yehl** (Skin., 1893)
4064 **viator** (Edw., 1865)
a. **zizaniae** Shapiro, 1971

PARATRYTONE Godm., 1900
4065 **melane** (Edw., 1869)
a. **vitellina** (H.-S., 1869)

CHORANTHUS Scudder, 1872
4066 **radians** (Luc., 1857)
magica (Plötz, 1883)
streckeri (Skin., 1893)
4067 **haitensis** (Skin., 1920)

MELLANA Hayward, 1948
4068 **eulogius** (Plötz, 1883)
mellona (Godm., 1900)
heberia (Dyar, 1914)
4069 **mexicana** (Bell, 1942), quest. occur.

EUPHYES Scudder, 1872
AROTIS Mabille, 1904
PERENEIA Linds., 1925
4070 **arpa** (Bdv. & Leconte, 1834)
4071 **pilatka** (Edw., 1867)
palatka (Edw., 1867), emend.?
floridensis (Plötz, 1883)
4072 **dion** (Edw., 1879)
palatka; (Scudder, 1889), not Edw.,
1867
4073 **alabamae** (Linds., 1923)
4074 **dukesi** (Linds., 1923)
4075 **conspicuus** (Edw., 1863)
pontiac (Edw., 1863)
orono (Scudder, 1872)
a. **buchholzi** (Ehrlich & Gillham,
1951)
4076 **berryi** (Bell, 1941)
4076.1 **macguirei** H. A. Freeman, 1975
4077 **bimacula** (G. & R., 1867)
acanootus (Scudder, 1868)
illinois (Dodge, 1872)
contradicta (Leussler, 1933)
4078 **ruricola** (Bdv., 1852)
vestris (Bdv., 1852)
osceola (Lint., 1878)
californica (Mabille, 1883)
a. **metacomet** (Harr., 1862)

rurea (Edw., 1862)
kiowah (Reak., 1866)
osyka (Edw., 1867), ♂
baeis (Scudder, 1889), nom. nud.?
immaculatus (R. C. Williams, 1914)

ASBOLIS Mabille, 1904
4079 **capucinus** (Luc., 1857)
sandarac (H.-S., 1865)
palaea (Hewitson, 1868)

ATRYTONOPSIS Godm., 1900
4080 **hianna** (Scudder, 1868)
grotei (Plötz, 1883)
a. **turneri** H. A. Freeman, 1948
4081 **deva** (Edw., 1876)
4082 **lunus** (Edw., 1884)
4083 **vierecki** (Skin., 1902)
4084 **loammi** (Whitney, 1876)
regulus (Edw., 1881)
apostologica (Strand, 1921)
4085 **pittacus** (Edw., 1882)
4086 **python** (Edw., 1882)
4087 **cestus** (Edw., 1884)
a. **margarita** (Skin., 1913)
4088 **ovinia** (Hewitson, 1866), extralim.
a. **edwardsi** B. & McD., 1916
zaovinia; auth., not Dyar, 1913
polingi Gunder, 1925

AMBLYSCIRTES Scudder, 1872
STOMYLES Scudder, 1872
MASTOR Godm., 1900
EPIPHYES Dyar, 1905
EPHIPHYES Evans, 1955, missp.
4089 **simius** Edw., 1881
4090 **exoteria** (H.-S., 1869)
nanno Edw., 1882
marcus (Strand, 1909)
4091 **cassus** Edw., 1883
simius; Wgt., 1905, not Edw., 1881
4092 **aenus** Edw., 1878
4093 **linda** H. A. Freeman, 1943
4094 **oslari** (Skin., 1899)
4095 **erna** H. A. Freeman, 1943
4096 **hegon** (Scudder, 1864)
samoset (Scudder, 1864)
nemoris (Edw., 1864)
argina (Plötz, 1884)
4097 **texanae** Bell, 1927
4098 **prenda** Evans, 1955
tolteca; auth., not Scudder, 1872
4099 **aesculapius** (F., 1793)
textor (Hbn., 1827–31)
oneko (Scudder, 1864)
wakulla (Edw., 1869)
oneka McD., 1938, missp.
4100 **carolina** (Skin., 1892)
4101 **reversa** F. M. Jones, 1926
4102 **nereus** (Edw., 1876)
4103 **nysa** Edw., 1877
similis (Stkr., 1878)
4104 **eos** (Edw., 1871)
comus (Edw., 1876)
nilus Edw., 1878
quinquemacula (Skin., 1911)
quinquimacula Bell, 1938, missp.
4105 **vialis** (Edw., 1862)
asella (H.-S., 1869)
4106 **celia** Skin., 1895

4107 **belli** H. A. Freeman, 1941
4108 **alternata** (G. & R., 1867)
meridionalis Dyar, 1905
4109 **phylace** (Edw., 1878)
4110 **fimbriata** (Plötz, 1882)
bellus (Edw., 1884)

LERODEA Scudder, 1872
4111 **eufala** (Edw., 1869)
osyka (Edw., 1867), ♀
dispersa (H.-S., 1869)
floridae (Mabille, 1876)
micylla (Burmeister, 1878)
obscura (Mabille, 1904)
4112 **arabus** (Edw., 1882)
4113 **dysaules** Godm., 1900

OLIGORIA Scudder, 1872
4114 **maculata** (Edw., 1865)
grossula (H.-S., 1869)
deleta (H.-S., 1869)
norus (Plötz, 1883)
orthomenes (Scudder, 1889), nom.
nud.?

CALPODES Hbn., 1819
4115 **ethlius** (Stoll, 1782)
chemnis (F., 1793)
olynthus (Bdv. & Leconte, 1834)
ethlinus auth., 1960, missp.

PANOQUINA Hemming, 1934
PRENES Scudder, 1872, preocc. by
Gistl, 1848
4116 **panoquin** (Scudder, 1864)
ophis (Edw., 1871)
wimico (Plötz, 1883)
cochles (Scudder, 1889), nom. nud.?
4117 **panoquinoides** (Skin., 1891)
4118 **errans** (Skin., 1892)
nereus; (Wgt., 1905), not Edw., 1876
4119 **ocola** (Edw., 1863)
stratyllis (Burmeister, 1878)
heterospila (Mabille, 1878)
ortygia (Mösch., 1882)
4120 **hecebola** (Scudder, 1872)
parilis (Mabille, 1891)
4121 **sylvicola** (H.-S., 1865)
nero; auth., not F., 1798
4122 **evansi** (H. A. Freeman, 1946)

NYCTELIUS Hayward, 1948
4123 **nyctelius** (Latr., 1824)
ares (C. Felder, 1862)
coscinia (H.-S., 1865)
aegialea (Plötz, 1883)

THESPIEUS Godm., 1900
4124 **macareus** (H.-S., 1869)
emacareus (Plötz, 1882)

MEGATHYMINAE

AGATHYMUS H. A. Freeman, 1959
4125 **neumoegeni** (Edw., 1882)
4126 **carlsbadensis** (D. Stallings & Turner,
1957)
4127 **florenceae** (D. Stallings & Turner,
1957)
4128 **judithae** (D. Stallings & Turner, 1957)

4129 **diabloensis** H. A. Freeman, 1962
4130 **mcalpinei** (H. A. Freeman, 1955)
macalpinei dos Passos, 1964, emend.
4131 **chisosensis** (H. A. Freeman, 1952)
4132 **aryxna** (Dyar, 1905)
drucei (Skin., 1911)
4133 **baueri** (D. Stallings & Turner, 1954)
4134 **freemani** D. Stallings & Turner, 1960
4135 **evansi** (H. A. Freeman, 1950)
4136 **mariae** (B. & Benj., 1924)
4137 **chinatiensis** H. A. Freeman, 1964
4138 **lajitaensis** H. A. Freeman, 1964
4139 **rindgei** H. A. Freeman, 1964
4140 **gilberti** H. A. Freeman, 1964
4141 **valverdiensis** H. A. Freeman, 1966
4142 **stephensi** (Skin., 1912)
4143 **polingi** (Skin., 1905)
4144 **alliae** (D. Stallings & Turner, 1957)

MEGATHYMUS Scudder, 1872
4145 **yuccae** (Bdv. & Leconte, 1834)
alabamae H. A. Freeman, 1943
a. **buchholzi** H. A. Freeman, 1952
4146 **coloradensis** Riley, 1877
a. **elidaensis** D. Stallings, Turner &
J. Stallings, 1966
b. **navajo** Skin., 1911
c. **browni** D. Stallings & Turner,
1960
d. **stallingsi** H. A. Freeman, 1943
dee H. A. Freeman, 1943, ♀
e. **reinthali** H. A. Freeman, 1963
f. **martini** D. Stallings & Turner,
1956
g. **maudae** D. Stallings, Turner &
J. Stallings, 1966
h. **arizonae** Tinkham, 1954
i. **albasuffusus** R., J. & D. Wielgus,
1974
j. **reubeni** D. Stallings, Turner &
J. Stallings, 1963
k. **winkensis** H. A. Freeman, 1965
l. **wilsonorum** D. Stallings &
Turner, 1958
m. **louiseae** H. A. Freeman, 1963
n. **kendalli** H. A. Freeman, 1965
4147 **cofaqui** (Stkr., 1876)
4148 **harrisi** H. A. Freeman, 1955
4149 **streckeri** (Skin., 1895)
4150 **texanus** B. & McD., 1912
albocincta Holl., 1930
a. **leussleri** Holl., 1931
4151 **ursus** Poling, 1902
a. **violae** D. Stallings & Turner, 1956
b. **deserti** R., J. & D. Wielgus, 1972

STALLINGSIA H. A. Freeman, 1959
4152 **maculosa** (H. A. Freeman, 1955)
smithi; auth., not Druce, 1896

Papilionoidea

PAPILIONIDAE

by LEE D. MILLER & F. MARTIN BROWN

PARNASSIINAE

PARNASSIUS Latr., 1804
 DORITIS F., 1807
 PARNASSIS Hbn., 1819, missp.?
 THERIUS Billberg, 1820
 TADUMIA Moore, 1902
 KAILASIUS Moore, 1902
 KORAMIUS Moore, 1902
 LINGAMIUS Bryk, 1935
 EUKORAMIUS Bryk, 1935
4153 eversmanni Mén., 1849, extralim.
 versmanni Bryk & Eisner, 1937,
  missp.
 a. thor Hy. Edw., 1881
  kohlsaati Gunder, 1932
  ochreoocellatus Bryk & Eisner,
   1932
  quincunx; Bryk & Eisner, 1934,
   not Bryk, 1914
  desubmarginatus Bryk, 1935
4154 clodius Mén., 1855
 castus Bryk, 1913
 sulfureus Gunder, 1932
 extinctoanalis Bryk & Eisner, 1932
 a. strohbeeni Sternitzky, 1945
  dodgei Gunder, 1932, ab.
 b. sol Bryk & Eisner, 1932
 c. baldur Edw., 1877
  lorquini Oberth., 1891
  lusca Stichel, 1907
  binigrimaculellus Gunder, 1926
  primoettertiopictaetornatus Bryk &
   Eisner, 1932
  medionigroocellatus; Bryk, 1935,
   not Bryk & Eisner, 1932
 d. claudianus Stichel, 1907
  baldus Ehrmann, 1918
  kallias Ehrmann, 1918
 e. pseudogallatinus Bryk, 1913
  hel Eisner, 1956
 f. incredibilis Bryk, 1932
 g. altaurus Dyar, 1903
  flavoocellatus Bryk, 1935
  shepardi Eisner, 1969
 h. gallatinus Stichel, 1907
  immaculatus Skin., 1911
 i. menetriesii Hy. Edw., 1877
  medionigroocellatus Bryk & Eisner,
   1932
  nigroanalis Bryk & Eisner, 1932
  albocentratus Bryk & Eisner, 1932
4155 phoebus (F., 1793), extralim.
NOTE: many palearctic synonyms not
 included
 a. behrii Edw., 1870
  niger Wgt., 1905
  astriotes Fruhstorfer, 1923
 b. sternitzkyi McD., 1936
 c. magnus Wgt., 1905
 d. olympianus Burdick, 1941
  guppyi Wyatt, 1969
 e. smintheus Doubleday, 1847
  sedakovii Mén., 1849

rocky Grum-Grshimaïlo, 1890
nanus Neum., 1890
minor Verity, 1907, preocc. by
 Stgr., 1881
minusculus Bryk, 1912
quincunx Bryk, 1914
verity Ehrmann, 1918, repl. name
ernestinae Bryk & Eisner, 1935
manitobaensis Bryk & Eisner, 1935
f. xanthus Ehrmann, 1918
 mendicus Stichel, 1907, ab.
 pseudocorybas Verity, 1907, ab.
 idahoensis Bryk & Eisner, 1931
 montanulus Bryk & Eisner, 1935
 maximus Bryk & Eisner, 1937
g. sayii Edw., 1863
 hermodur Hy. Edw., 1881
 nigerrimus Verity, 1907
 mariae Bryk, 1912
 quincunx; Bryk, 1915, not Bryk,
  1914
 polus Ehrmann, 1917
 montanus Ehrmann, 1918
 utahensis Roths., 1918
 fermatus Bryk, 1921
 melanophorus Bryk, 1921
 sordellus Fruhstorfer, 1923
 aristion Fruhstorfer, 1923
 catullius Fruhstorfer, 1923
 pholus B. & Benj., 1926, missp.
 dakotaensis Bryk & Eisner, 1935
 hollandi Bryk & Eisner, 1935
 rotgeri B.-H., 1938
 reducta B.-H., 1938
 discocircumcintus Eisner, 1955
 hermador F. M. Brown, Eff &
  Rotger, 1956, missp.
 rubiana Wyatt, 1961
 excelsior Eisner, 1969
h. pseudorotgeri Eisner, 1966
i. apricatus Stichel, 1906
j. golovinus Holl., 1930
k. alaskensis Eisner, 1956
 alaskaensis Eisner, 1957, emend.
l. elias Bryk, 1934
m. yukonensis Eisner, 1969

PAPILIONINAE

Troidini

PARIDES Hbn., 1819
 HECTORIDES Hbn., 1821
 ENDOPOGON Lacordaire, 1833
 ASCANIDES Gey., 1837
 BLAKEA Grt., 1875
4156 eurimedes (Stoll, 1780) extralim.
 arcas (Cram., 1777), preocc. by
  Drury, 1773
 a. mylotes (Bates, 1861)
  caleli (Reak., 1863)
  tonila (Reak., 1863)
  alcamedes (C. & R. Felder, 1865)
  aristomenes (C. & R. Felder, 1865)

BATTUS Scop., 1777
 LAERTIAS Hbn., 1819
 ITHOBALUS Hbn., 1819
 LAERTIADES Doubleday, 1846,
  missp.

4157 philenor (L., 1771)
 astinous (Drury, 1773)
 obsoleta (Ehrmann, 1900)
 wasmuthi (Weeks, 1901)
 a. hirsuta (Skin., 1908)
  inghami (Gunder, 1927)
  acauda; auth., not Oberth., 1879
  nezahualcoytl; auth., not Stkr.,
   1885
  corbis; auth., not Godm. & Salvin,
   1890
  orsua; auth., not Godm. & Salvin,
   1890
4158 polydamas (L., 1758)
 polydamus (McD., 1938), missp.
 a. lucayus (R. & J., 1906)

Papilionini

PAPILIO L., 1758
 PTEROURUS Scop., 1777
 PRINCEPS Hbn., 1807
 AMARYSSUS Dalm., 1816
 JASONIADES Hbn., 1819
 EUPHOEADES Hbn., 1819
 HERACLIDES Hbn., 1819
 ACHILLIDES Hbn., 1819
 ORPHEIDES Hbn., 1819, suppr.
  (ICZN Op. 179) to Princeps Hbn.,
  1807
 NESTORIDES Hbn., 1819
 CALAIDES Hbn., 1819
 PRIAMIDES Hbn., 1819
 ILIADES Hbn., 1819
 TROILIDES Hbn., 1825
 THOAS Swainson, 1833
 AERNAUTA Berge, 1842
 PYRRHOSTICTA Butler, 1872
 HARIMALA Moore, 1881
 CHARUS Moore, 1881
 SARBARIA Moore, 1882
 SAINIA Moore, 1882
 CHARES Swinhoe, 1885, missp.
 PANOSMIA Wood-Mason & de
  Nicéville, 1886
 TAMERA MOORE, 1888
 ACHIVUS Kby., 1896
 EQUES Kby., 1896
 PAENASMIA Kby., 1896, missp.
 TROS Kby., 1896
 PRIAMEDES Grt., 1899
 SADENGIA Moore, 1902
 HETEROCREON Kby., 1904
 CALIADES dos Passos, 1964, missp.
4159 polyxenes F., 1775, extralim.
 asterias; F., 1787, missp.
 asterioides; Eimer, 1895, not Reak.,
  1866
 a. asterius Stoll, 1782
  ajax L., 1758, part, rejected name
   (ICZN Op. 286)
  troilus; Drury, 1773, not. L., 1758
  ampliata Mén., 1857
  calverleyi Gr., 1864
  asterioides Reak., 1866
  viridis Ckll., 1889
  alunatus Skin. & Aaron, 1889
  mediocanda Eimer, 1895, incorr.
   orig. spell.
  mediocauda Eimer, 1895, emend.

astyanax; Scudder, 1898, not F., 1793
semialba Ehrmann, 1900
curvifascia Skin., 1902
ehrmanni Ehrmann, 1925
streckeri Holl., 1927
forsythae; Wood, 1937, not Gunder, 1933
pseudoamericus F. M. Brown, 1942
subampliata Dufrane, 1946
stabilis; dos Passos, 1964, not R. & J., 1906
4160 **joanae** J. R. Heitzman, 1974
4161 **rudkini** Comst., 1935
clarki F. & R. Chermock, 1937
comstocki F. & R. Chermock, 1937
4162 **kahli** F. & R. Chermock, 1937
4163 **brevicauda** Saund., 1869
asterius; Gosse, 1840, not Stoll, 1782
anticostiensis Stkr., 1873
medicandus Bryk, 1930
a. **gaspeensis** McD., 1934
b. **bretonensis** McD., 1939
4164 **bairdii** Edw., 1866
utahensis Stkr., 1878
hollandii Edw., 1892
brucei Edw., 1895
4165 **oregonius** Edw., 1876
machaon; Hagen, 1882, not L., 1758
a. **dodi** McD., 1939
4166 **machaon** L., 1758, extralim.
NOTE: many palearctic synonyms not included
a. **aliaska** Scudder, 1869
joannisi Verity, 1905
petersii A. H. Clark, 1932
b. **hudsonianus** A. H. Clark, 1932
avinoffi F. & R. Chermock, 1937
4167 **zelicaon** Luc., 1852
zolicaon Bdv., 1852, missp.
californica Mén., 1855
zelicayn Skin., 1898, missp.
coloro Wgt., 1905
impunctata Fischer, 1908
formosus Fischer, 1908
melanotaenia Fischer, 1908
mcdunnoughi Gunder, 1928
macdunnoughi dos Passos, 1964, emend.
a. **nitra** Edw., 1883
gothica Remington, 1968
4168 **indra** Reak., 1866
a. **minori** Cross, 1936
b. **pergamus** Hy. Edw., 1875
c. **kaibabensis** Bauer, 1955
d. **fordi** Comst. & Martin, 1956
e. **martini** T. & J. Emmel, 1966
f. **nevadensis** T. & J. Emmel, 1971
4169 **thoas** L., 1771, extralim.
a. **autocles** R. & J., 1906
nealces R. & J., 1906
b. **oviedo** Gundlach, 1866, quest. occur.
epithoas Oberth., 1897
4170 **cresphontes** Cram., 1777
oxilus (Hbn., 1819)
chresphontes Dury, 1878, missp.
maxwelli Franck, 1919
forsythae Gunder, 1933
a. **pennsylvanicus** F. & R. Chermock, 1945

4171 **aristodemus** Esper, 1794, extralim.
cresphontinus Martyn, 1797
daphnis G. R. Gray, 1852
a. **ponceanus** Schaus, 1911
4172 **andraemon** (Hbn., 1823), extralim.
a. **bonhotei** E. M. Sharpe, 1900
4173 **astyalus** God., 1819, extralim.
lycophron (Hbn., 1823)
mentor Dalm., 1823
pirithous Bdv., 1836
oebalus Bdv., 1836
a. **pallas** G. R. Gray, 1852
oebalus; G. R. Gray, 1852, not Bdv., 1836
4174 **ornython** Bdv., 1836
4175 **androgeus** Cram., 1775, extralim.
polycaon Cram., 1779
piranthus Cram., 1779
?amosis Stoll, 1787
androgeos Kby., 1871, missp.
a. **epidaurus** Godm. & Salvin, 1890
polycaon G. R. Gray, 1852, var.
androgeos; Stgr., 1884, part
4176 **glaucus** L., 1758
ajax L., 1758, part, rejected name (ICZN Op. 286)
antilochus L., 1758
turnus L., 1771
alcidamas Cram., 1775
fletcheri Kemp, 1900
dietzi Gunder, 1927
gerhardi Gunder, 1927
ehrmanni; McD., 1938, not Ehrmann, 1925
a. **canadensis** R. & J., 1906
arcticus Skin., 1906
deficiens Dufrane, 1946
b. **australis** Maynard, 1891
4177 **rutulus** L., 1852
nitulus Ckll., 1893, missp.
hospitonina LeCerf, 1912
fannyae Gunder, 1927
a. **ammoni** Behrens, 1887
b. **arizonensis** Edw., 1883
4178 **multicaudatus** Kby., 1884
daunus Bdv., 1836, preocc. by Cram., 1777
ragani Barnes, 1928
4179 **eurymedon** Luc., 1852
albanus C. & R. Felder, 1864
lewisii Kby., 1884
cocklei Gunder, 1925
columbiana Gunder, 1927
4180 **pilumnus** Bdv., 1836
4181 **troilus** L., 1758
radiatus Stkr., 1900
flavus Dufrane, 1946
obliteratus Dufrane, 1946
berioi Dufrane, 1946
addenda Dufrane, 1946
a. **ilioneus** J. E. Smith, 1797
texanus Ehrmann, 1900
4182 **palamedes** Drury, 1773
chalcas F., 1775
flavomaculatus Goeze, 1779
chalcus F., 1781, missp.
4183 **anchisiades** Esp., 1788, extralim.
anchises; Stoll, 1780, not L., 1758
hipponous Hbn., 1819
archelaus Godt., 1819

theramenes C. & R. Felder, 1861
isidorus; Bates, 1861, not Doubleday, 1846
pompeius; Kby., 1871, not. F., 1781
a. **idaeus** F., 1793
pandion C. & R. Felder, 1865

EURYTIDES Hbn., 1821
PROTESILAUS Swainson, 1832
COSMODESMUS Haase, 1891
GRAPHIUM; auth., not Scop., 1777
4184 **marcellus** (Cram., 1777)
ajax (L., 1758), part, rejected name (ICZN Op. 286)
telamonides (C. & R. Felder, 1864)
walshii (Edw., 1872)
abbotii (Edw., 1872)
floridensis (Holl., 1898)
lecontei (R. & J., 1906)
broweri (Gunder, 1927)
carolinianus (Holl., 1931) (ICZN Op. 286)
pricei (Field, 1936)
nigrosuffusa (Field, 1936)
4185 **philolaus** (Bdv., 1836)
xanticles; Rogenhofer, 1888, not Bates, 1863
ajax; Eimer, 1889, not L., 1758
nigrescens (Eimer, 1889)
niger (Eimer, 1889)
felicis (Fruhstorfer, 1904)

PIERIDAE

by LEE D. MILLER & F. MARTIN BROWN

DISMORPHIINAE

ENANTIA Hbn., 1819
LICINIA Swainson, 1820
4186 **melite** (Johansson, 1763)

PIERINAE

Pierini

NEOPHASIA Behr, 1869
4187 **menapia** (C. & R. Felder, 1859)
tau (Scudder, 1861)
ninonia (Bdv., 1869)
suffusa (Stretch, 1882), ♀
nigracosta Comst., 1918
4188 **terlootii** Behr, 1869
epyaxa Stkr., 1900
princetonia Poling, 1900

CATASTICTA Butler, 1870
EUTERPE; auth., not Swainson, 1831
4189 **nimbice** (Bdv., 1836)

APPIAS Hbn., 1819
CATOPHAGA Hbn., 1819
HIPOSCRITA Gey., 1832
TRIGONIA Gey., 1837, preocc. by Bruguière, 1789
TACHYRIS Wallace, 1867
GLUTOPHRISSA Butler, 1887
LADE de Nicéville, 1898
APPIUS F. M. Brown, 1942, missp.

4190 **drusilla** (Cram., 1777), extralim.
   a. **poeyi** (Butler, 1872)
   b. **neumoegenii** (Skin., 1894)
      hollandi Röber, 1909

PONTIA F., 1807
   MANCIPIUM Hbn., 1807, rejected
      (Name 214, ICZN Op. 137)
   SYNCHLOE Hbn., 1818
   PARAPIERIS de Nicéville, 1897
   LEUCOCHLOE Röber, 1907
   PIERIS; auth., part, not Schr., 1801
4191 **beckerii** (Edw., 1871)
   pseudochloridice (McD., 1928)
   gunderi (Ingham, 1933)
4192 **sisymbrii** (Bdv., 1852)
   flava; Edw., 1883, not Edw., 1881
   transversa (B. & Benj., 1926)
   a. **flavitincta** (Comst., 1924)
   b. **elivata** (B. & Benj., 1926)
4193 **protodice** (Bdv. & Leconte, 1829)
   vernalis (Edw., 1864)
   nasturtii (Edw., 1864)
   protodin (Edw., 1872), missp.
4194 **occidentalis** (Reak., 1866)
   calyce (Edw., 1870)
   a. **nelsoni** (Edw., 1883)

ARTOGEIA Verity, 1947
   PIERIS; auth., part, not Schr., 1801
4195 **napi** (L., 1761), extralim.
   napae (Edw., 1881), missp.
NOTE: many palearctic synonyms not included
   a. **pseudobryoniae** (Verity, 1908)
      arctica (Verity, 1911)
   b. **hulda** (Edw., 1869)
   c. **frigida** (Scudder, 1861)
      borealis (Grt., 1873)
      acadica (Edw., 1881), sum. f.
      pseudoleracea (Verity, 1908)
   d. **oleracea** (Harr., 1829)
      hyemalis (Harr., 1829)
      cruciferarum (Bdv., 1836), sum. f.
      casta (Kby., 1837)
      aestiva (Harr., 1850) =
         cruciferarum
   e. **venosa** (Scudder, 1861)
      castoria (Reak., 1866), sum. f.
      nasturtii; Bdv., 1869, not Edw.,
         1864 (=*castoria*)
      iberidis (Bdv., 1869)
      resedae (Bdv., 1869)
      flava (Edw., 1881)
      cottlei (Gunder, 1925)
      microstriata (Comst., 1924)
   f. **marginalis** (Scudder, 1861)
      pallida (Scudder, 1861)
   g. **mcdunnoughi** (Remington,
         1954)
      pseudonapi (B. & McD., 1916),
         preocc. by Verity, 1911
      macdunnoughi (dos Passos, 1964),
         emend.
   h. **pallidissima** (B. & McD., 1916)
   i. **mogollon** (Burdick, 1942)
4196 **virginiensis** (Edw., 1870)
4197 **rapae** (L., 1758)
NOTE: many palearctic synonyms not included
   yreka (Reak., 1866)
   novangliae (Scudder, 1872)

immaculata (Skin. & Aaron, 1889),
   preocc. by de Selys-Longchamps,
   1857

ASCIA Scop., 1777
4198 **monuste** (L., 1764), extralim.
   feronia (Steph., 1828)
   cleomes (Bdv. & Leconte, 1829)
   crameri Holl., 1931
   a. **phileta** (F., 1775)
   b. **raza** Klots, 1930, quest. occur.

GANYRA Billberg., 1820
4199 **josephina** (Godt., 1819), extralim.
   amaryllis (F., 1793), preocc. by Stoll,
      1782
   a. **josepha** (Salvin & Godm., 1868)

ANTHOCHARINAE

EUCHLOE Hbn., 1819
   PHYLLOCHARIS Schatz, 1886
4200 **ausonides** (Luc., 1852)
   ausonoides Edw., 1872, missp.
   ausoneides Hy. Edw., 1880, missp.
   aussonides Elrod & Maley, 1906,
      missp.
   flavidalis Comst., 1924
   semiflava Comst., 1924
   boharti Doudoroff, 1930
   hemiflava Field, 1936
   a. **coloradensis** (Hy. Edw., 1881)
   montana Verity, 1908
   b. **palaeoreios** K. Johnson, 1977
   c. **mayi** F. & R. Chermock, 1940
4201 **creusa** (Doubleday, 1847)
   crensa (Edw., 1881), missp.
   elsa Beutenmüller, 1898
   orientalides Verity, 1908, part
4202 **olympia** (Edw., 1871)
   a. **rosa** (Edw., 1871)
4203 **hyantis** (Edw., 1871)
   creusa; auth., part, not Doubleday,
      1847, part
   pseudoausonides Verity, 1908
   orientalides Verity, 1908, part
   pumilio Strand, 1914
   a. **andrewsi** Martin, 1936
   b. **lotta** (Beutenmüller, 1898)
      belioides (Verity, 1911)
      pumilio; Strand, 1927, not Strand,
         1914

ANTHOCHARIS Bdv., Rambur &
   Graslin, 1833
   TETRACHARIS Grt., 1898
   ANTHOCHARIS; auth., not Bdv.,
      Rambur & Graslin, 1833
   ANTHOCHRIS Cook, 1948, missp.
4204 **cethura** (C. & R. Felder, 1865)
   cooperii Behr., 1869
   angelina Bdv., 1869
   morrisoni Edw., 1881
   caliente Wgt., 1905
   deserti Wgt., 1905
   catalina Meadows, 1937
4205 **pima** Edw., 1888
4206 **sara** Luc., 1852
   reakirtii Edw., sum. f.
   flora Wgt., 1905

mollis Wgt., 1905
wrighti Comst., 1924
sternitzkyi Gunder, 1925
dammersi Comst., 1929
corcorani Gunder, 1931
broweri Gunder, 1932
flavicoloris Gunder, 1933
gunderi Ingham, 1933
   a. **stella** Edw., 1879
   b. **julia** Edw., 1872
      thoosa Scudder, 1878
      sulfuris Gunder, 1931
   c. **alaskensis** Gunder, 1932
   d. **browningi** Skin., 1905
   e. **inghami** Gunder, 1932
      duncani Gunder, 1932

FALCAPICA Klots, 1930
   MIDEA; H.-S., 1867, part, not
      Bruzelius, 1854
   ANTHOCARIS Bdv., Rambur &
      Graslin, 1833, part
4207 **midea** (Hbn., 1809)
   genutia; F., 1793, not Cram., 1780
   lherminieri (Godt., 1819)
   flavida (Skin., 1917)
   medea (Leussler, 1938), missp.
   a. **annickae** dos Passos & Klots,
      1969
4208 **lanceolata** (Luc., 1852)
   edwardsii (Behr, 1869)
   australis (Grinnell, 1908)

COLIADINAE

COLIAS F., 1807
   EURYMUS Horsfield, 1829
   SCALIDONEURA Butler, 1871
   ERIOCOLIAS Watson, 1895
   COLIASTES Hemming, 1931
4209 **philodice** Godt., 1819
   europome; Haw., 1803, not Esp.,
      1778
   dorippe Godt., 1819
   anthyale Hbn., 1823, spr. f.
   europome; Steph., 1828, not Esp.,
      1778
   santes Fitch, 1854
   philodin Edw., 1872, missp.
   eriphyle Edw., 1876, spr. f., fall f.
      (subspecies?)
   nig Stkr., 1878
   alba Stkr., 1878, not Stkr., 1878
   virida Stkr., 1878
   hagenii Edw., 1884
   maria Edw., 1884
   autumnalis Ckll., 1888, fall f. (this
      species?)
   suffusa Ckll., 1889
   misidice (Scudder, 1889), ♀
   pallidice (Scudder, 1889)
   nigridice (Scudder, 1889)
   albinic Skin., 1898
   melanic Skin., 1898
   nigrina Stkr., 1900
   alba; Röber, 1910, not Stkr., 1878
   rothkei Reiff, 1917
   nigrofasciata Reiff, 1917
   nigrofasciata; Reiff, 1918, not Reiff,
      1917

inversata Nakahara, 1925
plicaduta Nakahara, 1925
minor F. Chermock, 1927
ehrmanni F. Chermock, 1927
alba; F. Chermock, 1927, not Stkr., 1878
albida F. Chermock, 1928
serrata F. Chermock, 1929
notatus A. & L. Clark, 1941
reducta Dufrane, 1947
a. *vitabunda* Hovanitz, 1943
4210 *eurytheme* Bdv., 1852
amphidusa Bdv., 1852
californiana Mén., 1855
amphidona Edw., 1869, missp.
keewaydin Edw., 1869, sum. f.
ariadne Edw., 1870
enegthenu Edw., 1872, missp.
alba; Stkr., 1878, not Stkr., 1878
flava Stkr., 1878
hybrida Stkr., 1878, ?hybrid with *philodice*
pallida Ckll., 1888
typica Ckll., 1888
intermedia Ckll., 1888
autumnalis Ckll., 1888, fall f. (this species?)
luteitincta Wolcott, 1893
fumosa Stkr., 1900
kootenai Cockle, 1910, altitudinal f.
unictrina (Gunder, 1924)
laurae (F. Chermock, 1929)
nigricosta (F. Chermock, 1929)
rudkini (Gunder, 1932)
4211 *alexandra* Edw., 1863
alba Stkr., 1878
pallida; Ckll., 1889, not Ckll., 1888
a. *edwardsii* Edw., 1870
emilia Edw., 1870
hatui (B. & Benj., 1926)
b. *harfordii* Hy. Edw., 1877
barbara Hy. Edw., 1877
weaverae (Gunder, 1924)
martini (Gunder, 1931)
hartfordii auth., missp.
c. *columbiensis* Ferris, 1973
d. *astraea* Edw., 1872
e. *christina* Edw., 1863
pallida; Ckll., 1889, not Ckll., 1888
f. *krauthii* Klots, 1935
4212 *occidentalis* Scudder, 1862
a. *chrysomelaena* Hy. Edw., 1877
chryomelas Hy. Edw., 1877, missp.
shastae (B. & Benj., 1926)
4213 *meadii* Edw., 1871
medi (Gunder, 1934), ♀
a. *elis* Stkr., 1885
alberta Bowman, 1942, unavail., hybr. name
pallidissima Bowman, 1942, unavail., hybr. name, ♀
lambillioni Dufrane, 1947, ♀
4214 *hecla* Lefebre, 1836
pallida (Skin., 1892)
groenlandica Lampa, 1893
palamedes Hemming, 1934
a. *hela* Stkr., 1880
chrysothemoides Verity, 1911
b. *glacialis* M'Lachlan, 1878

4215 *boothii* Curt., 1835
chione Curt., 1835
4216 *nastes* Bdv., 1832
standfussi Röber, 1910, hybrid?
a. *rossii* Gn., 1864
gueneei Avinoff, 1935
b. *moina* Stkr., 1880
harperi (Gunder, 1932)
c. *streckeri* Grum-Grshimaïlo, 1895
obscurata Verity, 1911
palliflava McD., 1927
d. *cocandicides* Verity, 1911
subarctica McD., 1928
e. *aliaska* Bang-Haas, 1927
alaskae McD., 1938, missp.
4217 *thula* Hovanitz, 1955
4218 *palaeno* (L., 1761), extralim.
NOTE: many palearctic synonyms not included
a. *chippewa* Edw., 1870
helena Edw., 1863, preocc. by H.-S., 1843
kohlsaati (Gunder, 1931)
4219 *behrii* Edw., 1866
canescens (Comst., 1925), ♀
4220 *interior* Scudder, 1862
solivaga Edw., 1877
interier Kby., 1877, missp.
nepi (B. & Benj., 1926), ♀
a. *vividior* Berger, 1945
b. *laurentina* Scudder, 1876
raritus Gunder, 1928
4221 *pelidne* Bdv. & Leconte, 1829
labradorensis Scudder, 1862
anthyale; Mösch., 1870, not Hbn., 1823
moeschleri Grum-Grshimaïlo, 1893
mira Verity, 1911
a. *minisini* Bean, 1895
menisme Verity, 1908, missp.
isni (B. & Benj., 1926)
b. *skinneri* Barnes, 1897
neri (B. & Benj., 1926), ♀
4222 *gigantea* Stkr., 1900
pelidneides Stgr., 1901
a. *mayi* F. & R. Chermock, 1940
marjorie F. & R. Chermock, 1940, ♀
b. *harroweri* Klots, 1940
4223 *scudderii* Reak., 1865
flavotincta Ckll., 1901, ♀
a. *ruckesi* Klots, 1937

ZERENE Hbn., 1819
MEGONOSTOMA Reak., 1863
MEGANOSTOMA auth., missp.
4224 *cesonia* (Stoll, 1790)
rosa M'Neill, 1889
rosea (Röber, 1910)
immaculsecunda Gunder, 1928, ♀
stainkeae (Field, 1936), ♀
caesonia auth., missp.
4225 *eurydice* (Bdv., 1855)
lorquini Bdv., 1855, ♀
wosnesenskii (Mén., 1855)
helena (Reak., 1863)
amorphae (Hy. Edw., 1876)
bernardino (Edw., 1887), ♂
fanniae Gunder, 1924
newcombi Gunder, 1925
lineainta Gunder, 1928, ♂

masumbrosus Gunder, 1928, ♀
doudoroffi Gunder, 1932, ♀

ANTEOS Hbn., 1819
AMYNTHIA Swainson, 1831
RHODOCERA; auth., not Bdv. & Leconte, 1830
GONEPTERYX; auth., not Leach, 1815
4226 *clorinde* (Godt., 1824), extralim.
godarti (Perty, 1833)
swainsonia (Swainson, 1833)
a. *nivifera* Fruhstorfer, 1907
4227 *maerula* (F., 1775), extralim.
a. *lacordairei* (Bdv., 1836)

PHOEBIS Hbn., 1819
COLIAS; Hbn., 1819, not F., 1807
CALLIDRYAS Bdv. & Leconte, 1830
METURA Butler, 1873
PARURA Kby., 1896
4228 *sennae* (L., 1758)
a. *eubule* (L., 1767)
drya (F., 1775), spr. f.
pallida Ckll., 1889, ♀
browni Field, 1936, ♀
b. *marcellina* (Cram., 1777)
yamana Reak., 1863, ♀
4229 *philea* (Johansson, 1763)
aricye (Cram., 1776)
melanippe (Stoll, 1782), spr. f.
corday (Hbn., 1819)
obsoleta Niepelt, 1920, ♀
4230 *argante* (F., 1775)
cipris (Cram., 1777)
larra (F., 1798)
cnidia (Godt., 1819)
4231 *agarithe* (Bdv., 1836)
argante; auth.
a. *maxima* (Neum., 1891)
floridensis (Röber, 1910), preocc. by Neumoegen, 1891
albarithe F. M. Brown, 1929, ♀
4232 *neocypris* (Hbn., 1823), extralim.
a. *bracteolata* (Butler, 1865)
cipris; F., 1793, not Cram., 1777

APHRISSA Butler, 1873
4233 *statira* (Cram., 1777), extralim.
a. *floridensis* (Neum., 1891)
b. *jada* (Butler, 1870)
4234 *orbis* (Poey, 1832)

KRICOGONIA Reak., 1863
4235 *lyside* (Godt., 1819)
castalia; auth., not F., 1793
terissa (Luc., 1852), ♂
fantasia Butler, 1871, ♀
lanice Lint., 1884
unicolor Godm. & Salvin, 1889, ♀
xanthophila Röber, 1910

EUREMA Hbn., 1819
ABAEIS Hbn., 1819
XANTHIDIA Bdv. & Leconte, 1829
HEUREMA Agassiz, 1846, emend.
SPHAENOGONA Butler, 1870
PYRISITIA Butler, 1870
TERIAS; auth., not Swainson, 1821

4236 **proterpia** (F., 1775)
    gundlachia (Poey, 1851)
    longicauda (H. W. Bates, 1864)
4237 **lisa** Bdv. & Leconte, 1829
    euterpe; auth., not Mén., 1832
    alba (Stkr., 1878), ♀
    clappi (Maynard, 1891)
    immaculata Whittaker & D. Stallings,
      1944, sum. f.
4238 **nise** (Cram., 1776)
    neda (Godt., 1819)
    nelphe (R. Felder, 1869)
    linda (Edw., 1884)
    perimede; Klots, 1929, not (Prittwitz,
      1865)
4239 **messalina** (F., 1787), extralim.
    a. **blakei** (Maynard, 1891)
4240 **dina** (Poey, 1832), extralim.
    a. **westwoodi** (Bdv., 1836)
    b. **helios** M. Bates, 1934
4241 **chamberlaini** Butler, 1897, extralim.
4242 **nicippe** (Cram., 1780)
    flava (Stkr., 1878)
    dammersi Gunder, 1930
    callae Field, 1936
    pallens Whittaker & D. Stallings,
      1944, sum. f.
4243 **daira** (Godt., 1819)
    delia (Cram., 1780), preocc. by D. &
      S., 1775
    demoditus Hbn., 1819
    jucunda Bdv. & Leconte, 1829, sum.
      f.
    albina (Poey, 1851)
    pallida Klots, 1928
    fusca L. Harris, 1931
    delioides Haskin, 1933
    elathea; auth., not Cram., 1777
    lydia; dos Passos, 1964, not C. &
      R. Felder, 1861
4244 **palmira** (Poey, 1853)
    elathea; auth. not Cram., 1777
    palmyra auth., missp.
4245 **boisduvalianum** C. & R. Felder, 1865
    ingrata R. Felder, 1869
4246 **mexicanum** (Bdv., 1836)
    damaris (C. & R. Felder, 1865)
    depuiseti (Bdv., 1870)
    biedermanni Ehrmann, 1927
    recta Klots, 1928
    rosa Whittaker & D. Stallings, 1944,
      winter f.
4247 **salome** (C. Felder, 1861), extralim.
    a. **limoneum** (C. & R. Felder, 1861)

    NATHALIS Bdv., 1836
      NATALIS Doubleday, 1848, missp.
4248 **iole** Bdv., 1836
    irene Fitch, 1856
    luteolus Reak., 1863
    immaculata Field, 1936
    pallida Field, 1936
    viridis Whittaker & D. Stallings,
      1944, winter f.

LYCAENIDAE
by LEE D. MILLER & F. MARTIN BROWN

LIPHYRINAE

Spalgini

FENISECA Grt., 1869
4249 **tarquinius** (F., 1793)
    crataegi (Bdv. & Leconte, 1833)
    porsenna (Scudder, 1863)
    suffusa Dean, 1918
    a. **novascotiae** McD., 1935, emend.

LYCAENINAE

THARSALEA Scudder, 1876
4250 **arota** (Bdv., 1852)
    a. **virginiensis** (Edw., 1870)
    b. **nubila** Comst., 1926
    c. **schellbachi** Tilden, 1955

LYCAENA F., 1807
    LYCIA Sodoffsky, 1837
    MIGONITIS Sodoffsky, 1837
    RUMICIA Tutt, 1906
4251 **phlaeas** (L., 1761), extralim.
NOTE: many palearctic synonyms not included
    a. **americana** Harr., 1862
    fasciata (Stkr., 1878)
    fulliolus (Hulst, 1886)
    bacchus (Scudder, 1889)
    obliterata (Scudder, 1889)
    adrienne (Maynard, 1891)
    caeca (Reiff, 1913)
    obsoleta B. & McD., 1917, missp.
    octomaculata (Dean, 1918)
    banksi (Watson &
      W. P. Comstock, 1920)
    neui (Rummel, 1928)
    fulvus (Rummel, 1928)
    hypophlaeas; auth., not Bdv., 1852
    b. **hypophlaeas** (Bdv., 1852)
    c. **arethusa** (Wolley-Dod, 1907)
    d. **arctodon** Ferris, 1977
    e. **feildeni** (M'Lachlan, 1878)
    fieldeni Dyar, 1902, missp.
4252 **cuprea** (Edw., 1870)
    maculinita (Gunder, 1926)
    a. **snowi** (Edw., 1881)
    mcdunnoughi (Gunder, 1927)
    macdunnoughi dos Passos, 1964,
      emend.
    b. **henryae** (Cadbury, 1937)

GAEIDES Scudder, 1876
4253 **xanthoides** (Bdv., 1852)
    luctuosa (Watson & W. P. Comstock,
      1920)
    a. **dione** (Scudder, 1868)
    gibboni (Gunder, 1927)
4254 **edith** (Mead, 1878)
    vanduzeei (Gunder, 1927)
    a. **montana** (Field, 1936)
    meadi (Field, 1936), ♀
4255 **gorgon** (Bdv., 1852)

HYLLOLYCAENA L. Miller &
    F. M. Brown, 1978
    CHRYSOPHANUS; Scudder,
      1876, not Hbn., 1818

4256 **hyllus** (Cram., 1775)
    thoe (Guér.-Méneville, 1831)
    wyatti (Gunder, 1927)
    wormsbacheri (Gunder, 1927)

CHALCERIA Scudder, 1876
4257 **rubida** (Behr, 1866)
    a. **duofacies** (K. Johnson & Balogh,
      1977)
    b. **perkinsorum** (K. Johnson &
      Balogh, 1977)
    c. **longi** (K. Johnson & Balogh, 1977)
    d. **sirius** (Edw., 1871)
    e. **monachensis** (K. Johnson &
      Balogh, 1977)
4258 **ferrisi** (K. Johnson & Balogh, 1977)
4259 **heteronea** (Bdv., 1852)
    coloradensis (Gunder, 1925)
    gravenotata (Klots, 1930)
    klotsi (Field, 1936)
    a. **clara** (Hy. Edw., 1877)

EPIDEMIA Scudder, 1876
4260 **epixanthe** (Bdv. & Leconte, 1833)
    hypoxanthe (Kby., 1862)
    a. **phaedra** (Hall, 1924)
    b. **michiganensis** (Rawson, 1955)
4261 **dorcas** (Kby., 1837)
    anthelle Scudder, 1876
    a. **castro** (Reak., 1866)
    b. **florus** (Edw., 1883)
    hulbirti (Field, 1936), ♀
    sternitzkyi; Field, 1936, not
      Gunder, 1927
    c. **dospassosi** (McD., 1940), emend.
    d. **claytoni** (Brower, 1940)
    e. **megaloceras** Ferris, 1977
    f. **arcticus** Ferris, 1977
4262 **helloides** (Bdv., 1852)
    williamsi (Gunder, 1927)
    sternitzkyi (Gunder, 1927
    gunderi (Rudkin, 1933)
4263 **nivalis** (Bdv., 1869)
    ianthe (Edw., 1871)
    a. **browni** (dos Passos, 1938)
4264 **mariposa** (Reak., 1866)
    zeroe (Bdv., 1869)
    a. **charlottensis** (Holl., 1930)
    b. **penroseae** (Field, 1938)
    penrosae (F. M. Brown, Eff &
      Rotger, 1955), missp.

HERMELYCAENA L. Miller &
    F. M. Brown, 1978
    THARSALEA Scudder, 1876, part
4265 **hermes** (Edw., 1870)
    delsud (Wgt., 1905), emend.

THECLINAE

HYPAUROTIS Scudder, 1876
4266 **crysalus** (Edw., 1873)
    chrysalus auth., missp.
    a. **citima** (Hy. Edw., 1881)

HABRODAIS Scudder, 1876
    HABRODIAS McD., 1914, missp.
4267 **grunus** (Bdv., 1852)
    a. **lorquini** Field, 1938
    chloris Field, 1938
    b. **herri** Field, 1938

EUMAEINAE

Eumaeini

EUMAEUS Hbn., 1819
EUMENIA Godt., 1824
4268 atala (Poey, 1832), extralim.
  a. florida Röber, 1926
    grayi W. P. Comstock &
      Huntington, 1943
4269 minijas (Hbn., 1809), extralim.?
  minyas Hbn., 1819, emend.
  toxea (Godt., 1824)

Strymonini

ATLIDES Hbn., 1819
4270 halesus (Cram., 1777)
  dolichos (Hbn., 1823)
  juanita (Scudder, 1868)
  a. estesi Clench, 1942
    corcorani Gunder, 1934, ab.
    corcorani dos Passos, 1964

CHLOROSTRYMON Clench, 1961
4271 maesites (H.-S., 1864)
4272 telea (Hewitson, 1868)
4273 simaethis (Drury, 1773)
  a. sarita (Skin., 1895)

PHAEOSTRYMON Clench, 1961
4274 alcestis (Edw., 1871)
  a. oslari (Dyar, 1904)

HARKENCLENUS dos Passos, 1970
CHRYSOPHANUS Hbn., 1818,
  suppressed by ICZN (Op. 541)
4275 titus (F., 1793)
  a. mopsus (Hbn., 1818)
  b. watsoni (B. & Benj., 1926)
  c. immaculosus (W. P. Comstock,
    1913)

SATYRIUM Scudder, 1876
CALLIPSYCHE Scudder, 1876
4276 behrii (Edw., 1870)
  kali (Stkr., 1878)
  nigroinita (Gunder, 1924)
  a. crossi (Field, 1938)
  b. columbia (McD., 1944)
4277 fuliginosum (Edw., 1861)
  suasa (Bdv., 1869)
  immaculata Gunder, 1927
  a. semiluna Klots, 1930
4278 acadicum (Edw., 1862)
  souhegan (Whitney, 1868)
  muskoka (Watson & W. P. Comstock,
    1920)
  swetti (Watson & W. P. Comstock,
    1920)
  souhegon (McD., 1938), missp.
  acadia (F. M. Brown, Eff & Rotger,
    1955), missp.
  a. coolinense (Watson &
    W. P. Comstock, 1920)
  b. montanense (Watson &
    W. P. Comstock, 1920)
  c. watrini (Dufrane, 1939)
4279 californicum (Edw., 1862)
  borus (Bdv., 1869)
  cygnus (Edw., 1871)

4280 sylvinum (Bdv., 1852)
  a. dryope (Edw., 1870)
  b. desertorum (Grinnell, 1917)
  c. itys (Edw., 1882)
  d. putnami (Hy. Edw., 1876)
    putmani (F. M. Brown, Eff &
      Rotger, 1955), missp.
4281 edwardsii (G. & R., 1869)
  fabricii (Kby., 1871), mistake
4282 calanus (Hbn., 1809)
  wittfeldii (Edw., 1883)
  a. falacer (Godt., 1824)
    lorata (G. & R., 1868)
    inorata (G. & R., 1868)
    heathii (Fletcher, 1903)
    borealis (Lafontaine, 1969)
  b. godarti (Field, 1938)
4283 caryaevorum (McD., 1942)
4284 kingi (Klots & Clench, 1952)
4285 liparops (Leconte, 1833)
  a. strigosum (Harr., 1862)
    pruina (Scudder, 1889)
  b. fletcheri (Michener & dos
    Passos, 1942)
  c. aliparops (Michener & dos
    Passos, 1942)
4286 auretorum (Bdv., 1852)
  tacita (Hy. Edw., 1881)
  tetra; auth., not Edw., 1870
  a. spadix (Hy. Edw., 1881)
4287 tetra (Edw., 1870)
  adenostomatis (Hy. Edw., 1876)
4288 saepium (Bdv., 1852)
  chalcis (Edw., 1869)
  fulvescens (Hy. Edw., 1876)
  chlorophora (Watson &
    W. P. Comstock, 1920)
  provo (Watson & W. P. Comstock,
    1920)
  a. okanaganum (McD., 1944)

OCARIA Clench, 1970
4289 ocrisia (Hewitson, 1868)

MINISTRYMON Clench, 1961
4290 clytie (Edw., 1877)
  maevia (Godm. & Salvin, 1887),
    win. f.(?)
4291 leda (Edw., 1882)
  ines (Edw., 1882), fall f.

TMOLUS Hbn., 1819
4292 echion (L., 1758), extralim.
  a. echiolus (Draudt, 1924)
4293 azia (Hewitson, 1873)
  nipona (Hewitson, 1877)

OENOMAUS Hbn., 1819
4294 ortygnus (Cram., 1779)

THEREUS Hbn., 1819
HETEROSMAITIA Clench, 1964
4295 zebina (Hewitson, 1869)
4296 spurina (Hewitson, 1867), extralim.
4297 palegon (Stoll, 1780)
  myrtillus (Stoll, 1782)
  juicha (Reak., 1867)
  mytillus (Godm. & Salvin, 1887), missp.

ALLOSMAITIA Clench, 1964
4298 pion (Godm. & Salvin, 1887)

CALYCOPIS Scudder, 1876
4299 cecrops (F., 1793)
  poeas (Hbn., 1811)
  gottschalki (A. H. & L. Clark, 1938)
4300 isobeon (Butler & Druce, 1872)
  beon; auth., not Cram., 1782

CALLOPHRYS Billberg, 1820
LYCUS Hbn., 1819, preocc. by F., 1787
LICUS Hbn., 1823, emend.
4301 dumetorum (Bdv., 1852)
  a. perplexa B. & Benj., 1923
  b. oregonensis Gorelick, 1969
4302 comstocki Henne, 1940
4303 lemberti Tilden, 1963
4304 apama (Edw., 1882)
  a. homoperplexa B. & Benj., 1923
4305 affinis (Edw., 1862)
  a. washingtonia Clench, 1944
4306 viridis (Edw., 1862)
4307 sheridanii (Carpenter, 1877), emend.
  sheridonii (Carpenter, 1877), incorr.
    orig. spell.
  a. neoperplexa B. & Benj., 1923
  b. newcomeri Clench, 1963

CYANOPHRYS Clench, 1961
4308 miserabilis (Clench, 1946)
  longula; Hewitson, 1877
  pastor; B. & McD., 1913
4309 goodsoni (Clench, 1946)
  facuna; H. A. Freeman, 1950

MITOURA Scudder, 1872, emend.
MITOURI Scudder, 1872, incorr.
  orig. spell.
MITURA Rye, 1874, missp.
4310 spinetorum (Hewitson, 1867)
  ninus (Edw., 1871)
  cuyamaca (Wgt., 1922)
4311 johnsoni (Skin., 1904)
4312 rosneri K. Johnson, 1976
  a. plicataria K. Johnson, 1976
4313 barryi K. Johnson, 1976
  a. acuminata K. Johnson, 1976
4314 byrnei K. Johnson, 1976
4315 nelsoni (Bdv., 1869)
  exoleta (Hy. Edw., 1881)
  a. muiri (Hy. Edw., 1881)
4316 siva (Edw., 1874)
  rhodope (Godm. & Salvin, 1887)
  a. juniperaria Comst., 1925
  b. mansfieldi Tilden, 1951
4317 loki (Skin., 1907)
4318 grynea (Hbn., 1819)
  damon (Stoll, 1782), preocc. by D. &
    S., 1775
  damastus (Godt., 1824)
  auburniana (Harr., 1862)
  patersonia (Brehme, 1907)
  octoscripta Buchholz, 1951
  a. smilacis (Bdv. & Leconte, 1833)
  b. sweadneri F. Chermock, 1944
  c. castalis (Edw., 1871)
    discoidalis (Skin., 1897)
    brehmei B. & Benj., 1923
4319 hesseli Rawson & Ziegler, 1950

XAMIA Clench, 1961
4320 xami (Reak., 1866)
  blenina (Hewitson, 1868)

SANDIA Clench & Ehrlich, 1960
4321 **mcfarlandi** Ehrlich & Clench, 1960
    macfarlandi dos Passos, 1965 emend.

INCISALIA Scudder, 1872
4322 **augustus** (Kby., 1837)
    augustinus (Westwood, 1852)
    a. **helenae** dos Passos, 1943
    b. **croesioides** (Scudder, 1876)
    c. **iroides** (Bdv., 1852)
      immaculata (Cockle, 1910)
    d. **annetteae** dos Passos, 1943
4323 **fotis** (Stkr., 1878)
    a. **doudoroffi** dos Passos, 1940
    b. **windi** Clench, 1943
    c. **schryveri** Cross, 1937
    d. **mossii** (Hy. Edw., 1881)
    e. **bayensis** (R. M. Brown, 1969)
4324 **polia** Cook & Watson, 1907
    davisi Watson & W. P. Comstock,
      1920
    a. **obscura** Ferris & Fisher, 1973
4325 **irus** (Godt., 1824)
    balteata (Scudder, 1889)
    a. **arsace** (Bdv. & Leconte, 1833)
    b. **hadra** Cook & Watson, 1909
4326 **henrici** (G. & R., 1867)
    a. **margaretae** dos Passos, 1943
    b. **solata** Cook & Watson, 1909
    c. **turneri** Clench, 1943
4327 **lanaoriaeensis** Sheppard, 1934
4328 **niphon** (Hbn., 1819)
    nipha (Morris, 1860), missp.
    a. **clarki** Free., 1938
4329 **eryphon** (Bdv., 1852)
    a. **sheltonensis** F. Chermock &
      Frechin, 1949

ARAWACUS Kaye, 1904
    DOLYMORPHA Holl., 1931
4330 **jada** (Hewitson, 1867)

EURISTRYMON Clench, 1961
4331 **favonius** (J. E. Smith, 1797)
4332 **ontario** (Edw., 1868)
    a. **autolycus** (Edw., 1871)
    b. **violae** (D. Stallings & Turner,
      1947)
    c. **ilavia** (Beutenmüller, 1899)
      mirabelle (Barnes, 1900)
4333 **polingi** (B. & Benj., 1926)

HYPOSTRYMON Clench, 1961
4334 **critola** (Hewitson, 1874)

PARRHASIUS Hbn., 1819
    EUPSYCHE Scudder, 1876
    PANTHIADES; auth., not Hbn.,
      1819
4335 **m-album** (Bdv. & Leconte, 1833)
    psyche (Bdv. & Leconte, 1833), ♂

STRYMON Hbn., 1818
    CALLIPAREUS Scudder, 1872
    CALLICISTA Grt., 1873
    URANOTES Scudder, 1876
4336 **melinus** Hbn., 1818
    hyperici (Bdv. & Leconte, 1833)
    youngi Field, 1936
    a. **humuli** (Harr., 1841)
      meinersi Gunder, 1927

    b. **franki** Field, 1938
    c. **pudicus** (Hy. Edw., 1876)
    d. **setonia** McD., 1927
    e. **atrofasciatus** McD., 1921
4337 **avalona** (Wgt., 1905)
4338 **rufofuscus** (Hewitson, 1877)
4339 **bebrycia** (Hewitson, 1868)
    buchholzi H. A. Freeman, 1950
4340 **martialis** (H.-S., 1864)
4341 **yojoa** (Reak., 1866)
    daraba; auth., not Hewitson, 1867
    beroea; auth., not Hewitson, 1868
4342 **albatus** (C. & R. Felder, 1865),
      extralim.
    a. **sedecia** (Hewitson, 1874)
4343 **acis** (Drury, 1773), extralim.
    mars (F., 1777)
    a. **bartrami** (W. P. Comstock &
      Huntington, 1943)
4344 **alea** (Godm. & Salvin, 1887)
    laceyi (B. & Benj., 1910)
4345 **columella** (F., 1793), extralim.
    eurytalus (Butler, 1870)
    a. **modestus** (Maynard, 1873)
      ocellifera Grt., 1873
    b. **cybira** (Hewitson, 1874)
    c. **istapa** (Reak., 1866)
4346 **limenia** (Hewitson, 1867)
4347 **cestri** (Reak., 1866)
4348 **bazochii** (Godt., 1824)
    thius Gey., 1832
    agra (Hewitson, 1871)

ERORA Scudder, 1872
4349 **laetus** (Edw., 1862)
    clothilde (Edw., 1863)
4350 **quaderna** (Hewitson, 1868), extralim.
    a. **sanfordi** dos Passos, 1940

ELECTROSTRYMON Clench, 1961
4351 **endymion** (F., 1775), extralim.
    hugon (Godt., 1824)
    hugo (Westwood & Hewitson,
      1852)
    a. **cyphara** (Hewitson, 1874)
4352 **angelia** (Hewitson, 1874)

POLYOMMATINAE

Zizeerini

BREPHIDIUM Scudder, 1876
4353 **exile** (Bdv., 1852)
    fea (Edw., 1871)
    coolidgei Gunder, 1925
4354 **isophthalma** (H.-S., 1862), extralim.
    barbouri Clench, 1943
    a. **pseudofea** (Morr., 1873)

ZIZULA Chapman, 1910
4355 **cyna** (Edw., 1881)
    tulliola (Godm. & Salvin, 1887)
    mela (Stkr., 1900)
    gaika; auth., not Trimen, 1862

Lampidini

LEPTOTES Scudder, 1876
    SYNTAURUCOIDES Kaye, 1904
    SYNTAURUCHOIDES Sharp,
      1905, missp.

4356 **cassius** (Cram., 1775), extralim.
    a. **theonus** (Luc., 1857)
      catalina; auth., not F., 1793
      floridensis (Morr., 1873)
    b. **striata** (Edw., 1877)
4357 **marina** (Reak., 1868)
    violacea Gunder, 1925
    burdicki Henne, 1935
    reakirti Field, 1936, ♀

HEMIARGUS Hbn., 1818
    CYCLARGUS Nabokov, 1945
    ECHINARGUS Nabokov, 1945
4358 **thomasi** Clench, 1943, extralim.
    ammon; auth., not Luc., 1857
    a. **bethunebakeri** W. P. Comstock
      & Huntington, 1943, emend.
4359 **ceraunus** (F., 1793), extralim.
    a. **antibubastus** Hbn., 1818
      pseudoptiletes (Bdv. & Leconte,
      1833)
    b. **gyas** (Edw., 1871)
      astragala (Wgt., 1905)
      florencia (Clemence, 1914)
    c. **zachaeina** (Butler & Druce,
      1872)
4360 **isola** (Reak., 1866), extralim.
    zachaeina (Butler & Druce, 1872), ♀
    a. **alce** (Edw., 1871)

Everini

EVERES Hbn., 1819
    TIORA Evans, 1912
    UNUNCULA van Eecke, 1915
4361 **comyntas** (Godt., 1824)
    sissona (Wgt., 1905)
    watermani Nakahara, 1925
    meinersi Field, 1938, spr. f.
    a. **texanus** R. Chermock, 1944
4362 **amyntula** (Bdv., 1852)
    dodgei Gunder, 1927
    a. **valeriae** Clench, 1944
    b. **albrighti** Clench, 1944
    c. **herri** (Grinnell, 1901)
      arizonensis Gunder, 1927
      jemezensis Gunder, 1927

Lycaenopsini

CELASTRINA Tutt, 1906
    CYANIRIS; auth., not Dalman,
      1816
    LYCAENOPSIS; auth., not C. &
      R. Felder, 1865
4363 **ladon** (Cram., 1780)
    pseudargiolus (Bdv. & Leconte,
      1833)
    neglecta (Edw., 1862), sum. f.
    violacea (Edw., 1866)
    deutargiolus (Scudder, 1869), sum. f.
    intermedia (Stkr., 1870), ♀
    obsoletalunulata Tutt, 1908, sum. f.
    neglectamajor Tutt, 1908, sum. f.
    a. **lucia** (Kby., 1837)
      marginata (Edw., 1883)
      fumida (Scudder, 1889), nom. nud.?
      pseudora (Scudder, 1889), nom.
        nud.?
      brunnea Tutt, 1908
      subtusjuncta Tutt, 1908, emend.

  inaequalis Tutt, 1908, emend.
 b. **argentata** (Fletcher, 1903)
 c. **nigrescens** (Fletcher, 1903)
  quesnelii (Cockle, 1910)
  maculatasuffusa (Cockle, 1910),
   emend.
 d. **sidara** (Clench, 1944)
 e. **cinerea** (Edw., 1883)
  arizonensis (Edw., 1884), sum. f.
 f. **echo** (Edw., 1864)
  nunenmacheri (Strand, 1915)
  bakeri (Clench, 1944)
 g. **gozora** (Bdv., 1870)
  gazora (Holl., 1931), missp.
4364 **ebenina** Clench, 1972
 nig (Stkr., 1878), ab., unavailable
 nigra (Edw., 1887), ab., unavailable

Scolitantidini

PHILOTES Scudder, 1876
4365 **sonorensis** (C. & R. Felder, 1865)
 regia (Bdv., 1869)
 sonoralba Watson & W. P. Comstock,
  1920
 comstocki Gunder, 1925

EUPHILOTES Mattoni, 1978
 PHILOTES; auth., not Scudder,
  1876
 CALETA; auth., not Fruhstorfer,
  1922
 SHIJIMIAEOIDES; auth., not
  Beuret, 1958
4366 **battoides** (Behr, 1867)
 a. **oregonensis** (B. & McD., 1917)
 b. **glaucon** (Edw., 1871)
 c. **baueri** (Shields, 1975)
 d. **intermedia** (B. & McD., 1917)
  malcolmi (Gunder, 1927)
 e. **comstocki** (Shields, 1975)
 f. **centralis** (B. & McD., 1917)
 g. **ellisii** (Shields, 1975)
 h. **bernardino** (B. & McD., 1916)
  baldyensis (Gunder, 1925)
 i. **allyni** (Shields, 1975)
 j. **martini** (Mattoni, 1955)
4367 **enoptes** (Bdv., 1852)
 a. **dammersi** (Comst. & Henne,
  1933)
 b. **tildeni** (Langston, 1964)
 c. **langstoni** (Shields, 1975)
 d. **smithi** (Mattoni, 1955)
 e. **mojave** (Watson &
  W. P. Comstock, 1920)
  mohave auth., missp.
 f. **ancilla** (B. & McD., 1917)
 g. **columbiae** (Mattoni, 1955)
 h. **bayensis** (Langston, 1964)
4368 **pallescens** (Tilden & Downey, 1955)
 a. **elvirae** (Mattoni, 1966)
4369 **rita** (B. & McD., 1916)
 a. **mattoni** (Shields, 1975)
 b. **emmeli** (Shields, 1975)
 c. **coloradensis** (Mattoni, 1966)
 d. **spaldingi** (B. & McD., 1917)

PHILOTIELLA Mattoni, 1977
4370 **speciosa** (Hy. Edw., 1876)
 bohartorum (Tilden, 1968)

GLAUCOPSYCHE Scudder, 1872
 SCOLITANTIDES; auth., not
  Hbn., 1819
 NOMIADES; auth., not Hbn., 1819
 PHAEDROTES Scudder, 1876
 SHIJIMIA; dos Passos, 1964, not
  Matsumura, 1919
4371 **piasus** (Bdv., 1852)
 sagittigera (C. & R. Felder, 1865)
 lorquini (Behr, 1867), preocc. by
  H.-S., 1850
 viaca (Edw., 1871)
 a. **catalina** (Reak., 1866)
  rhaea (Bdv., 1869)
  gorgonioi (Gunder, 1925)
 b. **nevada** F. M. Brown, 1975
 c. **daunia** (Edw., 1871)
 d. **toxeuma** F. M. Brown, 1971
4372 **lygdamus** (Doubleday, 1842)
 nittanyensis F. Chermock, 1944
 boydi A. Clark, 1948
 a. **couperi** Grt., 1874
  mcdunnoughi Gunder, 1927
  macdunnoughi dos Passos, 1964,
   emend.
 b. **afra** (Edw., 1883)
 c. **mildredi** F. Chermock, 1944
 d. **oro** Scudder, 1876
  leussleri Gunder, 1927
 e. **jacki** D. Stallings & Turner, 1947
 f. **arizonensis** McD., 1936
 g. **australis** Grinnell, 1917
  ?orcus (Edw., 1869)
  sinepunctata Comst., 1926
  sinepuncta McD., 1938, missp.
 h. **incognita** Tilden, 1974
  behrii; auth., not Edw., 1862
  sternitzkyi Gunder, 1929, ab.,
   unavail.
 i. **columbia** Skin., 1917
4373 **xerces** (Bdv., 1852)
 antiacis (Bdv., 1852)
 behrii (Edw., 1862)
 mertila (Edw., 1866)
 polyphemus (Bdv., 1869)
 huguenini Gunder, 1925
 barnesi Gunder, 1927
 intermedia F. Chermock, 1929

Polyommatini

LYCAEIDES Hbn., 1819
4374 **argyrognomon** (Brgstr., 1779), extralim.
 idas; L., 1767, not L. 1758
 a. **anna** (Edw., 1861)
  cajona (Reak., 1866)
  argyrotoxus (Behr, 1867)
  philemon (Bdv., 1869)
 b. **ricei** (Cross, 1937)
  fretchini (F. Chermock, 1945)
 c. **lotis** (Lint., 1879)
 d. **alaskensis** (F. Chermock, 1945)
  kodiak; auth., not Edw., 1870
 e. **scudderii** (Edw., 1861)
  subarcticus (F. Chermock, 1945)
 f. **aster** (Edw., 1882)
 g. **empetri** (Free., 1938)
 h. **ferniensis** (F. Chermock, 1945)
 i. **atrapraetextus** (Field, 1939)
  sweadneri (F. Chermock, 1945)

 j. **sublivens** Nabokov, 1949
  sublivent F. M. Brown, Eff &
   Rotger, 1955, missp.
 k. **longinus** Nabokov, 1949
 l. **nabokovi** Masters, 1972
4375 **melissa** (Edw., 1873)
 fridayi (F. Chermock, 1945)
 a. **pseudosamuelis** Nabokov, 1949
 b. **paradoxa** (F. H. Chermock, 1945)
  inyoensis (Gunder, 1927), ab.,
   unavail.
 c. **annetta** (Edw., 1882)
 d. **samuelis** Nabokov, 1944
  scudderii; auth., not Edw., 1861

PLEBEJUS Kluk, 1802
 PLEBEIUS auth., missp.
 RUSTICUS Hbn., 1806, suppr.
  (ICZN Op. 278)
4376 **saepiolus** (Bdv., 1852)
 aehaja (Behr, 1867), altitudinal f.
 rufescens (Bdv., 1869), ♀
 a. **hilda** (Grinnell & Grinnell, 1907)
  garthi Gunder, 1928
  boharti Gunder, 1932
 b. **insulanus** Blkmre., 1919
 c. **amica** (Edw., 1863)
  kodiak (Edw., 1870)
 d. **gertschi** dos Passos, 1938
 e. **whitmeri** F. M. Brown, 1951
  leussleri Gunder, 1927, ab.
  caerulescens Ferris, 1970

PLEBULINA Nabokov, 1944
4377 **emigdionis** (Grinnell, 1905)
 melimona (Wgt., 1905)

ICARICIA Nabokov, 1944
4378 **icarioides** (Bdv., 1852)
 maricopa (Reak., 1866), ♀
 daedalus (Behr, 1867)
 phileros (Bdv., 1869)
 fulla (Edw., 1870)
 helios (Edw., 1870)
 spinimaculata (Gunder, 1926)
 a. **evius** (Bdv., 1869)
 b. **moroensis** (Sternitzky, 1930)
 c. **missionensis** (Hovanitz, 1937)
 d. **pardalis** (Behr, 1867)
  erymus (Bdv., 1869)
  fenderi (Macy, 1931)
  windi (Gunder, 1933)
 e. **pheres** (Bdv., 1852)
 f. **ardea** (Edw., 1871)
 g. **lycea** (Edw., 1864)
  rapahoe (Reak., 1866), ♀
 h. **buchholzi** (dos Passos, 1938)
 i. **pembina** (Edw., 1862)
 j. **blackmorei** (B. & McD., 1919)
 k. **montis** (Blkmre., 1929)
4379 **shasta** (Edw., 1862)
 zelmira (C. & R. Felder, 1865)
 calchas (Behr, 1867)
 nivium (Bdv., 1869)
 comstocki (C. Fox, 1924)
 calcas (McD., 1938), missp.
 a. **minnehaha** (Scudder, 1875)
  browni (Ferris, 1970)
 b. **pitkinensis** (Ferris, 1976)
4380 **acmon** (Westwood & Hewitson, 1852)

antaegon (Bdv., 1852)
cottlei (Grinnell, 1916)
labecula (Watson & W. P. Comstock, 1920)
kelseyi (Wgt., 1930)
a. **lutzi** dos Passos, 1938
b. **texana** Goodpasture, 1973
c. **spangelatus** (Burdick, 1942)
4381 **lupini** (Bdv., 1869)
immaculata (F. Chermock, 1929)
a. **monticola** (Clemence, 1909)
pallida (Gunder, 1925)
malcolmi (Gunder, 1925)
montanus (Gunder, 1929)
angelus (Gunder, 1929)
b. **chlorina** (Skin., 1902)
carolyna (Comst., 1922), ♂
4382 **neurona** (Skin., 1902)

VACCINIINA Tutt, 1909
VACCINIA Forster, 1938, missp.
4383 **optilete** (Knoch, 1781), extralim.
a. **yukona** (Holl., 1900)

AGRIADES Hbn., 1819
LATIORINA Tutt, 1908
4384 **franklinii** (Curt., 1835)
glandon; auth., not de Prunner, 1798
aquilo; auth., not Bdv., 1833
suttoni (Holl., 1931)
a. **lacustris** (Free., 1939)
b. **bryanti** (Leussler, 1935)
kohlsaati (Gunder, 1932), ab.
c. **megalo** (McD., 1927)
d. **rusticus** (Edw., 1865)
e. **podarce** (C. & R. Felder, 1865)
tehama (Reak., 1866)
cilla (Behr, 1867)
nestos (Bdv., 1869)

RIODINIDAE
by LEE D. MILLER & F. MARTIN BROWN

EUSELASIINAE

EUSELASIA Hbn., 1819
PSALIDOPTERIS Hbn., 1823
EURYGONA Bdv., 1836
PSADILOPTERIS Stichel, 1930, missp.
4385 **abreas** (Edw., 1881), quest. occur.
sergia (Godm. & Salvin, 1885)

RIODININAE

CALEPHELIS G. & R., 1869
NYMPHIDIUM; auth., not F., 1807
NYMPHIDIA Bdv. & Leconte, 1833, missp.
CALEPHILIS Kby., 1871, missp.
CALEPHALIS Wgt., 1908, missp.
LEPHELISCA B. & L., 1922
CALLEPHELIS Stichel, 1930, missp.
4386 **virginiensis** (Guér.-Méneville, 1831)
pumila (Bdv. & Leconte, 1833)
caeneus; auth., not. L., 1767
louisiana Holl., 1931
4387 **borealis** (G. & R., 1866)
geda Scudder, 1876

4388 **nemesis** (Edw., 1871)
a. **australis** (Edw., 1877)
guadeloupe (Stkr., 1878)
b. **dammersi** McAlpine, 1971
c. **californica** McAlpine, 1971
4389 **perditalis** B. & McD., 1918
4390 **wrighti** Holl., 1930
4391 **mutica** McAlpine, 1937
borealis; auth., not G. & R., 1866
4392 **rawsoni** McAlpine, 1939
guadeloupe; Powell, 1975, not (Stkr., 1878)
4393 **freemani** McAlpine, 1971
4394 **arizonensis** McAlpine, 1971
4395 **dreisbachi** McAlpine, 1971

CARIA Hbn., 1823
4396 **ino** Godm. & Salvin, 1886, extralim.
domitianus; auth., not F., 1793
a. **melicerta** Schaus, 1890

LASAIA Bates, 1868
TALITES Capronnier, 1874
LASAEA Glaeser, 1887, missp.
LASIA Hemming, 1967, missp.
4397 **sula** Stgr., 1888, extralim.
a. **peninsularis** Clench, 1972
sula; auth., not Stgr., 1888
narses; auth., not Stgr., 1888
sessilis; auth., not Schaus, 1890

MELANIS Hbn., 1819
LIMNAS Bdv., 1836, preocc., by Hbn., 1806
LYMNAS Blanchard, 1840
ERCHIA Wlk., 1854, preocc., by Wlk., 1854
LYNMAS Scudder, 1875, missp.
PSEUDERCHIA Kby., 1892
4398 **pixe** (Bdv., 1836)

EMESIS F., 1807
APHACITIS Hbn., 1819
POLYSTICHTIS Hbn., 1819
TAPINA Billberg, 1820
NIMULA Blanchard, 1840, syn. or missp.?
POLYSTICHTHIS Agassiz, 1846, missp.
NELONE Bdv., 1870
4399 **zela** Butler, 1870, extralim.
a. **cleis** (Edw., 1882)
ares (Edw., 1882), ♂
4400 **ares** (Edw., 1882), ♀
4401 **emesia** (Hewitson, 1867)

APODEMIA C. & R. Felder, 1865
CHRYSOBIA Bdv., 1869
POLYSTIGMA Godm. & Salvin, 1886, preocc. by Kraatz, 1880
4402 **mormo** (C. & R. Felder, 1859)
dumeti (Behr, 1865)
mormonia (Bdv., 1869)
a. **langei** Comst., 1938
b. **virgulti** (Behr, 1865)
sonorensis C. & R. Felder, 1865
vergulti auth., missp.
c. **tuolumnensis** Opler & Powell, 1962
d. **dialeuca** Opler & Powell, 1962, quest. occur.

e. **deserti** B. & McD., 1918
f. **cythera** (Edw., 1873)
g. **mejicana** (Behr, 1865), quest. occur.
h. **duryi** (Edw., 1882)
druryi Stichel, 1910, missp.
4403 **multiplaga** Schaus, 1902
4404 **hepburni** Godm. & Salvin, 1886
4405 **palmerii** (Edw., 1870)
marginalis (Skin., 1920)
4406 **walkeri** Godm. & Salvin, 1886
4407 **phyciodoides** B. & Benj., 1924
4408 **nais** (Edw., 1876)
4409 **chisosensis** H. A. Freeman, 1964

LIBYTHEIDAE
by LEE D. MILLER & F. MARTIN BROWN

LIBYTHEINAE

LIBYTHEANA Michener, 1943
LIBYTHEA F., 1807, part
4410 **bachmanii** (Kirtland, 1851)
kirtlandi (Field, 1938)
a. **larvata** (Stkr., 1878)
streckeri (Field, 1938)
4411 **carinenta** (Cram., 1777), extralim.
a. **mexicana** Michener, 1943
4412 **motya** (Bdv. & Leconte, 1883), quest. occur.

NYMPHALIDAE
by LEE D. MILLER & F. MARTIN BROWN

HELICONIINAE

AGRAULIS Bdv. & Leconte, 1833
DIONE; auth., not Hbn., 1819
4413 **vanillae** (L., 1758), extralim.
passiflorae (F., 1793)
a. **nigrior** Michener, 1942
b. **incarnata** (Riley, 1926)
comstocki (Gunder, 1925), ab.
fumosus (Gunder, 1927)
margineapertus (Gunder, 1928)
hewlettae (Gunder, 1930)

DIONE Hbn., 1819
4414 **moneta** Hbn., 1819, extralim.
a. **poeyi** (Butler, 1873)
moneta; (Doubleday, 1848), not Hbn., 1819

DRYADULA Michener, 1942
COLAENIS; auth., not Hbn., 1819
4415 **phaetusa** (L., 1758)
phaerusa (L., 1767), missp.

DRYAS Hbn., 1807
COLAENIS Hbn., 1819
4416 **iulia** (F., 1775), extralim.
juncta W. P. Comstock, 1944
julia auth., missp.
a. **largo** Clench, 1975
cillene; auth., not Cram., 1780
nudeola; auth., not Stichel, 1907
carteri; auth., not Riley, 1926
b. **moderata** Riley, 1926

EUEIDES Hbn., 1816
  MECHANITIS Illiger, 1807, suppr.
    (ICZN Op. 232)
  EIEIDES Hbn., 1821, missp.
  EVIDES Agassiz, 1846, emend.
  SEMELIA Erichson, 1848, preocc.
    by Doubleday, 1844
  EVEIDES Bdv., 1870, missp.
  EISIDES Scudder, 1875, emend.?
  EURIDES Godm. & Salvin, 1880,
    missp.
4417 **isabella** (Stoll, 1781) extralim.
  cleobaea; auth., not Gey., 1832
  a. **zorcaon** Reak., 1866

HELICONIUS Kluk, 1802
  HELICONIUS Latr., 1804, preocc.
    by Kluk, 1802
  MIGONITIS Hbn., 1816, preocc.
    by Rafinesque, 1815
  SUNIAS Hbn., 1816
  SICYONIA Hbn., 1816, suppr.
    (ICZN Op. 382)
  AJANTIS Hbn., 1816
  APOSTRAPHIA Hbn., 1816,
    suppr. (ICZN Op., 382)
  LAPARUS Billberg, 1820
  CRENIS Hbn., 1821
  PHLOGRIS Hbn., 1825
  SYCIONIA Hbn., 1826, missp.
  HELICONIA Godt., 1819, missp.
  HELICONA Guér.-Méneville, 1844,
    missp.
  BLANCHARDIA Buchecker, 1880
4418 **charitonius** (L., 1767), emend.,
  extralim.
  charithonia (L., 1767)
  a. **tuckeri** W. P. Comstock &
    F. M. Brown, 1950
  b. **vasquezae** W. P. Comstock &
    F. M. Brown, 1950
4419 **erato** (L., 1764), extralim.
NOTE: many neotropical synonyms not
  included
  a. **petiveranus** Doubleday, 1847,
    emend.
  demophoon Mén., 1855

NYMPHALINAE

Nymphalini

POLYGONIA Hbn., 1819
  EUGONIA Hbn., 1819
  COMMA Rennie, 1832
  GRAPTA Kby., 1837
4420 **interrogationis** (F., 1798)
  umbrosa (Lint., 1869)
  crameri (Scudder, 1870)
  fabricii (Edw., 1870), light f.
4421 **comma** (Harr., 1842)
  dryas (Edw., 1870), dark f.
  harrisii (Edw., 1873)
4422 **satyrus** (Edw., 1869)
  chrysoptera (Wgt., 1905)
  a. **neomarsyas** dos Passos, 1969
  marsyas; auth., not Edw., 1870
  hollandi Gunder, 1927, ab.
4423 **faunus** (Edw., 1862)
  virescens Scudder, 1875

  a. **smythi** A. H. Clark, 1937
  b. **rustica** (Edw., 1874)
  silvius (Edw., 1874), ♀
  c. **arctica** Leussler, 1935
4424 **hylas** (Edw., 1872)
  orpheus Cross, 1936, ♀
4425 **silvius** (Edw., 1874), ♂
4426 **zephyrus** (Edw., 1870)
  thiodamas Scudder, 1875
4427 **gracilis** (G. & R., 1867)
4428 **oreas** (Edw., 1869)
  a. **silenus** (Edw., 1870)
4429 **progne** (Cram., 1776)
  c-argenteum (Kby., 1837)
  l-argenteum Scudder, 1875, dark f.
  martineae Coleman, 1919

NYMPHALIS Kluk, 1802
  HAMADRYAS Hbn., 1806
  SCUDDERIA Grt., 1873
  EUVANESSA Scudder, 1889
  VANESSA; auth., not F., 1807
  AGLAIS; auth., not Dalm., 1816
4430 **vau-album** (D. & S., 1775), extralim.
  a. **j-album** (Bdv. & Leconte, 1833)
  pocahontas (Scudder, 1889)
  aureomarginata (Ckll., 1889)
  b. **watsoni** (Hall, 1924)
4431 **californica** (Bdv., 1852)
  a. **herri** Field, 1936
4432 **antiopa** (L., 1758)
  pompadour (Pollich, 1779)
  hygiaea (Heydenreich, 1851)
  lintnerii (Fitch, 1856)
  obscura (Ckll., 1890)
  hippolita (Skin., 1898)
  grandis (Ehrmann, 1900)
  a. **hyperborea** (Seitz, 1914)

AGLAIS Dalman, 1816
  NYMPHALIS; auth., part, not
    Kluk, 1802
  VANESSA; auth., part, not F.,
    1807
4433 **milberti** (Godt., 1819)
  melbirti (Edw., 1872), missp.
  rothkei Gunder, 1927
  a. **furcillata** (Say, 1825)
  subpallida (Ckll., 1889)
  b. **viola** dos Passos, 1938

VANESSA F., 1807
  NYMPHALIS Latr., 1804, preocc.
    by L., 1758
  CYNTHIA F., 1807
  PYRAMEIS Hbn., 1819
  AMMIRALIS Rennie, 1832
  PHANESSA Sodoffsky, 1837
  NEOPYRAMEIS Scudder, 1889
4434 **virginiensis** (Drury, 1773)
  huntera (F., 1775)
  jole (Cram., 1776)
  iole auth., missp.
  hunteri Hbn., 1819, missp.?
  fulvia (Dodge, 1900)
  ahwashtee C. Fox, 1921
  massachusettensis Gunder, 1925
  simmsi (Gunder, 1927)
4435 **cardui** (L., 1758)
  carduelis (Seba, 1765)

  elymi Rambur, 1829
  ate (Stkr., 1878)
  minor (Ckll., 1890)
  jacksoni (A. H. Clark, 1932)
4436 **annabella** (Field, 1971)
  carye; auth., not Hbn., 1812
  muelleri (Letcher, 1898), ab., unavail.
  intermedia Grinnell, 1918, ab.,
    unavail.
  nivosa (Gunder, 1927), ab., unavail.
  schraderi (Gunder, 1929), ab.,
    unavail.
4437 **atalanta** (L., 1758), extralim.
  admiralis (Retzius, 1783)
  a. **rubria** (Fruhstorfer, 1909)
  edwardsi Grinnell, 1918

HYPANARTIA Hbn., 1821
  EUREMA Doubleday, 1848, preocc.
    by Hbn., 1819
  HEUREMA H.-S., 1865, preocc. by
    Agassiz, 1846
  SIPROETA; auth., not Hbn., 1823
4438 **lethe** (F., 1793)
  demonica Hbn., 1821

Hypolimnini

HYPOLIMNAS Hbn., 1819, emend.
  HIPOLYMNAS Hbn., 1819, incorr.
    orig. spell.
  ESOPTRIA Hbn., 1819
  DIADEMA Bdv., 1832
  EURALIA Westwood, 1850
  EUCALIA C. Felder, 1861
4439 **misippus** (L., 1764)
NOTE: many palearctic synonyms not included
  bolina; Drury, 1773

JUNONIA Hbn., 1819
  PRECIS; auth., not Hbn., 1819
  ALCYONEIS Hbn., 1819
  ARESTA Billberg, 1820
4440 **coenia** (Hbn., 1822)
  schraderi Gunder, 1925
  coena B. & Benj., 1926, missp.
  wilhelmi Gunder, 1927
  rubrosuffusa (Field, 1936)
  rosa (Whittaker & D. Stallings,
    1944), winter f.
4441 **nigrosuffusa** B. & McD., 1916
4442 **evarete** (Cram., 1780)
  lavinia; auth., not F., 1775
  genoveva (Stoll, 1782)
  lavina W. P. Comstock, 1942, missp.
  a. **zonalis** C. & R. Felder, 1867
  weidenhammeri Polacek, 1925

ANARTIA Hbn., 1819
  CELAENA Doubleday, 1849
  CELOENA Bdv., 1870
  ANARTIELLA Fruhstorfer, 1907
4443 **jatrophae** (Johansson, 1763), extralim.
  a. **guantanamo** Mun., 1942
  b. **luteipicta** Fruhstorfer, 1907
4444 **chrysopelea** (Hbn., 1825)
  lytrea; auth., not Godt., 1819
  eurytis Fruhstorfer, 1907
4445 **fatima** (F., 1793)
  venusta Fruhstorfer, 1907

SIPROETA Hbn., 1823
   METAMORPHA; auth., not Hbn.,
    1819
   VICTORINA Blanchard, 1840
   AMPHIRENE Doubleday, 1844
   AMPHIRENE Bdv., 1870, preocc.
    by Doubleday, 1844
4446 stelenes (L., 1758), extralim.
   sthenele (Hbn., 1819)
   steneles (Blanchard, 1840), missp.
    a. **biplagiata** (Fruhstorfer, 1907)
    pallida (Fruhstorfer, 1907)
    insularis (Holl., 1916)

ARGYNNINAE

EUPTOIETA Doubleday, 1848
4447 claudia (Cram., 1776)
   daunius (Herbst, 1798)
   daunus auth., missp.
   dodgei Gunder, 1927
   fumosa Field, 1936
   albaclaudia Field, 1936
4448 hegesia (Cram., 1780), extralim.
   columbina (F., 1793)
    a. **hoffmanni** W. P. Comstock, 1944

SPEYERIA Scudder, 1872
   SEMNOPSYCHE Scudder, 1875
   ARGYNNIS; auth., not F., 1807
4449 diana (Cram., 1777)
4450 cybele (F., 1775)
   daphnis (Cram., 1775)
   baal (Stkr., 1878)
    a. **novascotiae** (McD., 1935)
    b. **krautwurmi** (Holl., 1931)
    c. **pseudocarpenteri** (F. &
     R. Chermock, 1940)
    d. **carpenterii** (Edw., 1876)
    e. **charlottii** (Barnes, 1897)
    f. **letona** dos Passos & Grey, 1945
    g. **pugetensis** F. Chermock &
     Frechin, 1947
    h. **leto** (Behr, 1862)
    letis (Wgt., 1905)
    valesinoidesalba (Reuss, 1926)
    lethe (Gunder, 1934)
4451 aphrodite (F., 1787)
   bartschi (Reiff, 1910)
   bakeri (A. H. Clark, 1932)
    a. **winni** (Gunder, 1932)
    b. **alcestis** (Edw., 1877)
    suffusa (Wolcott, 1916)
    c. **manitoba** (F.&R. Chermock, 1940)
    mayae (Gunder, 1932), ab.
    mayae dos Passos & Grey, 1947
    d. **whitehousei** (Gunder, 1932)
    e. **columbia** (Hy. Edw., 1877)
    f. **ethne** (Hemming, 1933)
    cypris (Edw., 1886), preocc. by
     Meigen, 1828
    g. **byblis** (B. & Benj., 1926)
4452 idalia (Drury, 1773)
   astarte (Fisher, 1858, preocc., by
    Doubleday, 1847
   ashtaroth (Fisher, 1859)
   infumata (Oberth., 1912)
   dolli (Gunder, 1927)
   pallida (Eisner, 1942)
4453 nokomis (Edw., 1862)

nigrocaerulea (W. & T. Ckll., 1900)
rufescens (Ckll., 1909)
valesinoidesalba (Reuss, 1926),
   preocc. by Reuss, 1926
   a. **nitocris** (Edw., 1874)
   nitrocris dos Passos, 1964, missp.
   b. **coerulescens** (Holl., 1900)
   caerulescens McD., 1938, missp.
   c. **apacheana** (Skin., 1918)
   nokomis; Edw., 1873
   hermosa (Comst., 1925)
4454 edwardsii (Reak., 1866)
   aglaia; Edw., 1864
   montana (Reuss, 1926)
   edonis (Gunder, 1934)
4455 coronis (Behr, 1864)
   californica (Skin., 1917)
    a. **hennei** (Gunder, 1934)
    b. **semiramis** (Edw., 1886)
    bernardensis (Gunder, 1933)
    c. **gunderi** (Comst., 1925)
    d. **simaetha** dos Passos & Grey, 1945
    e. **snyderi** (Skin., 1897)
    f. **halcyone** (Edw., 1869)
4456 zerene (Bdv., 1852)
   monticola (Behr, 1863)
    a. **conchyliata** (Comst., 1925)
    shastaensis (Comst., 1925)
    b. **gloriosa** Moeck, 1957
    c. **sordida** (Wgt., 1905)
    d. **malcolmi** (Comst., 1920)
    sineargentatus (Gunder, 1927)
    e. **carolae** (dos Passos & Grey, 1942)
    f. **hippolyta** (Edw., 1879)
    g. **behrensii** (Edw., 1869)
    h. **myrtleae** dos Passos & Grey, 1945
    i. **bremnerii** (Edw., 1872)
    j. **picta** (McD., 1924)
    k. **garretti** (Gunder, 1932)
    l. **sinope** dos Passos & Grey, 1945
    m.**platina** (Skin., 1897)
    n. **pfoutsi** (Gunder, 1933)
    o. **cynna** dos Passos & Grey, 1945
4457 callippe (Bdv., 1852)
    a. **comstocki** (Gunder, 1925)
    creelmani (Gunder, 1934)
    b. **liliana** (Hy. Edw., 1877)
    baroni (Edw., 1881)
    c. **semivirida** (McD., 1924)
    d. **elaine** dos Passos & Grey, 1945
    e. **rupestris** (Behr, 1863)
    f. **juba** (Bdv., 1869)
    inornata (Edw., 1872)
    g. **laura** (Edw., 1879)
    h. **sierra** dos Passos & Grey, 1945
    i. **nevadensis** (Edw., 1870)
    j. **macaria** (Edw., 1877)
    k. **laurina** (Wgt., 1905)
    wrighti (Wgt., 1905)
    l. **harmonia** dos Passos & Grey,
     1945
    m.**meadii** (Edw., 1872)
    gerhardi (Gunder, 1927)
    n. **gallatini** (McD., 1929)
    o. **calgariana** (McD., 1924)
4458 egleis (Behr, 1862)
   montivaga (Behr, 1863)
   astarte (Edw., 1864), preocc. by
    Doubleday, 1847
   mammothi (Gunder, 1924)

boharti (Gunder, 1929)
   a. **adiaste** (Edw., 1864)
   adiante (Bdv., 1869)
   adraste (Kby., 1871), missp.
   adianthe (B. & McD., 1917),
    missp.
   b. **clemencei** (Comst., 1925)
   c. **atossa** (Edw., 1890)
   tejonica (Comst., 1925)
   d. **tehachapina** (Comst., 1920)
   e. **oweni** (Edw., 1892)
   f. **linda** (dos Passos & Grey, 1942)
   g. **macdunnoughi** (Gunder, 1932)
   h. **albrighti** (Gunder, 1932)
   i. **utahensis** (Skin., 1919)
   j. **toiyabe** Howe, 1975
   k. **secreta** dos Passos & Grey, 1945
4459 atlantis (Edw., 1862)
   chemo (Scudder, 1889)
   a. **canadensis** (dos Passos, 1935)
   b. **hollandi** (F. & R. Chermock, 1940)
   c. **hesperis** (Edw., 1864)
   d. **nikias** (Ehrmann, 1917)
   e. **dorothea** Moeck, 1947
   f. **nausicaa** (Edw., 1874)
   arizonensis (Elwes, 1889)
   g. **schellbachi** Garth, 1949
   shellbachi dos Passos, 1964, missp.
   h. **chitone** (Edw., 1879)
   i. **wasatchia** dos Passos & Grey,
    1945
   j. **greyi** (Moeck, 1950)
   k. **tetonia** dos Passos & Grey, 1945
   l. **viola** dos Passos & Grey, 1945
   m.**dodgei** (Gunder, 1931)
   n. **irene** (Bdv., 1869)
   cottlei (Comst., 1925)
   o. **electa** (Edw., 1878)
   cornelia (Edw., 1892)
   p. **lurana** dos Passos & Grey, 1945
   q. **hutchinsi** dos Passos & Grey, 1947
   hutchinsi (Gunder, 1932), ab.
   r. **beani** (B. & Benj., 1926)
   s. **lais** (Edw., 1884)
   helena dos Passos & Grey, 1957
   t. **dennisi** dos Passos & Grey, 1947
   dennisi (Gunder, 1927), ab.
4460 hydaspe (Bdv., 1869)
   a. **viridicornis** (Comst., 1925)
   b. **purpurascens** (Hy. Edw., 1877)
   caliginosa (Comst., 1925)
   c. **minor** dos Passos & Grey, 1947
   minor (McD., 1927), infrasubsp.
   d. **rhodope** (Edw., 1874)
   gregsoni (Gunder, 1932)
   e. **sakuntala** (Skin., 1911)
   skinneri (Holl., 1931)
   f. **conquista** dos Passos & Grey,
    1945
4461 mormonia (Bdv., 1869)
   a. **bischoffii** (Edw., 1870)
   b. **opis** (Edw., 1874)
   c. **jesmondensis** dos Passos & Grey,
    1947
   jesmondensis (McD., 1940),
    infrasubsp.
   d. **washingtonia** (B. & McD., 1913)
   rainierensis (Gunder, 1932)
   e. **erinna** (Edw., 1883)
   cunninghami (Owen, 1893)

60

f. **arge** (Stkr., 1878)
astarte (Edw., 1862), preocc. by
Doubleday, 1847
montivaga (Behr, 1864), preocc. by
Behr, 1863
g. **artonis** (Edw., 1881)
rubyensis (Gunder, 1932)
h. **eurynome** (Edw., 1872)
eurynomes (Edw., 1873), missp.
clio (Edw., 1874)
eris (Igel, 1922), preocc. by
Meigen, 1828
benjamini (Gunder, 1927)
brucei (Gunder, 1927)
gunderi (Field, 1936), preocc. by
Comst., 1925
igeli dos Passos & Grey, 1947
fieldi dos Passos & Grey, 1947
i. **luski** (B. & McD., 1913)

BOLORIA Moore, 1900
BRENTHIS; auth., part, not F.,
1807
4462 **napaea** (Hoffmannsegg, 1804),
extralim.
isis (Hbn., 1799–1800), preocc. by
Drury, 1773
pales; auth., not. D. & S., 1775
a. **alaskensis** (Holl., 1900)
reiffi Reuss, 1925
b. **nearctica** Verity, 1932
pales; Lehmann, 1913
c. **halli** Klots, 1940

PROCLOSSIANA Reuss, 1926
BRENTHIS; auth., part, not F.,
1807
BOLORIA; auth., part, not Moore,
1900
4463 **eunomia** (Esper, 1787), extralim.
aphirape (Hbn., 1799–1800)
tomyris (Herbst, 1800)
a. **triclaris** (Hbn., 1821)
ossianus (Bdv., 1832), preocc. by
Herbst, 1800
lais (Scudder, 1875), unavail.,
publ. in syn.
b. **caelestis** (Hemming, 1933)
alticola (B. & McD., 1913),
preocc. by Sushkin &
Tsetverikov, 1907
c. **dawsoni** (B. & McD., 1916)
harperi (Gunder, 1934)
d. **nichollae** (B. & Benj., 1926)
e. **laddi** (Klots, 1940)
f. **ursadentis** (Ferris & Groothuis,
1971)
g. **denali** (Klots, 1940)

CLOSSIANA Reuss, 1920
BRENTHIS; auth., part, not F.,
1807
BOLORIA; auth., part, not Moore,
1900
4464 **selene** (D. & S., 1775), extralim.
NOTE: many palearctic synonyms not included
a. **myrina** (Cram., 1777)
myrissa (Godt., 1819)
nubes (Scudder, 1889)
nivea (Gunder, 1928)

marilandica (A. H. Clark, 1941)
b. **nebraskensis** (Holl., 1928)
c. **sabulicollis** (Kohler, 1977)
d. **tollandensis** (B. & Benj., 1925)
serratimarginata (Gunder, 1926)
e. **albequina** (Holl., 1928)
baxteri (Holl., 1928)
f. **atrocostalis** (Huard, 1927)
jenningsae (Holl., 1928)
g. **terraenovae** (Holl., 1928), emend.
4465 **bellona** (F., 1775)
myrina; Martyn, 1797
fasciata (Ckll., 1889)
kleeni (Watson, 1921)
ammiralis (Hemming, 1933)
pardopsis (Holl., 1928)
a. **toddi** (Holl., 1928)
b. **jenistae** (D. Stallings & Turner,
1947)
4466 **frigga** (Thunb., 1791), extralim.
a. **saga** (Stgr., 1861)
b. **alaskensis** (Lehmann, 1913)
gibsoni (B. & Benj., 1926)
lehmanni (Holl., 1928)
c. **sagata** (B. & Benj., 1923)
4467 **improba** (Butler, 1877)
a. **youngi** (Holl., 1900)
4468 **kriemhild** (Stkr., 1879)
laurenti (Skin., 1913)
luecki Reuss, 1923
kriemeld (McHenry, 1964), missp.
lucki (dos Passos, 1964), missp.
4469 **epithore** (Edw., 1864)
a. **chermocki** (E. & S. Perkins, 1966)
b. **borealis** (E. Perkins, 1973)
obscuripennis (Gunder, 1926), ab.
c. **sierra** (E. Perkins, 1973)
eldorado (Strand, 1915), ab.
wawonae (Gunder, 1924), ab.
4470 **polaris** (Bdv., 1828)
groenlandica (Skin., 1892)
americana (Strand, 1905)
a. **stellata** (Masters, 1972)
4471 **freija** (Thunb., 1791)
lapponica (Esp., 1793)
freya (Godt., 1819)
gunderi (Harper, 1933)
a. **tarquinius** (Curt., 1835)
b. **natazhati** (Gibs., 1920)
c. **nabokovi** (D. Stallings & Turner,
1947)
d. **browni** (Higgins, 1953)
4472 **alberta** (Edw., 1890)
banffensis (Gunder, 1932)
4473 **astarte** (Doubleday & Hewitson, 1847)
victoria (Edw., 1891)
a. **distincta** (Gibs., 1920)
4474 **titania** (Esp., 1793), extralim.
a. **boisduvalii** (Dup., 1832)
labradorensis (Gunder, 1928)
b. **rainieri** (B. & McD., 1913)
ranieri (B. & McD., 1913), missp.
c. **grandis** (B. & McD., 1916)
d. **montina** (Scudder, 1863)
e. **ingens** (B. & McD., 1918)
martini (Gunder, 1934)
f. **helena** (Edw., 1871)
4475 **chariclea** (Schneider, 1794), extralim.
carichlea (Zett., 1839), missp.
churiclea (Johansen, 1921), missp.

churchilea (dos Passos, 1964), missp.
a. **arctica** (Zett., 1839)
oenone (Scudder, 1875)
obscurata (M'Lachlan, 1878)
b. **butleri** (Edw., 1883)
butlerii (Edw., 1884), missp.

MELITAEINAE

ANTHANASSA Scudder, 1875
TRITANASSA Fbs., 1945
4476 **texana** (Edw., 1863)
smerdis (Hewitson, 1864)
a. **seminole** (Skin., 1911)
4477 **ptolyca** (Bates, 1864)

ERESIA Bdv., 1836
NEPTIS Illiger, 1807, suppr.
(ICZN Op. 232)
4478 **frisia** (Poey, 1832)
gyges (Hewitson, 1864)
a. **tulcis** (Bates, 1864)
genigueh Reak., 1865
archesilea R. Felder, 1869
punctata (Edw., 1870)
4479 **leucodesma** C. & R. Felder, 1861,
quest. occur.
cincta Edw., 1864

PHYCIODES Hbn., 1819
PHICIODES Hbn., 1826, missp.
4480 **phaon** (Edw., 1864)
gorgone (Hbn., 1810), ♀
aestiva (Edw., 1878), sum. f.
hiemalis (Edw., 1878), win. f.
thornei Gunder, 1934
4481 **tharos** (Drury, 1773)
morpheus (F., 1775), spr. f.
cocyta (Cram., 1777)
euclea (Brgstr., 1780)
tharossa (Godt., 1819)
selenis (Kby., 1837)
pulchella (Bdv., 1852)
marcia (Edw., 1868), spr. f.
packardii (Saund., 1869)
reaghi Reiff, 1913
dyari Gunder, 1928
a. **distinctus** Bauer, 1975
b. **pascoensis** Wgt., 1905
herse Hall, 1924, ♀
nigrescens Hall, 1924
c. **arcticus** dos Passos, 1935
4482 **batesii** (Reak., 1865)
harperi Gunder, 1932
4483 **campestris** (Behr, 1863)
pratensis (Behr, 1863), ♂
mcdunnoughi Gunder, 1928
macdunnoughi dos Passos, 1964,
emend.
a. **montanus** (Behr, 1863)
orsa (Bdv., 1869)
b. **camillus** Edw., 1871
emissa Edw., 1871
rohweri Ckll., 1913
tristis Ckll., 1913
4484 **pictus** Edw., 1865
jemezensis Brehme, 1913
a. **canace** Edw., 1871
4485 **vesta** (Edw., 1869)
hiemalis Edw., 1878, win. f.

aestiva Edw., 1878, sum. f.
boucardi Godm. & Salvin, 1882
4486 **orseis** Edw., 1871
edwardsi Gunder, 1927
a. **herlani** Bauer, 1975
4487 **pallidus** (Edw., 1864)
mata (Reak., 1866)
a. **barnesi** Skin., 1897
4488 **mylitta** (Edw., 1861)
collina (Behr., 1863)
epula (Bdv., 1869)
collinsi Gunder, 1930
macyi Fender, 1930
callina; auth.
a. **arizonensis** Bauer, 1975
thebais; auth., not Godm. & Salvin,
1878
arida; auth., not Skin., 1897

CHARIDRYAS Scudder, 1872
LIMNAECIA Scudder, 1872,
preocc. by Staint., 1851
MELITAEA; auth., part, not F.,
1807
PHYCIODES; auth., part, not Hbn.,
1819
CHLOSYNE; auth., part, not Butler,
1870
MILITAEA Edw., 1873, missp.
4489 **gorgone** (Hbn., 1810)
ismeria (Bdv. & Leconte, 1833),
ident. uncert.
a. **carlota** (Reak., 1866)
nigra (Cary, 1901)
nox (Gunder, 1928)
4490 **nycteis** (Doubleday, 1847)
oenone (Scudder, 1863)
harrisii; Skin., 1904
millburni (Rummel, 1926)
lacteus (Gunder, 1928)
greyi (Field, 1934)
a. **drusius** (Edw., 1884)
hewitsoni (Field, 1936)
b. **reversa** (F. & R. Chermock, 1940)
4491 **harrisii** (Scudder, 1862)
albimontana (Avinoff, 1930)
a. **hanhami** (Fletcher, 1904)
b. **liggetti** (Avinoff, 1930)
4492 **palla** (Bdv., 1852)
eremita (Wgt., 1905), ♀
wardi (Oberth., 1913)
blackmorei (Gunder, 1926)
stygiana (Comst., 1926)
hemifusa (Gunder, 1930)
whitneyi (Behr, 1863)
vanduzeei (Gunder, 1928)
b. **vallismortis** (F. Johnson, 1938),
emend.
c. **flavula** (B. & McD., 1918)
d. **calydon** (Stkr., 1878)
e. **sterope** (Edw., 1870)
hewesi (Leussler, 1931)
hopfingeri (Gunder, 1934)
dorothyi Bauer, 1975
4493 **acastus** (Edw., 1874)
pearlae (Gunder, 1926)
4494 **neumoegeni** (Skin., 1895)
fridayi (Gunder, 1932)
boharti (Gunder, 1933)
a. **sabina** (Wgt.; 1905)

4495 **gabbii** (Behr, 1863)
pola (Bdv., 1869)
sonorae (Bdv., 1869)
pasadenae (Gunder, 1924)
newcombi (Comst., 1926)
gunderi (Comst., 1926)
4496 **damoetas** (Skin., 1902)
damoetella (McD., 1927)
a. **malcolmi** (Comst., 1926)
4497 **hoffmanni** (Behr, 1863)
helicta (Bdv., 1869)
abnorma (Wgt., 1905)
hollandae (Gunder, 1928)
a. **segregata** (B. & McD., 1918)
bridgei (Comst., 1924)
b. **manchada** Bauer, 1975

CHLOSYNE Butler, 1870
SYNCHLOE Doubleday, 1844,
preocc. by Hbn., 1818
COATLANTONA Kby., 1871
4498 **californica** (Wgt., 1905)
chinoi Gunder, 1924
4499 **lacinia** (Gey., 1837), extralim.
tellias (Bates, 1864)
mediatrix (C. & R. Felder, 1867)
indigens Higgins, 1960
a. **crocale** (Edw., 1874)
nigrescens Ckll., 1893
nigra Ckll., 1893, part
bicolor Ckll., 1893, part
rufa (Ckll., 1893), part
rufescens Ckll., 1894, preocc. by
Bdv., 1869
flavida Higgins, 1960
b. **adjutrix** Scudder, 1875
nigra Ckll., 1893, part
bicolor Ckll., 1893, part
rufa Ckll., 1893, part
inghami Gunder, 1928
4500 **definita** (Aaron, 1884)
albiplaga (Aaron, 1884)
schausi (Godm., 1901)
4501 **endeis** (Godm. & Salvin, 1894)
4502 **erodyle** (Bates, 1864)
4503 **janais** (Drury, 1782)
4504 **rosita** Hall, 1924, extralim.
a. **browni** Bauer, 1961

THESSALIA Scudder, 1875
MELITAEA; auth., part, not F.,
1807
CHLOSYNE; auth., part, not Butler,
1870
4505 **leanira** (C. & R. Felder, 1860)
daviesi Wind, 1947
a. **oregonensis** Bauer, 1975
obsoleta (Hy. Edw., 1876), ab.
obliterata (Stkr., 1878), missp.
b. **wrighti** (Edw., 1886)
leona (Wgt., 1905)
carolynae (Gunder, 1926)
pelona (Gunder, 1930)
c. **cerrita** (Wgt., 1905)
d. **alma** (Stkr., 1878)
koebelei (Gunder, 1927)
4506 **fulvia** (Edw., 1879)
sinefascia (R. C. Williams, 1914)
4507 **cyneas** (Godm. & Salvin, 1882)
infrequens (Gunder, 1928)

4508 **theona** (Mén., 1855), extralim.
a. **thekla** (Edw., 1870)
benjamini (Gunder, 1928)
b. **bollii** (Edw., 1877)
4509 **chinatiensis** (Tinkham, 1944)

MICROTIA Bates, 1864
4510 **elva** Bates, 1864
horni Rebel, 1906
draudti Röber, 1914

DYMASIA Higgins, 1960
4511 **dymas** (Edw., 1877)
larunda (Stkr., 1878)
senrabii (Barnes, 1900)
4512 **chara** (Edw., 1883)
a. **imperialis** (Bauer, 1959)
jacintoi (Gunder, 1924), ab.
nitela (Comst., 1926), ab.

TEXOLA Higgins, 1959
4513 **elada** (Hewitson, 1868), extralim.
socia (C. & R. Felder, 1869)
a. **callina** (Bdv., 1869)
ulrica (Edw., 1877)
imitata (Stkr., 1878)
b. **perse** (Edw., 1882)

POLADRYAS Bauer, 1961
4514 **arachne** (Edw., 1869)
polingi (Gunder, 1926)
gunderiae (Holl., 1931)
gilensis (Holl., 1931)
pola; auth., not Bdv., 1869
a. **monache** (Comst., 1918)
b. **nympha** (Edw., 1884)
4515 **minuta** (Edw., 1861)
approximata (Stkr., 1900)

EUPHYDRYAS Scudder, 1872
LEMONIAS Hbn., 1806, suppr.
(ICZN Op. 97)
MELITAEA; auth., not F., 1807
4516 **phaeton** (Drury, 1773)
phaetaena (Hbn., 1819)
phaetontea (Godt., 1819)
phaedon (H.-S., 1865), missp.
superba (Stkr., 1878)
phaethusa (Hulst, 1881)
streckeri Ellsworth, 1902
schausi A. H. Clark, 1927
magnifica A. H. Clark, 1927
borealis F. & R. Chermock, 1940
a. **ozarkae** Masters, 1968

OCCIDRYAS Higgins, 1979
EUPHYDRYAS; auth., not
Scudder, 1872, part
4517 **chalcedona** (Doubleday, 1847)
cooperi (Behr, 1863)
fusimacula (Barnes, 1900)
mariana (Barnes, 1900)
grundeli (Coolidge, 1908)
lorquini (Oberth., 1913)
omniluteofuscus (Gunder, 1925)
hemiluteofuscus (Gunder, 1925)
suprafusa (Comst., 1926)
fusisecunda (Comst., 1926)
supranigrella (Comst., 1926)
hemimelanica (Comst., 1926)

a. **dwinnellei** (Hy. Edw., 1881)
  sperryi (F. & R. Chermock, 1945)
b. **quino** (Behr, 1863)
  hennei (Gunder, 1932)
c. **klotsi** (dos Passos, 1938)
  hermosa; dos Passos, 1964, part
d. **corralensis** (T. & J. Emmel, 1973)
e. **kingstonensis** (T. & J. Emmel, 1973)
f. **macglashanii** (Rivers, 1888)
  truckeensis (Gunder, 1928)
  georgei (Gunder, 1928)
  hilli (Gunder, 1928)
  mcglashanii (McD., 1938), missp.
g. **olancha** (Wgt., 1905)
  malcolmi (Gunder, 1927)
h. **sierra** (Wgt., 1905)
  magdalenae (Gunder, 1925)
  umbrobasana (Comst., 1925)
  irelandi (Gunder, 1929)
4518 **colon** (Edw., 1881)
  mcdunnoughi (Gunder, 1928)
  fenderi (Gunder, 1932)
  bakeri (Fender, 1945), preocc. by D.
    Stallings & Turner, 1945
  macdunnoughi (dos Passos, 1964), emend.
  a. **perdiccas** (Edw., 1880)
    perdiceas (Edw., 1880), missp.
    nigrisupernipennis (Gunder, 1926)
    svihlae (Gunder, 1932)
  b. **paradoxa** (McD., 1927)
  c. **wallacensis** (Gunder, 1928)
    huellemanni (Comst., 1926), ab.
    idahoensis (Gunder, 1929)
  d. **nevadensis** (Bauer, 1975)
4519 **anicia** (Doubleday, 1847)
  mayi (Gunder, 1932)
  a. **helvia** (Scudder, 1869)
  b. **howlandi** (D. Stallings &
    Turner, 1947)
  c. **capella** (Barnes, 1897)
    rubrolimbata (Comst., 1918)
    oslari (Gunder, 1925)
  d. **hermosa** (Wgt., 1905)
    venusta (Gunder, 1932)
  e. **magdalena** (B. & McD., 1918)
  f. **alena** (B. & Benj., 1926)
  g. **carmentis** (B. & Benj., 1926)
    charlotteae (Gunder, 1928)
  h. **windi** (Gunder, 1932)
  i. **eurytion** (Mead, 1875)
    nubigena; Mead, 1875
    brucei (Edw., 1888)
    melanodisca (Comst., 1918)
  j. **maria** (Skin., 1899)
    skinneri (Gunder, 1928)
    spaldingi (Gunder, 1928)
  k. **effi** (D. Stallings & Turner, 1945)
  l. **hopfingeri** (Gunder, 1934)
    andersoni (Gunder, 1934)
  m. **bernadetta** (Leussler, 1920)
    leussleri (Gunder, 1927)
    belli (Gunder, 1929)
  n. **veazieae** (Fender & Jewett, 1953)
  o. **bakeri** (D. Stallings & Turner, 1945)
  p. **macyi** (Fender & Jewett, 1953)
  q. **wheeleri** (Hy. Edw., 1881)
    duncani (Gunder, 1934)
  r. **morandi** (Gunder, 1928)
4520 **editha** (Bdv., 1852)
  cottlei (Gunder, 1928)

a. **wrighti** (Gunder, 1929)
  fieldi (Gunder, 1934)
  thornei (Gunder, 1934)
  rubyae (Hower, 1935)
b. **insularis** (T. & J. Emmel, 1975)
c. **bayensis** (Sternitzky, 1937)
d. **taylori** (Edw., 1888)
  victoriae (Gunder, 1926)
  barnesi (Gunder, 1928)
e. **beani** (Skin., 1897)
  blackmorei (Gunder, 1926)
f. **hutchinsi** (McD., 1928)
  montanus (McD., 1928)
g. **alebarki** (Ferris, 1971)
h. **gunnisonensis** (F. M. Brown, 1971)
i. **edithana** (Strand, 1914)
j. **baroni** (Edw., 1879)
  mirabilis (Wgt., 1905)
  dunni (Gunder, 1929)
  sternitzkyi (Gunder, 1929)
k. **rubicunda** (Hy. Edw., 1881)
l. **monoensis** (Gunder, 1928)
  fridayi (Gunder, 1931)
m. **lehmani** (Gunder, 1929)
n. **augusta** (Edw., 1890)
  augustina (Wgt., 1905)
o. **colonia** (Wgt., 1905)
p. **aurilacus** (Gunder, 1928)
  foxi (Gunder, 1924), ab.
  albiradiata (Gunder, 1926), ab.
q. **remingtoni** (Burdick, 1959)
  thielsenensis (Gunder, 1931), ab.
  thielsonensis (McD., 1938), missp.
r. **lawrencei** (Gunder, 1931)
  diamondensis (Gunder, 1931)
s. **nubigena** (Behr, 1863)
  nubigeria (Edw., 1873), missp.
  rubrosuffusa (Comst., 1926)
  tiogaensis (Gunder, 1929)
  boharti (Gunder, 1929)
  tiogoensis (McD., 1938), missp.

HYPODRYAS Higgins, 1979
4521 **gillettii** (Barnes, 1897)
  glacialis (Skin., 1921)
  herri (Gunder, 1929)
  gracialis (McD., 1938), missp.

LIMENITIDINAE

Limenitidini

BASILARCHIA Scudder, 1872
  CALLIANIRA Hbn., 1819, preocc.
    by Péron & Lesueur, 1810
  NYMPHALIS C. & R. Felder,
    1861, preocc. by L., 1758
  LIMENITIS; auth., not F., 1807
  NAJAS Hbn., 1806, in part,
    rejected (ICZN Op. 97)
  NYMPHA; auth., not Krause, 1839,
    preocc. by Fitzinger, 1826
  NYMPHALUS; auth., not Boitard,
    1828
4522 **arthemis** (Drury, 1773)
  lamina (F., 1793)
  proserpina (Edw., 1865)
  rufescens (Ckll., 1889)
  arthechippus Scudder, 1889, unavail.
    hybr. name

cerulea (Ehrmann, 1900)
albofasciata (Newcomb, 1907)
benjamini (Nakahara, 1924)
virithemis Field, 1936
prosperpina McD., 1938, missp.
a. **rubrofasciata** B. & McD., 1916
  rubrofasechippus Gunder, 1934,
    unavail., hybr. name
b. **astyanax** (F., 1775)
  ephestion (Stoll, 1790)
  ursula (F., 1793)
  ephestiaena (Hbn., 1819), missp.?
  viridis (Stkr., 1878)
  atlantis (Nakahara, 1923)
  inornata (Nakahara, 1924)
  purpuratus Gunder, 1934
  tildeni Field, 1934
  rubidus (Stkr., 1878), unavail.,
    hybr. name
c. **arizonensis** (Edw., 1882)
  doudoroffi Gunder, 1934
4523 **archippus** (Cram., 1776)
  disippe (Godt., 1824)
  disippus (Bdv. & Leconte, 1834)
  pseudorippus (Stkr., 1878)
  rubidus (Stkr., 1878), unavail., hybr.
    name
  arthechippus Scudder, 1889, unavail.,
    hybr. name
  lanthanis Cook & Watson, 1909
  advena (Ellsworth, 1918)
  cayuga Nakahara, 1923
  nivosus Gunder, 1929
  rubrofasechippus Gunder, 1934,
    unavail., hybr. name
  weidechippus Cross, 1936, unavail.,
    hybr. name
a. **floridensis** (Stkr., 1878)
  nig (Stkr., 1878)
  eros (Edw., 1880)
  nigricans (Stkr., 1900)
  halli (Watson & W. P. Comstock,1920)
b. **watsoni** dos Passos, 1938
c. **obsoleta** (Edw., 1882)
  hulstii (Edw., 1882)
d. **lahontani** (Herlan, 1971)
4524 **weidemeyerii** (Edw., 1861)
  nigerrima Ckll., 1927
  weidechippus Cross, 1936, unavail.,
    hybr. name
a. **angustifascia** B. & McD., 1912
  sinefascia (Edw., 1882), ab.,
    emend. (B. & McD., 1924)
b. **nevadae** B. & Benj., 1924
c. **oberfoelli** (F. M. Brown, 1960)
d. **latifascia** (E. & S. Perkins, 1967)
  fridayi Gunder, 1932, unavail.,
    hybr. name
4525 **lorquini** (Bdv., 1852)
  eavesii (Hy. Edw., 1877)
  comstocki Gunder, 1925
  fridayi Gunder, 1932, unavail., hybr.
    name
  gunderi Field, 1936
  powelli Field, 1936
a. **burrisonii** (Maynard, 1891)
  maynardi Field, 1936

ADELPHA Hbn., 1819
  HETEROCHROA Bdv., 1836

LIMENITIS; auth., not F., 1807
4526 **fessonia** (Hewitson, 1847)
4527 **bredowii** (Gey., 1837), extralim.
    a. **eulalia** (Doubleday, 1848)
    b. **californica** (Butler, 1865)

Epicaliini

MYSCELIA Doubleday, 1844
  SAGARITIS Hbn., 1821, preocc.
    by Billberg, 1820
4528 **ethusa** (Bdv., 1836)
  cyanecula C. & R. Felder, 1866
4529 **skinneri** Mengel, 1894, quest. occur.
4530 **streckeri** (Skin., 1889), quest. occur.
4531 **cyananthe** C. & R. Felder, 1866

EUNICA Hbn., 1819
  EVONYME Hbn., 1819
  EUNICE Gey., 1832, preocc. by
    Rafinesque, 1815
  CRENIS Bdv., 1833, preocc. by
    Hbn., 1821
  CALLIANIRA Doubleday &
    Hewitson, 1847, preocc. by Péron
    & Lesueur, 1810
  FAUNIA Poey, 1847
  AMYCLA Doubleday, 1849, preocc.
    by Rafinesque, 1815
  SALLYA Hemming, 1964
4532 **monima** (Stoll, 1782)
  hyperipte (Hbn., 1823)
  myrto (Godt., 1824)
  modesta Bates, 1864
  myrta Skin., 1898, missp.
4533 **tatila** (H.-S., 1853), extralim.
  coerulea Godm. & Salvin, 1877
    a. **tatilista** Kaye, 1926

DYNAMINE Hbn., 1819
  SIRONIA Hbn., 1823
  EUBAGIS Bdv., 1832
  ARISBA Doubleday, 1847
  SERONIA Kby, 1871, missp.
4534 **dyonis** Gey., 1837

DIAETHRIA Billberg, 1820
  CORECALLA Röber, 1916
  CALLICORE; auth., not Hbn., 1819
4535 **clymena** (Cram., 1776)
4536 **asteria** (Godm. & Salvin, 1894)

Eurytelini

MESTRA Hbn., 1825
  CYSTINEURA Bdv., 1836
4537 **amymone** (Mén., 1857)
  dorcas (Skin., 1904)
4538 **cana** (Erichson, 1848), extralim.
    a. **floridana** (Stkr., 1900),
      quest. occur.

BIBLIS F., 1807
  ZONAGA Billberg, 1820
  DIDONIS; Scudder, 1875, not Hbn.,
    1819
  ZONAGRA Sherborn, 1932, missp.
4539 **hyperia** (Cram., 1780), extralim.
  biblis (F., 1775), preocc. by Drury,
    1773

thadana (Godt., 1819)
    a. **aganisa** Bdv., 1836

Ageroniini

HAMADRYAS Hbn., 1806
  APATURA Illiger, 1807, suppr.
    (ICZN Op. 232)
  AGERONIA Hbn., 1819
  PHILOCALA Billberg, 1820
  PERIDROMIA Bdv., 1836
  AMPHICHLORA C. Felder, 1861
4540 **fornax** (Hbn., 1823), extralim.
    a. **fornacalia** (Fruhstorfer, 1907)
4541 **feronia** (L., 1758), extralim.
    a. **farinulenta** (Fruhstorfer, 1916)
4542 **februa** (Hbn., 1823), extralim.
    a. **gudula** (Fruhstorfer, 1916)
4543 **ferox** (Stgr., 1888), quest. occur.
4544 **amphinome** (L., 1767), extralim.
    a. **mexicana** (Luc., 1853)

Coloburini

HISTORIS Hbn., 1819
  COEA Hbn., 1819
  AGANISTHOS Bdv. & Leconte,
    1834
  MEGISTANIS Doubleday, 1844
  MEGISTANIS Bdv., 1870, preocc.
    by Doubleday, 1844
  AGANISTOS Bdv., 1870, missp.
  AGANISTHUS Kby., 1904, missp.
4545 **odius** (F., 1775)
  orion (F., 1775), preocc. by Pallas, 1771
  danae (Cram., 1776), preocc. by F.,
    1775
4546 **acheronta** (F., 1775), quest. occur.
  cadmus (Cram., 1776)
  pherecydes (Stoll, 1782)

SMYRNA Hbn., 1823
4547 **karwinskii** Gey., 1833

MARPESIINAE

MARPESIA Hbn., 1818
  ATHENA Hbn., 1819
  EUGLYPHUS Billberg, 1820
  MARIUS Swainson, 1830
  PETREUS Swainson, 1833
  MEGALURA Blanchard, 1840
  TIMETES Doubleday, 1844
  TYMETES Bdv., 1846, missp.
  TIMETES Bdv., 1870, preocc. by
    Doubleday, 1848
  EUMARGARETA Grt., 1898
4548 **coresia** (Godt., 1820)
  zerynthia Hbn., 1823
  sylla (Perty, 1833)
4549 **chiron** (F., 1775)
  marius (Cram., 1779)
  chironias (Hbn., 1819)
4550 **petreus** (Cram., 1776)
  peleus (Sulz., 1776), preocc. by L., 1764
  thetys (F., 1777), preocc. by
    Rottemburg, 1763
  thetis Godt., 1819
4551 **eleuchea** Hbn., 1818
  eleucha auth., missp.

APATURIDAE
by LEE D. MILLER & F. MARTIN BROWN

CHARAXINAE

ANAEA Hbn., 1819
  PYRRHANAEA Röber, 1888
  PAPHIA; auth., not Bolton, 1798
4552 **aidea** (Guér.-Méneville, 1844)
  didea (Weidemeyer, 1864), missp.?
  morrisonii (Edw., 1883)
  morrisonia (Sm., 1884), missp.
  appiciata Röber, 1916
4553 **floridalis** F. Johnson &
    W. P. Comstock, 1941
  portia; Schaus, 1898
  floraesta F. Johnson &
    W. P. Comstock, 1941, sum. f.
4554 **andria** Scudder, 1875
  glycerium, Edw., 1871
  ops (Druce, 1877)
  troglodyta; Stkr., 1878
  glycerinus Stichel, 1939
  andriaesta F. Johnson &
    W. P. Comstock, 1941, sum. f.

MEMPHIS Hbn., 1819
  CORYCIA Hbn., 1825
  CYMATOGRAMMA Doubleday &
    Hewitson, 1849
  EUSCHATZIA Grt., 1898
  ANAEA; auth., not Hbn., 1819
4555 **glycerium** (Doubleday, 1850)
  glyzerium Fassl, 1918, missp.
4556 **pithyusa** (R. Felder, 1869)
  daguana Bargmann, 1929

APATURINAE

ASTEROCAMPA Röber, 1916
  CELTIPHAGA B. & L., 1922
  APATURA; auth., not F., 1807
4557 **celtis** (Bdv. & Leconte, 1833)
  alb (Stkr., 1878)
  inornata (Wolcott, 1916)
4558 **antonia** (Edw., 1877)
4559 **montis** (Edw., 1883)
4560 **subpallida** (B. & McD., 1913)
4561 **leilia** (Edw., 1874)
    a. **cocles** (Lint., 1884)
4562 **alicia** (Edw., 1868)
4562.1 **clyton** (Bdv. & Leconte, 1833)
  proserpina (Scudder, 1868)
  ocellata (Edw., 1876)
  nig (Stkr., 1878)
4563 **flora** (Edw., 1876)
4564 **texana** (Skin., 1911)
4565 **louisa** D. Stallings & Turner, 1947

DOXOCOPA Hbn., 1819
  CATARGYRIA Hbn., 1823
  CHLORIPPE Doubleday, 1844
  CHLORIPPE Bdv., 1870, preocc.
    by Doubleday, 1844
4566 **pavon** (Latr., 1809)
  pavonii (H.-S., 1856), missp.
4567 **laure** (Drury, 1773)

SATYRIDAE

by LEE D. MILLER & F. MARTIN BROWN

ELYMNIINAE

Parargini

ENODIA Hbn., 1819
4568 **portlandia** (F., 1781)
  andromacha (Hbn., 1809)
  androcardia Hbn., 1821
  a. **floralae** (J. R. Heitzman & dos
    Passos, 1974)
  b. **missarkae** (J. R. Heitzman & dos
    Passos, 1974)
4568.1 **anthedon** A. H. Clark, 1936
  a. **borealis** A. H. Clark, 1936
4568.2 **creola** (Skin., 1897)

SATYRODES Scudder, 1875
4568.3 **eurydice** (Johansson, 1763)
  cantheus (Godt., 1823)
  canthus; auth., not L., 1767
  transmontana (Gosse, 1840), nom.
    nud.
  boisduvalli (Harr., 1862)
  boweri F. Chermock, 1927
  rawsoni Field, 1926
  a. **fumosus** Leussler, 1916
4569 **appalachia** (R. Chermock, 1947)
  a. **leeuwi** (Gatrelle & Arbogast, 1974)

SATYRINAE

Euptychiini

CYLLOPSIS R. Felder, 1869
  EUPTYCHIA; auth., part, not
    Hbn., 1818
  NEONYMPHA; auth., part, not
    Hbn., 1818
4570 **pyracmon** (Butler, 1866), extralim.
  a. **nabokovi** L. Miller, 1974
4571 **henshawi** (Edw., 1876)
4572 **pertepida** (Dyar, 1912), extralim.
  a. **dorothea** (Nabokov, 1942)
  edwardsi (Nabokov, 1942)
  b. **maniola** (Nabokov, 1942)
  c. **avicula** (Nabokov, 1942)
  texana (Wind, 1946)
4573 **gemma** (Hbn., 1808)
  ?cornelius (F., 1793)
  a. **freemani** (D. Stallings & Turner,
    1946)
  inductura (D. Stallings & Turner,
    1946), win. f.

HERMEUPTYCHIA Forster, 1964
  EUPTYCHIA; auth., part, not
    Hbn., 1818
4574 **hermes** (F., 1775)
  canthe (Hbn., 1811)
4575 **sosybius** (F., 1793)

NEONYMPHA Hbn., 1818
  EUPTYCHIA; auth., part, not
    Hbn., 1818
4576 **areolata** (J. E. Smith, 1797)
  phocion (F., 1787), preocc. by F.,
    1775

?helicta (Hbn., 1808)
  a. **septentrionalis** Davis, 1924
4577 **mitchellii** French, 1889

MEGISTO Hbn., 1819
  EUPTYCHIA; auth., part, not
    Hbn., 1818
4578 **cymela** (Cram., 1777)
  eurytus (F., 1775), preocc. by L.,
    1758
  eurytris (F., 1793)
  eurythris (Godt., 1821), missp.
  eurytis (dos Passos, 1964), missp.
  a. **viola** (Maynard, 1891)
4579 **rubricata** (Edw., 1871)
  a. **smithorum** (Wind, 1946)
  b. **cheneyorum** (R. Chermock, 1949)

PARAMACERA Butler, 1868
  PARAMECERA Butler, 1868
4580 **allyni** L. Miller, 1972
  xicaque; auth., not Reak., 1867

Coenonymphini

COENONYMPHA Hbn., 1819
  CHORTOBIUS Dunning &
    Pickard, 1858
  CHORTOBIUS Doubleday, 1859,
    preocc. by Dunning & Pickard,
    1858
  SICCA Verity, 1953
4581 **haydenii** (Edw., 1872)
4582 **kodiak** Edw., 1869
  a. **yukonensis** Holl., 1900
  b. **mixturata** Alpheraky, 1897
4583 **inornata** Edw., 1861
  quebecensis B. & Benj., 1926
  a. **mcisaaci** dos Passos, 1935
  maciasaaci dos Passos, 1964,
    emend.
  b. **benjamini** McD., 1928
  c. **nipisiquit** McD., 1939
  d. **heinemani** F. M. Brown, 1959
4584 **ochracea** Edw., 1861
  albescens Field, 1936
  ochracia F. & R. Chermock, 1938,
    missp.
  phantasma Burdick, 1956
  a. **brenda** Edw., 1869
  b. **mackenziei** Davenport, 1936
  c. **mono** Burdick, 1942
  d. **furcae** B. & Benj., 1926
  e. **subfusca** B. & Benj., 1926
4585 **ampelos** Edw., 1871
  a. **eunomia** Dornfeld, 1967
  eunomia Field, 1937, spr. f.
  b. **elko** Edw., 1881
  sweadneri F. & R. Chermock, 1941
  c. **columbiana** McD., 1928
  d. **insulana** McD., 1928
4586 **california** Westwood, 1851
  galactinus (Bdv., 1852)
  ceres Butler, 1866
  pulla Hy. Edw., 1881
  californica Hy. Edw., 1881, missp.
  a. **eryngii** Hy. Edw., 1877
  siskiyouensis Comst., 1925

Maniolini

CERCYONIS Scudder, 1875
  MINOIS; auth., not Hbn., 1819
4587 **pegala** (F., 1775)
  maritima (Edw., 1880)
  pegale Sm., 1884, missp.
  a. **abbotti** F. M. Brown, 1969
  b. **alope** (F., 1793)
  ochracea (F. & R. Chermock,
    1942)
  carolina (F. & R. Chermock, 1942)
  c. **nephele** (Kby., 1837)
  d. **olympus** (Edw., 1880)
  borealis F. Chermock, 1929
  e. **texana** (Edw., 1880)
  f. **ino** Hall, 1924
  g. **boopis** (Behr, 1864)
  baroni (Edw., 1880)
  incana (Edw., 1880)
  h. **ariane** (Bdv., 1852)
  gabbii (Edw., 1870)
  i. **stephensi** (Wgt., 1905)
  j. **wheeleri** (Edw., 1873)
  hoffmani (Stkr., 1873)
  k. **damei** B. & Benj., 1926
  l. **blanca** T. Emmel & Mattoon,
    1972
4588 **meadii** (Edw., 1872)
  a. **melania** Wind, 1946
  b. **mexicana** R. Chermock, 1948
  c. **alamosa** T. & J. Emmel, 1969
4589 **sthenele** (Bdv., 1852)
  behrii (Grinnell, 1905)
  a. **paulus** (Edw., 1879)
  b. **masoni** Cross, 1937
4590 **oetus** (Bdv., 1869)
  a. **charon** (Edw., 1872)
  b. **silvestris** (Edw., 1861)
  phocus (Edw., 1874)
  c. **pallescens** T. & J. Emmel, 1971

Erebiini

EREBIA Dalm., 1816
  SYNGEA Hbn., 1819
  EPIGEA Hbn., 1819
  PHORCIS Hbn., 1819
  MARICA Hbn., 1819
  GORGO Hbn., 1819
  TRIARIIA Verity, 1953
  TRUNCAEFALCIA Verity, 1953
  MEDUSIA Verity, 1953
  SIMPLICIA Verity, 1953
4591 **vidleri** Elwes, 1898
4592 **rossii** (Curt., 1835)
  a. **ornata** Leussler, 1935
  b. **kuskoquima** Holl., 1931
  c. **gabrieli** dos Passos, 1949
4593 **disa** (Thunb., 1791), extralim.
  griela (F., 1793)
  a. **mancinus** Doubleday & Hewitson,
    1849
  macinus B. & Benj., 1926, missp.
  b. **steckeri** Holl., 1930
  c. **subarctica** McD., 1937
4594 **magdalena** Stkr., 1880
  a. **mackinleyensis** Gunder, 1932
4595 **fasciata** Butler, 1868
  suffusa Warr., 1936

a. **avinoffi** Holl., 1900
b. **semo** Grum-Grshimaïlo, 1899, quest. occur.

4596 **discoidalis** (Kby., 1837)
a. **mcdunnoughi** dos Passos, 1940
  macdunnoughi dos Passos, 1964, emend.

4597 **theano** (Tauscher, 1809), extralim.
a. **ethela** (Edw., 1891)
b. **demmia** Warr., 1936
c. **sofia** Stkr., 1881
  canadensis Warr., 1931
  churchillensis Warr., 1936
d. **alaskensis** Holl., 1900

4598 **youngi** Holl., 1900
a. **herscheli** Leussler, 1935
b. **rileyi** dos Passos, 1947

4599 **epipsodea** Butler, 1868
  sineocellata Skin., 1889
a. **rhodia** Edw., 1871
  brucei Elwes, 1889
b. **remingtoni** Ehrlich, 1952
c. **hopfingeri** Ehrlich, 1954
d. **freemani** Ehrlich, 1954
  sineocellata; dos Passos, 1964, not Skin., 1889

4600 **callias** Edw., 1871
  tyndarus; auth., not Esp., 1777

4601 **dabanensis** Erschoff, 1871

Pronophilini

GYROCHEILUS Butler, 1867
  GEIROCHEILUS Edw., 1874, missp.
  GYROCHILUS Kby., 1882, missp.
  GIROCHEILUS F. M. Brown, 1964, missp.

4602 **patrobas** (Hewitson, 1862), extralim.
a. **tritonia** (Edw., 1874)

Satyrini

NEOMINOIS Scudder, 1875
  EUMENIS; auth., not Hbn., 1819

4603 **ridingsii** (Edw., 1865)
a. **stretchii** (Edw., 1870)
b. **dionysus** Scudder, 1878
  ashtaroth (Stkr., 1878)
  dionysius B. & McD., 1917, missp.

OENEIS Hbn., 1819
  CHIONOBAS Bdv., 1833

4603.1 **ivallda** (Mead, 1878)
  invallda Tilden, 1959, missp.

4604 **nevadensis** (C. & R. Felder, 1866)
  californica (Bdv., 1868)
a. **iduna** (Edw., 1874)
b. **gigas** Butler, 1868

4605 **macounii** (Edw., 1885)

4606 **chryxus** (Doubleday & Hewitson, 1849)
a. **stanislaus** Hovanitz, 1937
b. **valerata** Burdick, 1958
c. **strigulosa** McD., 1934
d. **calais** (Scudder, 1865)
e. **caryi** Dyar, 1904

4607 **uhleri** (Reak., 1866)
  obscura (Edw., 1892)
a. **reinthali** F. M. Brown, 1953

b. **varuna** (Edw., 1882)
  dennisi Gunder, 1927
c. **nahanni** Dyar, 1904
d. **cairnesi** Gibs., 1920

4608 **alberta** Elwes, 1893
a. **oslari** Skin., 1911
b. **capulinensis** F. M. Brown, 1970
c. **daura** (Stkr., 1894)

4609 **taygete** Gey., 1830
  bootes (Bdv., 1832)
a. **gaspeensis** dos Passos, 1949
b. **fordi** dos Passos, 1949
c. **edwardsi** dos Passos, 1949

4610 **bore** (Schneider, 1792), extralim.
a. **hanburyi** Watkins, 1928
b. **mckinleyensis** dos Passos, 1949
  mackinleyensis dos Passos, 1964, emend.

4611 **jutta** (Hbn., 1805–6), extralim.
  balder (Guér.-Ménéville, 1831)
a. **alaskensis** Holl., 1900
b. **leussleri** Bryant, 1935
c. **harperi** dos Passos, 1976
d. **chermocki** Wyatt, 1965
e. **reducta** McD., 1929
f. **ridingiana** F. & R. Chermock, 1940
g. **ascerta** Masters & Sorenson, 1968
h. **terraenovae** dos Passos, 1935, emend.

4612 **melissa** (F., 1775)
  oeno (Bdv., 1832)
a. **semidea** (Say, 1828)
  eritiosa (Bdv., 1832)
  nigra (Edw., 1894)
b. **semplei** Holl., 1931
c. **assimilis** Butler, 1868
  arctica Gibs., 1920
  simulans Gibs., 1920
d. **gibsoni** Holl., 1931
e. **beanii** Elwes, 1893
f. **lucilla** B. & McD., 1918

4613 **polixenes** (F., 1775)
  crambis (Freyer, 1845)
  polyxenes Riley & Gabriel, 1924, missp.
a. **katahdin** (Newcomb, 1901)
b. **subhyalina** (Curt., 1835)
c. **peartiae** (Edw., 1897)
d. **yukonensis** Gibs., 1920
e. **brucei** (Edw., 1891)

DANAIDAE

by LEE D. MILLER & F. MARTIN BROWN

DANAINAE

DANAUS Kluk, 1802
  DANAIDA Latr., 1804
  LIMNAS Hbn., 1806, suppr. (ICZN Op. 97)
  DANAIS Latr., 1807
  DANAUS Latr., 1809, preocc. by Kluk, 1802
  ANOSIA Hbn., 1816, subgenus?
  FESTIVUS Crotch, 1872
  TIRUMALA Moore, 1880
  SALATURA Moore, 1880
  PARANTICA Moore, 1880

  CHITTIRA Moore, 1880
  CADUGA Moore, 1882
  MELINDA Moore, 1883, preocc. by Robineau-Desvoidy, 1830
  LINTORATA Moore, 1883
  NASUMA Moore, 1883
  TASITIA Moore, 1883
  RAVADEBA Moore, 1883
  PHIRDANA Moore, 1883
  BAHORA Moore, 1883
  ASTHIPA Moore, 1883
  VADEBRA Moore, 1883
  CHANAPA Moore, 1883
  ANDASENA Moore, 1883
  PENOA Moore, 1883
  VADEBA Schatz, 1886, missp.
  BADACARA Moore, 1890
  CHLOROCHROPSIS Roths., 1892
  RAVADEBRA Roths., 1892, missp.
  ELSA Honrath, 1892
  SALATURIA Swinhoe, 1893, missp.
  REVADEBRA Grose-Smith, 1895, missp.
  CHITIRA Grünberg, 1908, missp.
  ASHTIPA Fruhstorfer, 1910, missp.
  DANAOMORPHA Kremky, 1925
  PANLYMNAS Bryk, 1937
  DIOGAS d'Almeida, 1938

4614 **plexippus** (L., 1758)
  archippus (F., 1793), preocc. by Cram., 1776
  fumosus (Hulst, 1886)
  pulchra (Stkr., 1900)
  americanus Gunder, 1927
  nivosus Gunder, 1927
  menippe; auth., not Hbn., 1816
  megalippe; auth., not Hbn., 1826

4615 **gilippus** (Cram., 1776), extralim.
  gilippe (Godt., 1819), missp.
a. **berenice** (Cram., 1780)
b. **strigosus** (Bates, 1864)
  kerri Comst., 1925

4616 **eresimus** (Cram., 1777), extralim.
  eresime (Godt., 1819)
a. **montezuma** Talbot, 1943
b. **tethys** Fbs., 1943
  cleothera; auth., not Godt., 1819

LYCOREINAE

LYCOREA Doubleday & Hewitson, 1847
  LYCORELLA Hemming, 1933

4617 **cleobaea** (Godt., 1819), extralim.
  ceres (Cram., 1776), preocc. by F., 1775
a. **demeter** (C. & R. Felder, 1865)
b. **atergatis** (Doubleday, 1847)

Zygaenoidea

ZYGAENIDAE

by DONALD R. DAVIS

HARRISINA Pack., 1864
4618 **aversa** (Hy. Edw., 1884)
4619 **cyanea** (B. & McD., 1910)

4620 lustrans (Beutenmüller, 1894)
4621 brillians B. & McD., 1910
4622 coracina (Clem., 1860)
    nigrina Graef, 1887
4623 metallica Stretch, 1885
4624 americana (Guér., 1829)
    a. **texana** Stretch, 1872
    b. **australis** Stretch, 1885

SERYDA Wlk., 1856
4625 basirei (Druce, 1896)
4626 constans (Hy. Edw., 1881)
    sancta (N. & D., 1894), var.

ACOLOITHUS Clem., 1860
4627 rectarius Dyar, 1898
4628 novaricus B. & McD., 1913
4629 falsarius Clem., 1860
    sanborni Pack., 1864
    ruficollis Druce, 1884

TRIPROCRIS Grt., 1873
4630 yampai Barnes, 1905
4631 smithsoniana (Clem., 1860)

TETRACLONIA Jordan, 1913
4632 dyari Jordon, 1913
    laterculae (Dyar, 1900)
4633 latercula (Hy. Edw., 1882)

PYROMORPHA H.-S., 1854
    PYRRHOMORPHA Hy. Edw.,
    1884, missp.
    MALTHACA Clem., 1860
4634 martenii (French, 1883)
    marteni (Dyar, 1903), missp.
    barnea Druce, 1891
4635 fusca Hy. Edw., 1884
    landia (Druce, 1891)
4646 rata (Hy. Edw., 1882)
4637 centralis (Wlk., 1854)
    notha (Hy. Edw., 1885)
4638 caelebs Blanchard, 1972
4639 dimidiata H.-S., 1854
    perlucidula (Clem., 1860)

MEGALOPYGIDAE
by DONALD R. DAVIS

TROSIA Hbn., 1820
    SCIATHOS Wlk., 1855
    EDEBESSA Wlk., 1856
    ISOCHROMA Felder, 1875
    SAROTHROMA H.-S., 1856
    ENDOBRACHYS Felder, 1875
4640 obsolescens Dyar, 1899

LAGOA Harr., 1841
4641 immaculata (Cass., 1928)
4642 pyxidifera (J. E. Smith, 1797)
4643 lacyi B. & McD., 1910
    laceyi B. & McD., 1917, missp.
4644 crispata (Pack., 1864)
    grisea (B. & McD., 1910), ab.

MEGALOPYGE Hbn., 1820
    ALPHIS Wlk., 1855
    CHRYSOPYGA H.-S., 1855
    GASINA Wlk., 1855

    OCHROSOMA H.-S., 1856
    PODALIA Wlk., 1856
    OYLOTHRIX Clem., 1860
    PIMELA Clem., 1860
    ZEBONDA Wlk., 1865
    CYCLARA Schaus, 1896
4645 lapena Schaus, 1896
    a. **heteropuncta** B. & McD., 1918
4646 bissesa Dyar, 1911
4647 opercularis (J. E. Smith, 1797)
    lanuginosa (Clem., 1860)
    subcitrina (Wlk., 1869)

NORAPE Wlk., 1855
    CARAMA Wlk., 1855
    MALLOTODESMA Wallgr., 1858
    ULOSOTA Grt., 1864
    SULYCHRA Butler, 1878
    ANARCHYLUS Dyar, 1906
4648 tenera (Druce, 1897)
    archrigelos (Dyar, 1910)
4649 virgo (Butler, 1877)
4650 ovina (Sepp, 1848–52)
    cretata (Grt., 1864)

LIMACODIDAE
by DONALD R. DAVIS

EPIPEROLA Dyar, 1898
    PALAEOPHOBETRON Dyar, 1898
4651 perornata Dyar, 1905

TORTRICIDIA Pack., 1864
4652 testacea Pack., 1864
    a. **crypta** Dyar, 1902
4653 pallida (H.-S., 1854)
    a. **flavula** (H.-S., 1854)
4654 flexuosa (Grt., 1880)
    caesonia (Grt., 1880), form

SLOSSONELLA Dyar, 1904
4655 tenebrosa Dyar, 1904

KRONAEA Reak., 1864
4656 minuta (Reak., 1864)

HETEROGENEA Knoch, 1783
4657 shurtleffi Pack., 1864
    shurtleffii Dyar, 1898, missp.

PACKARDIA G. & R., 1867
    CYRTOSIA Pack., 1864, preocc. by
    Perris, 1839
4658 albipunctata (Pack., 1864)
    goodellii Grt., 1880
    goodelli McD., 1939, missp.
    a. **ocellata** (Grt., 1865)
4659 geminata (Pack., 1864)
4660 ceanothi Dyar, 1908
4661 elegans (Pack., 1864)
    nigripunctata Goodell, 1881
    a. **fusca** (Pack., 1864)

LITHACODES Pack., 1864
    LITHOCODIA Pack., 1887, missp.
4662 graefii Pack., 1887
    graefi McD., 1939, missp.
4663 fiskeanus (Dyar, 1900)
4664 gracea Dyar, 1921

4665 fasciola (H.-S., 1854)
    laticlavia (Clem., 1860)
    divergens (Wlk., 1862)
    a. **belfragei** Dyar., 1925

APODA Haw., 1809
    COCHLIDION Hbn., 1806, suppr.
    (ICZN Op. 97)
    COCHLIDIUM Hbn., 1822
    LIMACODES Berthold, 1827
4666 maxima Dyar, 1927
4667 y-inversum (Pack., 1864)
    a. **parallela** (Hy. Edw., 1886)
4668 rectilinea (G. & R., 1868)
    a. **latomia** (Harv., 1877)
4669 biguttata (Pack., 1864)
    tetraspilaris (Wlk., 1865)

PROLIMACODES Schaus, 1896
4670 trigona (Hy. Edw., 1882)
    telligii (Barnes, 1900)
    filifera Dyar, 1925
4671 badia (Hbn., 1822)
    scapha (Harr., 1841)
    undifera (Wlk., 1855)
    a. **argentimacula** B. & McD., 1913

CNIDOCAMPA Dyar, 1905
    MONEMA Wlk., 1855, preocc. by
    Greville, 1829
4672 flavescens (Wlk., 1855)

ALARODIA Mösch., 1886
    PHRYNE Grt., 1865, preocc. by
    H.-S., 1843
    CALYBIA Kby., 1892
    EUPOEYA Pack., 1893
4673 slossoniae (Pack., 1893)

CRYPTOPHOBETRON Dyar, 1905
4674 oropeso (Barnes, 1905)

ISOCHAETES Dyar, 1899
4675 beutenmuelleri (Hy. Edw., 1887)

PHOBETRON Hbn., 1825
    ECNOMIDEA Westwood, 1841
    EURYDA H.-S., 1854
    NEMETA Wlk., 1855
    PHOBETRUM Pack., 1864, missp.
    ECNOMIDIA Dyar, 1895, missp.
4676 dyari B. & Benj., 1926
4677 pithecium (J. E. Smith, 1797)
    abbotana Hbn., 1825
    nigricans (Pack., 1864)
    hyalinus Walsh, 1864
    tetradactylus (Walsh, 1864)
    nondescriptus (Wetherby, 1875)

NATADA Wlk., 1855
    BOMBYCOCERA R. & A. F.
    Felder, 1874
    RHINAXIME Berg, 1882
4678 nigripuncta B. & McD., 1910
4679 nasoni (Grt., 1876)
    rude (Hy. Edw., 1882)
    daona (Druce, 1887)

ISA Pack., 1864
    SISYROSEA Grt., 1876

SOSIOSA Kby., 1892, missp.
SICYROSEA Kby., 1892, missp.
4680 schaefferana Dyar, 1905
4681 textula (H.-S., 1854)
  textula (Pack., 1854), missp.
  inornata (G. & R., 1867)

ADONETA Clem., 1860
  CYCLOPTERYX Pack., 1864
4682 gemina Dyar, 1906
4683 pygmaea G. & R., 1868
4684 bicaudata Dyar, 1904
4685 spinuloides (H.-S., 1854)
  voluta Clem., 1860
  ferrigera (Wlk., 1865)
  ruptilinea (Wlk., 1865)
  nebulosa (Wetherby, 1875)
  a. leucosigma (Pack., 1865)

MONOLEUCA G. & R., 1869
4686 disconcolorata B. & Benj., 1925
4687 fieldi B. & Benj., 1925
4688 occidentalis B. & McD., 1912
4689 erectifascia Dyar, 1925
4690 obliqua Hy. Edw., 1886
4691 semifascia (Wlk., 1855)
  a. sulfurea Grt., 1880
4692 angustilinea Dyar, 1927
4693 subdentosa Dyar, 1891

EUCLEA Hbn., 1819
  METRAGA Wlk., 1855
  NOCHELIA Clem., 1860
4694 dolliana Dyar, 1905
  a. spadicis (Grossb., 1906)
4695 flava B. & McD., 1910
4696 incisa (Harv., 1876
  mira Dyar, 1905
4697 delphinii (Bdv., 1832)
  cippus; auth., not Cram., 1775
  strigata (Bdv., 1832)
  quercicola (H.-S., 1854)
  querceti (H.-S., 1854), form
  quereti (H.-S., 1854), missp.
  viridiclava Wlk., 1855, form
  paenulata (Clem., 1860), form
  tardigrada (Clem., 1860)
  bifida Pack., 1864
  ferruginea Pack., 1864
  monitor Pack., 1864
  excisa (Wlk., 1865)
  argentata (Wetherby, 1875)
  ellioti Pears., 1887, form
  interjecta Dyar, 1891, form
4697.1 nanina Dyar, 1899
  nana Dyar, 1891, preocc. by H.-S.,
  1856

PARASA Moore, 1858
  NEAERA H.-S., 1854, preocc. by
  Robineau-Desvoidy, 1830
  CALLOCHLORA Pack., 1864
4698 chloris (H.-S., 1854)
  viridis (Reak., 1864)
  fraterna Grt., 1881
  a. huachuca Dyar, 1905
4699 indetermina (Bdv., 1832)
  vernata (Pack., 1864)

SIBINE H.-S., 1855
  NYSSIA Wlk., 1855, preocc. by
  Dup., 1829
  EMPRETIA Clem., 1860
  EUPALIA Wlk., 1866
  NEOMIRESA Butler, 1878
  STREBLOTA Berg, 1882
  EPISIBINE Dyar, 1898
4700 stimulea (Clem., 1860)
  ephippiatus (Harr., 1869)
4700.1 extensa Schaus, 1896

EPIPYROPIDAE

by DONALD R. DAVIS

FULGORAECIA Newman, 1851
  EPIPYROPS Westwood, 1876
  PSEUDOPSYCHE Hy. Edw.,
  1882, preocc. by Oberth., 1880
  OEDONIA Kby., 1892, n. syn.
4701 exigua (Hy. Edw., 1882), n. comb.
  barberiana (Dyar, 1902), n. syn.

DALCERIDAE

by DONALD R. DAVIS

DALCERIDES N. & D., 1893
4702 ingenitus (Hy. Edw., 1882)

Pyraloidea

PYRALIDAE

by EUGENE MUNROE (except CRAMBINAE)

(Series Crambiformes)

SCOPARIINAE

GESNERIA Hbn., 1824–25
  SCOPARONA Chapman, 1912
4703 centuriella (D. & S., 1775), extralim.
  centurionalis (Hbn., 1824–25),
  emend.
  numeralis (Zett., 1839)
  quadratella (Zett., 1839)
  centurialis (Gn., 1854), emend.
  confluella (Krulikowsky, 1909)
  a. borealis (Dup., 1835)
  b. caecalis (Wlk., 1858)
  caliginosalis (Wlk., 1866)
  c. beringiella Mun., 1972
  d. ninguidalis (Hulst, 1886)
4704 rindgeorum Mun., 1972

COSIPARA Mun., 1972
4705 tricoloralis (Dyar, 1904)
  rufitinctalis (Dyar, 1929)
4706 modulalis Mun., 1972
  delphusa; auth., not Druce, 1896
  smithi; auth., not Druce, 1896
  sabura; auth., not Druce, 1896
  flexuosa; auth., not Dyar, 1918
4707 chiricahuae Mun., 1972

SCOPARIA Haw., 1812
  SCOPEA Haw., 1812

EUDOREA Curt., 1827
  CHOLIUS Gn., 1845
  TETRAPROSOPUS Butler, 1882
  XEROSCOPA Meyr., 1884
4708 rigidalis B. & McD., 1912
4709 denigata Dyar, 1929
4710 normalis Dyar, 1904
4711 palloralis Dyar, 1906
  obispalis Dyar, 1906
  cervalis McD., 1927
4712 californialis Mun., 1972
4713 apachealis Mun., 1972
  a. pinalensis Mun., 1972
  b. utalis Mun., 1972
4714 ruidosalis Mun., 1972
4715 blanchardi Mun., 1972
4716 biplagialis Wlk., 1866
  libella Grt., 1878
  a. bellaeislae Mun., 1972
  b. fernaldalis Dyar, 1904
  c. pacificalis Dyar, 1921
  alaskalis B. & Benj., 1922
  d. afognakalis Mun., 1972
4717 penumbralis Dyar, 1906
4718 cinereomedia Dyar, 1904
  truncatalis McD., 1922
4719 basalis Wlk., 1866
4720 dominicki Mun., 1972
4721 huachucalis Mun., 1972

EUDONIA Billberg., 1820
  EUDORIA Chapman, 1912
  WITLESIA Chapman, 1912
4722 rectilinea (Zell., 1874)
  refugalis (Hulst, 1886)
  nominatalis (Hulst, 1886)
4723 commortalis (Dyar, 1921)
4724 expallidalis (Dyar, 1906)
  rufitinctalis (Hamp., 1907)
4725 franciscalis Mun., 1972
4726 torniplagalis (Dyar, 1904)
  a. alialis (B. & McD., 1912)
  b. perfectalis Mun., 1972
4727 albertalis (Dyar, 1929)
4728 vivida Mun., 1972
4729 spaldingalis (B. & McD., 1912)
4730 spenceri Mun., 1972
4731 rotundalis Mun., 1972
4732 franclemonti Mun., 1972
4733 schwarzalis (Dyar, 1906)
4734 leucophthalma (Dyar, 1929)
  a. petaluma Mun., 1972
4735 echo (Dyar, 1929)
  a. gartrelli Mun., 1972
4736 bronzalis (B. & Benj., 1922)
4737 lugubralis (Wlk., 1866)
  phycitinalis (Dyar, 1929)
  persimilalis (McD., 1961)
  a. madgei Mun., 1972
4738 strigalis (Dyar, 1906)
4739 heterosalis (McD., 1961)

NYMPHULINAE

Ambiini

UNDULAMBIA Lange, 1956
4740 striatalis (Dyar, 1906)
4741 polystichalis Capps, 1965
4742 rarissima Mun., 1972

68

Nymphulini

NEOCATACLYSTA Lange, 1956
4743 **magnificalis** (Hbn., 1796)
  lamialis (Wlk., 1859)
  heliopalis (Clem., 1860)

CHRYSENDETON Grt., 1881
4744 **medicinalis** Grt., 1881
  claudialis; auth., part, not Wlk., 1859
4745 **kimballi** Lange, 1956
4746 **imitabilis** (Dyar, 1917)

NYMPHULA Schr., 1802
  ELOPHILA Hbn., 1822
  HYDROCAMPUS Berthold, 1827
  HYDROCAMPA Steph., 1829
  HYDROCAMPE Latr., 1829
4747 **ekthlipsis** (Grt., 1876)

MUNROESSA Lange, 1956
4748 **icciusalis** (Wlk., 1859)
  formosalis (Clem., 1860)
  genuialis (Led., 1863)
  a. **albiplaga** Mun., 1972
  b. **avalona** Mun., 1972
4749 **faulalis** (Wlk., 1859)
  pacalis (Grt., 1881)
4750 **nebulosalis** (Fern., 1887)
4751 **gyralis** (Hulst, 1886)
  dentilinea (Hmps., 1897)
  a. **serralinealis** (B. & Benj., 1924)

CONTIGER Lange, 1956
4752 **vittatalis** (Dyar, 1906)

NYMPHULIELLA Lange, 1956
4753 **daeckealis** (Haim., 1915)
  broweri (Heinr., 1940)

SYNCLITA Led., 1863
4754 **tinealis** Mun., 1972
4755 **obliteralis** (Wlk., 1859)
  proprialis (Fern., 1888)
  obscuralis; auth., part, not Mösch., 1881
4756 **atlantica** Mun., 1972
4757 **occidentalis** Lange, 1956

LANGESSA Mun., 1972
4758 **nomophilalis** (Dyar, 1906)

PARAPOYNX Hbn., 1825
  EUSTALES Clem., 1860
  SIRONIA Clem., 1860
  NYMPHAEELLA Grt., 1880
4759 **maculalis** (Clem., 1860)
  seminivella (Wlk., 1866)
  dispar (Grt., 1880)
  foeminalis (Dyar, 1906), form, infrasubsp.
  masculinalis (Dyar, 1906), form, infrasubsp.
4760 **obscuralis** (Grt., 1881)
4761 **badiusalis** (Wlk., 1859)
  albalis (Rob., 1869)
4762 **curviferalis** (Wlk., 1866)
4763 **seminealis** (Wlk., 1859)
  tedyuscongalis (Clem., 1860)
4764 **allionealis** Wlk., 1859
  aptalis Led., 1863

  a. **itealis** (Wlk., 1859)
  cretacealis Led., 1863
  plenilinealis Grt., 1881
4765 **diminutalis** Snell., 1880

OLIGOSTIGMOIDES Lange, 1956
4766 **cryptale** (Druce, 1896)

Argyractini

USINGERIESSA Lange, 1956
4767 **onyxalis** (Hamp., 1897)
  cancellalis (Dyar, 1917)
4768 **brunnildalis** (Dyar, 1906)

NEARGYRACTIS Lange, 1956
4769 **slossonalis** (Dyar, 1906)

PETROPHILA Guilding, 1830
  PARARGYRACTIS Lange, 1956, n. syn.
4770 **drumalis** (Dyar, 1906), n. comb.
4771 **daemonalis** (Dyar, 1907), n. comb.
4772 **cappsi** (Lange, 1956), n. comb.
4773 **kearfottalis** (B. & McD., 1917), n. comb.
4774 **bifascialis** (Rob., 1869), n. comb.
4775 **jaliscalis** (Schaus, 1906), n. comb.
  satanalis (Dyar, 1917), n. comb.
4776 **hodgesi** (Mun., 1972), n. comb.
4777 **fulicalis** (Clem., 1860), n. comb.
  ?angulatalis (Led., 1863), n. comb.
4778 **santafealis** (Heppner, 1976), n. comb.
4779 **canadensis** (Mun., 1972), n. comb.
  opulentalis; auth., part, not Led., 1863
4780 **confusalis** (Wlk., 1866), n. comb.
  truckeealis (Dyar, 1917), n. comb.
4781 **avernalis** (Grt., 1878), n. comb.
  confusalis (B. & McD., 1913), n. comb., preocc. by Wlk., 1866
4782 **cronialis** (Druce, 1896), n. comb.
4783 **longipennis** (Hamp., 1906), n. comb.
4784 **schaefferalis** (Dyar, 1906), n. comb.
  castusalis (Schaus, 1924), n. comb.

EOPARARGYRACTIS Lange, 1956
4785 **irroratalis** (Dyar, 1917)
4786 **floridalis** Lange, 1956
4787 **plevie** (Dyar, 1917)

OXYELOPHILA Fbs., 1922
4788 **callista** (Fbs., 1922)

ODONTIINAE

Dichogamini

METREA Grt., 1882
4789 **ostreonalis** Grt., 1882
  urticaloides (Fyles, 1894)

DICHOGAMA Led., 1863
4790 **redtenbacheri** Led., 1863
4791 **amabilis** Mösch., 1891
4792 **colotha** Dyar, 1912

ALATUNCUSIA Amsel, 1956
4793 **bergii** (Mösch., 1891)

EUSTIXIA Hbn., 1823
  THELCTERIA Led., 1863
4794 **pupula** Hbn., 1823

Odontiini

MICROTHEORIS Meyr., 1932
4795 **vibicalis** (Zell., 1873)
4796 **ophionalis** (Wlk., 1859)
  sesquialteralis (Zell., 1873)
  nasonialis (Zell., 1873)
  a. **lacustris** Mun., 1961
  b. **eremica** Mun., 1961
  c. **baboquivariensis** Mun., 1961
  d. **occidentalis** Mun., 1961

RHODOCANTHA Mun., 1961
4797 **diagonalis** Mun., 1961

FRECHINIA Mun., 1961
4798 **helianthiales** (Murt., 1897)
  thyanalis (Druce, 1899)
  murmuralis (Dyar, 1917)
4799 **lutosalis** (B. & McD., 1914)
4800 **laetalis** (B. & McD., 1914)
4801 **criddlealis** (Mun., 1951)
4802 **texanalis** Mun., 1961

PROCYMBOPTERYX Mun., 1961
4803 **belialis** (Druce, 1899)

CYMBOPTERYX Mun., 1961
4804 **fuscimarginalis** Mun., 1961
4805 **unilinealis** (B. & McD., 1918)
  phaeopasta (Hamp., 1919)

NEOCYMBOPTERYX Mun., 1973
4806 **heitzmani** Mun., 1973

EDIA Dyar, 1913
4807 **semiluna** (Sm., 1905)
  bidentalis (B. & McD., 1912)
  macrostagma Dyar, 1913
4808 **minutissima** (Sm., 1906)
  coolidgei Dyar, 1921

DICHOZOMA Mun., 1961
4809 **parvipicta** (B. & McD., 1918)

CUNEIFRONS Mun., 1961
4810 **coloradensis** Mun., 1961

GYROS Hy. Edw., 1881
  ORIBATES Hy. Edw., 1881, preocc. by Dugès, 1834
  MONOCONA Warr., 1892
4811 **muirii** (Hy. Edw., 1881)
  a. **rubralis** (Warr., 1892)
4812 **atripennalis** B. & McD., 1914
4813 **powelli** Mun., 1959

ANATRALATA Mun., 1961
4814 **versicolor** (Warr., 1892)

EREMANTHE Mun., 1972
4815 **chemsaki** Mun., 1972

METAXMESTE Hbn., 1825
4816 **nubicola** Mun., 1954

POGONOGENYS Mun., 1961
4817 **proximalis** (Fern., 1894)
4818 **frechini** Mun., 1961
4819 **masoni** Mun., 1961

CHRISMANIA B. & McD., 1914
4820 **pictipennalis** B. & McD., 1914

PLUMIPALPIA Mun., 1961
4821 **martini** Mun., 1961

NANNOBOTYS Mun., 1961
4822 **commortalis** (Grt., 1881)
    minima (Dyar, 1917)

PORPHYRORHEGMA Mun., 1961
4823 **fortunata** Mun., 1961

PSAMMOBOTYS Mun., 1961
4824 **fordi** Mun., 1961
4825 **alpinalis** Mun., 1972

MIMOSCHINIA Warr., 1892
4826 **rufofascialis** (Steph., 1834), extralim.
    fascialis (Haw., 1803), preocc. by
      Hbn., 1796
    thalialis (Wlk., 1859)
    perviana (Wlk., 1866)
    gelidalis (Wlk., 1866)
    costaemaculalis (Snell., 1887)
    a. **novalis** (Grt., 1876)
    b. **decorata** (Druce, 1898)
    c. **nuchalis** (Grt., 1878)

JATIVA Mun., 1961
4827 **castanealis** (Hulst, 1886)
    jativa (Barnes, 1905)

PSEUDOSCHINIA Mun., 1961
4828 **elautalis** (Grt., 1881)
    magnalis (Hulst, 1886)

ODONTIVALVIA Mun., 1973
4829 **radialis** (Mun., 1972)

NOCTUELIOPSIS Mun., 1961
4830 **brunnealis** Mun., 1972
4831 **puertalis** (B. & McD., 1912)
4832 **palmalis** (B. & McD., 1918)
4833 **atascaderalis** (Mun., 1951)
4834 **aridalis** (B. & Benj., 1922)
4835 **pandoralis** (B. & McD., 1914)
4836 **rhodoxanthinalis** Mun., 1974
4837 **bububattalis** (Hulst, 1886)
    tectalis (B. & McD., 1914)
4838 **virula** (B. & McD., 1918)

MOJAVIA Mun., 1961
4839 **achemonalis** (B. & McD., 1914)
    pulcharalis (B. & Benj., 1924)

MOJAVIODES Mun., 1972
4840 **blanchardae** Mun., 1972

HELIOTHELOPSIS Mun., 1961
4841 **arbutalis** (Snell., 1875)
    rhea (Druce, 1894)
4842 **costipunctalis** (B. & McD., 1914)
4843 **unicoloralis** (B. & McD., 1914)

CHLOROBAPTA B. & McD., 1914
4844 **rufistrigalis** B. & McD., 1914

GLAUCODONTIA Mun., 1972
4845 **pyraustoides** Mun., 1972

GLAPHYRIINAE

HELLULA Gn., 1854
    OEBIA auth., suppr.
    OEOBIA auth., suppr.
    PHYRATOCOSMA Meyr.
4846 **rogatalis** (Hulst, 1886)
    undalis; auth., part, not F., 1794
4847 **phidilealis** (Wlk., 1859)
    trypheropa (Meyr., 1936)
4848 **kempae** Mun., 1972
4849 **aqualis** B. & McD., 1914
4850 **subbasalis** (Dyar, 1923)

UPIGA Capps, 1964
4851 **virescens** (Hulst, 1900)

PAREGESTA Mun., 1964
4852 **californiensis** Mun., 1964

SCYBALISTODES Mun., 1964
4853 **periculosalis** (Dyar, 1908)
4854 **vermiculalis** Mun., 1964
4855 **regularis** Mun., 1964
4856 **fortis** Mun., 1972

NEPHROGRAMMA Mun., 1964
4857 **reniculalis** (Zell., 1872)
4858 **separata** Mun., 1972

STEGEA Mun., 1964
    EGESTA Rag., 1891, preocc. by
      Conrad, 1845
4859 **mexicana** Mun., 1964
4860 **sola** Mun., 1972
4861 **simplicialis** (Kft., 1907)
4862 **minutalis** (Walter, 1928)
4863 **powelli** Mun., 1972
4864 **eripalis** (Grt., 1878)
4865 **salutalis** (Hulst, 1886)
    a. **ochralis** (Haim., 1908)
    b. **grisealis** Mun., 1972
    c. **riparialis** Mun., 1972

ABEGESTA Mun., 1964
4866 **reluctalis** (Hulst, 1886)
4867 **remellalis** (Druce, 1899)
4868 **concha** Mun., 1964

GLAPHYRIA Hbn., 1823
    HOMOPHYSA Gn., 1854
4869 **glaphyralis** (Gn., 1854)
    stipatalis (Wlk., 1866)
    albolineata (G. & R., 1867)
4870 **sequistrialis** Hbn., 1823
    dimotalis; auth., part, not Wlk., 1866
4871 **basiflavalis** B. & McD., 1913
4872 **peremptalis** (Grt., 1878)
4873 **fulminalis** (Led., 1863)
4874 **cappsi** Mun., 1972

AETHIOPHYSA Mun., 1964
4875 **delicata** Mun., 1964
4876 **dualis** (B. & McD., 1914)

CHLOROBAPTA B. & McD., 1914

4877 **lentiflualis** (Zell., 1872)
    invisalis; auth., part, not Gn., 1854
4878 **consimilis** Mun., 1964

XANTHOPHYSA Mun., 1964
4879 **psychialis** (Hulst, 1886)
    psychicalis auth., missp.

PLUMEGESTA Mun., 1972
4880 **largalis** Mun., 1972

LIPOCOSMA Led., 1863
    LIPOCOSMOPSIS Mun., 1964
4881 **sicalis** (Wlk., 1859)
    perfusalis (Wlk., 1866)
4882 **diabata** Dyar, 1917
4883 **adelalis** (Kft., 1903)
4884 **intermedialis** B. & McD., 1912
4885 **septa** Mun., 1972
4886 **albibasalis** B. & McD., 1911
4887 **polingi** Mun., 1972

LIPOCOSMODES Mun., 1964
4888 **fuliginosalis** (Fern., 1888)

DICYMOLOMIA Zell., 1872
    BIFALCULINA Amsel, 1956
4889 **julianalis** (Wlk., 1859)
    decora Zell., 1872
4890 **metalophota** (Hamp., 1897)
    consortalis (Dyar, 1914)
    argentipunctalis (Amsel, 1956)
4891 **opuntialis** Dyar, 1908
4892 **metalliferalis** (Pack., 1873)
    sauberi v. Hedemann, 1883
4893 **grisea** Mun., 1964
4894 **micropunctalis** Mun., 1964

CHALCOELA Zell., 1872
4895 **iphitalis** (Wlk., 1859)
    aurifera Zell., 1872
4896 **pegasalis** (Wlk., 1859)
    principalis (Wlk., 1866)
    egressalis (Wlk., 1866)
    robinsonii (Grt., 1871)
    discedalis Mösch., 1890

EVERGESTINAE

EVERGESTIS Hbn., 1825
    MESOGRAPHE Hbn., 1825
    HOMOCHROA Hbn., 1825
    SCOPOLIA Hbn., 1825
    PIONEA Gn., 1844
    OROBENA Gn., 1854
    AEDIS Grt., 1878
    PARAEDIS Grt., 1882
    PAROEDIS auth., missp.
    PACHYZANCLOIDES
      Matsumura, 1925
4897 **pallidata** (Hufn., 1767)
    straminalis (Hbn., 1793)
    elutalis (Hbn., 1796)
    stramentalis (Hbn., 1825)
    eunusalis (Wlk., 1859)
4898 **rimosalis** (Gn., 1854)
4899 **consimilis** Warr., 1892
    extimalis; auth., part, not Scop.,
      1763
4900 **aridalis** B. & McD., 1914

4901 **unimacula** (G. & R., 1867)
4902 **lunulalis** B. & McD., 1914
4903 **nolentis** Heinr., 1940
4904 **simulatilis** (Grt., 1880)
    brunneogrisea (Hy. Edw., 1886)
4905 **angustalis** (B. & McD., 1918)
    a. **catalinae** Mun., 1974
    b. **arizonae** Mun., 1974
4906 **vinctalis** B. & McD., 1914
    a. **muricoloralis** Mun., 1974
4907 **obscuralis** B. & McD., 1914
    a. **palousalis** Mun., 1974
4908 **comstocki** Mun., 1974
4909 **funalis** (Grt., 1878)
    a. **insulalis** B. & McD., 1914
    b. **columbialis** Mun., 1974
    c. **angelina** Mun., 1974
    d. **wallacensis** Mun., 1974
4910 **subterminalis** B. & McD., 1914
4911 **eurekalis** B. & McD., 1914
    a. **fuscistrigalis** Mun., 1974
4912 **obliqualis** (Grt., 1883)
4913 **dimorphalis** Mun., 1974
4914 **triangulalis** B. & McD., 1914
4915 **borregalis** Mun., 1974

PRORASEA Grt., 1878
4916 **simalis** Grt., 1878
4917 **gracealis** Mun., 1974
4918 **praeia** (Dyar, 1917)
4919 **fernaldi** Mun., 1974
4920 **sideralis** (Dyar, 1917)
4921 **pulveralis** (Warr., 1892)

CORNIFRONS Led., 1858
4922 **actualis** B. & McD., 1918
4923 **phasma** Dyar, 1917
    chlorophasma Dyar, 1917

CYLINDRIFRONS Mun., 1951
4924 **succandidalis** (Hulst, 1886)
    simplex (Warr., 1895)

ORENAIA Dup., 1844
4925 **trivialis** B. & McD., 1914
    subargentalis (B. & McD., 1918)
4926 **coloradalis** B. & McD., 1914
4927 **arcticalis** Mun., 1974
4928 **sierralis** Mun., 1974
4929 **alticolalis** (B. & McD., 1914)
4930 **pallidivittalis** Mun., 1956
4931 **macneilli** Mun., 1974

EVERGESTELLA Mun., 1974
4932 **evincalis** (Mösch., 1890)

TRISCHISTOGNATHA Warr., 1892
4933 **pyrenealis** (Wlk., 1859)
    palindialis; auth., part, not Gn., 1854
    medonalis (Wlk., 1859)
    dyaralis (Fern., 1901)

PYRAUSTINAE

Pyraustini

MUNROEODES Amsel, 1957
    MUNROEIA Amsel, 1956, preocc.
      by Marion, 1954

4934 **thalesalis** (Wlk., 1859)

SAUCROBOTYS Mun., 1976
4935 **fumoferalis** (Hulst, 1886)
4936 **futilalis** (Led., 1863)
    erectalis (Grt., 1876)
    a. **inconcinnalis** (Led., 1863)
    crocotalis (Grt., 1881)
    festalis (Hulst, 1886)

NASCIA Curt., 1835
4937 **acutella** (Wlk., 1866)
    venalis (Grt., 1878)

EPICORSIA Hbn., 1818
4938 **oedipodalis** (Gn., 1854)
    cerata; auth., part, not F., 1795
    mellinalis; auth., part, not Hbn., 1818
    butyrosa (Butler, 1878)

PSEUDOPYRAUSTA Amsel, 1956
4939 **santatalis** (B. & McD., 1914)
    acutangulalis; auth., part, not Snell.,
    1875

OENOBOTYS Mun., 1976
4940 **vinotinctalis** (Hamp., 1895)
4941 **texanalis** Mun. & A. Blanchard, 1976

TRIUNCIDIA Mun., 1976
4942 **eupalusalis** (Wlk., 1859)

CROCIDOPHORA Led., 1863
4943 **pustuliferalis** Led., 1863
4944 **serratissimalis** Zell., 1872
    subdentalis (Grt., 1873)
4945 **tuberculalis** Led., 1863

OSTRINIA Hbn., 1825
    MICRACTIS Warr., 1892
    EUPOLEMARCHA Meyr., 1937
    ZEAPHAGUS Agenjo, 1952
4946 **penitalis** (Grt., 1876)
    nelumbialis (Sm., 1890)
    pyraustalis (Dyar, 1925)
4947 **obumbratalis** (Led., 1863)
    obliteralis (Wlk., 1866)
    ainsliei (Heinr., 1919)
4948 **marginalis** (Wlk., 1866)
    stenopteralis (Grt., 1878)
4949 **nubilalis** (Hbn., 1796)
    silacealis (Hbn., 1796)
    lupulinalis; auth., part, not Cl., 1759

FUMIBOTYS Mun., 1976
4950 **fumalis** (Gn., 1854)
    orasusalis (Wlk., 1859)
    badipennis (Grt., 1873)

PERISPASTA Zell., 1875
4951 **caeculalis** Zell., 1875
    immixtalis Grt., 1881

EURRHYPARA Hbn., 1825
4952 **hortulata** (L., 1758)
    urticata (L., 1761)
    urticalis (D. & S., 1775)

PHLYCTAENIA Hbn., 1825
4953 **coronata** (Hufn., 1767), extralim.

    sambucalis (D. & S., 1775)
    a. **tertialis** (Gn., 1854)
    plectilis (G. & R., 1867)
    syringicola (Pack., 1870)
4954 **quebecensis** Mun., 1954
4955 **leuschneri** Mun., 1976

NEALGEDONIA Mun., 1976
4956 **extricalis** (Gn., 1854)
    intricatalis (Led., 1863)
    oppilalis (Grt., 1880)
    a. **dionalis** (Wlk., 1859)
    nisoeecalis (Wlk., 1859)
    beddeci (Dyar, 1913)

MUTUURAIA Mun., 1976
4957 **mysippusalis** (Wlk., 1859)
    terrealis; auth., part, not Tr., 1829
    humilalis (Led., 1863)

ANANIA Hbn., 1823
4958 **funebris** (Ström, 1768), extralim.
    octomaculata (L., 1771)
    atralis (F., 1775)
    guttalis (D. & S., 1775)
    trigutta (Esp., 1791)
    octomaculalis (Tr., 1829)
    a. **glomeralis** (Wlk., 1859)
4959 **labeculalis** (Hulst, 1886)

HAHNCAPPSIA Mun., 1976
4960 **fordi** (Capps, 1967)
4961 **alpinensis** (Capps, 1967)
4962 **marculenta** (G. & R., 1867)
    obliteralis; auth., part, not Wlk., 1866
4963 **neomarculenta** (Capps, 1967)
4964 **pseudobliteralis** (Capps, 1967)
4965 **neobliteralis** (Capps, 1967)
4966 **jaralis** (Schaus, 1920)
4967 **mancalis** (Led., 1863)
4968 **pergilvalis** (Hulst, 1886)
4969 **cochisensis** (Capps, 1967)
4970 **coloradensis** (G. & R., 1867)
4971 **ramsdenalis** (Schaus, 1920)
4972 **huachucalis** (Capps, 1967)
4973 **mellinialis** (Druce, 1899)
    phrixalis (Dyar, 1914)

ACHYRA Gn., 1849
    EURYCREON Led., 1863
    TRITAEA Meyr., 1884
4974 **bifidalis** (F., 1794)
    inornatalis (Wlk., 1866)
    evanidalis (Berg, 1875)
    obsoletalis (Berg, 1875)
    stolidalis (Schaus, 1940)
4975 **rantalis** (Gn., 1854)
    similalis; auth., part, not Gn., 1854
    siriusalis (Wlk., 1859)
    licealis (Wlk., 1859)
    murcialis (Wlk., 1859)
    nestusalis (Wlk., 1859)
    diotimealis (Wlk., 1859)
    crinisalis (Wlk., 1859)
    intractella (Wlk., 1863)
    crinitalis (Led., 1863)
    posticata (G. & R., 1867)
    subfulvalis (H.-S., 1871)
    communis (Grt., 1876)
    collucidalis (Mösch., 1890)

caffreii (Flint & Malloch, 1920)
4976 **occidentalis** (Pack., 1873)

NEOHELVIBOTYS Mun., 1976
4977 **neohelvialis** (Capps, 1967)
4978 **arizonensis** (Capps, 1967)
4979 **polingi** (Capps, 1967)

HELVIBOTYS Mun., 1976
4980 **helvialis** (Wlk., 1859)
thycesalis (Wlk., 1859)
apertalis (Wlk., 1866)
citrina (G. & R., 1867)
4981 **pseudohelvialis** (Capps, 1967)
4982 **freemani** Mun., 1976
4983 **subcostalis** (Dyar, 1912)
4984 **pucilla** (Druce, 1895), n. comb.

SITOCHROA Hbn., 1825
SPILODES Gn., 1854
4985 **aureolalis** (Hulst, 1886)
cyralis (Druce, 1895)
4986 **dasconalis** (Wlk., 1859)
4987 **chortalis** (Grt., 1873)

ARENOCHROA Mun., 1976
4988 **flavalis** (Fern., 1894)
unipunctalis (Walter, 1928)

XANTHOSTEGE Mun., 1976
4989 **roseiterminalis** (B. & McD., 1914)
4990 **plana** (Grt., 1883)

SERICOPLAGA Warr., 1892
4991 **externalis** Warr., 1892
maclurae (Riley, 1893)

URESIPHITA Hbn., 1825
THOLERIA; auth., part, not Hbn.,
1823
MECYNA; auth., part, not
Doubleday, 1849
4992 **reversalis** (Gn., 1854)

LOXOSTEGE Hbn., 1825
BOREOPHILA Gn., 1844
COSMOCREON Warr., 1892
MAROA B. & McD., 1914
POLINGIA B. & McD., 1914
PARASITOCHROA Hannemann,
1964
4993 **albiceralis** (Grt., 1878)
4994 **floridalis** B. & McD., 1913
4995 **lepidalis** (Hulst, 1886)
4996 **indentalis** (Grt., 1883)
4997 **kearfottalis** Walter, 1928
4998 **terpnalis** B. & McD., 1918
4999 **unicoloralis** (B. & McD., 1914)
5000 **allectalis** (Grt., 1877)
perplexalis (Fern., 1885)
5001 **typhonalis** B. & McD., 1914
5002 **oberthuralis** Fern., 1894
5003 **egregialis** Mun., 1976
5004 **sticticalis** (L., 1761)
lupulina (Clerck, 1759), preocc. by
L., 1758
fuscalis (Hbn., 1796), preocc. by D.
& S., 1775
tetragonalis (Haw., 1811)
lupulinalis (Gn., 1854), emend.

sordida; auth., part, not Butler, 1886
5005 **mojavealis** Capps, 1967
5006 **kingi** Mun., 1976
5007 **annaphilalis** (Grt., 1881)
5008 **immerens** (Harv., 1875)
triumphalis (Grt., 1902)
5009 **quaestoralis** (B. & McD., 1914)
5010 **anartalis** (Grt., 1877)
a. **lulualis** (Hulst, 1886)
b. **albertalis** B. & McD., 1918
c. **saxicolalis** B. & McD., 1918
d. **rainierensis** Mun., 1976
5011 **ephippialis** (Zett., 1839)
dubitaria (Zett., 1839)
scandinavialis (Gn., 1854)
frigidalis (Gn., 1854)
5012 **thallophilalis** (Hulst, 1886)
thrallophilalis (Hulst, 1886), incorr.
orig. spell.
flavifimbrialis (Warr., 1892)
thallophyllalis (B. & McD., 1917),
missp.
thrallophyllalis (McD., 1939), missp.
5013 **brunneitincta** Mun., 1976
5014 **offumalis** (Hulst, 1886)
5015 **sierralis** Mun., 1976
a. **internationalis** Mun., 1976
b. **tularealis** Mun., 1976
c. **sanpetealis** Mun., 1976
5016 **commixtalis** (Wlk., 1866)
indotatellus (Wlk., 1866)
septentrionalis (Tengström, 1869)
5017 **cereralis** (Zell., 1872)

PYRAUSTA Schr., 1802
BOTYS Latr., 1802–03
HAEMATIA Hbn., 1818
BOTIS Swainson, 1821
SYLLYTHRIA Hbn., 1825
PANSTEGIA Hbn., 1825
ENNYCHIA Tr., 1828
RHODARIA Gn., 1844
HERBULA Gn., 1854
SYNCHROMIA Gn., 1854
CINDAPHIA Led., 1863
SCIORISTA Warr., 1890
AUTOCOSMIA Warr., 1892
TRIGONUNCUS Amsel, 1952
5018 **demantrialis** (Druce, 1895)
monotonigra Amsel, 1956
5019 **nexalis** (Hulst, 1886)
concinna Warr., 1892
5020 **sartoralis** B. & McD., 1914
5021 **roseivestalis** Mun., 1976
5022 **zonalis** B. & McD., 1918
5023 **napaealis** (Hulst, 1886)
5024 **linealis** (Fern., 1894)
5025 **ochreicostalis** B. & McD., 1918
5026 **pilatealis** B. & McD., 1914
5027 **lethalis** (Grt., 1881)
5028 **corinthalis** B. & McD., 1914
5029 **volupialis** (Grt., 1877)
5030 **morenalis** (Dyar, 1908)
5031 **atropurpuralis** (Grt., 1877)
5032 **nicalis** (Grt., 1878)
uxorculalis (Hulst, 1886)
subnicalis (Warr., 1892)
5033 **grotei** Mun., 1976
augustalis (Grt., 1881), preocc. by
Felder & Rogenhofer, 1874

5034 **signatalis** (Wlk., 1866)
vinulenta (G. & R., 1867)
5035 **pythialis** B. & McD., 1918
5036 **inveterascalis** B. & McD., 1918
5037 **inornatalis** (Fern., 1885)
rosa (Druce, 1985)
5038 **shirleyae** Mun., 1976
5039 **coccinea** Warr., 1892
5040 **bicoloralis** (Gn., 1854)
julialis; auth., part, not Wlk., 1859
incensalis; auth., part, not Led., 1863
amiculatalis; auth., part, not Berg,
1876
facitalis; auth., part, not *facetalis*
Berg, 1884, missp.
pulchripictalis; auth., part, not Hamp.,
1895
5041 **augustalis** (Felder & Rogenhofer, 1874)
angustalis Hamp., 1899, missp.
5042 **onythesalis** (Wlk., 1859)
5043 **pseudonythesalis** Mun., 1976
5044 **insignitalis** (Gn., 1854)
5045 **aurea** (Hamp., 1913)
5046 **flavibrunnea** (Hamp., 1913)
5047 **klotsi** Mun., 1976
5048 **flavofascialis** (Grt., 1882)
5049 **phoenicealis** (Wlk., 1859)
flegialis (Wlk., 1859)
noraxalis (Wlk., 1859)
5050 **panopealis** (Wlk., 1859)
coecilialis (Wlk., 1859)
probalis (Wlk., 1859)
ocellusalis (Wlk., 1859)
catenalis (Wlk., 1866)
juncturalis (Wlk., 1866)
concatenalis (Wlk., 1866)
heliamma (Meyr., 1885)
5051 **rubricalis** (Hbn., 1796)
nescalis (Wlk., 1859)
similalis (Led., 1863)
5052 **californicalis** (Pack., 1873)
a. **sierranalis** Mun., 1976
5053 **pseuderosnealis** Mun., 1976
5054 **dapalis** (Grt., 1881)
5055 **homonymalis** Mun., 1976, repl. name
submarginalis (Wlk., 1866: 1288),
preocc. by Wlk., 1866: 1286
5056 **generosa** (G. & R., 1867)
5057 **subgenerosa** Mun., 1976
5058 **orphisalis** Wlk., 1859
ochosalis Dyar, 1903
5059 **tuolumnalis** B. & McD., 1918
5060 **subsequalis** (Gn., 1854)
insequalis (Gn., 1854)
madetesalis (Wlk., 1859)
repletalis (Wlk., 1866)
efficitalis (Wlk., 1866)
matronalis (Grt., 1875)
a. **borealis** Pack., 1867
b. **plagalis** Haim., 1908
c. **petaluma** Mun., 1976
5061 **tatalis** (Grt., 1877)
5062 **retidiscalis** Mun., 1976
5063 **andrei** Mun., 1976
5064 **perrubralis** (Pack., 1873)
a. **saanichalis** Mun., 1951
b. **shastanalis** Mun., 1976
5065 **scurralis** (Hulst, 1886)
postrubralis Hamp., 1899
a. **awemealis** Mun., 1976

5066 **arizonicalis** Mun., 1976
   arizonensis Mun., 1976, preocc. by
     Mun., 1957
5067 **semirubralis** (Pack., 1873)
5068 **unifascialis** (Pack., 1873)
   obnigralis (Hulst, 1886)
   a. **subolivalis** (Pack., 1873)
    hircinalis (Grt., 1875)
   b. **rindgei** Mun., 1957
   c. **arizonensis** Mun., 1957
5069 **tyralis** (Gn., 1854)
   erosnealis Wlk., 1859
   agathalis (Wlk., 1859)
   diffissa (G. & R., 1867)
   bellulalis (Hulst, 1886)
   idessa (Druce, 1895)
5070 **laticlavia** (G. & R., 1867)
   cinerosa (G. & R., 1867)
5071 **acrionalis** (Wlk., 1859)
   acuphisalis (Wlk., 1859)
   proceralis (Led., 1863)
   rubicundalis (Led., 1863), unavail.,
    publ. in syn.
   sumptuosalis Wlk., 1866
   haruspica (G. & R., 1867)
   rufifimbrialis (Grt., 1881)
5072 **obtusanalis** Druce, 1899
5073 **niveicilialis** (Grt., 1875)
5074 **fodinalis** (Led., 1863)
   a. **monticola** Mun., 1976
   b. **septentrionicola** Mun., 1976
5075 **socialis** (Grt., 1877)
   a. **perpallidalis** Mun., 1976
5076 **antisocialis** Mun., 1976

HYALORISTA Warr., 1892
  PYRAUSTOPSIS Amsel, 1956
5077 **taeniolalis** (Gn., 1854)

PORTENTOMORPHA Amsel, 1956
  APOECETES Mun., 1956
5078 **xanthialis** (Gn., 1854)
   superbalis (Wlk., 1866)
   incalis (Snell., 1875)
   rosealis (Mösch., 1890)

Spilomelini

UDEA Gn., 1845
  OEOBIA Hbn., 1825, suppr.
  OEBIA Hbn., 1825, suppr.
  MELANOMECYNA Butler, 1883
  PROTOCOLLETIS Meyr., 1888
  MNESICTENA Meyr., 1890, n. syn.
  PROTAULACISTIS Meyr., 1899
  NOTOPHYTIS Meyr., 1932
5079 **rubigalis** (Gn., 1854)
   ferrugalis; auth., part, not Hbn., 1796
   oblunalis (Led., 1863)
   harveyana (Grt., 1877)
5080 **profundalis** (Pack., 1873)
5081 **washingtonalis** (Grt., 1882)
   invinctalis (Hulst, 1886)
   a. **hollandi** Mun., 1966
   b. **nomensis** Mun., 1966
   c. **pribilofensis** Mun., 1966
5082 **octosignalis** (Hulst, 1886)
   straminea (Warr., 1892)
5083 **vacunalis** (Grt., 1881)
   galactalis Dyar, 1925

5084 **torvalis** (Mösch., 1864)
   gelida (M'Lachlan, 1878)
5085 **alaskalis** (Gib., 1920)
5086 **inquinatalis** (Zell., 1846)
   glacialis (Pack., 1867)
5087 **rusticalis** (B. & McD., 1914)
5088 **nordeggensis** (McD., 1930)
5089 **berberalis** (B. & McD., 1918)
5090 **indistinctalis** Warr., 1892
   a. **johnstoni** Mun., 1966
5091 **sheppardi** (McD., 1930)
5092 **saxifragae** (McD., 1935)
5093 **brevipalpis** Mun., 1966
5094 **cacuminicola** Mun., 1966
5095 **beringialis** Mun., 1966
5096 **derasa** Mun., 1966
5097 **livida** Mun., 1966
5098 **turmalis** (Grt., 1881)
   a. **catronalis** Mun., 1966
   b. **tularensis** Mun., 1966
   c. **griseor** Mun., 1966
5099 **itysalis** (Wlk., 1859)
   variegata (Wlk., 1863)
   hyperborealis (Mösch., 1874)
   a. **mertensialis** Mun., 1966
   b. **tillialis** (Dyar, 1904)
   c. **rindgeorum** Mun., 1966
   d. **kodiakensis** Mun., 1966
   e. **albimontanalis** Mun., 1966
   f. **durango** Mun., 1966
   g. **wasatchensis** Mun., 1966
   h. **clarkensis** Mun., 1966
   i. **marinensis** Mun., 1966
5100 **abstrusa** Mun., 1966
   a. **subarctica** Mun., 1966
   b. **cordilleralis** Mun., 1966
   c. **pullmanensis** Mun., 1966
5101 **radiosalis** (Mösch., 1883)

NEOLEUCINODES Capps, 1958
5102 **prophetica** (Dyar, 1914)

LAMPROSEMA Hbn., 1823
5103 **lunulalis** Hbn., 1823
5104 **victoriae** Dyar, 1923
5105 **sinaloanensis** Dyar, 1923

LINEODES Gn., 1854
  SCOPTONOMA Zell., 1873
5106 **fontella** Wlsm., 1913
   contortalis; auth., part, not Gn.,
    1854
5107 **integra** (Zell., 1873)
5108 **interrupta** (Zell., 1873)
5109 **triangulalis** Mösch., 1890

ZELLERINA de la Torre y Callejas,
  1958, repl. name
  STENOPTYCHA Zell., 1863,
   preocc. by Agassiz, 1862
5110 **solanalis** (B. & McD., 1913). n. comb.

ERCTA Wlk., 1859
5111 **vittata** (F., 1794)
   hemialis (Gn., 1859)
   tipulalis Wlk., 1859

APILOCROCIS Amsel, 1956
5112 **brumalis** (B. & McD., 1914)
5113 **pimalis** (B. & Benj., 1926)

DIAPHANTANIA Mösch., 1890
5114 **impulsalis** (H.-S., 1871), n. comb.

LOXOSTEGOPSIS Dyar, 1917,
  emend.
  LOXOTEGOPSIS Dyar, 1917,
   incorr. orig. spell.
5115 **polle** Dyar, 1917
5116 **xanthocrypta** (Dyar, 1913)
5117 **merrickalis** (B. & McD., 1918)
5118 **emigralis** (B. & McD., 1918)
5119 **curialis** B. & McD., 1918

SUFETULA Wlk., 1859
  MIROBRIGA Wlk., 1863
  LOETRINA Wlk., 1863
  PSEUDOCHOREUTIS Snell.,
   1880
5120 **diminutalis** (Wlk., 1866)
   dematrialis (Druce, 1895)

FALCIMORPHA Amsel, 1957
  FALX Amsel, 1956, preocc. by
   Gouan, 1770
5121 **philogelos** (Dyar, 1922), n. comb.
   sinuosalis (Amsel, 1956), n. syn.

EURRHYPARODES Snell., 1880
  MOLYBDANTHA Meyr., 1884
5122 **lygdamis** Druce, 1902

DEUTEROPHYSA Warr., 1889
  GONOPIONEA Hamp., 1913, n.
   syn.
5123 **fernaldi** Mun., n. name
   costimaculalis (Fern., 1901), n.
    comb., preocc. by Warr., 1889,
    sec. hom.

HYDROPIONEA Hamp., 1917
5124 **oblectalis** (Hulst, 1886)
   eumoros (Dyar, 1917)
5125 **fenestralis** (B. & McD., 1914)

GESHNA Dyar, 1906
5126 **cannalis** (Quaintance, 1899)

HYDRIRIS Meyr., 1885
  SPANISTA Led., 1863, preocc. by
   Forster, 1862
  ANTIERCTA Amsel, 1956
5127 **ornatalis** (Dup., 1831)
   saturnalis (Tr., 1835)
   saturalis auth., missp.
   deciusalis (Wlk., 1859)
   invenustalis (Wlk., 1866)
   fraterna (Butler, 1875)
   elutalis; Strand, 1918, part, not Wlk.,
    1859
   bifascialis (Heeger, 1938)

CHORISTOSTIGMA Warr., 1892
  NAMANGANIA Amsel, 1952. n.
   syn.
5128 **plumbosignalis** (Fern., 1888)
5129 **zephyralis** (B. & McD., 1914), n.
   comb.
5130 **roseopennalis** (Hulst, 1886), n. comb.
5131 **perpulchralis** (Hamp., 1898), n.
   comb.

chromaphila (Dyar, 1914), n. comb., n. syn.

microchroia (Dyar, 1916), n. comb., n. syn.

5132 **elegantalis** Warr., 1892

argalis (Fern., 1894), n. comb.

5133 **disputalis** (B. & McD., 1917), n. comb.

5134 **leucosalis** (B. & McD., 1914), n. comb.

MECYNA Doubleday, 1849

5135 **submedialis** (Grt., 1876)

dissectalis (Grt., 1880), n. comb.

pilalis (Hulst, 1886), n. comb.

5136 **fuscimaculalis** (Grt., 1878), n. comb.

flavicoloralis (Grt., 1878), n. comb.

confovealis (Hulst, 1886), n. comb.

5137 **mustelinalis** (Pack., 1873)

catenulalis (Grt., 1877), n. comb.

monulalis (Hulst, 1886), n. comb.

5138 **luscitialis** (B. & McD., 1914)

MIMORISTA Warr., 1890

5139 **subcostalis** (Hamp., 1913), n. comb.

5140 **trimaculalis** (Grt., 1878), n. comb.

5141 **tristigmalis** (Hamp., 1898), n. comb.

DIACME Warr., 1892

5142 **elealis** (Wlk., 1859), n. comb.

taedialis (Wlk., 1859), n. comb.

5143 **adipaloides** (G. & R., 1867), n. comb.

5144 **phyllisalis** (Wlk., 1859)

5145 **mopsalis** (Wlk., 1859), n. comb.

mettiusalis (Wlk., 1859), n. comb.

griseicinctalis (Hamp., 1913), n. comb.

griseicinctus (Kimball, 1965), missp.

EPIPAGIS Hbn., 1825

STENOPHYES Led., 1863

5146 **forsythae** Mun., 1955

5147 **huronalis** (Gn., 1854)

fenestralis (Hbn., 1796), unused sr. syn.

serinalis (Wlk., 1859)

5148 **disparilis** (Dyar, 1910)

SAMEODES Snell., 1880

PESSOCOSMA Meyr., 1884

5149 **albiguttalis** (Warr., 1889), mispl.

SAMEA Gn., 1854

5150 **ecclesialis** Gn., 1854

castellalis Gn., 1854

luccusalis Wlk., 1859

disertalis Wlk., 1866

5151 **multiplicalis** (Gn., 1854)

discessalis Wlk., 1866

nicaeusalis; auth., part, not Wlk., 1859

5152 **baccatalis** (Hulst, 1886), n. comb.

SOMATANIA Mösch., 1890

SOMATAMIA Kimball, 1965, missp.

5153 **pellucidalis** Mösch., 1890

LOXOMORPHA Amsel, 1956

CHRYSOBOTYS Mun., 1956, n. syn.

5154 **cambogialis** (Gn., 1854)

5155 **flavidissimalis** (Grt., 1878), n. comb.

NOMOPHILA Hbn., 1825

STENOPTERYX Gn., 1844

MACRONOMEUTIS Meyr., 1936

5156 **nearctica** Mun., 1973

noctuella; auth., part, not D. & S., 1775

PILEMIA Mösch., 1881

RHECTOTHYRIS Warr., 1890

RAPOONA v. Hedemann, 1894

RAPONA Schaus, 1940, missp.

5157 **periusalis** (Wlk., 1859)

deformalis Mösch., 1881

tristis (v. Hedemann, 1894)

ATEGUMIA Amsel, 1956

5158 **ebulealis** (Gn., 1854), n. comb.

DESMIA Westwood, 1831

AEDIODES Gn., 1854

ARNA Wlk., 1875

5159 **funeralis** (Hbn., 1796)

5160 **maculalis** Westwood, 1831, rev. stat.

5161 **subdivisalis** Grt., 1871, rev. stat.

nominabilis E. Hering, 1906, rev. syn.

5162 **ufeus** (Cram., 1777)

orbalis (Gn., 1854)

prognealis Wlk., 1859

bulisalis Wlk., 1859

5163 **divisalis** Wlk., 1866, rev. stat.

5164 **tages** (Cram., 1777)

tagesalis (Gn., 1854), emend.

propinqualis Mösch., 1881

5165 **stenizonalis** Hamp., 1912

5166 **deploralis** Hamp., 1912

5167 **ploralis** (Gn., 1854)

5168 **desmialis** (B. & McD., 1914), n. comb.

kaeberalis (Haim., 1915). n. comb.

HYMENIA Hbn., 1825

ZINCKENIA Zell., 1852

5169 **perspectalis** (Hbn., 1796)

phrasiusalis (Wlk., 1859)

SPOLADEA Gn., 1854

5170 **recurvalis** (F., 1794)

animalis (Gn., 1854)

fascialis; auth., part, not Cram., 1782

DIASEMIOPSIS Mun., 1957

5171 **leodocusalis** (Wlk., 1859)

ramburialis; auth., part, not Dup., 1834

DIASEMIODES Mun., 1957

5172 **janassialis** (Wlk., 1859)

hariolalis (Hulst, 1886)

5173 **nigralis** (Fern., 1892)

DIATHRAUSTA Led., 1863

TRIPODAULA Meyr., 1933

TRIPLODAULA Mun., 1956, missp.

5174 **reconditalis** (Wlk., 1859)

minualis (Wlk., 1866)

octomaculalis Fern., 1887

5175 **harlequinalis** Dyar, 1913

a. **montana** Haim., 1915

b. **amaura** Mun., 1956

c. **lauta** Mun., 1956

ANAGESHNA Mun., 1956

5176 **primordialis** (Dyar, 1907)

a. **vividior** Mun., 1956

b. **pallidior** Mun., 1956

APOGESHNA Mun., 1956

EUVALVA Amsel, 1956, n. syn.

5177 **stenialis** (Gn., 1854)

acestealis (Wlk., 1859), n. comb.

phaerusalis (Wlk., 1859), n. comb.

STENIODES Snell., 1875

HERINGIA v. Hedemann, 1894, preocc. by Rondani, 1856, n. syn.

HERINGIELLA Berg, 1898, repl. name, n. syn.

SCAEOCERANDRA Meyr., 1936, n. syn.

5178 **mendica** (v. Hedemann, 1894), n. comb.

indianalis (Dyar, 1915), n. syn.

gelliasalis; auth., part, not Wlk., 1859

PENESTOLA Mösch., 1890

5179 **bufalis** (Gn., 1854), n. comb.

praeficalis Mösch., 1890, n. syn.

stercoralis; auth., part, not Mösch., 1881

5180 **simplicialis** (B, & McD., 1913), n. comb.

ANTIGASTRA Led., 1863

5181 **catalaunalis** (Dup., 1833)

venosalis (Wlk., 1866)

BLEPHAROMASTIX Led., 1863

5182 **ranalis** (Gn., 1854)

archasialis (Wlk., 1859)

ofellusalis (Wlk., 1859)

olliusalis (Wlk., 1859)

strictalis (Wlk., 1866)

gracilis (G. & R., 1867)

a. **datisalis** Druce, 1895

occidentalis Haim., 1908

5183 **pseudoranalis** (B. & McD., 1914)

5184 **potentalis** (B. & McD., 1914), mispl.

5185 **achroalis** (Hamp., 1913), mispl.

5186 **haedulalis** (Hulst, 1886), mispl.

5187 **magualis** (Gn., 1854), mispl.

magnalis auth., missp.

medealis (Wlk., 1859)

belusalis (Wlk., 1859)

curtalis (Wlk., 1866)

5188 **aplicalis** (Gn., 1854), mispl.

xeniolalis (Hulst, 1886)

5189 **rehamalis** (Dyar, 1914), mispl.

5190 **differentialis** (Dyar, 1914), mispl.

5191 **schistisemalis** (Hamp., 1912), mispl.

5192 **hampsoni** (B. & McD., 1913), mispl.

5193 **eudamidasalis** (Druce, 1899), mispl.

ARASCHNOPSIS Amsel, 1956

5194 **subulalis** (Gn., 1854)

preciosalis (Mösch., 1881), n. comb.

pretiosalis auth., missp.

EULEPTE Hbn., 1825
5195 **anticostalis** (Grt., 1871), n. comb.,
rev. stat.
levalis (Hulst, 1886), n. comb.
inguinalis; auth., part, not Gn., 1854

SYNCLERA Led., 1863
5196 **jarbusalis** (Wlk., 1859), rev. stat.
traducalis; auth., part, not Zell., 1852
retinalis; auth., part, not Led., 1857
univocalis; auth., part, not Wlk., 1859
cottalis (Wlk., 1859)
chlorophasma; auth., part, not Butler,
1878

GLYPHODES Gn., 1854
5197 **pyloalis** Wlk., 1859
sylpharis (Butler, 1878)
5198 **sibillalis** Wlk., 1859
batesi Felder & Rogenhofer, 1864
atlitalis Hulst, 1886
bivitralis; auth., part, not Gn., 1854
5199 **floridalis** (Fern, 1901), n. comb.

COLOMYCHUS Mun., 1956
5200 **talis** (Grt., 1878)
florepicta; auth., part, not Dyar, 1914

DIAPHANIA Hbn., 1808–18
EUDIOPTIS Hbn., 1823
PHAKELLURA Guilding, 1830
PHACELLURA Led., 1863, missp.
5201 **olealis** (Felder & Rogenhofer, 1874)
5202 **nitidalis** (Stoll, 1781)
vitralis (Hbn., 1818)
fumosalis; auth., part, not Gn., 1854
praxialis; auth., part, not Druce,
1895
5203 **arguta** (Led., 1863)
5204 **hyalinata** (L., 1767)
marginalis (Stoll, 1781)
lucernalis (Hbn., 1796)
hyalinatalis (Gn., 1854), emend.
zapillitalis (Weyenbergh, 1873)
niveocilia (Hamp., 1898)
5205 **modialis** (Dyar, 1912)
5206 **infimalis** (Gn., 1854)
5207 **indica** (Saund., 1851)
hyalinalis; (Bdv., 1833), part, not
hyalinata L., 1767, emend,
capensis (Zell., 1852)
gazorialis (Gn., 1854)
zygaenalis (Gn., 1854)
cucurbitalis (Gn., 1863)
5208 **lualis** (H.-S., 1871)

HOTERODES Gn., 1854
5209 **ausonia** (Cram., 1779)
canastralis (Hbn., 1825), repl. name
ausonialis Gn., 1854, emend.

LEUCOCHROMA Gn., 1854
5210 **corope** (Stoll, 1781)
splendidalis (Stoll, 1781)
corrivalis (Hbn., 1825), repl. name
selectalis (Wlk., 1866)
minoralis Warr., 1889
mineralis Hamp., 1898, missp.

OMIODES Gn., 1854, rev. stat.
SPARGETA Led., 1863, rev. syn.

COENOSTOLA Led., 1863, rev.
syn.
HEDYLEPTA Led., 1863, n. syn.
DEBA Wlk., 1866, n. syn.
PELECYNTIS Meyr., 1884, n. syn.
CHAREMA Moore, 1888, rev. syn.
5211 **simialis** Gn., 1854
jasonalis (Wlk., 1859), n. comb.
orontesalis (Wlk., 1859), n. comb.
eruptalis (Led., 1863), n. comb.
5212 **indicata** (F., 1775), n. comb.
vulgalis (Gn., 1854), n. comb.
sabalis (Wlk., 1859), n. comb.
moeliusalis (Wlk., 1859), n. comb.
connexalis (Wlk., 1866), n. comb.
reductalis (Wlk., 1866), n. comb.
vulgaris (Dognin, 1909), missp.
pigralis (Dognin, 1909), ab.,
infrasubsp.
5213 **rufescens** (Hamp, 1912), n. comb.
miamialis (Schaus, 1917), n. comb.
5214 **stigmosalis** (Warr., 1892), n. comb.

CONDYLORRHIZA Led., 1863
5215 **vestigialis** (Gn., 1854)
tritealis (Wlk., 1859), form
mestoralis (Wlk., 1859)
ortalis (Hulst, 1886)

STEMORRHAGES Led., 1863
5216 **costata** (F., 1794)
aurocostalis (Gn., 1854)

PALPITA Hbn., 1808
HAPALIA Hbn., 1808–18
CONCHIA Hbn., 1821
MARGARONIA Hbn., 1825
PARADOSIS Zell., 1852
SAROTHRONOTA Led., 1863, n.
syn.
SEBUNTA Wlk., 1863
APYRAUSTA Amsel, 1951, n. syn.
5217 **flegia** (Cram., 1777)
virginalis (Hbn., 1825), repl. name,
n. comb.
flegyalis (Poey, 1832), emend., n.
comb.
villosalis (Zell., 1852), n. comb.
phantasmalis (Gn., 1854), n. comb.
5218 **quadristigmalis** (Gn., 1854)
5219 **kimballi** Mun., 1959
5220 **gracilalis** (Hulst, 1886)
atrisquamalis (Hamp., 1912), n.
comb.
5221 **cincinnatalis** Mun., 1952
5222 **arsaltealis** (Wlk., 1859)
5223 **illibalis** (Hbn., 1818)
5224 **euphaesalis** (Wlk., 1859)
subjectalis (Led., 1863), repl. name
5225 **freemanalis** Mun., 1952
5226 **magniferalis** (Wlk., 1861)
fascialis (Wlk., 1862)
guttulosa (Wlk., 1863)
5227 **aenescentalis** Mun., 1952

POLYGRAMMODES Gn., 1854
ASTURA Gn., 1854, n. syn.
HILAOPSIS Led., 1863, n. syn.
DICHOCROPSIS Dyar, 1910, n.
syn.
5228 **flavidalis** (Gn., 1854)

lacoalis (Wlk., 1859)
cinctipedalis (Wlk., 1866)
a. *oxydalis* (Gn., 1854)
5229 **langdonalis** (Grt., 1877)
5230 **elevata** (F., 1794)
elevalis (Gn., 1854), emend.
grandimacula (Dognin, 1912), form
5231 **sanguinalis** Druce, 1895

AZOCHIS Wlk., 1859
5232 **rufidiscalis** Hamp., 1904

COMPACTA Amsel, 1956
5233 **capitalis** (Grt., 1881)
5234 **hirtalis** (Gn., 1854)
lybialis (Wlk., 1859), n. comb.
amatalis (Wlk., 1859), n. comb.
5235 **hirtaloides** (Dyar, 1912), n. comb.

LANIIFERA Hamp., 1899
5236 **cyclades** (Druce, 1895)

MIMOPHOBETRON Mun., 1950
5237 **pyropsalis** (Hamp., 1904), n. comb.
rhodope (Hamp., 1913), n. syn.
liopasialis (Dyar, 1914), n. syn.
rhodopides (Strand, 1917), ab.,
infrasubsp.

LIOPASIA Mösch., 1881
TERASTIODES Warr., 1892
DICHOTIS Warr., 1892
5238 **teneralis** (Led., 1863)

TERASTIA Gn., 1854
5239 **meticulosalis** Gn., 1854
coeligenalis (Hulst, 1886)

AGATHODES Gn., 1854
5240 **designalis** Gn., 1854
floridalis (Hulst, 1886)
a. *monstralis* Gn., 1854

PANTOGRAPHA Led., 1863
PANTOGRAPTA auth., missp.
5241 **limata** (G. & R., 1867)

PLEUROPTYA Meyr., 1890
5242 **penumbralis** (Grt., 1877), n. comb.
5243 **silicalis** (Gn., 1854), n. comb.
cypraealis (Wlk., 1859), n. comb.
sublutalis (Warr., 1889), n. comb.
5244 **fluctuosalis** (Led., 1863), ident.
uncert.

SYLLEPTE Hbn., 1823
SYLEPTA Hbn., 1825, emend.
SYLLEPTA Hbn., 1826, missp.
5245 **diacymalis** (Hamp., 1912), mispl.
butyrosa; auth., part, not Butler,
1878

PHAEDROPSIS Warr., 1890
TRICHOGNATHOS Amsel, 1956,
n. syn.
5246 **chromalis** (Gn., 1854)
principialis (Led., 1863), n. comb.
principalis (Led., 1863), incorr. orig.
spell.
5247 **stictigramma** (Hamp., 1912), n.
comb.

LYGROPIA Led., 1863
5248 **tripunctata** (F., 1794), n. comb.
    campalis (Gn., 1854), n. comb.
    cubanalis (Gn., 1854), n. comb.
    memmialis (Wlk., 1859), n. comb.
5249 **plumbicostalis** (Grt., 1871), n. comb.
5250 **rivulalis** Hamp., 1898, mispl.
    nymphulalis (Haim., 1908)
5251 **octonalis** (Zell., 1873), mispl.
    sexmaculalis (Grt., 1876)

LYPOTIGRIS Hbn., 1825
5252 **reginalis** (Stoll, 1781)

DIASTICTIS Hbn., 1818
    ANOMOSTICTIS Warr., 1892,
      repl. name
5253 **argyralis** Hbn., 1818
5254 **pseudargyralis** Mun., 1956
5255 **ventralis** (G. & R., 1867)
    a. seamansi Mun., 1956
5256 **fracturalis** (Zell., 1872)
5257 **holguinalis** Mun., 1956
5258 **viridescens** Mun., 1956
5259 **robustior** Mun., 1956
5260 **sperryorum** Mun., 1956
5261 **caecalis** (Warr., 1892)

FRAMINGHAMIA Strand, 1920
5262 **helvalis** (Wlk., 1859)
    oscitalis (Grt., 1880)
    gyralis (Hulst, 1886)
    botys Strand, 1920

MICROTHYRIS Led., 1863
    CROSSOPHORA Mösch., 1890
    GROSSOPHORA Mun., 1956,
      missp.
5263 **anormalis** (Gn., 1854)
    alvinalis (Gn., 1854)
    helcitalis (Wlk., 1859), n. comb.
    orphnealis (Wlk., 1859), n. comb.
    dracusalis (Wlk., 1859), n. comb.
    subaequalis (Wlk., 1866), n. comb.
5264 **prolongalis** (Gn., 1854)
    sectalis (Gn., 1854)
    eurytalis (Wlk., 1859)
    scotalis (Led., 1863), missp.
    subaequalis (Wlk., 1866)

PHOSTRIA Hbn., 1819
5265 **tedea** (Stoll, 1780)
    temira (Stoll, 1782), n. syn.
5266 **oajacalis** (Wlk., 1866), rev. stat.
    pelialis (Felder & Rogenhofer, 1874),
      rev. syn.

ASCIODES Gn., 1854
5267 **gordialis** Gn., 1854
    quietalis (Wlk., 1859)
    confusalis (Hulst, 1886)

PSARA Snell., 1875
5268 **obscuralis** (Led., 1863), n. comb.
5269 **dryalis** (Wlk., 1859), n. comb.
    glaucusalis (Wlk., 1859), n. comb.

SATHRIA Led., 1863
5270 **internitalis** (Gn., 1854)
    stercoralis Led., 1863
    serenalis (Wlk., 1866), n. comb.

BICILIA Amsel, 1956
5271 **iarchasalis** (Wlk., 1859), n. comb.
    differalis (Wlk., 1865), n. comb.
    iarchisalis (Schaus, 1940), missp.
    vogli Amsel, 1956, n. syn.

HERPETOGRAMMA Led., 1863
    PACHYZANCLA Meyr., 1884
    ACHARANA Moore, 1885
    PILOPTILA Swinhoe, 1884
    PANTOEOCOME Warr., 1896
    PTILOPTILA Hamp., 1899, missp.
    MACROBOTYS Mun., 1950
5272 **bipunctalis** (F., 1794)
    detritalis (Gn., 1854)
    lycialis (Wlk., 1859), n. comb.
    philealis (Wlk., 1859), n. comb.
    repetitalis (Grt., 1882), n. comb.
5273 **ipomoealis** (Capps, 1964), n. comb.
5274 **phaeopteralis** (Gn., 1854)
    vecordalis (Gn., 1854), n. comb.
    vestalis (Wlk., 1859), n. comb.
    additalis (Wlk., 1862), n. comb.
    plebejalis (Led., 1863: 373), n. comb.
    cellatalis (Wlk., 1866), n. comb.
5275 **pertextalis** (Led., 1863), n. comb.
    thesealis; (Zell., 1872), part, not
      Led., 1863
    gentilis (Grt., 1873), n. comb.
    genitalis auth., missp.
5276 **abdominalis** (Zell., 1872), n. comb.
    fissalis (Grt., 1881), n. comb., n. syn.
5277 **thestealis** (Wlk., 1859), n. comb.
    magistralis (Grt., 1873), n. comb.
    gulosalis (Hulst, 1886), n. comb.
5278 **centrostrigalis** (Steph., 1834)
5279 **theseusalis** (Wlk., 1859), n. comb.
    thesealis (Led., 1863), emend., n.
      comb.
    feudalis (Grt., 1875), n. comb.
5280 **aeglealis** (Wlk., 1859)
    quinquelinealis (Grt., 1875), n. comb.

PILOCROCIS Led., 1863
5281 **ramentalis** Led., 1863
    perfuscalis (Hulst, 1886)

CRYPTOBOTYS Mun., 1956
5282 **zoilusalis** (Wlk., 1859)
    masculinalis (B. & McD., 1913)

SYLLEPIS Poey, 1832
5283 **hortalis** (Wlk., 1859)

SYNGAMIA Gn., 1854
    ANANIA; auth., not Hbn., 1823
5284 **florella** (Stoll, 1781)
    quinqualis (Hbn., 1823)
    florellalis Gn., 1854, emend.
    quisqualis Hamp., 1898, missp.

SALBIA Gn., 1854
5285 **tytiusalis** (Wlk., 1859), n. comb.
5286 **mizaralis** (Druce, 1899), n. comb.
5287 **haemorrhoidalis** Gn., 1854
    dircealis (Wlk., 1859), n. comb.
    futilalis (B. & McD., 1914), n. comb.
    haemorrhoidalis (Klima, 1939), missp.

MARASMIA Led., 1863
    DOLICHOSTICHA Meyr., 1884

EPIMIMA Meyr., 1886
    LASIACME Warr., 1896
5288 **trapezalis** (Gn., 1854)
    creonalis (Wlk., 1859)
    neoclesalis (Wlk., 1859)
    suspicalis (Wlk., 1859)
    convectalis (Wlk., 1866)
    bifurcalis (Snell., 1880)
5289 **cochrusalis** (Wlk., 1859)
    azionalis (Wlk., 1859)
    ruptalis (Wlk., 1865)

CONCHYLODES Gn., 1854
5290 **diphteralis** (Gey., 1832)
5291 **salamisalis** Druce, 1895
5292 **ovulalis** (Gn., 1854), ident. uncert.
5293 **concinnalis** Hamp., 1898

OMMATOSPILA Led., 1863
5294 **narcaeusalis** (Wlk., 1859)
    nummulalis Led., 1863
    venustalis (Wlk., 1866)

DAULIA Wlk., 1859
    GIRTEXTA Swinhoe, 1890
5295 **magdalena** (Fern., 1892)
5296 **arizonensis** Mun., 1957

PALPUSIA Amsel, 1956
5297 **goniopalpia** (Hamp., 1912), n. comb.

MARACAYIA Amsel, 1956
5298 **chlorisalis** (Wlk., 1859)

SCHOENOBIINAE

ACENTRIA Steph., 1829
    ZANCLE Steph., 1833
    ACENTROPUS Curt., 1834
5299 **nivea** (Olivier, 1791)
    hansoni (Steph., 1833)
    garnonsii (Curt., 1834)

PATISSA Moore, 1886
5300 **xantholeucalis** (Gn., 1854)
    fasciella (Fern., 1887)
5301 **flavicostella** (Fern., 1887)
5302 **flavifascialis** B. & McD., 1913
5303 **parthenialis** Dyar, 1917
5304 **chrysozona** Dyar, 1917
5305 **sordidalis** B. & McD., 1913
5306 **vestaliella** (Zell., 1872)

SCIRPOPHAGA Tr., 1832
    TOPEUTIS; auth., not
      Thopeutis Hbn., 1818, or
      Topeutis Hbn., 1825
5307 **perstrialis** (Hbn., 1825)
    serriradiellus (Wlk., 1863)
    semiradiellus auth., missp.
    macrinellus (Zell., 1866)
5308 **repugnatalis** (Wlk., 1863)
    consortalis Dyar, 1909

CARECTOCULTUS A. Blanchard,
    1975
5309 **dominicki** A. Blanchard, 1975

RUPELA Wlk., 1863
    STORTERIA B. & McD., 1913
5310 **segrega** Heinr., 1937

albinella; auth., part, not Cram., 1781
nivea; auth., part, not Wlk., 1863
5311 **tinctella** (Wlk., 1863)
albinella; auth., part, not Cram., 1781
nivea; auth., part, not Wlk., 1863
zelleri (Mösch., 1882)
holophaealis (Hamp., 1904)
unicolor (B. & McD., 1913)
5312 **sejuncta** Heinr., 1937
albinella; auth., part, not Cram., 1781
nivea; auth., part, not Wlk., 1863

DONACAULA Meyr., 1890
5313 **sordidella** (Zinck., 1821), n. comb.
5314 **unipunctella** (Rob., 1870), n. comb.
5315 **tripunctella** (Rob., 1870), n. comb.
5316 **melinella** (Clem., 1860), n. comb.
　　a. **dispersella** (Rob., 1870), n. comb.
　　b. **albicostella** (Fern., 1888), n. comb.
　　c. **uniformella** (Dyar, 1917), n. comb.
5317 **aquilella** (Clem., 1860), n. comb.
clemensella (Rob., 1870), n. comb.
5318 **pallulella** (B. & McD., 1912), n. comb.
5319 **longirostrella** (Clem., 1860), n. comb.
forficella; auth., part, not Thunb., 1794
5320 **amblyptepennis** (Dyar, 1917), n. comb.
5321 **roscidella** (Dyar, 1917), n. comb.
5322 **nitidella** (Dyar, 1917), n. comb.
5323 **uxorialis** (Dyar, 1921), n. comb.
5324 **maximella** (Fern., 1891), n. comb.

CYBALOMIINAE

CYBALOMIA Led., 1863
CYBOLOMIA auth., missp.
5325 **extorris** Warr., 1892
quadristrigalis (Fern., 1894)

ANCYLOLOMIINAE

SURATTHA Wlk., 1863
CALARINA Wlk., 1866
PLATYTESIA Strand, 1919
5326 **santella** Kft., 1908
5327 **indentella** Kft., 1908

MESOLIA Rag., 1888
EUGROTEA Fern., 1896
5328 **baboquivariella** (Kft., 1907)
5329 **oraculella** Kft., 1908
5330 **huachucella** Kft., 1908
5331 **incertella** (Zinck., 1821)
olivella (Grt., 1882)
dentella (Fern., 1896)

PRIONAPTERYX Steph., 1834
NAURACE Wlk., 1863
PRIONOPTERYX auth., missp.
5332 **yavapai** (Kft., 1908)
5333 **nebulifera** Steph., 1834
5334 **achatina** Zell., 1863
delectalis (Hulst, 1886)
5335 **cuneolalis** (Hulst, 1886)
5336 **serpentella** Kft., 1908

PSEUDOSCHOENOBIUS Fern., 1896
5337 **opalescalis** (Hulst, 1886)
griseosparsa (Hamp., 1896)

EUFERNALDIA Hulst, 1900
5338 **cadarella** (Druce, 1896)
argenteonervella Hulst, 1900

CRAMBINAE
by ALEXANDER B. KLOTS

Crambini

CRAMBUS F., 1798
PALPARIA Haw., 1811
TETRACHILA Hbn., 1822
ARGYROTEUCHIA Hbn., 1825
5339 **pascuellus** (L. 1798), extralim.
　　a. **floridus** Zell., 1872
hastiferellus; auth., not Wlk., 1863
5340 **hamellus** (Thunb., 1794)
hastiferellus Wlk., part, ♀
　　a. **carpenterellus** Pack., 1872
5341 **alienellus** (Zinck., 1817), extralim.
　　a. **labradoriensis** Christoph, 1858
luctuellus; auth., not. H.-S., 1856
moestellus Wlk., 1863
indotatellus Wlk., 1866
　　b. **dissectus** Grt., 1880, uncert. stat.
5342 **bidens** Zell., 1872
5343 **perellus** (Scop., 1763)
　　a. **innotatellus** Wlk., 1863
sericinellus Zell., 1863
inornatellus Clem., 1864
5344 **unistriatellus** Pack., 1867
exesus Grt., 1880
5345 **whitmerellus** Klots, 1942
dumetellus; auth., part, not Hbn., 1824
pratellus; auth., part, not L., 1758
　　a. **browni** Klots, 1942
5346 **tutillus** McD., 1921
5347 **awemellus** McD., 1921
5348 **lyonsellus** Haim., 1915
5349 **youngellus** Kft., 1908
5350 **daeckellus** Haim., 1907
5351 **gausapalis** Hulst, 1886
5352 **trichusalis** Hulst, 1886
5353 **cockleellus** Kft., 1908
5354 **ainsliellus** Klots, 1942
5355 **praefectellus** (Zinck., 1821)
involutellus Clem., 1860
hastiferellus; Felt, 1894, part, not Wlk., 1863
pulchellus; Felt, 1894, part, not Wlk., 1863
　　a. **oslarellus** Haim., 1908
5356 **bigelovi** Klots, 1967
5357 **leachellus** (Zinck., 1818)
pulchellus Zell., 1863
involutellus; Zell., 1863, not Clem., 1860
lativittellus Zell., 1863
hastiferellus Wlk., 1863
5358 **cypridalis** Hulst, 1886
5359 **occidentalis** Grt., 1880
agricolellus Dyar, 1923
5360 **rickseckerellus** Klots, 1940
5361 **albellus** Clem., 1860
5362 **agitatellus** Clem., 1860
alboclavellus Zell., 1863, n. syn.
carolinellus Haim., 1915., n. syn.
5363 **saltuellus** Zell., 1863
agitatellus; auth., part, not Clem., 1860

5364 **multilinellus** Fern., 1887
5365 **girardellus** Clem., 1860
nivihumellus Wlk., 1863
5366 **watsonellus** Klots, 1942
5367 **sanfordellus** Klots, 1942
5368 **braunellus** Klots, 1940
5369 **quinquareatus** Zell., 1877
quinqueareatus auth., missp.
extorralis Hulst, 1886
argentictus Hamp., 1919
hastiferellus; auth., not Walk., 1863
5370 **sperryellus** Klots, 1940
5371 **leuconotus** Zell., 1882
5372 **satrapellus** (Zinck., 1821)
elegantellus Wlk., 1863
aculeilellus Wlk., 1863
aureorufus Hamp., 1919, uncert. stat.
5373 **cyrilellus** Klots, 1942
5374 **harrisi** Klots, 1967
5375 **johnsoni** Klots, 1942
5376 **sargentellus** Klots, 1942
5377 **angustexon** Bleszynski, 1962
5378 **laqueatellus** Clem., 1860
semifusellus Wlk., 1863
5379 **luteolellus** Clem., 1860, mispl.
duplicatus Grt., 1880
　　a. **refotalis** Hulst, 1886, n. stat.
ulae Ckll., 1888, n. syn.
5380 **zeellus** Fern., 1885, mispl.
5381 **caliginosellus** Clem., 1860, mispl.
5382 **murellus** Dyar, 1904, mispl.
simpliciellus Kft., 1908, n. syn.
5383 **modestellus** B. & McD., 1918, mispl.
5384 **albilineellus** Fern., 1893, mispl.
5385 **quadrinotellus** Zell., 1877, mispl.
5386 **hulstellus** Fern., 1885, mispl.
5387 **coloradellus** Fern., 1893, mispl.
5388 **dimidiatellus** Grt., 1883, mispl.
leucorhabdon Hamp., 1919, n. stat.

FERNANDOCRAMBUS Auriv., 1922
5389 **harpipterus** (Dyar, 1916)
5390 **ruptifascia** (Hamp., 1919)

CHRYSOTEUCHIA Hbn., 1825
5391 **topiaria** (Zell., 1866)
hortuella; auth., part, not Hbn., 1796
　　a. **vachellella** (Kft., 1903)

AREQUIPA Wlk., 1863
5392 **turbatella** Wlk., 1863
bipunctella (Zell., 1863)

RAPHIPTERA Hamp., 1896
5393 **argillaceella** (Pack., 1867)
　　a. **minimella** (Rob., 1870), n. stat.

PLATYTES Gn., 1845
5394 **vobisne** Dyar, 1920, mispl.

AGRIPHILA Hbn., 1825
5395 **biarmica** (Tengström, 1865)
　　a. **paganella** (McD., 1925)
5396 **straminella** (D. & S., 1775)
culmella; auth., part, not L., 1758
5397 **plumbifimbriella** (Dyar, 1904)
5398 **costalipartella** (Dyar, 1921)
5399 **ruricolella** (Zell., 1863)
canadella (Haim., 1930), n. syn.
5400 **undata** (Grt., 1881)

5401 **anceps** (Grt., 1880)
5402 **biothanatalis** (Hulst, 1886)
    behrensella (Fern., 1887)
5403 **vulgivagella** (Clem., 1860)
    aurifimbrialis (Wlk., 1863)
    chalybirostris (Zell., 1863)
5404 **attenuata** (Grt., 1880)
5405 **angulata** (B. & McD., 1918), mispl.
    diegonella (Dyar, 1923)

CATOPTRIA Hbn., 1825
    EXORIA Hbn., 1825
5406 **trichostoma** (Christoph, 1858)
    albisinuatella (Pack., 1867)
5407 **maculalis** (Zett., 1840)
5408 **latiradiella** (Wlk., 1863)
    interrupta (Grt., 1877)
    myella; auth., part, not Hbn., 1796
    luctuella; auth., part, not H.-S., 1856
    conchella; auth., part, not D. & S.,
      1775
5409 **oregonica** (Grt., 1880)
    bartella (B. & McD., 1918), n. syn.

PEDIASIA Hbn., 1825
    CARVANCA Wlk., 1856
5410 **aridella** (Thunb., 1788)
    a. **edmontella** (Mc.D., 1923)
5411 **truncatella** (Zett., 1840)
    obtrusella (Wlk., 1863)
    lienigiella (Zell., 1863)
    rufinalis (Wlk., 1865)
5412 **browerella** (Klots, 1942)
    a. **katahdini** (Klots, 1942)
5413 **trisecta** (Wlk., 1856)
    interminella (Wlk., 1863)
    exsiccata (Zell., 1863)
    fuscisquamella (Zell., 1863)
    biliturella (Zell., 1874)
5414 **laciniella** (Grt., 1880)
5415 **ericella** (B. & McD., 1918), n. comb.
5416 **abnaki** (Klots, 1942)
5417 **dorsipunctella** (Kft., 1908)
    geminatella (Zell., 1863), quest. syn.

MICROCRAMBUS Bleszynski, 1963
5418 **copelandi** Klots, 1968
    pusionellus; auth., not Zell., 1863
5419 **biguttellus** (Fbs., 1920)
5420 **elegans** (Clem., 1860)
    terminellus (Zell., 1863)
5421 **polingi** (Kft., 1908)
5422 **minor** (Fbs., 1920)
    immunellus; Fbs., 1920, part, not
      Zell., 1872
    polingi; Fbs., 1923, part, not Kft.,
      1908
5423 **discludellus** (Mösch., 1890)
    micralis (Hamp., 1919), quest. syn.
    domingellus (Schaus, 1922)
    discobolus Bleszynski, 1963
5424 **kimballi** Klots, 1968
    minor (Fbs., 1923), part
5425 **matheri** Klots, 1968
5426 **croesus** Bleszynski, 1967

LOXOCRAMBUS Fbs., 1920
5427 **canellus** Fbs., 1920
5428 **mohaviellus** Fbs., 1920
5429 **awemensis** McD., 1929

FISSICRAMBUS Bleszynski, 1963
5430 **fissiradiellus** (Wlk., 1863)
    curtellus (Wlk., 1863)
    gestatellus (Mösch., 1890)
5431 **profanellus** (Wlk., 1866)
5432 **intermedius** (Kft., 1908)
5433 **haytiellus** (Zinck., 1821)
5434 **hemiochrellus** (Zell., 1877)
5435 **mutabilis** (Clem., 1860), n. comb.
    fuscicostella (Zell., 1863)
5436 **hospition** Bleszynski, 1963
5437 **minuellus** (Wlk., 1863), mispl.
    santiagellus (Schaus, 1922)
    habanellus (Schaus, 1922)

THAUMATOPSIS Morr., 1874
    PROPEXUS Grt., 1880
5438 **edonis** (Grt., 1880)
5439 **pexella** (Zell., 1863)
    macropterella (Zell., 1863)
    longipalpa Morr., 1874
    a. **gibsonella** Kft., 1908, n. stat.
    b. **coloradella** Kft., 1908, n. stat.
    c. **strictalis** (Dyar, 1914), n. comb.
    idion (Dyar, 1919), n. stat.
5440 **magnifica** (Fern., 1891)
5441 **fernaldella** Kft., 1905
    a. **nortella** Kft., 1905
    b. **lagunella** Dyar, 1912, n. stat.
5442 **atomosella** Kft., 1908
5443 **floridalis** B. & McD., 1913
    fernaldella Kft., 1905, part, n. syn.
5444 **fieldella** B. & McD., 1912
5445 **repanda** (Grt., 1880)
5446 **crenulatella** Kft., 1908
5447 **pectinifer** (Zell., 1877)
5448 **actuella** B. & McD., 1918
5449 **solutella** (Zell., 1863)
    striatella Fern., 1896, n. syn.
    daeckeella Kft., 1903, n. syn.

PARAPEDIASIA Bleszynski, 1966
5450 **decorella** (Zinck., 1821)
    polyactinella (Zell., 1863)
    goodelliana (Grt., 1880)
    bonusculalis (Hulst, 1886)
5451 **teterrella** (Zinck., 1821), mispl.
    camurella (Clem., 1860)
    terrella (Zell., 1863)
5452 **bolterella** (Fern., 1887), mispl.

TEHAMA Hulst, 1888
5453 **bonifatella** (Hulst, 1887)
    inornatella (Wlk., 1863), quest. syn.
    nevadella (Kft., 1908)

EUCHROMIUS Gn., 1845
    OMMATOPTERYX Kby., 1897
    EROMENE Hbn., 1825, preocc. by
      Hbn., 1821
5454 **ocelleus** (Haw., 1811)
    texanus (Rob., 1870)
5455 **californicalis** (Pack., 1873)

MICROCAUSTA Hamp., 1895
5456 **flavipunctalis** B. & McD., 1913
5457 **bipunctalis** B. & McD., 1914

DIPTYCHOPHORA Zell., 1866
    DITOMOPTERA Hamp., 1893

    SCISSOLIA B. & McD., 1914
    COLIMEA Dyar, 1925
5458 **harlequinalis** (B. & McD., 1914)
5459 **incisalis** (Dyar, 1925)

Argyriini

ARGYRIA Hbn., 1818
    CATHARYLLA Zell., 1863
    UROLA; auth., not Wlk., 1863
5460 **nummulalis** Hbn., 1818
    argentana Fern., 1896, n. syn.
5461 **subaenescens** (Wlk., 1863)
    fuscipes (Zell., 1863)
5462 **rufisignella** (Zell., 1872)
    rileyella Dyar, 1914
5463 **lacteella** (F., 1794)
    albana (F., 1798)
    pusillalis Hbn., 1818

UROLA Wlk., 1863
5464 **nivalis** (Drury, 1773)
    argentata (Emmons, 1854)
    microchrysella Wlk., 1863
    nummulalis; Zell., 1863, part, not
      Hbn., 1818
    quadristigmalis (Gn., 1854), quest.
      syn.

VAXI Bleszynski, 1962
    ARGYRIA; auth., part, not Hbn.,
      1818
    UROLA; auth., part, not Wlk., 1863
5465 **auratella** (Clem., 1860), n. comb.
    pulchella (Wlk., 1863), n. comb.
5466 **critica** (Fbs., 1820), n. comb.
5467 **tripsacas** (Dyar, 1921), n. comb.

Chilini

EPINA Wlk., 1866
    DIATRAENOPSIS Dyar & Heinr.,
      1927
    PLATYTES; auth., part, not Gn.,
      1845
5468 **dichromella** (Wlk., 1866)
    differentialis (Fern., 1888)
    matanzalis (Schaus, 1922)
5469 **alleni** (Fern., 1888)

CHILO Zinck., 1817
    DIPHRYX Grt., 1881
    SILVERIA Dyar, 1925
    DIATRAENOPSIS Dyar & Heinr.,
      1927, part
5470 **plejadellus** Zinck., 1821
    sabuliferus (Wlk., 1863)
    prolatellus (Grt., 1881)
    oryzaeellus Riley, 1882
5471 **erianthalis** Capps, 1963
5472 **demotellus** Wlk., 1866
    idalis (Fern., 1896)
    fernaldalis Dyar & Heinr., 1927

THOPEUTIS Hbn., 1818
    CEPHIS Rag., 1892
    STENOCHILO Hamp., 1896
    HOMBERGIA de Joannis, 1910
5473 **forbesellus** (Fern., 1896)

ACIGONA Hbn., 1825
  OCCIDENTALIA Dyar & Heinr., 1927
  HAIMBACHIA; auth., part, not Dyar, 1909
  EOREUMA; auth., part, not Ely, 1910
  XUBIDA; auth., part, not Schaus, 1922
5474 **comptulatalis** (Hulst, 1886)

DIATRAEA Guilding, 1832
  IESTA Dyar, 1909
  DIATRAERUPA Schaus, 1913
  ZEADIATRAEA Box, 1955
5475 **saccharalis** (F., 1794)
  sacchari (F., 1798)
  crambidoides; auth., part, not Grt., 1880
5476 **crambidoides** (Grt., 1880)
  zeacolella Dyar, 1911
  tripsacicola Dyar, 1921
5477 **venosalis** (Dyar, 1917)
5478 **evanescens** Dyar, 1917
  sobrinalis Schaus, 1922
5479 **grandiosella** Dyar, 1911
  lineolata; auth., part, not Wlk., 1856
5480 **lineolata** (Wlk., 1856)
5481 **lisetta** (Dyar, 1909)

HAIMBACHIA Dyar, 1909
5482 **squamulella** (Zell., 1881)
5483 **arizonensis** Capps, 1965
5484 **pallescens** Capps, 1965
5485 **indistinctalis** Capps, 1965
5486 **discalis** Dyar & Heinr., 1927
5487 **floridalis** Capps, 1965
5488 **albescens** Capps, 1965
5489 **placidella** (Haim., 1907)
5490 **cochisensis** Capps, 1965
5491 **diminutalis** Capps, 1965

EOREUMA Ely, 1910
5492 **densella** (Zell., 1881)
5493 **loftini** (Dyar, 1917)
  opinionella (Dyar, 1917)
5494 **evae** Klots, 1970
5495 **confederata** Klots, 1970
5496 **multipunctella** (Kft., 1908)
5497 **callista** Klots, 1970
5498 **crawfordi** Klots, 1970

XUBIDA Schaus, 1922
5499 **linearella** (Zell., 1863)
  multilineatella (Hulst, 1887)
5500 **panalope** (Dyar, 1917)
  acerata (Dyar, 1917)
5501 **relovae** Klots, 1970
5502 **punctilineella** (B. & McD., 1913)
5503 **lipan** Klots, 1970
5504 **dentilineatella** (B. & McD., 1913)
5505 **puritella** (Kft., 1908)
  dinephelalis (Dyar, 1917)
5506 **chiloidella** (B. & McD., 1913)

HEMIPLATYTES B. & Benj., 1924
  ALAMOGORDIA Dyar & Heinr., 1927, n. syn.
5507 **epia** (Dyar, 1912)
  damon (B. & McD., 1918)
5508 **prosenes** (Dyar, 1912)
5509 **parallela** (Kft., 1908)

(Series Pyraliformes)

PYRALINAE

PYRALIS L., 1758
  ASOPIA Tr., 1828
  SACATIA Wlk., 1863
  EUTRICHODES Warr., 1891
  THERAPNE Rag., 1891
5510 **farinalis** L., 1758
  fraterna Butler, 1879
5511 **costiferalis** Wlk., 1866
  costigeralis Wlk., 1866, preocc. by Wlk., 1862
5512 **disciferalis** Dyar, 1908
5513 **electalis** Hulst, 1886
5514 **cacamica** Dyar, 1913
5515 **manihotalis** Gn., 1854
  vetustalis Wlk., 1859
  gerontesalis Wlk., 1859
  laudatella (Wlk., 1863)
  despectalis Wlk., 1866
  miseralis Wlk., 1866
  achatina Butl., 1877
  gerontialis Meyr., 1888, emend.

AGLOSSA Latr., 1796
5516 **pinguinalis** (L., 1758)
  streatfieldii Curt., 1833
5517 **caprealis** (Hbn., 1800–09)
  capreolatus (Haw., 1809)
  cuprealis auth., missp.
  domalis Gn., 1854
  enthealis (Hulst, 1886)
  cuprialis Heinr., 1931, missp.
5518 **cuprina** Zell., 1872
5519 **acallalis** Dyar, 1908
5520 **baba** Dyar, 1914
5521 **gigantalis** B. & Benj., 1925
5522 **furva** Heinr., 1931
5523 **oculalis** Hamp., 1906

HYPSOPYGIA Hbn., 1825
5524 **costalis** (F., 1775)
  fimbrialis (D. & S., 1775)
  aurotaenialis (Christoph, 1881), quest. syn.
  hyllalis (Wlk., 1859)
  rubrocilialis (Stgr., 1870)

HERCULIA Wlk., 1859
  PSEUDASOPIA Grt., 1873
  DOLICHOMIA Rag., 1891
5525 **planalis** (Grt., 1880)
  enniculalis (Hulst, 1886)
  occidentalis (Hulst, 1886)
5526 **intermedialis** (Wlk., 1862)
  sodalis (Wlk., 1869)
  squamealis (Grt., 1873)
5527 **phoezalis** Dyar, 1908
5528 **cohortalis** (Grt., 1878)
  florencealis Blkmre., 1920, n. syn.
5529 **thymetusalis** (Wlk., 1859)
  devialis (Grt., 1875)
5530 **binodulalis** (Zell., 1872)
5531 **sordidalis** B. & McD., 1913
5532 **infimbrialis** Dyar, 1910
5533 **olinalis** (Gn., 1854)
  trentonalis (Led., 1863)
  himonialis (Zell., 1872)

USCODYS Dyar, 1910
5534 **cestalis** (Hulst, 1886)
5535 **atalis** Dyar, 1910

NEODAVISIA B. & McD., 1914, repl. name
  DAVISIA B. & McD., 1913, preocc. by Preston, 1910
5536 **singularis** (B. & McD., 1913)

CHRYSAUGINAE

CAPHYS Wlk., 1863
  UGRA Wlk., 1863
  EUEXIPPE Rag., 1891
5537 **arizonensis** Mun., 1970
  bilinea; auth., not Wlk., 1863

PARACHMA Wlk., 1866
  ZAZACA Wlk., 1866
  PERSEIS Rag., 1891
  ARTOPSIS Dyar, 1908
5538 **ochracealis** Wlk., 1866
  auratalis (Wlk., 1866)
  culiculalis (Hulst, 1886)
  nua (Dyar, 1914)
5539 **borregalis** (Dyar, 1908)
5540 **tarachodes** Dyar, 1914

ACALLIS Rag., 1891
  POLLOCCIA Dyar, 1910
5541 **gripalis** (Hulst, 1886)
  gryphalis auth., missp.
  fernaldi Rag., 1891
  angustipennis (Warr., 1891)
  griphalis Dyar, 1910, missp.
5542 **centralis** Dyar, 1910
5543 **alticolalis** (Dyar, 1910)

ZABOBA Dyar, 1914
5544 **mitchelli** (Dyar, 1914)
5545 **unicoloralis** Mun., 1970

ANEMOSELLA Dyar, 1914
  BALIDARCHA Dyar, 1914
5546 **nevalis** (B. & Benj., 1925)
5547 **obliquata** (Hy. Edw., 1886)
  albistrigalis (B. & McD., 1913)
5548 **polingalis** B. & Benj., 1926
5549 **viridalis** (B. & McD., 1912)
  cuis (Dyar, 1914), n. syn.

LEPIDOMYS Gn., 1852
  CHALINITIS Rag., 1891
5550 **irrenosa** Gn., 1852
  olealis (Rag., 1891)

NEGALASA B. & McD., 1913
5551 **fumalis** B. & McD., 1913
  a. **rubralis** B. & McD., 1913

GALASA Wlk., 1886
  CORDYLOPEZA Zell., 1873
5552 **nigrinodis** (Zell., 1873)
  rubrana Fitch, 1891, unavail., publ. in syn.
  palmipes G. & R., 1891, unavail., publ. in syn.
5553 **nigripunctalis** (B. & McD., 1913)
  fulvusana Haim., 1915

TETRASCHISTIS Hamp., 1897
5554 **leucogramma** Hamp., 1904

PENTHESILEA Rag., 1891
5555 **sacculalis** Rag., 1891

TOSALE Wlk., 1863
   FABATANA Wlk., 1866
   SIPAROCERA Rob., 1876
5556 **oviplagalis** (Wlk., 1866)
   anthoecioides (G. & R., 1867)
   nobilis (Rob., 1876)
5557 **similalis** B. & Benj., 1924
5558 **aucta** Hamp., 1897

SALOBRENA Wlk., 1863
   SALOBRANA auth., missp.
   OECTOPERIA Zell., 1875
5559 **sincera** (Zell., 1875)
5560 **rubiginea** (Hamp., 1897)

EPITAMYRA Rag., 1891
5561 **birectalis** Hamp., 1897

SATOLE Dyar, 1908
5562 **ligniperdalis** Dyar, 1908

CLYDONOPTERON Riley, 1880
5563 **tecomae** Riley, 1880

BONCHIS Wlk., 1862
   ETHNISTIS Led., 1863
   VURNA Wlk., 1866
   ZARANIA Wlk., 1866
   GAZACA Wlk., 1866
5564 **munitalis** (Led., 1863)
   instructalis (Wlk., 1866)
   cossalis (Wlk., 1866)
   dirutalis (Wlk., 1866)

STREPTOPALPIA Hamp., 1895
5565 **minusculalis** (Mösch., 1890)
   deera (Druce, 1895)
   ustalis (Hamp., 1895)

ARTA Grt., 1875
   XANTIPPIDES Dyar, 1908
5566 **statalis** Grt., 1875
5567 **epicoenalis** Rag., 1891
   descansalis (Dyar, 1908), n. comb., n.
     syn.
5568 **olivalis** Grote, 1878

XANTIPPE Rag., 1891
5569 **beatifica** Dyar, 1921
5570 **uranides** Dyar, 1921

CONDYLOLOMIA Grt., 1873
5571 **participalis** Grt., 1873

BLEPHAROCERUS Blanch., 1852
5572 **rosellus** Blanch., 1852
   rufalis (Led., 1863)
5573 **ignitalis** Hamp., 1906

HELIADES Rag., 1891
5574 **mulleolella** (Hulst, 1887)
   huachucalis (Haim., 1915)

EPIPASCHIINAE

MACALLA Wlk., 1859
   HOMURA Led., 1863
5575 **thyrsisalis** Wlk., 1859
   mixtalis (Wlk., 1866)
5576 **phaeobasalis** Hamp., 1916

EPIPASCHIA Clem., 1860
   MOCHLOCERA Grt., 1876
5577 **superatalis** Clem., 1860
   borealis (Grt., 1870)
   olivalis (Hulst, 1886)
5578 **albomedialis** B. & Benj., 1924
5579 **zelleri** (Grt., 1876)

CACOZELIA Grt., 1877
5580 **basiochrealis** Grt., 1877
5581 **alboplagialis** Dyar, 1905

JOCARA Wlk., 1863
   DEUTEROLLYTA Led., 1863
   TORIPALPUS Grt., 1877
   WINONA Hulst, 1888
5582 **incrustalis** (Hulst, 1887)
5583 **perseella** B. & McD., 1913
5584 **breviornatalis** (Grt., 1877)
5585 **trabalis** (Grt., 1881)
   adulatalis (Hulst, 1887)
5586 **interruptella** (Rag., 1888)
   dentilineella Hulst, 1900
5587 **elegans** (Schaus, 1912)

ONEIDA Hulst, 1889
5588 **lunulalis** (Hulst, 1887)
5589 **luniferella** Hulst, 1895
   a. **pallidalis** B. & Benj., 1924

POCOCERA Zell., 1848
   PHIDOTRICHA Rag., 1888
5590 **atramentalis** Led., 1863
   erigens (Rag., 1888)

TALLULA Hulst, 1888
5591 **atrifascialis** (Hulst, 1886)
5592 **watsoni** B. & McD., 1917
5593 **baboquivarialis** B. & Benj.,
   1926
5594 **fieldi** B. & McD., 1913

TETRALOPHA Zell., 1848
   LANTHAPHE Clem., 1860
   BENTA Wlk., 1863
   SALUDA Hulst, 1888
   KATONA Hulst, 1888
   TIOGA Hulst, 1888
   LOMA Hulst, 1888
   WANDA Hulst, 1889
   ATTACAPA Hulst, 1889
5595 **robustella** Zell., 1848
   diluculella Grt., 1880
5596 **scortealis** (Led., 1863)
   slossoni (Hulst, 1895)
5597 **melanogrammos** Zell., 1872
5598 **texanella** Rag., 1888
5599 **callipeplella** Hulst, 1888
5600 **speciosella** (Hulst, 1901)
5601 **floridella** (Hulst, 1901)
5602 **subcanalis** (Wlk., 1863)
   taleolalis (Hulst, 1886)

   talleolalis auth., missp.
   querciella B. & McD., 1913
5603 **maritimalis** McD., 1939
5604 **militella** Zell., 1848
   platanella (Clem., 1860)
5605 **aplastella** (Hulst, 1888)
5606 **asperatella** (Clem., 1860)
5607 **vacciniivora** Mun. 1963
5608 **expandens** (Wlk., 1863)
   nephelotella (Hulst, 1888)
   clemensalis Dyar, 1905, infrasubsp.,
     n. syn.
5609 **spaldingella** B. & Benj., 1924
5610 **dolorosella** B. & Benj., 1924
5611 **provoella** B. & Benj., 1924
5612 **arizonella** B. & Benj., 1924
5613 **thoracicella** B. & Benj., 1924
5614 **griseella** B. & Benj., 1924
5615 **fuscolotella** Rag., 1888
5616 **tiltella** (Hulst, 1888)
5617 **humerella** Rag., 1888
   formosella Hulst, 1900
5618 **tertiella** (Dyar, 1905)
5619 **baptisiella** Fern., 1887
5620 **euphemella** (Hulst, 1888)
   variella (Rag., 1888)
   melanographella (Rag., 1888)

COENODOMUS Wlsm., 1889
   DYARIA Neum., 1893
   ALIPPA Auriv., 1894
5621 **hockingii** Wlsm., 1889
   singularis (Neum., 1893)
   anomala (Auriv., 1894)

GALLERIINAE

Galleriini

GALLERIA F., 1798
   CERIOCLEPTA Sodoffsky, 1837
   VINDANA Wlk., 1866
5622 **mellonella** (L., 1758)
   cereana (L., 1767)
   cerella (F., 1775), emend.
   cerea (Haw., 1811), emend.
   cerealis Hbn., 1825, emend.
   obliquella (Wlk., 1866)
   austrinia (Felder & Rogenhofer, 1874)
   crombrugheella Dufrane, 1930,
     infrasubsp.

ACHROIA Hbn., 1819
   ACHROEA auth., missp.
   MELIPHORA Gn., 1845
   VOBRIX Wlk., 1864
5623 **grisella** (F., 1794)
   aluearia (F., 1798)
   alvearia auth., missp.
   cinereola (Hbn., 1802)
   alvea (Haw., 1811), emend.
   alveariella (Gn., 1848), emend.
   anticella (Wlk., 1863)
   obscurevittella Rag., 1901
   major Dufrane, 1930, infrasubsp.

TRACHYLEPIDIA Rag., 1887
   AGANACTESIS Dyar, 1921
5624 **fructicassiella** Rag., 1887
   indecora (Dyar, 1921)

Megarthridiini

OMPHALOCERA Led., 1863
5625 **cariosa** Led., 1863
    dentosa Grt., 1881
5626 **occidentalis** B. & Benj., 1924
5627 **munroei** Martin, 1956
    cariosa; auth., part, not Led.,
      1863

THYRIDOPYRALIS Dyar, 1901
5628 **gallaerandialis** Dyar, 1901

Tirathabini

APHOMIA Hbn., 1825
    ILITHYIA Berthold, 1827
    MELIA Curt., 1828
    ILYTHIA Steph., 1829, missp.
    MELISSOBLAPTES Zell., 1839
5629 **sociella** (L., 1758)
    colonella (L., 1758)
    tribunella (D. & S., 1775)
    socia (F., 1798), emend.
    colonum (F., 1798), emend.
    colonata (Haw., 1809), emend.
    rufinella Krulikowsky, 1909,
      infrasubsp.
    virescens Skala, 1929, infrasubsp.
    minor Dufrane, 1930, infrasubsp.
    lanceolata Dufrane, 1930, infrasubsp.
5630 **terrenella** Zell., 1848
    furella (Zell., 1873)
    terenella Whalley, 1964, missp.
5631 **fuscolimbella** (Rag., 1887)

PARALIPSA Butler, 1879
    PARALISPA Spuler, 1910, emend.
5632 **gularis** (Zell., 1877)
    modesta Butler, 1879
    tenebrosa (Butler, 1879)
5633 **decorella** Hulst, 1892

CORCYRA Rag., 1885
    TINEOPSIS Dyar, 1913
5634 **cephalonica** (Staint., 1866)
    oeconomella (Mann, 1872)
    translineella Hamp., 1901
    theobromae (Dyar, 1913)

Macrothecini, n. stat

MACROTHECA Rag., 1891
    CACOTHERAPIA Dyar, 1904
5635 **interalbicalis** Rag., 1891
    vulnifera Dyar, 1917
    vulnifica McD., 1939, missp.
5636 **bilinealis** B. & McD., 1918
5637 **angulalis** B. & McD.
5638 **unicoloralis** B. & McD., 1913
5639 **unipuncta** Dyar, 1913
5640 **ponda** (Dyar, 1907)
5641 **nigrocinereella** (Hulst, 1900)
5642 **flexilinealis** (Dyar, 1905)
5643 **leucocope** Dyar, 1917
5644 **lecerfialis** B. & Benj., 1925

ALPHEIAS Rag., 1891
    AMESTRIA Rag., 1891
5645 **vicarilis** Dyar, 1913

5646 **transferrens** Dyar, 1917
    ponda; B. & McD., 1912, part, not
      Dyar, 1907
5647 **querula** Dyar, 1913
5648 **oculiferalis** (Rag., 1891)

ALPHEIOIDES B. & McD., 1912
5649 **parvulalis** B. & McD., 1912

DECATURIA B. & McD., 1912
5650 **pectinalis** B. & McD., 1912

PHYCITINAE

ACROBASIS Zell., 1839
    MINEOLA Hulst, 1890
    SENECA Hulst, 1890
    ACROCAULA Hulst, 1900
    CATACROBASIS Gozmány, 1958
5651 **indigenella** (Zell., 1848)
    nebulo (Walsh, 1860)
    nebulella (Riley, 1872)
    zelatella (Hulst, 1887)
    indiginella auth., missp.
5652 **grossbecki** (B. & McD., 1917)
5653 **vaccinii** Riley, 1884
5654 **amplexella** Rag., 1887
5655 **tricolorella** Grt., 1878
    scitulella (Hulst, 1900)
5656 **comptella** Rag., 1887
5657 **minimella** Rag., 1889
    nigrosignella Hulst, 1890
5658 **feltella** Dyar, 1910
5659 **palliolella** Rag., 1887
    albocapitella Hulst, 1888
5660 **caryalbella** Ely, 1913
5661 **juglandis** (LeBaron, 1872)
5662 **sylviella** Ely, 1908
5663 **kearfottella** Dyar, 1905
5664 **caryae** Grt., 1881
5665 **carpinivorella** Neunzig, 1970
5666 **elyi** Neunzig, 1970
5667 **nuxvorella** Neunzig, 1970
5668 **evanescentella** Dyar, 1908
5669 **stigmella** Dyar, 1908
5670 **aurorella** Ely, 1910
5671 **peplifera** Dyar, 1925
5672 **exsulella** (Zell., 1848)
    septentrionella Dyar, 1925
5673 **angusella** Grt., 1880
    eliella Dyar, 1908
5674 **demotella** Grt., 1881
5675 **latifasciella** Dyar, 1908
5676 **irrubriella** Ely, 1908
5677 **normella** Dyar, 1908
5678 **malipennella** Dyar, 1908
5679 **dyarella** Ely, 1910
5680 **ostryella** Ely, 1913
5681 **secundella** Ely, 1913
5682 **coryliella** Dyar, 1908
5683 **hebescella** Hulst, 1890
5684 **cirroferella** Hulst, 1892
5685 **cunulae** Dyar & Heinr., 1929
5686 **caryivorella** Rag., 1887
    caryaevorella Hulst, missp.
    conjivorella Hulst, missp.
5687 **comacornella** (Hulst, 1900)
5688 **betulella** Hulst, 1890
5689 **betulivorella** Neunzig, 1975
5690 **rubrifasciella** Pack., 1873

    rubifasciella Beutenmüller, missp.
    alnella McD., 1922
5691 **comptoniella** Hulst, 1890
5692 **myricella** B. & McD., 1917
5693 **tumidulella** (Rag., 1887)
5694 **blanchardorum** Neunzig, 1973

RHODOPHAEA Gn., 1845
    AURANA Wlk., 1863
    GAANA Wlk., 1866
    RHODOPHAEOPSIS Amsel, 1950
5695 **caliginella** (Hulst, 1889)
    caliginoidella (Dyar, 1905)
5696 **cruza** Opler, 1977
5697 **kofa** Opler, 1977
5698 **aurata** Opler, 1977
5699 **fria** Opler, 1977
5700 **neva** Opler, 1977
5701 **yuba** Opler, 1977
5702 **suavella** (Zinck., 1818)
    porphyrea (Steph., 1834)
    supposita (Heinr., 1940)

TRACHYCERA Rag., 1893
5703 **pallicornella** (Rag., 1887)

ANABASIS Heinr., 1956
5704 **ochrodesma** (Zell., 1881)
    crassisquamella (Hamp., 1901)

HYPSIPYLA Rag., 1888
5705 **grandella** (Zell., 1848)
    cnabella Dyar, 1914

CROCIDOMERA Zell., 1848
5706 **turbidella** Zell., 1848

CUNIBERTA Heinr., 1956
5707 **subtinctella** (Rag., 1887)

ADANARSA Heinr., 1956
5708 **intransitella** (Dyar, 1905)

BERTELIA B. & McD., 1913
5709 **grisella** B. & McD., 1913
5710 **dupla** A. Blanchard, 1976

HYPARGYRIA Rag., 1888
5711 **slossonella** (Hulst, 1900)
    tenuella (B. & McD., 1913)

CHARARICA Heinr., 1956
5712 **annuliferella** (Dyar, 1905)
5713 **hystriculella** (Hulst, 1889)
5714 **bicolorella** (B. & McD., 1917)

MYELOIS Hbn., 1825
    KYRA Gozmány, 1958
5715 **grossipunctella** Rag., 1888, mispl.

MYELOPSIS Heinr., 1956
5716 **coniella** (Rag., 1887)
    nefas (Dyar, 1922)
5717 **immundella** (Hulst, 1890)
5718 **subtetricella** (Rag., 1889)
    zonulella (Rag., 1889)
    obnupsella (Hulst, 1890)
5719 **minutularia** (Hulst, 1887)
    minutulella (Hulst, 1890)
5720 **alatella** (Hulst, 1887)

rectistrigella (Rag., 1887)
fragilella (Dyar, 1904)
piazzella (Dyar, 1925)

APOMYELOIS Heinr., 1956
5721 **bistriatella** (Hulst, 1887)
bilineatella (Rag., 1887)

ECTOMYELOIS Heinr., 1956
5722 **decolor** (Zell., 1881)
ephestiella (Hamp., 1901)
5723 **ceratoniae** (Zell., 1839)
ceratoniella (F. v. Röslerstamm, 1839)
pryerella (Vaughan, 1870)
tuerkheimiella (Sorhagen, 1881)
zellerella (Sorhagen, 1881)
psarella (Hamp., 1903)
oporedestella (Dyar, 1911)
phoenicis (Durrant, 1915)

AMYELOIS Amsel, 1956
PARAMYELOIS Heinr., 1956
5724 **transitella** (Wlk., 1863)
notatalis (Wlk., 1863)
solitella (Zell., 1881)
duplipunctella (Rag., 1881)
venipars (Dyar, 1914)
cassiae (Dyar, 1917)

FUNDELLA Zell., 1848
BALLOVIA Dyar, 1913
5725 **pellucens** (Zell., 1848)
cistipennis (Dyar, 1913)
5726 **argentina** (Dyar, 1919)
eucasis (Dyar, 1919)

PROMYLEA Rag., 1887
5727 **lunigerella** Rag., 1887
a. **glendella** (Dyar, 1906)

ANADELOSEMIA Dyar, 1919
5728 **texanella** (Hulst, 1892)
dulciella (Hulst, 1900)
5729 **condigna** Heinr., 1956

DASYPYGA Rag., 1887
5730 **alternosquamella** Rag., 1887
stictophorella Rag., 1887
5731 **salmocolor** A. Blanchard, 1970

DAVARA Wlk., 1859
HOMALOPALPIA Dyar, 1914
EUCARDINIA Dyar, 1918
5732 **caricae** (Dyar, 1913)
dalera (Dyar, 1914)

SARASOTA Hulst, 1900
CUBA Dyar, 1919
5733 **plumigerella** Hulst, 1900

ATHELOCA Heinr., 1956
5734 **subrufella** (Hulst, 1887)
filiolella (Hulst, 1888)
ptychis (Dyar, 1919)

TRIOZOSNEURA A. Blanchard, 1973
5735 **dorsonotata** A. Blanchard, 1973

MONOPTILOTA Hulst, 1900
5736 **pergratialis** (Hulst, 1886)

grotella (Rag., 1887)
nubilella Hulst, 1900

ZAMAGIRIA Dyar, 1914
5737 **australella** (Hulst, 1900)
bumeliella (B. & McD., 1913)
5738 **kendalli** A. Blanchard, 1970
5739 **laidion** (Zell., 1881)
deia Dyar, 1919
striella Dyar, 1919

ANEGCEPHALESIS Dyar, 1917
5740 **arctella** (Rag., 1893)
cathaeretes Dyar, 1917
catheretes Dyar, 1919, missp.

ANCYLOSTOMIA Rag., 1893
5741 **stercorea** (Zell., 1848)
ignobilis (Butler, 1878)
diffissella (Zell., 1881)

CARISTANIUS Heinr., 1956
5742 **decoloralis** (Wlk., 1863)
metagrammalis (Wlk., 1863)
furfurellus (Hulst, 1890)
floridellus (Hulst, 1893)
decoralis (Hulst, 1890), missp.
decorellus (Hulst, 1891), missp.
5743 **minimus** Neunzig, 1977

ETIELLA Zell., 1839
RHAMPHODES Gn., 1845
MELLA Wlk., 1859
ALATA Wlk., 1863
ARUCHA Wlk., 1863
MODIANA Wlk., 1863
CERATAMNA Butler, 1880
5744 **zinckenella** (Tr., 1832)
etiella (Tr., 1835), repl. name
colonnella (Costa, 1836)
majorella (Costa, 1836)
dymnusalis (Wlk., 1859)
heraldella (Gn., 1862)
anticalis (Wlk., 1863)
indicatalis (Wlk., 1863)
hastiferella (Wlk., 1866)
decipiens Stgr., 1870, infrasubsp.
spartiella Rondani, 1876
sabulina (Butler, 1879)
madagascariensis Saalmüller, 1879–80
schisticolor Zell., 1881
villosella Hulst, 1887
rubribasella Hulst, 1890
scitivittalis; auth., not Wlk., 1896

GLYPTOCERA Rag., 1889
5745 **consobrinella** (Zell., 1872)
busckella (Dyar, 1904)

PIMA Hulst, 1888
5746 **boisduvaliella** (Gn., 1845)
farrella (Curt., 1850)
lefauryella (Constant, 1865)
5747 **albiplagiatella** (Pack., 1874)
a. **occidentalis** Heinr., 1956
5748 **fosterella** Hulst, 1888
5749 **vividella** (McD., 1935)
5750 **albocostalialis** (Hulst, 1886)
albocostalis (Hulst, 1890), emend.
a. **subcostella** (Rag., 1887)

5751 **fulvirugella** (Rag., 1887)
5752 **granitella** (Rag., 1887)
piperella (Dyar, 1904)
5753 **parkerella** (Schaus, 1924)

PIMODES A. Blanchard, 1976
5754 **insularis** A. Blanchard, 1976

INTERJECTIO Heinr., 1956
5755 **denticulella** (Rag., 1887)
lallatalis; auth., part, not Hulst, 1886
5756 **columbiella** (McD., 1935)
5757 **ruderella** (Rag., 1887)
5758 **niviella** (Hulst, 1888)

AMBESA Grt., 1880
5759 **laetella** Grt., 1880
5760 **walsinghami** (Rag., 1887)
monodon (Dyar, 1913)
a. **mirabella** Dyar, 1908
5761 **lallatalis** (Hulst, 1886)

CATASTIA Hbn., 1825
DIOSIA Dup., 1832
5762 **bistriatella** (Hulst, 1895)
5763 **incorruscella** (Hulst, 1895)
5764 **actualis** (Hulst, 1886)

GLYPHOCYSTIS A. Blanchard, 1973
5765 **viridivallis** A. Blanchard, 1973

IMMYRLA Dyar, 1906
5766 **nigrovittella** Dyar, 1906

OREANA Hulst, 1888
5767 **unicolorella** (Hulst, 1887)
leucophaeella (Hulst, 1892)
nebulella; McD., 1939, not Riley, 1872

OLYBRIA Heinr., 1956
5768 **aliculella** (Hulst, 1887)
oberthuriella (Rag., 1887)
5769 **furciferella** (Dyar, 1903)

SALEBRIACUS Heinr., 1956
5770 **odiosellus** (Hulst, 1887)
bakerella (Dyar, 1904)
yumaella (Dyar, 1905)

SALEBRIARIA Heinr., 1956
5771 **turpidella** (Rag., 1888)
ademptanella (Dyar, 1908)
5772 **nubiferella** (Rag., 1887)
5773 **engeli** (Dyar, 1906)
5774 **annulosella** (Rag., 1903)
robustella (Dyar, 1908)
5775 **tenebrosella** (Hulst, 1887)
quercicolella (Rag., 1887)
heinrichalis (Dyar, 1917)
5776 **pumilella** (Rag., 1887)
georgiella (Hulst, 1895)
5777 **fructetella** (Hulst, 1892)
rectistrigella (Dyar, 1908)

SALEBRIA Zell., 1846
5778 **nigricans** Hulst, 1900, mispl.

QUASISALEBRIA Heinr., 1956
5779 **admixta** Heinr., 1956

ORTHOLEPIS Rag., 1887
   METRIOSTOLA Rag., 1893
5780 **jugosella** Rag., 1887
5781 **myricella** McD., 1958
5782 **rhodorella** McD., 1958
5783 **pasadamia** (Dyar, 1917)

POLOPEUSTIS Rag., 1893
5784 **arctiella** (Gibson, 1920)

MEROPTERA Grt., 1882
   EMMERITA Hamp., 1930
5785 **mirandella** (Rag., 1893)
5786 **cviatella** Dyar, 1905
5787 **pravella** (Grt., 1878)
5788 **abditiva** Heinr., 1956

NEPHOPTERIX Hbn., 1825
   NEPHOPTERYX auth., missp.
   GYRTONA Wlk., 1863
   SCIOTA Hulst, 1888
   SCOTIA Hannemann, 1964, missp.
5789 **subfuscella** (Rag., 1887)
   semiobscurella (Hulst, 1890)
   pravella; auth., part, not Grt., 1878
5790 **delassalis** Hulst, 1886
   purpurella (Hulst, 1892)
   pudibundella (Rag., 1893)
5791 **rubescentella** (Hulst, 1900)
5792 **fernaldi** (Rag., 1887)
   delassalis; auth., part, not Hulst,
     1886
5793 **dammersi** Heinr., 1956
   a. **floridensis** Heinr., 1956
5794 **vetustella** (Dyar, 1904)
5795 **inconditella** (Rag., 1893)
5796 **subcaesiella** (Clem., 1860)
   contatella (Grt., 1880)
5797 **virgatella** (Clem., 1860)
   quinquepunctella (Grt., 1880)
5798 **carneella** Hulst, 1887
   inquilinella Rag., 1887
5799 **basilaris** Zell., 1872
5800 **termitalis** (Hulst, 1886)
   levigatella (Hulst, 1892)
   a. **yuconella** (Dyar, 1925)
5801 **bifasciella** Hulst, 1887
   nogalesella (Dyar, 1905)
5802 **uvinella** (Rag., 1887)
   afflictella (Hulst, 1900)
   liquidambarella (Dyar, 1904)
5803 **celtidella** (Hulst, 1890)
5804 **rubrisparsella** (Rag., 1887)
   rufibasella (Rag., 1887)
   croceella (Hulst, 1888)
   texanella (Hulst, 1900)
5805 **gilvibasella** Hulst, 1890
   lacteella (Hulst, 1900)
5806 **crassifasciella** Rag., 1887
   decipientella Dyar, 1905
   crataegella B. & McD., 1917
5807 **bisra** Dyar, 1919

TLASCALA Hulst, 1890
5808 **reductella** (Wlk., 1863)
   gleditschiella (Fern., 1881)

TULSA Heinr., 1956
5809 **finitella** (Wlk., 1863)
   melanella (Hulst, 1890)

5810 **umbripennis** (Hulst, 1895)
   gillettella (Dyar, 1904)
5811 **oregonella** (B. & McD., 1918)

TELETHUSIA Heinr., 1956
5812 **ovalis** (Pack., 1873)
   latisfasciatella (Pack., 1873)
   geminipunctella (Rag., 1893)
   modestella (Hulst, 1900)
5813 **rhypodella** (Hulst, 1887), ident.
   uncert.

PHOBUS Heinr., 1956
5814 **brucei** (Hulst, 1895)
   lallatalis; auth., part, not Hulst, 1886
5815 **funerellus** (Dyar, 1905)
5816 **curvatellus** (Rag., 1887)
   rhypodella; auth., part, not Hulst,
     1887
5817 **incertus** Heinr., 1956

ACTRIX Heinr., 1956
5818 **nyssaecolella** (Dyar, 1904)
5819 **dissimulatrix** Heinr., 1956

STYLOPALPIA Hamp., 1901
5820 **scobiella** (Grt., 1880)
   decimerella (Hulst, 1888)

HYPOCHALCIA Hbn., 1825
   ARAXES Steph., 1834
5821 **hulstiella** Rag., 1887

PYLA Grt., 1882
5822 **fasciolalis** (Hulst, 1886)
5823 **impostor** Heinr., 1956
5824 **aequivoca** Heinr., 1956
5825 **gaspeensis** McD., 1958
5826 **insinuatrix** Heinr., 1956
5827 **aenigmatica** Heinr., 1956
5828 **criddlella** Dyar, 1907
5829 **fusca** (Haw., 1828)
   spadicella; Zinck., 1818, part, not
     Hbn., 1796
   carbonariella (Dup., 1836)
   janthinella (Dup., 1836)
   moestella (Wlk., 1853)
   procellariana (Wlk., 1863)
   frigidella (Pack., 1866)
   cacabella (Hulst, 1887)
   triplagiatella (Dyar, 1904)
5830 **hypochalciella** (Rag., 1887)
   blackmorella Dyar, 1921
5831 **hanhamella** Dyar, 1904
5832 **scintillans** (Grt., 1881)
   feella Dyar, 1921
   sylphiella Dyar, 1921
5833 **rainierella** Dyar, 1904
5834 **aeneella** Hulst, 1895
5835 **aeneoviridella** Rag., 1887
5836 **metalicella** Hulst, 1895
5837 **fasciella** B. & McD., 1917
5838 **nigricula** Heinr., 1956
5839 **viridisuffusella** B. & McD., 1917

PHYCITOPSIS Rag., 1887
5840 **flavicornella** Rag., 1887, ident. uncert.

DIORYCTRIA Zell., 1846
   PINIPESTIS Grt., 1878

5841 **abietivorella** (Grt., 1878)
   abietella; auth., part, not D. & S.,
     1775
   reniculella (Grt., 1880)
   elegantella (Hulst, 1892)
5842 **taedae** Schaber & Wood, 1971
5843 **reniculelloides** Mutuura & Mun.,
     1973
   reniculella; auth., part, not Grt., 1880
5844 **pseudotsugella** Mun., 1959
5845 **rossi** Mun., 1959
5846 **auranticella** (Grt., 1883)
   miniatella Rag., 1887
   xanthoenobares Dyar, 1911
5847 **disclusa** Heinr., 1953
   auranticella; auth., part, not Grt.,
     1883
5848 **erythropasa** (Dyar, 1914)
5849 **pygmaeella** Rag., 1887
5850 **ponderosae** Dyar, 1914
5851 **okanaganella** Mutuura, Mun. & Ross,
     1969
5852 **zimmermani** (Grt., 1877)
   delectella (Hulst, 1895)
   austriana (Cosens, 1906)
5853 **amatella** (Hulst, 1887)
5854 **cambiicola** (Dyar, 1914)
5855 **tumicolella** Mutuura, Mun. & Ross, 1969
5856 **contortella** Mutuura, Mun. & Ross,
     1969
5857 **monticolella** Mutuura, Mun. & Ross,
     1969
5858 **banksiella** Mutuura, Mun. & Ross,
     1969
5859 **albovittella** (Hulst, 1900)
5860 **baumhoferi** Heinr., 1956
5861 **pentictonella** Mutuura, Mun. & Ross,
     1969
   a. **vancouverella** Mutuura, Mun. &
     Ross, 1969
5862 **gulosella** (Hulst, 1890)
5863 **subtracta** Heinr., 1956
5863.1 **clarioralis** (Wlk., 1863)
   brunneella (Dyar, 1904)

ORYCTOMETOPIA Rag., 1888
5864 **fossulatella** Rag., 1888
   moeschleri (Rag., 1888)

SARATA Rag., 1887
5865 **edwardsialis** (Hulst, 1886)
   polyphemella (Rag., 1893)
5866 **pullatella** (Rag., 1887)
5867 **punctella** (Dyar, 1915)
   a. **septentrionaria** Heinr., 1956
5868 **incanella** (Hulst, 1895)
   aridella (Dyar, 1905)
5869 **atrella** (Hulst, 1890)
5870 **caudellella** (Dyar, 1904)
5871 **dnopherella** Rag., 1887
5872 **nigrifasciella** Rag., 1887
5873 **cinereella** Hulst, 1900
5874 **rubrithoracella** (B. & McD., 1913)
5875 **tephrella** Rag., 1893
5876 **alpha** Heinr., 1956, unassoc. ♀
5877 **beta** Heinr., 1956, unassoc. ♀
5878 **gamma** Heinr., 1956, unassoc. ♀
5879 **iota** Heinr., 1956, unassoc. ♀
5880 **perfuscalis** (Hulst, 1886), unassoc. ♀
   excantalis (Hulst, 1886)

5881 **epsilon** Heinr., 1956, unassoc. ♀
5882 **phi** Heinr., 1956, unassoc. ♀
5883 **kappa** Heinr., 1956, unassoc. ♀
5884 **delta** Heinr., 1956, unassoc. ♀

PHILODEMA Heinr., 1956
5885 **rhoiella** (Dyar, 1903)

LIPOGRAPHIS Rag., 1893
5886 **fenestrella** (Pack., 1873)
    humilis Rag., 1887
5887 **leoninella** (Pack., 1873)
    pallidella (Dyar, 1903)
5888 **truncatella** (W. S. Wright, 1916)
5889 **umbrella** (Dyar, 1908)

ADELPHIA Heinr., 1956
5890 **petrella** (Zell., 1846)
    rubiginella (Wlk., 1863)
    rufinalis (Wlk., 1863)
    hapsella (Hulst, 1887)
5891 **ochripunctella** (Dyar, 1908)

UFA Wlk., 1863
5892 **lithosella** (Rag., 1887)
    luteella (Hulst, 1900)
5893 **roseitinctella** (Dyar, 1912)
5894 **senta** Heinr., 1956
5895 **rubedinella** (Zell., 1848)
    translucida (Wlk., 1863)
    rufescentalis (Wlk., 1863)
    minualis (Wlk., 1863)
    deprivalis (Wlk., 1863)
    venezuelalis Wlk., 1863
    pyrrhochrella (Rag., 1888)

ELASMOPALPUS Blanchard, 1852
5896 **lignosellus** (Zell., 1848)
    angustellus Blanchard, 1852
    tartarellus (Zell., 1872)
    incautellus (Zell., 1872)
    major (Zell., 1874)
    anthracellus Rag., 1888
    carbonellus (Hulst, 1888)
    puer Dyar, 1919

ACRONCOSA B. & McD., 1917
5897 **albiflavella** B. & McD., 1917
    a. **castrella** B. & McD., 1917
5898 **similella** B. & McD., 1917

PASSADENA Hulst, 1900
5899 **flavidorsella** (Rag., 1887)
    canescentella (Hulst, 1890)
    constantella Hulst, 1900
    cinctella Hulst, 1900

ULOPHORA Rag., 1890
    ACROMESERES Dyar, 1919
5900 **groteii** Rag., 1890
    tephrosiella Dyar, 1904

TACOMA Hulst, 1888
5901 **feriella** Hulst, 1888
    submedianella Dyar, 1913

RAGONOTIA Grt. 1888
    CIRIS Rag., 1887, preocc. by Koch,
      1848
    PSAMMIA Hamp., 1930

5902 **dotalis** (Hulst, 1886)
    discigerella (Rag., 1887)
    olivella (Hulst, 1888)
    saganella (Hulst, 1890), n. comb.
    indianella (Dyar, 1923)
    flavipicta (Hamp., 1930)

MARTIA Rag., 1887
    URULA Hulst, 1900
5903 **arizonella** Rag., 1887
    incongruella (Hulst, 1900)

EUMYSIA Dyar, 1925
5904 **mysiella** (Dyar, 1905)
5905 **maidella** (Dyar, 1905)
5906 **fuscatella** (Hulst, 1900)
5907 **semicana** Heinr., 1956
5908 **idahoensis** Mackie, 1958

DIVITIACA B. & McD., 1913
5909 **ochrella** B. & McD., 1913
5910 **simulella** B. & McD., 1913
5911 **parvulella** B. & McD., 1913

MACRORRHINIA Rag., 1887
    DOLICHORRHINIA Rag., 1888,
      repl. name
5912 **aureofasciella** Rag., 1887
5913 **signifera** A. Blanchard, 1976

OCALA Hulst, 1892
5914 **dryadella** Hulst, 1892
    platanella (Grossb., 1917)

VALDIVIA Rag., 1888
    MARICOPA Hulst, 1890
5915 **lativittella** (Rag., 1887)
    aureomaculella (Dyar, 1903)

HETEROGRAPHIS Rag., 1885
    MONA Hulst, 1888, preocc. by
      Reichenbach, 1863
5916 **morrisonella** Rag., 1887
    coloradensis Rag., 1887
    olbiella (Hulst, 1890)
    ignistrigella Rag., 1901
    palloricostella Walter, 1928

STAUDINGERIA Rag., 1887
5917 **albipenella** (Hulst, 1887)
    olivacella Dyar, 1904
    perluteella Dyar, 1904

HULSTIA Rag., 1901
    HULSTEA Hulst, 1903, missp.
5918 **undulatella** (Clem., 1860)
    rubiginalis (Wlk., 1865)
    obsipella (Hulst, 1888)
    oblitella; Rag., 1889, not Zell., 1848
    fumosella (Hulst, 1900)

HONORA Grt., 1878
5919 **mellinella** Grt., 1878
    ochrimaculella Rag., 1887
5920 **subsciurella** Rag., 1887
5921 **sciurella** Rag., 1887
5922 **dotella** Dyar, 1910
5923 **montinatatella** (Hulst, 1887)
    canicostella (Rag., 1887)
5924 **perdubiella** (Dyar, 1905)

5925 **dulciella** Hulst, 1900, mispl.

CANARSIA Hulst, 1890
5926 **ulmiarrosorella** (Clem., 1860)
    pneumatella (Hulst, 1887)
    ulmella (Rag., 1887)
    fuscatella (Hulst, 1888)
    gracilella Hulst, 1900
    feliculella Dyar, 1907

EURYTHMIDIA Rag., 1901
5927 **ignidorsella** (Rag., 1887)

WUNDERIA Grossb., 1917
5928 **neaeriatella** Grossb., 1917

DIVIANA Rag., 1888
    DANNEMORA Hulst, 1890
5929 **eudoreella** Rag., 1888
    edentella (Hulst, 1890)
    eudoriella Hulst., 1902, missp.

PALATKA Hulst, 1892
5930 **nymphaeella** (Hulst, 1892)
    verecuntella Grossb., 1917

PSOROSINA Dyar, 1904
5931 **hammondi** (Riley, 1872)
    angulella Dyar, 1904

PATRICIOLA Heinr., 1956
5932 **semicana** Heinr., 1956

ANDERIDA Heinr., 1956
5933 **sonorella** (Rag., 1887), emend.
    senorella (Rag., 1887), missp.
    placidella (Dyar, 1908)

MESCINIA Rag., 1901
5934 **estrella** B. & McD., 1913

HOMOEOSOMA Curt., 1833
    PHYCIDEA Zell., 1839
    LOTRIA Gn., 1845
5935 **electellum** (Hulst, 1887)
    opalescellum (Hulst, 1887)
    texanellum Rag., 1887
    tenuipunctellum Rag., 1887
    differtellum B. & McD., 1913
5936 **stypticellum** Grt., 1878
    uncanale Hulst, 1886
5937 **striatellum** Dyar, 1905
5938 **oslarellum** Dyar, 1905
    a. **breviplicitum** Heinr., 1956
5939 **illuviellum** Rag., 1888
    candidellum Hulst, 1888
    a. **emendator** Heinr., 1956
5940 **imitator** Heinr., 1956
5941 **albescentellum** Rag., 1887
    elongellum Dyar, 1903
5942 **impressale** Hulst, 1886
    uncanale; Rag., 1901, not Hulst, 1886
5943 **inornatellum** (Hulst, 1900)
5944 **deceptorium** Heinr., 1956
5945 **peregrinum** Heinr., 1956

PHYCITODES Hamp., 1917
    ROTRUDA Heinr., 1956
5946 **albatella** (Rag., 1887), extralim.
    parva (Gerasimov, 1930)

84

nimbella; Heinr., 1956, not Dup.,
1836
a. **mucidella** (Rag., 1887)
b. **reliquella** (Dyar, 1904)

UNADILLA Hulst, 1890
STRYMAX Dyar, 1914
5947 **floridensis** Heinr., 1956
5948 **nasutella** Hulst, 1890

LAETILIA Rag., 1889
LAOSTICHA Hulst, 1902
5949 **coccidivora** (J. H. Comstock, 1879)
pallida (J. H. Comstock, 1879)
dilatifasciella (Rag., 1887)
hulstii Ckll., 1897
a. **quadricolorella** (Dyar, 1904)
5950 **zamacrella** Dyar, 1925
5951 **myersella** Dyar, 1910
5952 **ephestiella** (Rag., 1887)
lustrella (Dyar, 1903)
5953 **fiskella** Dyar, 1904

ROSTROLAETILIA A. Blanchard &
Fgn., 1975
5954 **placidella** (B. & McD., 1918)
5955 **minimella** A. Blanchard & Fgn., 1975
5956 **placidissima** A. Blanchard & Fgn.,
1975
5957 **utahensis** A. Blanchard & Fgn., 1975
5958 **coloradella** A. Blanchard & Fgn.,
1975
5959 **eureka** A. Blanchard & Fgn., 1975
5960 **nigromaculella** (Hulst, 1901)
5961 **ardiferella** (Hulst., 1888)
5962 **texanella** A. Blanchard & Fgn., 1975
5963 **pinalensis** A. Blanchard & Fgn., 1975

WELDERELLA A. Blanchard, 1978
5964 **parvella** (Dyar, 1906)

BAPHALA Heinr., 1956
5965 **basimaculatella** (Rag., 1887)
basimaculella (Hulst, 1890), missp.
eremiella (Dyar, 1910)

RHAGEA Heinr., 1956
5966 **packardella** (Rag., 1887)
orobanchella (Dyar, 1904)
5967 **stigmella** (Dyar, 1910)
maculicula (Dyar, 1913)
maculiella (Dyar, 1925), missp.

ZOPHODIA Hbn., 1825
DAKRUMA Grt., 1878
5968 **convulutella** (Hbn., 1796)
grossulariella (Hbn., 1807–09)
grossularialis Hbn., 1825, emend.
grossulariae (Riley, 1829)
turbatella (Grt., 1878)
franconiella (Hulst, 1890)
bella Hulst, 1892
ihouna Dyar, 1925
dilativitta Dyar, 1925
magnificans Dyar, 1925
5969 **epischnioides** Hulst, 1900, ident.
uncert.

MELITARA Wlk., 1863
MEGAPHYCIS Grt., 1882

5970 **prodenialis** Wlk., 1863
bollii (Zell., 1872)
bolli (Grt. 1882), missp.
5971 **dentata** (Grt., 1876)
junctolineella; Hulst, 1900, part
doddalis (Dyar, 1925)
bollii; auth., part, not Zell., 1872

OLYCELLA Dyar, 1928
5972 **junctolineella** (Hulst, 1900)
5973 **subumbrella** (Dyar, 1925)
nephelepasa; auth., part, not Dyar,
1919

ALBERADA Heinr., 1939
5974 **parabates** (Dyar, 1913)
5975 **bidentella** (Dyar, 1908)
5976 **holochlora** (Dyar, 1925)

CAHELA Heinr., 1939
5977 **ponderosella** (B. & McD., 1918)
purgatoria (Dyar, 1925)
interstitialis (Dyar, 1925)
phoenicis (Dyar, 1925)

RUMATHA Heinr., 1939
5978 **glaucatella** (Hulst, 1888)
5979 **bihinda** (Dyar, 1922)
5980 **polingella** (Dyar, 1906)

YOSEMITIA Rag., 1901
YOSEMETIA Hulst, 1903, missp.
5981 **graciella** (Hulst, 1887)
5982 **longipennella** (Hulst, 1888)
graciella; auth., part, not Hulst, 1887
5983 **fieldiella** (Dyar, 1913)

EREMBERGA Heinr., 1939
5984 **leuconips** (Dyar, 1925)
5985 **creabates** (Dyar, 1923)
5986 **insignis** Heinr., 1939

OZAMIA Rag., 1901
5987 **fuscomaculella** (W. S. Wright, 1916)
heliophila Dyar, 1925
a. **clarefacta** Dyar, 1919
odiosella; auth., part, not Hulst, 1887
5988 **thalassophila** Dyar, 1925

CACTOBROSIS Dyar, 1915
5989 **fernaldialis** (Hulst, 1886)
gigantella (Rag., 1901)
cinerella (Hulst, 1901)
fernaldalis (Dyar, 1905), missp.
5990 **longipennella** (Hamp., 1901)
elongatella (Hamp., 1901)
fernaldialis; auth., part, not Hulst, 1886
5991 **strigalis** (B. & McD., 1912)

LASCELINA Heinr., 1956
5992 **canens** Heinr., 1956

METEPHESTIA Rag., 1901
5993 **simplicula** (Zell., 1881)

SELGA Heinr., 1956
5994 **arizonella** (Hulst, 1900)

EUZOPHERA Zell., 1867
STENOPTYCHA Heinemann,

1865, preocc. by Agassiz, 1862,
Zell, 1863
MELIA Heinemann, 1865, preocc.
by Bosc [in Risso], 1813
PISTOGENES Meyr., 1937
AHWAZIA Amsel, 1949
LONGIGNATHIA Roesler, 1965
EUZOPHORA auth., missp.
5995 **semifuneralis** (Wlk., 1863)
aglaeella Rag., 1887
pallulella (Hulst, 1887)
5996 **magnolialis** Capps, 1964
5997 **ostricolorella** Hulst, 1890
5998 **nigricantella** Rag., 1887
griselda Dyar, 1913

EULOGIA Heinr., 1956
5999 **ochrifrontella** (Zell., 1876)
ferruginella (Rag., 1887)
ochifrontella (McD., 1939), missp.

EPHESTIODES Rag., 1887
6000 **gilvescentella** Rag., 1887
nigrella Hulst, 1900
6001 **infimella** Rag., 1887
6002 **erythrella** Rag., 1887
coloradella (Hulst, 1900)
benjaminella Dyar, 1905
6003 **mignonella** Dyar, 1908
6004 **erasa** Heinr., 1956

MOODNA Hulst, 1890
6005 **ostrinella** (Clem., 1860)
obtusangulella (Rag., 1887)
pelviculella Hulst, 1890
6006 **bisinuella** Hamp., 1901

VITULA Rag., 1887
ECCOPSIA Hulst, 1903
6007 **edmandsii** (Pack., 1864)
dentosella Rag., 1887
edmandsae Heinr., 1956, emend.
a. **bombylicolella** Amsel, 1955
serratilineella; auth., part, not Rag.,
1887
6008 **lugubrella** (Rag., 1887)
6009 **pinei** Heinr., 1956

MANHATTA Hulst, 1890
HORNIGIA Rag., 1887, preocc. by
Rag., 1885
6010 **setonella** (McD., 1927)
6011 **broweri** Heinr., 1956

CAUDELLIA Dyar, 1904
6012 **apyrella** Dyar, 1904
6013 **albovittella** Dyar, 1904
6014 **nigrella** (Hulst, 1890)
arizonella (Walter, 1928)

SOSIPATRA Heinr., 1956
6015 **rileyella** (Rag., 1887)
6016 **anthophila** (Dyar, 1925)
6017 **thurberiae** (Dyar, 1917)
6018 **nonparilella** (Dyar, 1904)

PLODIA Gn., 1845
6019 **interpunctella** (Hbn., 1810–13)
interpunctalis (Hbn., 1825)
zeae (Fitch, 1856)

castaneella Reutti, 1898
latercula (Hamp., 1901)
glycinivora (Matsumura, 1917)
glycinivorella (Matsumura, 1920)

ANAGASTA Heinr., 1956
6020 **kuehniella** (Zell., 1879)
fuscofasciella (Rag., 1887)
sericarium; auth., part, not Scott, 1859
ceratoniae; Thompson, 1887, not Zell., 1848
kühmiella Poulton, 1888, missp.
kuhniella Adkin, 1892, missp.
kurhnirela Johnson, 1895, missp.
gitonella (Druce, 1896)
kuehniela D'Utra, 1901, missp.
lunella Noel, 1904, missp.
huchinella Craveri, 1915, missp.
kuchinella Bolle, 1921, missp.

EPHESTIA Gn., 1845
HYPHANTIDIUM Scott, 1859
6021 **elutella** (Hbn., 1796)
elutea (Haw., 1811)
semirufa (Haw., 1811)
angusta (Haw., 1811)
rufa (Haw., 1811)
sericarium (Scott, 1859)
roxburghii Gregson, 1873
unicolorella; auth., part, not Stgr., 1879
infumatella Rag., 1887
affusella (Rag., 1888)
icosiella Rag., 1888
roxburgii Stgr. & Rebel, 1901, missp.
amarella Dyar, 1904
roxburghi Spuler, 1913, missp.
uniformata Dufrane, 1942, infrasubsp.

CADRA Wlk., 1864
XENEPHESTIA Gozmány, 1958
6022 **cautella** (Wlk., 1863)
defectella Wlk., 1864
desuetella (Wlk., 1866)
cahiritella; auth., part, not Zell., 1867
passulella (Barrett, 1875)
formosella (Wileman & South, 1918)
rotundatella (Turati, 1930)
irakella (Amsel, 1959)
6023 **figulilella** (Gregson, 1871)
ficulella (Barrett, 1875), emend.
milleri (Zell., 1876)
gypsella (Rag., 1887)
figuliella (Fbs., 1923), missp.
venosella (Turati, 1926)
figulella (Curran, 1926), missp.
ernestinella (Turati, 1927)

BANDERA Rag., 1887
NASUTES Hamp., 1930
6024 **binotella** (Zell., 1872)
subluteella Rag., 1887
6025 **cupidinella** Hulst, 1888
conspersella (Rag., 1901)
venata (Hamp., 1930)
6026 **virginella** Dyar, 1908

WAKULLA Shaffer, 1968
6027 **carneella** (B. & McD., 1913)

TAMPA Rag., 1887
6028 **dimediatella** Rag., 1887

VARNERIA Dyar, 1904
6029 **postremella** Dyar, 1904
6030 **atrifasciella** B. & McD., 1913

EURYTHMIA Rag., 1887
6031 **hospitella** (Zell., 1875)
spaldingella Dyar, 1905
a. *yavapaella* Dyar, 1906
6032 **angulella** Ely, 1910
diffusella Ely, 1910
6033 **fumella** Ely, 1910

ERELIEVA Heinr., 1956
6034 **quantulella** (Hulst, 1887)
santiagella (Dyar, 1919)
hospitella; auth., part, not Zell., 1875
6035 **parvulella** (Ely, 1910)

BARBERIA Dyar, 1905
6036 **affinitella** Dyar, 1905

CABNIA Dyar, 1904
6037 **myronella** Dyar, 1904

ANERASTIA Hbn., 1825
PRINANERASTIA Hamp., 1901
6038 **lotella** (Hbn., 1810–13)
miniosella (Zinck., 1818)
lobella B. & McD., 1917, missp.

COENOCHROA Rag., 1887
PETALUMA Hulst, 1888
ALAMOSA Hamp., 1901
6039 **californiella** Rag., 1887
inspergella Rag., 1887
6040 **illibella** (Hulst, 1887)
puricostella Rag., 1887
piperatella (Hamp., 1901)
6041 **bipunctella** (B. & McD., 1913)

PEORIINAE

PEORIA Rag., 1887
AURORA Rag., 1887
STATINA Rag., 1887
CEARA Rag., 1888
CALERA Rag., 1888
ALTOONA Hulst, 1888
CAYUGA Hulst, 1888
VOLUSIA Hulst, 1890, preocc. by Robineau-Desvoidy, 1830, Adams, 1861
WEKIVA Hulst, 1890
OSCEOLA Hulst, 1891, preocc. by Baird & Girard, 1853
CHIPETA Hulst, 1892
CHIPOTA Hulst, 1902, missp.
TRIVOLUSIA Dyar, 1903
OLLIA Dyar, 1904
6042 **longipalpella** (Rag., 1887)
6043 **bipartitella** Rag., 1887
roseopennella (Hulst, 1890)
bipunctella; auth., part, not Rag., 1888
6044 **tetradella** (Zell., 1872)
6045 **opacella** (Hulst, 1887)
dichroeella (Rag., 1889)
6046 **floridella** Shaffer, 1968

6047 **rostrella** (Rag., 1887)
6048 **gemmatella** (Hulst, 1887)
bistriatella (Hulst, 1890)
pamponerella (Dyar, 1908)
6049 **roseotinctella** (Rag., 1887)
punctilimbella (Rag., 1888)
bifasciella (Hamp., 1901)
6050 **johnstoni** Shaffer, 1968
6051 **santaritella** (Dyar, 1904)
6052 **holoponerella** (Dyar, 1908)
6053 **approximella** (Wlk., 1866)
haematica (Zell., 1872)
roseatella (Pack., 1873)
cremoricosta (Hamp., 1918)
6054 **luteicostella** (Rag., 1887)
nodosella (Hulst, 1890)
perlepidella (Hulst, 1892)
6055 **punctata** Shaffer, 1976
6056 **gaudiella** (Hulst, 1890), mispl.

TOLIMA Rag., 1888
6057 **cincaidella** Dyar, 1904, mispl.

NAVASOTA Rag., 1887
6058 **hebetella** Rag., 1887

ANACOSTIA Shaffer, 1968
6059 **tribulella** Shaffer, 1968

ARIVACA Shaffer, 1968
6060 **pimella** (Dyar, 1906)
6061 **linella** Shaffer, 1968
6062 **ostreella** (Rag., 1887)
discostrigella (Dyar, 1904)
6063 **poohella** Shaffer, 1968
6064 **albidella** (Hulst, 1900)
6065 **artella** Shaffer, 1968
6066 **albicostella** (Grossb., 1917)

ATASCOSA Hulst, 1890
6067 **glareosella** (Zell., 1872)
bicolorella Hulst, 1890
albocostella (Hulst, 1900)

HOMOSASSA Hulst, 1890
6068 **ella** (Hulst, 1887)
6069 **platella** Shaffer, 1968
6070 **incudella** Shaffer, 1968
6071 **blanchardi** Shaffer, 1976

REYNOSA Shaffer, 1968
6072 **floscella** (Hulst, 1890)

GOYA Rag., 1888
AROPOTHOURES A. Blanchard, 1975
6073 **stictella** (Hamp., 1918)
6074 **ovaliger** (A. Blanchard, 1975)

UINTA Hulst, 1888, mispl.
6075 **oreadella** Hulst, 1888

THYRIDIDAE
by EUGENE MUNROE

THYRIDINAE

THYRIS Laspeyres, 1803
6076 **maculata** Harr., 1839
perspicua (Wlk., 1856)

6077 **sepulchralis** Guér., 1832
  lugubris Bdv., 1852, rev. syn.
  margaritana (Clem., 1862)
  nevadae; auth., part, not Oberth.,
    1884

  DYSODIA Clem., 1860
  VARNIA Wlk., 1864
  PLATYTHYRIS G. & R., 1867
  PACHYTHYRIS Felder &
    Rogenhofer, 1868
6078 **oculatana** Clem., 1860
  plena (Wlk., 1865)
  fasciata (G. & R., 1867)
  montana (Hy. Edw., 1875)
  aurea (Pagenstecher, 1892)
6079 **granulata** (Neum., 1883)
  a. *igualensis* Dyar, 1913
6080 **speculifera** (Sepp, 1830)
  aequalis (Wlk., 1865)
6081 **flagrata** (Wlk., 1865)
  floridana (Hulst, 1886)

  HEXERIS Grt., 1875
  OTTOLENGUIA Beutenmüller, 1896
6082 **enhydris** Grt., 1875
  reticulina (Beutenmüller, 1896)

SICULODINAE

  BELONOPTERA H.-S., 1858
  BELENOPTERA auth., missp.
  BELNOPTERA auth., missp.
6083 **fratercula** (Pagenstecher, 1892)

  RHODONEURA Gn., 1857
  SICULODES Gn., 1859
  IZA Wlk., 1865
6084 **terminalis** (Wlk., 1865)
  subapicula (Dyar, 1913)

  MESKEA Grt., 1877
6085 **dyspteraria** Grt., 1877

STRIGLININAE

  BANISIA Wlk., 1864
  DURDARA Moore, 1882
  VERNIFILIA Schultze, 1907
  TROPHOESSA Turner, 1911
  NEOBANISIA Whalley, 1967
6086 **myrsusalis** (Wlk., 1859)
  immaculata (Mösch., 1890)
  radiata (Pagenstecher, 1892)
6087 **furva** (Warr., 1905), extralim.
  flavalis (Dognin, 1911)
  a. *fracta* Whalley, 1976

HYBLAEIDAE
by EUGENE MUNROE

  HYBLAEA F., 1793
  AENIGMA Stkr., 1876, preocc. by
    Newman, 1836, Koch, 1846
6088 **puera** (Cr., 1777)
  saga (F., 1787)
  unxia (Hbn., 1810)
  apricans (Bdv., 1834)
  mirificum (Stkr., 1876)

Pterophoroidea

PTEROPHORIDAE
by EUGENE MUNROE

AGDISTINAE

  AGDISTIS Hbn., 1825
  ADACTYLA Curt., 1834
  ERNESTIA Tutt, 1906
  HERBERTIA Tutt, 1906
6089 **americana** B. & L., 1921
  adactyla; auth., part, not Hbn., 1823

PLATYPTILIINAE

  SPHENARCHES Meyr., 1886
6090 **ontario** (McD., 1927)

  GEINA Tutt, 1907
6091 **periscelidactyla** (Fitch, 1854)
6092 **tenuidactyla** (Fitch, 1854)
  nigrociliata (Zell., 1873)
  cygnus (B. & L., 1921)
6093 **buscki** (McD., 1933)

  CAPPERIA Tutt, 1905
6094 **ningoris** (Wlsm., 1880)
  bernardinus (Grinnell, 1908), n.
    comb.
6095 **evansi** (McD., 1923)
6096 **raptor** (Meyr., 1908)

  OXYPTILUS Zell., 1841
6097 **delawaricus** Zell., 1873
  finitimus Grinnell, 1908, infrasubsp.

  TRICHOPTILUS Wlsm., 1880
6098 **parvulus** B. & L., 1921
6099 **potentellus** Lange, 1940
6100 **californicus** (Wlsm., 1880)
  wrightii Grinnell, 1908
6101 **pygmaeus** Wlsm., 1880
6102 **lobidactylus** (Fitch, 1854)
6103 **indentatus** (Meyr., 1930), mispl.?

  MEGALORRHIPIDA Amsel, 1935
6104 **defectalis** (Wlk., 1864)
  congrualis (Wlk., 1864)
  oxydactyla (Wlk., 1864)
  hawaiiensis (Butl., 1881)
  ochrodactyla (Fish, 1881)
  centetes (Meyr., 1886)
  compsochares (Meyr., 1886)
  ralumensis (Pagenstecher, 1900)
  palästinalis [sic] Amsel, 1935

  CNAEMIDOPHORUS Wallgr., 1862
  EUCNAEMIDOPHORUS Wallgr.,
    1881, repl. name
6105 **rhododactylus** (D. & S., 1775)

  PLATYPTILIA Hbn., 1825
  PLATYPTILUS Zell., 1841,
    emend.
  GILLMERIA Tutt, 1905
  FREDERICINA Tutt, 1905
6106 **tesseradactyla** (L., 1761)
  fischeri (Zell., 1841)

6107 **pallidactyla** (Haw., 1811)
  marginidactyla (Fitch, 1854)
  nebulaedactyla (Fitch, 1854)
  bertrami (Rössler, 1864)
  bischoffi (Zell., 1867)
  cervinidactyla (Pack., 1873)
  adusta (Wlsm., 1880)
6108 **johnstoni** Lange, 1940
6109 **carduidactyla** (Riley, 1869)
  cardui Zell., 1873
  hesperis Grinnell, 1908
6110 **percnodactyla** (Wlsm., 1880)
6111 **comstocki** Lange, 1939
6112 **williamsii** Grinnell, 1908
6113 **ardua** McD., 1927
6114 **washburnensis** McD., 1929
6115 **albicans** (Fish, 1881)
6116 **albertae** B. & L., 1921

  ANSTENOPTILIA Zimmerman, 1958
6117 **marmarodactyla** (Dyar, 1902)
  fuscicornis; auth., part, not Zell.,
    1877
  pasadenensis (Grinnell, 1908), n.
    comb.

  AMBLYPTILIA Hbn., 1825
  AMBLYPTILUS Wallgr., 1859,
    emend.
6118 **pica** (Wlsm., 1880), n. comb.
  punctidactyla; auth., part, not Haw.,
    1811
  acanthadactyla; auth., part, not Hbn.,
    1805–13
  cosmodactyla; auth., part, not Hbn.,
    1818–19
  ulodactyla; auth., part, not Zett., 1839
  a. *calisequoiae* (Lange, 1950), n.
    comb.
  b. *marina* (Lange, 1950), n. comb.
  c. *sierrae* (Lange, 1950), n. comb.
  d. *monticola* (Grinnell, 1908), n.
    comb.
  e. *crataea* (T. B. Fletcher, 1940), n.
    comb.

  LANTANOPHAGA Zimmerman,
    1958
6119 **pusillidactyla** (Wlk., 1864)
  technidion (Zell., 1877)
  hemimetra (Meyr., 1886)
  lantana (Bsk., 1914)

  LIOPTILODES Zimmerman, 1958
6120 **parvus** (Wlsm., 1880)

  STENOPTILODES Zimmerman, 1958
6121 **crenulata** (B. & McD., 1913), n.
    comb.
  taprobanes; auth., part, not Felder &
    Rogenhofer, 1875
  brachymorpha; auth., part, not Meyr.,
    1888
  seeboldi; auth., part, not Hofmann,
    1898
  phoenicodactyla; auth., part, not
    Chrétien, 1915
  terlizzii; auth., part, not Turati, 1926
6122 **brevipennis** (Zell., 1883), n. comb.
6123 **antirrhina** (Lange, 1939), n. comb.

6124 **grandis** (Wlsm., 1880), n. comb.
6125 **carolina** (Kft., 1907), n. comb.
6126 **immaculata** (McD., 1939), n. comb.
6127 **auriga** (B. & L., 1921), n. comb.
6128 **edwardsii** (Fish, 1881), n. comb.
6129 **baueri** (Lange, 1950), n. comb.
6130 **albiciliata** (Wlsm., 1880), n. comb.
    a. **canadensis** (McD., 1927), n. comb.
    b. **rubricans** (Lange, 1950), n. comb.
    c. **orthocarpi** (Wlsm., 1880), n. comb.
6131 **lutescens** (Lange, 1950), n. comb.
6132 **albida** (Wlsm., 1880), n. comb.
6133 **shastae** (Wlsm., 1880), n. comb.
6134 **nana** (McD., 1927), n. comb.
6135 **albidorsella** (Wlsm., 1880), n. comb.
6136 **fragilis** (Wlsm., 1880), n. comb.
6137 **maea** (B. & L., 1921), n. comb.
6138 **cooleyi** (Fern., 1898), n. comb.
    schwarzi (Dyar, 1903), n. comb.
6139 **xylopsamma** (Meyr., 1908), n. comb.
    gorgoniensis (Grinnell, 1908), n. comb.
6140 **modesta** (Wlsm., 1880), n. comb.
6141 **bifida** (Lange, 1950), n. comb.
6142 **petrodactyla** (Wlk., 1864), n. comb.

STENOPTILIA Hbn., 1825
    MIMAESEOPTILUS Wallgr., 1862
    MIMESEOPTILUS Zell., 1867, emend.
    ADKINIA Tutt, 1905
6143 **pterodactyla** (L., 1761)
    fusca (Retzius, 1763)
    fuscodactyla (Villers, 1789)
    ptilodactyla (Hbn., 1805-13)
    paludicola (Wallgr., 1862)
6144 **zophodactyla** (Dup., 1838)
    loewii (Zell., 1847)
    canalis (Wlk., 1864)
    semicostata (Zell., 1873)
6145 **pallistriga** B. & McD., 1913
6146 **mengeli** Fern., 1898
6147 **bowmani** McD., 1923
6148 **exclamationis** (Wlsm., 1880)
6149 **coloradensis** Fern., 1898
6150 **columbia** McD., 1927
6151 **grandipuncta** McD., 1939
6152 **philocremna** Meyr., 1930

PTEROPHORINAE

PTEROPHORUS Schäffer, 1766
    ACIPTILIA Hbn., 1825
    MERRIFIELDIA Tutt, 1905
    PORRITTIA Tutt, 1905
    WHEELERIA Tutt, 1905
    ALUCITA; auth., not L., 1758
6153 **walsinghami** (Fern., 1898)

PSELNOPHORUS Wallgr., 1881
    GYPSOCHARES Meyr., 1890
    CRASIMETIS Meyr., 1890
6154 **belfragei** (Fish, 1881)

ADAINA Tutt, 1905
6155 **bipunctata** (Mösch., 1890)
    simplicia (Grossb., 1917)
6156 **zephyria** B. & L., 1921

6157 **montana** (Wlsm., 1880)
    declivis (Meyr., 1913)
6158 **cinerascens** (Wlsm., 1880)
6159 **buscki** B. & L., 1921
6160 **ambrosiae** (Murt., 1880)
    perplexa (Grossb., 1917)

OIDAEMATOPHORUS Wallgr., 1862
    LEIOPTILUS Wallgr., 1862
    OEDEMATOPHORUS Zell., 1867, emend.
    LIOPTILUS Zell., 1867, emend.
    OVENDENIA Tutt, 1905
    HELLINSIA Tutt, 1905
6161 **occidentalis** Wlsm., 1880
    cretidactylus; auth., part, not Fitch, 1854
    californicus (Grinnell, 1908)
6162 **balsamorrhizae** McD., 1939
6163 **cretidactylus** (Fitch, 1854)
    gypsodactylus Wlsm., 1880
6164 **downesi** McD., 1927
6165 **guttatus** Wlsm., 1880
6166 **mathewianus** (Zell., 1874)
    gorgoniensis (Grinnell, 1908)
    hilda (Grinnell, 1908)
6167 **fishii** (Fern., 1893)
    fishi (B. & McD., 1917), emend.
6168 **eupatorii** (Fern., 1891)
6169 **alaskensis** B. & L., 1921
6170 **phaceliae** McD., 1938
6171 **grisescens** Wlsm., 1880
    acrias (Meyr., 1908)
    behrii (Grinnell, 1908)
6172 **cineraceus** Fish, 1881
    lugubris Fish, 1881
6173 **rileyi** (Fern., 1898)
6174 **lindseyi** McD., 1923
6175 **baroni** Fish, 1881
6176 **castor** B. & L., 1921
6177 **pollux** B. & L., 1921
6178 **mizar** B. & L., 1921
6179 **meyricki** B. & L., 1921
6180 **gratiosus** Fish, 1881
6181 **fieldi** (W. S. Wright, 1921)
6182 **confusus** Braun, 1930
6183 **citrites** (Meyr., 1908)
6184 **brucei** (Fern., 1898)
    chionastes (Meyr., 1908)
6185 **albilobatus** McD., 1939
6186 **inquinatus** Zell., 1873
6187 **eros** B. & L., 1921
6188 **pan** B. & L., 1921
6189 **phoebus** B. & L., 1921
6190 **thor** McD., 1939
6191 **triton** B. & L., 1921
6192 **integratus** (Meyr., 1913)
6193 **auster** B. & L., 1921
6194 **medius** B. & L., 1921
6195 **lienigianus** (Zell., 1852)
    melinodactylus (H.-S., 1855)
    linus B. & L., 1921
6196 **cadmus** B. & L., 1921
6197 **iobates** B. & L., 1921
6198 **cochise** B. & L., 1921
6199 **ares** B. & L., 1921
6200 **tinctus** (Wlsm., 1915)
6201 **thoracicus** McD., 1939
6202 **helianthi** (Wlsm., 1880)

6203 **homodactylus** (Wlk., 1864)
6204 **elliottii** (Fern., 1893)
    elliotti (B. & McD., 1917), emend.
6205 **stramineus** (Wlsm., 1880)
6206 **angustus** (Wlsm., 1880)
6207 **paleaceus** (Zell., 1873)
    sericadactylus (Murt., 1880)
6208 **venapunctus** B. & L., 1921
6209 **luteolus** B. & L., 1921
6210 **balanotes** (Meyr., 1908)
    aquila (Meyr., 1908)
6211 **grandis** (Fish, 1881)
    baccharides (Grinnell, 1908)
6212 **kellicottii** (Fish, 1881)
    chlorias (Meyr., 1908)
    kellicotti (Meyr., 1910)
6213 **lacteodactylus** (Cham., 1873)
    subochraceus; auth., part, not Wlsm., 1880
6214 **glenni** Cashatt, 1972
6215 **subochraceus** (Wlsm., 1880)
6216 **sulphureodactylus** (Pack., 1873)
    sulphureus (Wlsm., 1880)
    sulphureidactylus (Meyr., 1910), missp.
6217 **serenus** (Meyr., 1913)
6218 **australis** (Grinnell, 1908)
6219 **costatus** B. & L., 1921
6220 **falsus** B. & L., 1921
6221 **varius** B. & L., 1921
6222 **varioides** McD., 1939
6223 **corvus** B. & L., 1921
6224 **perditus** B. & L., 1921
6225 **simplicissimus** McD., 1938
6226 **unicolor** (B. & McD., 1938)
6227 **inconditus** (Wlsm., 1880)
6228 **caudelli** (Dyar, 1903)
6229 **rigidus** McD., 1938
6230 **contortus** McD., 1938
6231 **catalinae** (Grinnell, 1908)
    agraphodactylus; auth., part, not Wlk., 1864
6232 **arion** B. & L., 1921
6233 **longifrons** (Wlsm., 1915)

EMMELINA Tutt, 1905
6234 **monodactyla** (L., 1758)
    pterodactyla; auth., part, not L., 1761
    bidactyla (Hochen., 1785)
    galactodactyla; auth., part, not D. & S., 1775
    albodactyla (F., 1794)
    tephradactyla; auth., part, not Hbn., 1805-13
    cineridactyla (Fitch, 1854)
    naevosidactyla (Fitch, 1854)
    pergracilidactyla (Pack., 1873)
    barberi (Dyar, 1903)
    pictipennis (Grinnell, 1908)

Drepanoidea

THYATIRIDAE
by DOUGLAS C. FERGUSON

THYATIRINAE

Habrosynini

HABROSYNE Hbn., 1821
  GONOPHORA Bruand, 1845,
    preocc. by Dejean, 1835
6235 **scripta** (Gosse, 1840)
  abrasa (Gn., 1852)
    a. **chatfieldii** Grt., 1895
      chatfeldi McD., 1938, missp.
      chatfeldti Werny, 1966, missp.
    b. **abrasoides** B. & Benj., 1929
6236 **gloriosa** (Gn., 1852)
  rectangulata Ottol., 1897
    a. **arizonensis** B. & McD., 1912

PSEUDOTHYATIRA Grt., 1865
  LACINIA Grt., 1863, preocc. by
    Conrad, 1853
6237 **cymatophoroides** (Gn., 1852)
  expultrix (Grt., 1863), rev. stat., form

Thyatirini

THYATIRA Ochs., 1816
  CALLEIDA Sodoffsky, 1837,
    preocc. by Latr., 1824
6238 **mexicana** Hy. Edw., 1884
  superba (Barnes, 1901)

Macrothyatirini

EUTHYATIRA Sm., 1891
  PERSISCOTA Grt., 1895
6239 **lorata** (Grt., 1881)
6240 **pudens** (Gn., 1852)
  anticostiensis (Grt., 1886)
  pennsylvanica Sm., 1902, rev. stat.,
    form
6241 **semicircularis** (Grt., 1881)
  tema (Stkr., 1898)
    a. **griseor** (B. & McD., 1910)

POLYPLOCINAE

Ceranemotini

CERANEMOTA Clarke, 1938
6242 **improvisa** (Hy. Edw., 1873)
6243 **fasciata** (B. & McD., 1910)
6244 **crumbi** Benj., 1938
6245 **semifasciata** Benj., 1938
6246 **tearlei** (Hy. Edw., 1886)
6247 **partida** Clarke, 1938
6248 **albertae** Clarke, 1938
6249 **amplifascia** Clarke, 1938

BYCOMBIA Benj., 1938
6250 **verdugoensis** (Hill, 1927)

DREPANIDAE
by DOUGLAS C. FERGUSON

DREPANINAE

DREPANA Schr., 1802
  PLATYPTERYX Laspeyres, 1803
  FALCARIA Haw., 1809
  PLATYPTERIX Ochs., 1816,
    missp.
  PRIONIA Hbn., 1819
  EDAPTERYX Pack., 1864
6251 **arcuata** Wlk., 1855
  fabula (Grt., 1862)
  genicula (Grt., 1862), sum. f.
    a. **siculifer** Pack., 1872
    b. **alaskensis** B. & Benj., 1922
      grotei B. & Benj., 1922, spr. f.
6252 **bilineata** (Pack, 1864)
  levis (Hudson, 1893), sum. f.
    a. **rampartensis** B. & Benj., 1922, n.
      stat.
      hudsoni B. & Benj., 1922

EUDEILINIA Pack., 1876
6253 **herminiata** (Gn., 1857)
  albata (Gn., 1857)
  biseriata (Pack., 1873)
6254 **luteifera** Dyar, 1917

ORETINAE

ORETA Wlk., 1855
  DRYOPTERIS Grt., 1862
6255 **rosea** (Wlk., 1855)
  marginata Wlk., 1855
  americana (H.-S., 1855)
  formula (Grt., 1862)
  irrorata (Pack., 1864), form

Geometroidea

GEOMETRIDAE
by DOUGLAS C. FERGUSON (except
STERRHINAE)

ARCHIEARINAE

ARCHIEARIS Hbn., 1823
  BREPHOS Hbn., 1806, suppr.
    (ICZN Op. 97)
  BREPHIA Hbn., 1822, emend.
  CATOXANTHIA Sodoffsky, 1837
6256 **infans** (Mösch., 1862), n. comb.
  hamadryas (Harr., 1869), n. comb.
    a. **oregonensis** (Swett, 1917), n.
      comb.

LEUCOBREPHOS Grt., 1874
6257 **brephoides** (Wlk., 1857)
  resoluta (Zell., 1863)
  hoyi (Grt., 1880)

OENOCHROMINAE

ALSOPHILA Hbn., 1825
6258 **pometaria** (Harr., 1841)
  restituens (Wlk., 1860)
  autumnata (Pack., 1876)

AMETRIS Hbn., 1822
  MECOCERAS Gn., 1857
6259 **nitocris** (Cram., 1780)
  nitocritaria Hbn., 1822
  bitactaria (Wlk., 1861), n. syn.
  peninsularia (Grt., 1883)
  schausaria (Hy. Edw., 1884)

ALMODES Gn., 1857
  POLYSEMIA Gn., 1857
6260 **terraria** Gn., 1857
  stellidaria (Gn., 1857)
  squamigera (Felder, 1875), n. syn.
  rivularia Grt., 1883
  assecoma (Druce, 1892), n. syn.
  subaustralis (Hulst, 1898)
  pedicellata (Hulst, 1898)

ENNOMINAE

Abraxini

HELIOMATA G. & R., 1866
6261 **cycladata** G. & R., 1866
6262 **fulliola** B. & McD., 1917
6263 **infulata** (Grt., 1863)
6264 **elaborata** (Grt., 1863)

PROTITAME McD., 1939
6265 **matilda** (Dyar, 1904)
6266 **subalbaria** (Pack., 1873)
6267 **hulstiaria** (Tayl., 1906)
  subalbaria (Hulst, 1896), preocc. by
    Pack., 1873
6268 **discalis** McD., 1939
6269 **albescens** McD., 1939
6270 **virginalis** (Hulst, 1900)
6271 **pallicolor** (Dyar, 1923)

Semiothisini

EUMACARIA Pack., 1873
6272 **latiferrugata** (Wlk., 1863)
    a. **brunneata** Pack., 1873

ITAME Hbn., 1823
  ITAMA Hbn., 1826, emend.
  SPERANZA Curt., 1828
  GRAMMATOPHORA Steph.,
    1829
  HALIA Dup., 1829, preocc. by
    Risso, 1826
  THAMNONOMA Led., 1853
  EUFITCHIA Pack., 1876
  CATASTICTIS Gump., 1887
  PHYSOSTEGANIA Warr., 1894
  DYSMIGIA Warr., 1895
  PROUTICTIS Bryk, 1938
6273 **pustularia** (Gn., 1857)
6274 **ribearia** (Fitch, 1848)
  sigmaria (Gn., 1857)
  annisaria (Wlk., 1860)
  aniusaria (Wlk., 1863)
6275 **fascioferaria** (Hulst, 1887)
6276 **flavicaria** (Pack., 1876)
6277 **subfalcata** (Hulst, 1896), n. stat.
  helena Cass., 1928, n. syn.
6278 **evagaria** (Hulst, 1900)
6279 **occiduaria** (Pack., 1874)
6280 **andersoni** (Swett, 1916)
    a. **orientis** Fgn., 1953

6281 **simplex** (Dyar, 1907)
6282 **argillacearia** (Pack., 1874)
    modestaria (Hulst, 1895)
6283 **sulphurea** (Pack., 1873)
    sulphuraria (Pack., 1876)
    olivalis (Hulst, 1898), n. syn.
6284 **amboflava** Fgn., 1953, n. stat.
6285 **inextricata** (Wlk., 1861)
    floridensis (Hulst, 1898)
6286 **brunneata** (Thunb., 1784)
    fulvaria (Villers, 1789)
    pinetaria (Hbn., 1796–99)
    quinquaria (Hbn., 1819–22)
    sylvaria (Curt., 1828)
    ferruginaria (Pack., 1873)
6287 **anataria** (Swett, 1913)
6288 **quadrilinearia** (Pack., 1873)
    inquinaria (Hulst, 1887)
6289 **coloradensis** (Hulst, 1896)
    aegaria (Stkr., 1899)
    disparata (Warr., 1904)
6290 **loricaria** (Evers., 1837), extralim.
    a. **julia** (Hulst, 1896)
        nubilata (Warr., 1904)
6291 **semivolata** (Dyar, 1923)
6292 **exauspicata** (Wlk., 1861)
6293 **umbriferata** (Hulst, 1887)
    umbrifasciata (Hulst, 1896), nom. nud.
6294 **abruptata** (Wlk., 1862)
6295 **confederata** B. & McD., 1917
6296 **plumosata** B. & McD., 1917
6297 **pallidula** (Hulst, 1896)
6298 **extemporata** B. & McD., 1917
6299 **coortata** (Hulst, 1887)
    a. **enigmata** B. & McD., 1917
6300 **perornata** B. & McD., 1916
6301 **guenearia** (Pack., 1876)
6302 **trilinearia** (Grossb., 1910)
6303 **subcessaria** (Wlk., 1861)
    perarcuata (Wlk., 1862)
6304 **bitactata** (Wlk., 1862)
    astrosignata (Wlk., 1862)
    packardaria (Mösch., 1883), n. syn.
    a. **epigenata** B. & McD., 1917
6305 **denticulodes** (Hulst, 1896)
6306 **decorata** (Hulst, 1896)
6307 **pallescens** (Grossb., 1907)
6308 **colata** (Grt., 1881)
    sericeata (Hulst, 1898)
    correllatum (Hulst, 1896)
6309 **benigna** (Hulst, 1898)
6310 **schatzeata** Cass., 1927
    dimidiata Cass., 1927, form
6311 **graphidaria** (Hulst, 1887)
6312 **sobriaria** B. & McD., 1917
6313 **minata** Cass., 1928
6314 **varadaria** (Wlk., 1860)
    inaptata (Wlk., 1861)
    abbreviata (Wlk., 1863)
    donataria (Wlk., 1863), ident. uncert.
    florida (Hulst, 1896)
6315 **grossbecki** (B. & McD., 1913)
6316 **simpliciata** B. & McD., 1918
6317 **pallipennata** (B. & McD., 1912)
6318 **particolor** (Hulst, 1898)
6319 **crocearia** (Hulst, 1887)
6320 **deleta** (Hulst, 1900)

EPELIS Hulst, 1896
    ISTURGIA; auth., not Hbn., 1823
6321 **truncataria** (Wlk., 1862)

MELLILLA Grt., 1873
    GONILYTHRIA Gump., 1887
6322 **xanthometata** (Wlk., 1862)
    chamaechrysaria Grt., 1873
    snoviaria (Pack., 1876), sum. f.
    rilevaria (Pack., 1876)

ELPISTE Gump., 1887
    SYMPHERTA Hulst, 1896
    GLADELA Grossb., 1909
6323 **marcescaria** (Gn., 1857)
    cineraria (Pack., 1871)
6324 **lorquinaria** (Gn., 1857)
    tripunctaria (Pack., 1873)
6325 **metanemaria** (Hulst, 1887)
    castalia (Druce, 1893), n. syn.

SEMIOTHISA Hbn., 1818
    MACARIA Curt., 1826
    EUTROPA Hbn., 1827–31
    PHILOBIA Dup, 1829
    GODONELA Bdv., 1840
    ACADRA H.-S., 1854
    PSAMATODES Gn., 1857, n. syn.
    SCIAGRAPHIA Hulst, 1896
    LIGDIFORMIA Wehrli, 1937
    THYRIDESIA Wehrli, 1940
6326 **aemulataria** (Wlk., 1861)
    sectomaculata (Morr., 1874)
6327 **perplexata** (Pears., 1913), n. comb.
6328 **aspirata** (Pears., 1913), n. comb.
6329 **versitata** (Pears., 1913), n. comb.
6330 **ulsterata** (Pears., 1913)
6331 **promiscuata** Fgn., 1974
6332 **punctolineata** (Pack., 1873)
    simulata Hulst, 1887
6333 **errata** McD., 1939
6334 **infimata** (Gn., 1857)
6335 **aequiferaria** (Wlk., 1861)
    postrema (Wlk., 1861)
    morosaria (Wlk., 1861)
    subpunctaria (Wlk., 1861), n. syn.
    festa (Hulst, 1896)
6336 **distribuaria** (Hbn., 1825)
    oppositaria (Gn., 1857), unavail.,
        publ. in syn.
    proxanthata (Wlk., 1863)
    antaurata (Wlk., 1863)
6337 **sanfordi** Rindge, 1958
6338 **adonis** (B. & McD., 1918)
6339 **transitaria** (Wlk., 1861)
6340 **minorata** (Pack., 1873)
6341 **bicolorata** (F., 1798)
    praeatomata (Haw., 1809)
    consepta (Wlk., 1861)
    consimilata (Zell., 1872)
    grassata (Hulst, 1881)
6342 **bisignata** (Wlk., 1866)
    galbineata (Zell., 1872)
6343 **sexmaculata** (Pack., 1867)
    unimodaria (Morr., 1874)
    labradoriata Mösch., 1883
    a. **incolorata** (Dyar, 1904)
        purcellata (Tayl., 1908)
6344 **signaria** (Hbn., 1800–09), extralim.
    a. **dispuncta** (Wlk., 1860)
        inordinaria (Wlk., 1860)
        subapiciaria (Wlk., 1861)
        haliata (Wlk., 1861)
        irregulata (Wlk., 1861)
        exnotata (Wlk., 1862)

    subapicaria (Wlk., 1863)
    quadrisignata (Wlk., 1866)
6345 **fraserata** Fgn., 1974
6346 **unipunctaria** (W. S. Wright, 1916)
    a. **perplexa** (McD., 1927)
6347 **pinistrobata** Fgn., 1972
6348 **fissinotata** (Wlk., 1863)
    retinotata (Wlk., 1863)
6349 **banksianae** Fgn., 1974
    marmorata Fgn., 1972, preocc. by
        Warr., 1897
6350 **submarmorata** (Wlk., 1861)
6351 **oweni** (Swett, 1907)
6352 **granitata** (Gn., 1857)
    succosata (Zell., 1872)
6353 **multilineata** (Pack., 1873)
    patriciata Grt., 1883
6354 **trimaculata** (Warr., 1906)
    pallidata (Warr., 1897), preocc. by
        Pack., 1873, n. syn.
    atrimacularia (B. & McD., 1913), n. syn.
6355 **sublacteolata** Hulst, 1887
    lapitaria (Stkr., 1899)
6356 **maculifascia** (Hulst, 1896)
6357 **eremiata** (Gn., 1857)
    retectata (Wlk., 1861)
    gradata (Wlk., 1861)
    retentata (Wlk., 1861)
    subcinctaria (Wlk., 1863)
6358 **ordinata** (Wlk., 1862)
    aucillaria (Stkr., 1899)
6359 **quadrifasciata** (Tayl., 1906)
6360 **quadrinotaria** (H.-S., 1855)
    quadriguttaria (Wlk., 1861), n. syn.
    determinataria (Wlk., 1863), n. syn.
    septemfluaria (Grt., 1881)
    septemlinearia (Grt., 1882)
6361 **orillata** (Wlk., 1863)
6362 **continuata** (Wlk., 1862)
    strigularia (Wlk., 1863)
6363 **excurvata** (Pack, 1874)
    cinereola (Hulst, 1896)
    spodopterata (Hulst, 1898)
6364 **setonana** (McD., 1927)
6365 **pertinata** McD., 1939
6366 **napensis** McD., 1939
6367 **nigroalbana** (Cass., 1928)
6368 **atrofasciata** (Pack., 1876)
6369 **maricopa** (Hulst, 1898)
6370 **curvata** (Grt., 1880)
    cruciata (Grt., 1883)
6371 **nubiculata** (Pack., 1876)
6372 **pictipennata** (Hulst, 1898)
6373 **denticulata** Grt., 1883
    sexpunctata Bates, 1886
6374 **delectata** Hulst, 1887
6375 **septemberata** (B. & McD., 1917)
6376 **burneyata** McD., 1939
6377 **muscariata** (Gn., 1857)
6378 **respersata** (Hulst, 1880)
    subacuta (Hulst, 1896)
6379 **teucaria** (Stkr., 1899)
6380 **californiaria** (Pack., 1871)
6381 **colorata** Grt., 1883
    conarata (Grossb., 1908)
6382 **davisata** (Cass., 1928)
6383 **pervolata** (Hulst, 1880)
6384 **azureata** (Cass., 1928)
6385 **triviata** (B. & McD., 1917)
6386 **ocellinata** (Gn., 1857)
    duplicata (Pack., 1873)

6387 **aliceata** (Cass., 1928)
6388 **woodgateata** (Cass., 1928)
6389 **arubrescens** McD., 1939, n. stat.
    sinuata (Warr., 1904), preocc. by
      Pack., 1874
    decorata (Grossb., 1907), preocc. by
      Warr., 1906
    suffusata McD., 1939, form
6390 **vernata** McD., 1939
6391 **spinata** McD., 1939
6392 **indeterminata** McD., 1939
6393 **yavapai** (Grossb., 1907)
6394 **hebetata** (Hulst, 1881)
    ponderosa (B. & McD., 1917), form
    demaculata (B. & McD., 1917)
    a. **tulareata** (C. & S., 1923), n. stat.
      innotata (C. & S., 1923), form
6395 **irrorata** (Pack., 1876)
    a. **venosata** McD., 1939
6396 **neptaria** (Gn., 1857)
    flavofasciata (Pack., 1871)
    sinuata (Pack., 1874), n. syn.
    cinereata (Bates, 1886)
    a. **trifasciata** (Pack., 1874)
6397 **mellistrigata** (Grt., 1873)
6398 **snoviata** (Pack., 1876)
    subminiata; Pack., 1874, not Pack., 1873
6399 **subminiata** (Pack., 1873)
6400 **gilletteata** (Dyar, 1904)
6401 **meadiaria** (Pack., 1874)
6402 **fieldi** (Swett, 1916)
    grossbecki (Swett, 1916), rev. syn.
    comstocki Sperry, 1949, n. syn.
6403 **minuta** (Hulst, 1896)
6404 **puertata** (Grossb., 1912)
6405 **gnophosaria** (Gn., 1857)
    infectata (Wlk., 1862)
    reductaria (Wlk., 1863)
    caesiaria Hulst, 1888
    da (Dyar, 1916), n. syn.
6406 **parcata** (Grossb., 1908)
6407 **nigricomma** Warr., 1904
    nigrocomina McD., 1938, missp.
6408 **piccoloi** Rindge, 1976
6409 **stipularia** (B. & McD., 1913)
6410 **pallidata** (Pack., 1873)
6411 **sirenata** McD., 1939
6412 **flaviterminata** (B. & McD., 1913)
6413 **subterminata** (B. & McD., 1913)
6414 **s-signata** (Pack., 1873)
6415 **cyda** (Druce, 1893)
6416 **ballandrata** (W. S. Wright, 1923), n. stat.
    melanderi Sperry, 1948, n. syn.
6417 **hypaethrata** (Grt., 1881)
6418 **kuschea** Guedet, 1939, mispl.

    ENCONISTA Led., 1853
6419 **dislocaria** (Pack., 1876)
    a. **malefactaria** (B. & McD. 1917)

    NARRAGA Wlk., 1862
    FERNALDELLA Hulst, 1896
6420 **fimetaria** (G. & R., 1870)
    halesaria (Zell., 1875)
    a. **partitaria** (Grt., 1883)
6421 **stalachtaria** (Stkr., 1878)
    alternaria (Grt., 1883)

    SPERRYA Rindge, 1958
6422 **cervula** Rindge, 1958

    XENOECISTA Warr., 1897
    PROPHASIANE McD., 1939, n.
      syn.
6423 **octolineata** (Hulst, 1887), n. comb.
6424 **mendicata** (Hulst, 1887), n. comb.
6425 **tenebrosata** (Hulst, 1887), n. comb.

Boarmiini

    DASYFIDONIA Pack., 1876
6426 **avuncularia** (Gn., 1857)
6427 **macdunnoughi** Guedet, 1935

    ORTHOFIDONIA Pack., 1876
6428 **tinctaria** (Wlk., 1860)
6429 **exornata** (Wlk., 1862)
    albifusata (Wlk., 1863), rev. syn.
    deceptata (Hulst, 1896), rev. syn.
6430 **flavivenata** (Hulst, 1898)

    HESPERUMIA Pack., 1873
    ULTRALCIS McD., 1920
6431 **sulphuraria** Pack., 1873
    ochreata Pack., 1873
    baltearia (Hulst, 1880), form
    unicoloraria (Hulst, 1886), form
    duaria C. & S., 1922, form
6432 **fumosaria** Comst., 1937
    a. **impensa** Rindge, 1974
6433 **latipennis** (Hulst, 1896)
6434 **fumida** (Warr., 1904)

    NEOALCIS McD., 1920
6435 **californiaria** (Pack., 1871)
    latifasciaria (Pack., 1873), form

    EMATURGA Led., 1853
6436 **amitaria** (Gn., 1857)
    faxonii (Minot, 1869)

    HYPOMECIS Hbn., 1821
    PSEUDOBOARMIA McD., 1920,
      n. syn.
6437 **luridula** (Hulst, 1896), n. comb.
6438 **buchholzaria** (Lemmer, 1937), n.
    comb.
6439 **umbrosaria** (Hbn., 1813)
6440 **gnopharia** (Gn., 1857), n. comb.

    PIMAPHERA C. & S., 1927
6441 **percata** C. & S., 1927
6442 **sparsaria** (Wlk., 1863)

    GLENOIDES McD., 1920
6443 **texanaria** (Hulst, 1888)
6444 **lenticuligera** A. Blanchard, 1973

    GLENA Hulst, 1896
    MONROA Warr., 1904
    HETERERANNIS Warr., 1904
6445 **grisearia** (Grt., 1883)
    pexata (Swett, 1907)
6446 **furfuraria** (Hulst, 1888)
    alpenata Cass., 1927, missp.
    a. **minor** Sperry, 1952
6447 **arcana** Rindge, 1958
6448 **nigricaria** (B. & McD., 1913)
    rusticaria (B. & McD., 1917)
6449 **cribrataria** (Gn., 1857)
    fuliginaria (Hulst, 1888)

6450 **cognataria** (Hbn., 1824–31)
    acidaliaria (Wlk., 1863)
    infixaria (Wlk., 1863)
    crassata (Hulst, 1896)
    muricolor (Hulst, 1896)
    umatillaria (Stkr., 1899)
    insaria (Dyar, 1909)
6451 **interpunctata** (B. & McD., 1917)
    a. **thomasaria** Sperry, 1952
6452 **plumosaria** (Pack., 1874)
6453 **quinquelinearia** (Pack., 1874)
6454 **macdunnougharia** Sperry, 1952
    a. **kirkwoodaria** Rindge, 1965

    STENOPORPIA McD., 1920
6455 **pulchella** (Grossb., 1909)
    a. **coolidgearia** Dyar, 1923
6456 **margueritae** Rindge, 1968
    a. **farina** Rindge, 1968
6457 **asymmetra** Rindge, 1959
6458 **dionaria** (B. & McD., 1918)
6459 **polygrammaria** (Pack., 1876)
6460 **mediatra** Rindge, 1958
6461 **dissonaria** (Hulst, 1896)
    elena Cass., 1927
    a. **campa** Rindge, 1968
6462 **anastomosaria** (Grossb., 1908)
6463 **pulmonaria** (Grt., 1881)
    a. **dejecta** (Hulst, 1896)
    b. **lita** Rindge, 1968
    c. **albescens** (Hulst, 1896)
    d. **satisfacta** (B. & McD., 1917)
    e. **vicaria** Rindge, 1968
6464 **purpuraria** (B. & McD., 1913)
6465 **vernata** (B. & McD., 1917)
    jemesata Cass., 1927
    a. **variana** Rindge, 1968
6466 **vernalella** McD., 1940
6467 **insipidaria** McD., 1945
6468 **anellula** (B. & McD., 1917)
6469 **badia** Rindge, 1968
6470 **macdunnoughi** Sperry, 1938
6471 **blanchardi** Rindge, 1968
6472 **glaucomarginaria** McD., 1945
6473 **separataria** (Grt., 1883)
    umbraria McD., 1925
6474 **excelsaria** (Stkr., 1899)
    a. **pullata** Rindge, 1968
6475 **larga** Rindge, 1968
6476 **graciella** McD., 1940
6477 **lea** Rindge, 1968

    EXELIS Gn., 1857
    PATRIDAVA Wlk., 1863
6478 **pyrolaria** Gn., 1857
    tensaria (Wlk., 1863)
    approximaria (Pack., 1876)
6479 **dicolus** Rindge, 1952
6480 **ophiurus** Rindge, 1952

    TORNOS Morr., 1875
6481 **punctatus** (Druce, 1898)
6482 **hoffmanni** Rindge, 1954
6483 **benjamini** C. & S., 1925
6484 **erectarius** Grossb., 1909
    pimensarius C. & S., 1923
    a. **fieldi** Grossb., 1912
6485 **cinctarius** Hulst, 1887
6486 **scolopacinarius** (Gn., 1857)
    pervelata (Wlk., 1863)

robiginosus Morr., 1875
piazzata C. & S., 1923
  a. **spodius** Rindge, 1954
  b. **forsythae** Rindge, 1954
6487 **abjectarius** Hulst, 1887
kerrvillaria C. & S., 1922
  a. **calcasiatus** C. & S., 1923
  b. **ravus** Rindge, 1954
  c. **kimballi** Rindge, 1954

GLAUCINA Hulst, 1896
COENOCHARIS Hulst, 1896
6488 **erroraria** Dyar, 1907
pearsalli Grossb., 1912
abdominalis Grossb., 1912
bilineata Grossb., 1912
6489 **biartata** Rindge, 1959
6490 **utahensis** (C. & S., 1924)
6491 **cilla** Rindge, 1959
6492 **macdunnoughi** (Grossb., 1912)
albidaria C. & S., 1922
6493 **epiphysaria** Dyar, 1908
6494 **baea** Rindge, 1959
6495 **escaria** (Grt., 1882)
puellaria Dyar, 1907
6496 **eupetheciaria** (Grt., 1883)
pygmeolaria (Grt., 1883)
tuconensis (C. & S., 1924)
  a. **lucida** Rindge, 1959
  b. **osiana** (Druce, 1893)
  c. **escariola** Rindge, 1959
6497 **elongata** (Hulst, 1896)
6498 **golgolata** (Stkr., 1899)
golgata (Dyar, 1903), missp.
mormonaria Dyar, 1907
6499 **magnifica** Grossb., 1912
6500 **ampla** Rindge, 1959
6501 **bifida** Rindge, 1959
6502 **interruptaria** (Grt., 1882)
jemezata (C. & S., 1923)
  a. **alboceptata** (Dyar, 1915)
6503 **spaldingata** (C. & S., 1923)
6504 **gonia** Rindge, 1959
  a. **microgonia** Rindge, 1959
6505 **platia** Rindge, 1959
6506 **nephos** Rindge, 1959
6507 **infumataria** (Grt., 1877)
6508 **ignavaria** (Pears., 1906)
6509 **lowensis** (C. & S., 1925)
6510 **dispersa** Rindge, 1959
6511 **denticularia** (Dyar, 1907)
6512 **imperdata** (Dyar, 1915)
6513 **mayelisaria** A. Blanchard, 1966
6514 **nota** Rindge, 1959
6515 **eureka** (Grossb., 1912)
  a. **agnesae** Rindge, 1959
6516 **ochrofuscaria** (Grt., 1882)
obscura (Grossb., 1912)
  a. **indistincta** (Grossb., 1912)
albata (C. & S., 1923)
6517 **loxa** Rindge, 1959
6518 **anomala** Rindge, 1959

SYNGLOCHIS Hulst, 1896
6519 **perumbraria** Hulst, 1896

EUBARNESIA Ckll., 1917
BARNESIA Grossb., 1910, preocc.
by Bertoni, 1901

CEDARIA Grossb., 1910, unavail.,
publ. in syn.
BARNESIATA Strand, 1926
6520 **ritaria** (Grossb., 1910)
  a. **arida** Rindge, 1959

PARAGLAUCINA Rindge, 1959
6521 **hulstinoides** (Grossb., 1912)

NEPTEROTAEA McD., 1920
6522 **diagonalis** Cass., 1927
6523 **marjorae** Rindge, 1973
6524 **ozarkensis** Rindge, 1973
6525 **dorotheata** Sperry, 1949
6526 **furva** Rindge, 1973
6527 **memoriata** (Pears., 1906)
6528 **obliviscata** (B. & McD., 1918)

CHESIADODES Hulst, 1896
MORINA Grossb., 1912
JENANA Clarke, 1939
6529 **simularia** (B. & McD., 1918)
6530 **cinerea** Rindge, 1973
6531 **morosata** Hulst, 1896
6532 **tubercula** Rindge, 1973
6533 **polingi** (Cass., 1927)
6534 **bicolor** Rindge, 1973
6535 **coniferaria** (Grossb., 1912)
6536 **fusca** Rindge, 1973
6537 **curvata** (B. & McD., 1916)
6538 **longa** Rindge, 1973
6539 **dissimilis** Rindge, 1973

HULSTINA Dyar, 1903
6540 **formosata** (Hulst, 1896)
terlineata Dyar, 1903
6541 **imitatrix** Rindge, 1970
  a. **fulva** Rindge, 1970
6542 **tanycraeros** Rindge, 1970
  a. **deserta** Rindge, 1970
6543 **aridata** B. & Benj., 1929
6544 **xera** Rindge, 1970
6545 **grossbecki** Rindge, 1970
6546 **exhumata** (Swett, 1918)
6547 **wrightiaria** (Hulst, 1888)
inconspicua (Hulst, 1896)
aethalodaria (Dyar, 1908)

PTEROTAEA Hulst, 1896
6548 **crickmeri** (Sperry, 1946)
6549 **crinigera** Rindge, 1970
6550 **sperryae** McD., 1938
6551 **leuschneri** Rindge, 1970
6552 **plagia** Rindge, 1970
6553 **lamiaria** (Stkr., 1899)
agrestaria (Grossb., 1909)
  a. **tytthos** Rindge, 1970
6554 **campestraria** McD., 1941
6555 **succurva** Rindge, 1970
6556 **glauca** Rindge, 1970
6557 **cavea** Rindge, 1970
6558 **miscella** Rindge, 1970
6559 **systole** Rindge, 1970
6560 **euroa** Rindge, 1970
6561 **lira** Rindge, 1970
6562 **obscura** Rindge, 1970
6563 **comstocki** Rindge, 1970
6564 **macrocercos** Rindge, 1970
6565 **powelli** Rindge, 1970
6566 **albescens** McD., 1941

6567 **melanocarpa** (Swett, 1916)
tremularia B. & McD., 1916
serrataria B. & McD., 1916
  a. **opaca** Rindge, 1970
6568 **cariosa** Hulst, 1896
  a. **incompta** Rindge, 1970
  b. **aporema** Rindge, 1970
6569 **newcombi** (Swett, 1914)
  a. **orinomos** Rindge, 1970

AETHALURA McD., 1920
AETHALOPTERA Hulst, 1896,
preocc. by Brauer, 1875
6570 **intertexta** (Wlk., 1860), n. comb.
intextata auth., missp.
anticaria (Wlk., 1860), n. syn.
submuraria (Wlk., 1860), rev. syn.
  a. **fumata** (B. & McD., 1917), rev.
stat.

ANACAMPTODES McD., 1920
6571 **cypressaria** (Grossb., 1917)
6572 **jacumbaria** (Dyar, 1908)
6573 **dataria** (Grt., 1882)
6574 **angulata** Rindge, 1966
6575 **clivinaria** (Gn., 1857)
  a. **impia** Rindge, 1966
  b. **profanata** (B. & McD., 1917)
6576 **sanctissima** (B. & McD., 1917)
6577 **obliquaria** (Grt., 1883)
rufaria (Grt., 1883)
6578 **providentia** Rindge, 1966
6579 **sancta** Rindge, 1966
6580 **pergracilis** (Hulst, 1900)
6581 **perfectaria** McD., 1940
6582 **vellivolata** (Hulst, 1881)
6583 **ephyraria** (Wlk., 1860)
expressaria (Wlk., 1863)
takenaria (Pears., 1909)
6584 **humaria** (Gn., 1857)
intraria (Gn., 1857)
intractaria (Wlk., 1860)
illaudata (Wlk., 1860)
  a. **pallida** Rindge, 1966
6585 **fragilaria** (Grossb., 1909)
6586 **defectaria** (Gn., 1857)
albigenaria (Wlk., 1860)
6587 **gemella** Rindge, 1966

IRIDOPSIS Warr., 1894
6588 **larvaria** (Gn., 1857)
6589 **emasculata** (Dyar, 1904)

ANAVITRINELIA McD., 1922
VITRINELLA McD., 1920, preocc.
by Adams, 1850
6590 **pampinaria** (Gn., 1857)
frugaliaria (Gn., 1857)
collecta (Wlk., 1860)
psilogrammaria (Zell., 1872)
fraudulentaria (Zell., 1872)
nubiferaria (Swett, 1913)
stygia (Franc., 1938), melanic f.
  a. **erosiata** (Wlk., 1863)
6591 **atristrigaria** (B. & McD., 1913)
6592 **addendaria** (Grossb., 1908)
6593 **ocularia** (B. & McD., 1917)

CLEORA Curt., 1825
6594 **sublunaria** (Gn., 1857)

transfixaria (Wlk., 1860)
atrolinearia (Hulst, 1888)
areataria Broadwell, 1907
6595 **projecta** (Wlk., 1860)
manitoba (Grossb., 1911)

GNOPHOS Tr., 1825
SCOTOPTERIX Hbn., 1825
CHARISSA Curt., 1826
SCIADION Prout, 1904
DYSGNOPHOS Wehrli, 1951
6596 **macguffini** Smiles, 1978

ECTROPIS Hbn., 1825
TEPHROSIA Bdv., 1840
6597 **crepuscularia** (D. & S., 1775)
virginiaria (Cram., 1780), n. comb., n. syn.
occiduaria (Gn., 1857)
signaria (Wlk., 1860)
spatiosaria (Wlk., 1860)
intrataria (Wlk., 1860)
cineraria (Wlk., 1860)
abraxaria (Wlk., 1860), form
divisaria (Wlk., 1860)
fumataria (Minot, 1869), form
fernaldaria (Hulst, 1888)

PROTOBOARMIA McD., 1920
6598 **porcelaria** (Gn., 1857)
filaria (Wlk., 1860)
maestosa (Hulst, 1898)
a. **indicataria** (Wlk., 1860)

EPIMECIS Hbn., 1825
BRONCHELIA Gn., 1857
6599 **hortaria** (F., 1794)
virginiaria; auth., not **virginiaria** Cram., 1780, misident. and missp.
liriodendraria (J. E. Smith, 1797)
disserptaria (Wlk., 1860)
amplaria (Wlk., 1860)
dendraria (Gn., 1857), form
gravilinearia (Andrews, 1878)
carbonaria Haim., 1915, melanic f.
6600 **subaustralis** (Hulst, 1898)
6601 **matronaria** (Gn., 1857)
6602 **anonaria** (Felder, 1874)
6603 **fraternaria** (Gn., 1857)
6604 **detexta** (Wlk., 1860)

MERICISCA Hulst, 1896
6605 **gracea** Hulst, 1896
6606 **perpictaria** (B. & McD., 1916)
6607 **scobina** Rindge, 1958

PARAPHEROMIA McD., 1920
6608 **cassinoi** McD., 1927
6609 **lichenaria** (Pears., 1906)
6610 **ficta** Rindge, 1972
6611 **configurata** (Hulst, 1898)
6612 **falsata** McD., 1920

PRIONOMELIA Warr., 1895
MERISMA McD., 1920, preocc. by Broun, 1909
6613 **spododea** (Hulst, 1896), n. comb.
6614 **ceraea** (Rindge, 1958), n. comb.

TRACHEOPS Hulst, 1896
6615 **bolteri** Hulst, 1896

MELANOCHROIA Hbn., 1819
6616 **chephise** (Cram., 1782)
6617 **geometroides** Wlk., 1854
mors (Lucas, 1857)

Melanolophiini

MELANOLOPHIA Hulst, 1896
6618 **imitata** (Wlk., 1860)
subgenericata Dyar, 1904
subimitata Fbs., 1948, nom. nud.
a. **barbara** McD., 1940
b. **cana** Rindge, 1964
6619 **centralis** McD., 1920
6620 **canadaria** (Gn., 1857)
a. **crama** Rindge, 1964
carbonata C. & S., 1923, ab., infrasubsp.
b. **choctawae** Rindge, 1964
6621 **signataria** (Wlk., 1860)
ejectaria (Wlk., 1860)
patularia (Wlk., 1861)
a. **timucuae** Rindge, 1964
6622 **imperfectaria** (Wlk., 1860)

CARPHOIDES McD., 1920
6623 **setigera** Rindge, 1958
6624 **incopriaria** (Hulst, 1887)
lineata (Hulst, 1898)
6625 **inconspicuaria** (B. & McD., 1916)

PARAPHOIDES Rindge, 1964
6226 **errantaria** (McD., 1940)

GALENARA McD., 1920
6627 **consimilis** Heinr., 1931
6628 **lallata** (Hulst, 1898)
6629 **glaucaria** (Grossb., 1912)
6630 **lixaria** (Grt., 1883)
6631 **lixarioides** McD., 1945
6632 **stenomacra** Rindge, 1958
6633 **olivacea** Rindge, 1958

VINEMINA McD., 1920
6634 **perdita** Guedet, 1939
6635 **opacaria** (Hulst, 1881)
nigaria Cass., 1928
6636 **catalina** McD., 1945

EUFIDONIA Pack., 1876
6637 **convergaria** (Wlk., 1860)
famulata (Hulst, 1887), rev. syn.
6638 **notataria** (Wlk., 1860)
fidoniata (Wlk., 1862)
bicoloraria (Minot, 1869)
6639 **discospilata** (Wlk., 1862)
quadripunctata (Morr., 1874), rev. syn.

Bistonini

BISTON Leach, 1815
PACHYS Hbn., 1822
AMPHIDASIS Tr., 1825
EUBYJA Hbn., 1825
EUBOEA Gump., 1887, emend.
6640 **betularia** (L., 1758), extralim.
a. **cognataria** (Gn., 1857)

swettaria (B. & McD., 1917), melanic f.
fortitaria (B. & McD., 1917)
mesle (Fbs., 1928)
b. **contrasta** (B. & Benj., 1923)
6641 **multidentata** (Guedet, 1941), mispl.

COCHISEA B. & McD., 1916
6642 **rigidaria** B. & McD., 1916
6643 **barnesi** C. & S., 1922
6644 **undulata** Rindge, 1975
6645 **paula** Rindge, 1975
6646 **unicoloris** Rindge, 1975
6647 **sonomensis** McD., 1941
abrunnea McD., 1941
6648 **recisa** Rindge, 1975
6649 **curva** Rindge, 1975
6650 **sinuaria** B. & McD., 1916

LYCIA Hbn., 1825
NYSSIA Dup., 1829
MELANOCOMA Harrison, 1910
POECILOPSIS Harrison, 1910
6651 **ursaria** (Wlk., 1860)
6652 **ypsilon** (S. A. Forbes, 1885)
a. **carlotta** (Hulst, 1896)
6653 **rachelae** (Hulst, 1896)

HYPAGYRTIS Hbn., 1818
PARAPHIA Gn., 1857
AMILAPIS Gn., 1857
6654 **unipunctata** (Haw., 1809)
pustularia Hbn., 1818, n. syn.
subatomaria (Wood, 1839), spr. f., n. syn.
deplanaria (Gn., 1857)
nubecularia (Gn., 1857), rev. syn.
mamurraria (Gn., 1857)
triplicipunctata (Fitch, 1860), rev. syn.
impropriata (Wlk., 1861)
fidoniaria (Wlk., 1862)
exsuperata (Wlk., 1863)
6655 **esther** (Barnes, 1928)
lemmeri (Barnes, 1928)
6656 **piniata** (Pack., 1870)
guttata (Hulst, 1901)
6657 **brendae** R. L. Heitzman, 1975

PHIGALIA Dup., 1829
CONIODES Hulst, 1896
RHAPHIDODEMAS Hulst, 1896
6658 **titea** (Cram., 1782)
titearia (Hbn., 1825)
revocata Wlk., 1863
cinctaria French, 1878
nevadaria (Hulst, 1896)
mephistaria Reiff, 1913, melanic f.
deplorans Franc., 1938, melanic f.
6659 **denticulata** Hulst, 1900
6660 **strigataria** (Minot, 1869)
olivacearia (Morr., 1874)
6661 **plumogeraria** (Hulst, 1888)
plumigeraria (Hulst, 1888), emend.
immacularia (C. & S., 1922), form

PALEACRITA Riley, 1876
6662 **vernata** (Peck, 1795)
sericeiferata (Wlk., 1863)
speciosa Hulst, 1898

6663 **merriccata** Dyar, 1903
6664 **longiciliata** Hulst, 1898

    ERANNIS Hbn., 1825
      HYBERNIA Berthold, 1827
      HIBERNIA missp.
      LAMPETIA Steph., 1827
6665 **tiliaria** (Harr., 1841)
    coloradata Hulst, 1896
    a. **vancouverensis** Hulst, 1896
      brunneata C. & S., 1923

Baptini

    LOMOGRAPHA Hbn., 1825
      BAPTA Steph., 1829
      CORYCIA Dup., 1829, preocc. by
        Hbn., 1823
6666 **semiclarata** (Wlk., 1866)
    viatica (Harv., 1874)
6667 **vestaliata** (Gn., 1857)
6668 **glomeraria** (Grt., 1881)
    merricki (C. & S., 1922)
    virginalis (C. & S., 1923)
6669 **elsinora** (Hulst, 1900)

    PALYAS Gn., 1857
6670 **auriferaria** (Hulst, 1887)

    PHRYGIONIS Hbn., 1825
      BYSSODES Gn., 1857
6671 **argentata** (Drury, 1773)
    argentistriata Stkr., 1876
    cerussaria (Grt., 1882)
    obrussata (Grt., 1882)

Caberini

    SERICOSEMA Warr., 1895
      EUEMERA Hulst, 1896
6672 **juturnaria** (Gn., 1857)
    californiaria (Pack., 1871)
    inturnaria (Gump., 1892)
6673 **immaculata** (B. & McD., 1913)
    a. **argentata** C. & S., 1922
6674 **wilsonensis** C. & S., 1922
    a. **meadowsaria** Sperry, 1948
    b. **macdunnoughi** Rindge, 1950
6675 **simularia** (Tayl., 1906)

    CABERA Tr., 1825
      DEILINIA Hbn., 1825
      CABIRA Sodoffsky, 1837
      DILINA Agassiz, 1847, preocc. by
        Dalman, 1816
6676 **exanthemata** (Scop., 1763), extralim.
    exanthemaria (Bkh., 1794)
    striaria (Hbn., 1796–99)
    approximaria (Haw., 1809)
    arenosaria (Haw., 1809)
    a. **bryantaria** (Tayl., 1906)
6677 **erythemaria** Gn., 1857
    incoloraria (Wlk., 1863)
    pacificaria (Pack., 1876)
    a. **undularia** (B. & McD.,
      1913)
6678 **variolaria** Gn., 1857
    intentaria Wlk., 1861
6679 **borealis** (Hulst, 1896)
    solamata (Hulst, 1900)

6680 **quadrifasciaria** (Pack., 1873)
    elimaria (Hulst, 1887)

    EUDREPANULATRIX Rindge, 1949
6681 **rectifascia** (Hulst, 1896)
    a. **erubescens** (Warr., 1904)

    DREPANULATRIX Gump., 1887
      AETHYCTERA Hulst, 1896
6682 **unicalcararia** (Gn., 1857)
    behrensaria (Hulst, 1887)
    ida (Hulst, 1896)
    cervivicolor (Hulst, 1898)
6683 **hulstii** (Dyar, 1903)
    lenitaria (Grossb., 1912)
    a. **carneolata** B. & McD., 1917
    b. **verdiaria** (Grossb., 1912)
      rindgearia Sperry, 1948
6684 **bifilata** (Hulst, 1880)
    perpallidaria (Grt., 1882)
    carneata (Warr., 1904)
    a. **ella** (Hulst, 1896)
    b. **ruthiaria** Sperry, 1948
6685 **quadraria** (Grt., 1882)
    a. **usta** Rindge, 1949
6686 **foeminaria** (Gn., 1857)
    pulveraria (Hulst, 1898)
6687 **nevadaria** (Hulst, 1888)
6688 **carnearia** (Hulst, 1888)
    lutearia (B. & McD., 1916)
    a. **columbiaria** McD., 1939
6689 **falcataria** (Pack., 1873)
    electa (Hulst, 1896)
6690 **secundaria** B. & McD., 1916
6691 **baueraria** Sperry, 1948
6692 **monicaria** (Gn., 1857)
    californiaria (Pack., 1870)
    ferruginosaria (Pack., 1870)
    celataria (Hulst, 1888)
    mercedulata (Stkr., 1899)
    indurata (Dyar, 1908)

    APODREPANULATRIX Rindge,
      1949
6693 **liberaria** (Wlk., 1860)
    integraria (Wlk., 1861)
    lintneraria (Pack., 1874)
    helena (Hulst, 1896)
6694 **litaria** (Hulst, 1887)
    fumosa (Hulst, 1896)

    IXALA Hulst, 1896
6695 **desperaria** (Hulst, 1887)
    a. **unicoloraria** (Hulst, 1888)
6696 **proutearia** Cass., 1928
6697 **klotsi** Sperry, 1940
6698 **adventaria** Pears., 1906

    NUMIA Gn., 1857
6699 **terebintharia** Gn., 1857
    buxaria Gn., 1857
    diffusa (Wlk., 1861)
    factaria (Wlk., 1861)
    subcelata (Wlk., 1861)
    subvectaria (Wlk., 1861)

    CHLORASPILATES Pack., 1876
6700 **bicoloraria** Pack., 1876
    a. **arizonaria** Grt., 1882
6701 **minima** (Hulst, 1898)

    ERASTRIA Hbn., 1813
      TROSTHIS Hbn., 1821
      SYRRHODIA Hbn., 1823, n. syn.
      CATOPYRRHA Hbn., 1823, n. syn.
      EUCHLIDON Hbn., 1823, n. syn.
      ACROLEUCA H.-S., 1855
6702 **decrepitaria** (Hbn., 1823), n. comb.,
    extralim.
    a. **esperanza** (B. & McD., 1917), n.
      comb.
6703 **viridirufaria** (Neum., 1881), n. comb.
    incarnata (B. & McD., 1917), n.
      comb.
6704 **coloraria** (F., 1798), n. comb.
    dissimilaria Hbn., 1813, form
    accessaria (Hbn., 1827–31), n. comb.
    olenusaria (Wlk., 1863), n. comb.
6705 **cruentaria** (Hbn., 1796–99), n. comb.
    sphaeromacharia (Harv., 1875), n.
      comb.
    perolivata (Hulst, 1896), n. comb.

    PTEROSPODA Dyar, 1903
      SPODOPTERA Hulst, 1896,
        preocc. by Gn., 1852
6706 **nigrescens** (Hulst, 1898)
6707 **opuscularia** (Hulst, 1887)
6708 **kunzei** (Hulst, 1898)

    STERGAMATAEA Hulst, 1896
6709 **inornata** Hulst, 1896
6710 **delicata** (Hulst, 1900)
    a. **dolliata** Grossb., 1908

    THYSANOPYGA H.-S., 1855
6711 **intractata** (Wlk., 1863), mispl.
    gausaparia (Grt., 1881)
    fulva (Warr., 1900), n. comb.
6712 **proditata** (Wlk., 1861), mispl.
    nicetaria; auth., part, not Gn., 1857

    EPISEMASIA Hulst, 1896
6713 **solitaria** (Wlk., 1861)
    stabilita (Wlk., 1861), n. syn.
    repugnata (Wlk., 1863), n. syn.
    morbosa Hulst, 1896
6714 **cervinaria** (Pack., 1873)

Angeronini

    ASPITATES Tr., 1825
      ASPILATES Tr., 1827, emend.
      NAPUCA Wlk., 1863
      PSEUDOSIONIA Butler, 1893
6715 **aberratus** (Hy. Edw., 1884)
    a. **assiniboiarius** Mun., 1963
6716 **forbesi** Mun., 1963
6717 **orciferarius** (Wlk., 1863)
    a. **baffinensis** Mun., 1963
    b. **churchillensis** Mun., 1963
    c. **occidentalis** Mun., 1963
6718 **conspersarius** Stgr., 1901
6719 **taylorae** (Butl., 1893)
    taylori (Butl., 1893), incorr. gender

    LYTROSIS Hulst, 1896
6720 **unitaria** (H.-S., 1854)
6721 **sinuosa** Rindge, 1971
6722 **heitzmanorum** Rindge, 1971
6723 **permagnaria** (Pack., 1876)

EUCHLAENA Hbn., 1823
6724 serrata (Drury, 1770)
   concisaria (Wlk., 1866)
6725 muzaria (Wlk., 1860)
6726 obtusaria (Hbn., 1809–13)
   decisaria (Wlk., 1860)
   incisaria (Wlk., 1866)
6727 silacea Rindge, 1958
6728 effecta (Wlk., 1860)
   effectaria (Wlk., 1863)
   clementina (Gump., 1895)
6729 johnsonaria (Fitch, 1869)
   bilinearia (Pack., 1870)
   a. minoraria (Hulst, 1886)
6730 mollisaria (Hulst, 1886)
   occantaria (Hulst, 1886)
6731 madusaria (Wlk., 1860)
   oponearia (Wlk., 1860)
   tiviaria (Wlk., 1860)
   vinulentaria (G. & R., 1867)
   vinosaria (G. & R., 1867)
   a. ochrearia McD., 1940
6732 deplanaria (Wlk., 1863)
6733 amoenaria (Gn., 1857)
   arefactaria (G. & R., 1867)
   a. astylusaria (Wlk., 1860)
6734 marginaria (Minot, 1869)
   marginata, missp.
   albertanensis Swett, 1917, n. syn.
6735 pectinaria (D. & S., 1775)
   deductaria (Wlk., 1860)
6736 manubiaria (Hulst, 1886)
6737 tigrinaria (Gn., 1857)
   propriaria (Wlk., 1860)
   a. sirenaria (Stkr., 1899)
   abnormalis Hulst, 1900
6738 milnei McD., 1945
6739 irraria (B. & McD., 1917)
   majoraria; Pack., 1876, not Gn., 1857

XANTHOTYPE Warr., 1894
6740 urticaria Swett, 1918
   watsoni Swett, 1918
   vagaria Swett, 1918
   turbidaria Swett, 1918
6741 barnesi Swett, 1918
6742 rufaria Swett, 1918
6743 sospeta (Drury, 1773)
   crocataria (F., 1798)
   citrinaria (Hbn., 1824)
   citrina, missp.
   caelaria (Hulst, 1886)
   manitobensis Swett, 1918
   marylandensis Swett, 1918
6744 attenuaria Swett, 1918

CYMATOPHORA Hbn., 1812
   STENOTRACHELYS Gn., 1857,
   n. syn.
6745 approximaria Hbn., 1812

Azelinini

STENASPILATODES Franc. and
   Poole, 1972
6746 antidiscaria (Wlk., 1863)
   lentaria (Hulst, 1886)

PERO H.-S., 1855
   PERGAMA H.-S., 1855

AZELINA Gn., 1857, n. syn.
SYNEMIA Gn., 1857, n. syn.
METIDULODES Gn., 1857, n. syn.
EGABRA Wlk., 1858, n. syn.
EUSENEA Wlk., 1860, n. syn.
STENASPILATES Pack., 1876, n.
   syn.
AZELINOPSIS Warr., 1896, n.
   syn.
MARMAREA Hulst, 1896, n. syn.
STENODONTA Warr., 1905, n.
   syn.
6747 meskaria (Pack., 1876), n. comb.
   albomacularia (Hy. Edw., 1882), n.
   comb.
   arizonaria (Hy. Edw., 1882), n.
   comb.
   muricolor (Hulst, 1896), n. comb.
   albidula (Hulst, 1900), n. comb., n.
   syn.
6748 astapa (Druce, 1892), n. comb.
   egregiata (Pears, 1906), n. comb.
6749 radiosaria (Hulst, 1886), n. comb.
   muricolor (Warr., 1900), preocc. by
   Hulst, 1896, n. comb., n. syn.
   fulvata (Warr., 1905), n. comb., n.
   syn.
   apapinaria (Dyar, 1908), n. comb., n.
   syn.
   metzaria (Dyar, 1909), n. comb.
   rectissima (Dyar, 1910), n. comb., n.
   syn.
6750 inviolata (Hulst, 1898), n. comb.
   levisaria (Grossb., 1909), n. comb.,
   n. syn.
6751 flavisaria (Grossb., 1906), n. comb.
6752 zalissaria (Wlk., 1860), n. comb.
6753 honestaria (Wlk., 1860), n. comb.
   stygiaria (Wlk., 1866), n. comb.
   dyari (C. & S., 1922), n. comb.
6754 hubneraria (Gn., 1857), n. comb.
   marmoratus (Grossb., 1910), n.
   comb., n. syn.
   barnesi (C. & S., 1922), n. comb., n.
   syn.
6755 morrisonaria (Hy. Edw., 1881), n. comb.
   morrisonatus (Grossb., 1910), n. comb.
6756 gigantea (Grossb., 1910), n. comb.
   provoensis (C. & S., 1922), n. comb.
6757 mizon (Rindge, 1955), n. comb.
6758 macdunnoughi (C. & S., 1922), n.
   comb.
6759 modesta (Grossb., 1910), n. comb.
   grossbecki (Pears., 1911), n. comb., n.
   syn.
6760 behrensaria (Pack., 1871), n. comb.
   behrensata (Pack., 1876), n. comb.
   a. vanduzeeata (W. S. Wright,
   1921), n. comb.
   b. smithii (Grossb., 1906), n. comb.
   blackmorei (C. & S., 1922), n.
   comb.
   c. modocata (C. & S., 1922), n.
   comb.
   splendorata (C. & S., 1922), n.
   comb.
   d. daula (Rindge, 1955), n. comb.
   e. sperryi (Rindge, 1955), n. comb.
6761 occidentalis (Hulst, 1896), n. syn.
   a. packardi (C. & S., 1922), n. comb.

   b. canastra (Rindge, 1955), n. comb.
   c. helva (Rindge, 1955), n. comb.
   d. peplarioides (Hulst, 1898), n.
   comb.
   colorado (Grossb., 1910), n. comb.
   patriciata (C. & S., 1925), n. comb.
6762 nerisaria (Wlk., 1860), n. comb.
   vetustaria (Wlk., 1866), n. comb., n. syn.
   curvistrigaria (H.-S., 1870), n.
   comb., n. syn.
   atrocolorata (Hulst, 1886), n. comb.,
   n. syn.

Nacophorini

NACOPHORA Hulst, 1896
   PHAEOURA Hulst, 1896
6763 quernaria (J. E. Smith, 1797)
   phigaliaria (Gn., 1857)
   sperataria (Wlk., 1860)
   paenulataria (Grt., 1863)
   cupidaria (Grt., 1864)
   atrescens Hulst, 1898
6764 cristifera (Hulst, 1896)
6765 kirkwoodi (Rindge, 1961), n. comb.
6766 mexicanaria (Grt., 1883)
   triaria (B. & McD., 1917)
   magnificans (Dyar, 1923)
6767 belua (Rindge, 1961), n. comb.
6768 perfidaria (B. & McD., 1917).
6769 utahensis (C. & S., 1923)
6770 aetha (Rindge, 1961), n. comb.
6771 cana (Rindge, 1961), n. comb.

THYRINTEINA Mösch., 1890
6772 arnobia (Stoll, 1782), extralim.
   a. phala Rindge, 1961

HOLOCHROA Hulst, 1896
   GLODURIA Dyar, 1924
6773 dissociaria (Hulst, 1887)
   a. varia Rindge, 1961

AETHALOIDA McD., 1920
   AETHALODES Hulst, 1896,
   preocc. by Gahan, 1888
6774 packardaria (Hulst, 1888)
   lachrymosa (Hulst, 1898)
   homopteroides (Hulst, 1900)

HEMIMORINA McD., 1941
6775 dissociata McD., 1941

HEMNYPIA McD., 1941
6776 baueri McD., 1941

PAREXCELSA Pears., 1912
6777 ultraria Pears., 1912

CERATONYX Gn., 1857
   STENOCHARIS Grossb., 1912
6778 arizonensis (Capps, 1950)
6779 permagnaria (Grossb., 1912)
6780 satanaria Gn., 1857
   candida (Sm., 1890)

GABRIOLA Tayl., 1904
6781 dyari Tayl., 1904
   a. pruina Rindge, 1974
   b. bakeri Rindge, 1974

6782 **sierrae** McD., 1945
    a. **australis** Rindge, 1974
    b. **baliola** Rindge, 1974
6783 **minima** (Hulst, 1896)
    bidisata (Dyar, 1903)
6784 **minor** Rindge, 1974
6785 **regularia** McD., 1945

    YERMOIA McD., 1940
6786 **perplexata** McD., 1940
6787 **glaucina** Rindge, 1961

    ANIMOMYIA Dyar, 1908
    GRAEFIA Pears., 1910
6788 **smithii** (Pears., 1910)
    a. **magna** Rindge, 1974
    b. **nigris** (C. & S., 1923)
6789 **dilatata** Rindge, 1974
6790 **hardwicki** Rindge, 1974
6791 **morta** Dyar, 1908
    increscens Dyar, 1923
6792 **turgida** Rindge, 1974
6793 **nuda** Rindge, 1974
6794 **minuta** Rindge, 1974

Gonodontini

    COLOTOIS Hbn., 1823
    METRA Steph., 1827
    HIMERA Dup., 1829
6795 **pennaria** (L., 1761)
    bifidaria (Haw., 1809)

Campaeini

    CAMPAEA Lamarck, 1816
    METROCAMPUS Berthold, 1827
    METROCAMPA Bruand, 1846
6796 **perlata** (Gn., 1857)
    praegrandaria (Gn., 1857), ident. uncert.
    viridoperlata (Pack., 1873)
    perlaria (Pack., 1876)

Ennomini

    ENNOMOS Tr., 1825
    EUGONIA Hbn., 1823, preocc. by
      Hbn., 1819
    ODOPTERA Sodoffsky, 1837
    DEUTERONOMOS Prout, 1914
6797 **magnaria** Gn., 1857
    lutaria Wlk., 1866
    a. **ochreatus** Hulst, 1898
6798 **subsignaria** (Hbn., 1823)
    niveosericeatus (Harris, 1855)

Epirranthini

    EPIRRANTHIS Hbn., 1823
    PLOSERIA Bdv., 1840
    SPODOLEPIS Hulst, 1896
    JUBARELLA Hulst, 1898
6799 **substriataria** (Hulst, 1896)
    a. **danbyi** (Hulst, 1898)
    demorsaria (Stkr., 1899)

Sphacelodini

    SPHACELODES Gn., 1857
6800 **vulneraria** (Hbn., 1819–21)

    fusilineata (Wlk., 1860)
    floridensis Holl., 1884
6801 **haitiaria** Oberth., 1923

Lithinini

    PHILEDIA Hulst, 1896
6802 **punctomacularia** (Hulst, 1888)
    a. **connecta** Guedet, 1939

    PETROPHORA Hbn., 1811
    ORTHOLITHA Hbn., 1821
    LITHINA Hbn., 1825
    LOZOGRAMMA Steph., 1829
6803 **divisata** Hbn., 1811
    extremaria (Wlk., 1861), n. comb.
    ordinata (Wlk., 1862), n. comb.
6804 **subaequaria** (Wlk., 1860), n. comb.
    defluaria (Wlk., 1861), n. comb.
    defluata auth., missp.

    TACPARIA Wlk., 1860
    APAECASIA Hulst, 1896
6805 **zalissaria** Wlk., 1860
    deductaria (Wlk., 1860)
    darlingtoni (Lemmer, 1937)
6806 **atropunctata** (Pack., 1874)
    fernaldi (Grt., 1878)
6807 **detersata** (Gn., 1857)

    THALLOPHAGA Hulst, 1896
    ANTHELIA Hulst, 1896, preocc. by
      Lamarck, 1816
6808 **taylorata** (Hulst, 1896)
    perfalcata (Sm., 1903)
    angulata (Warr., 1905)
6809 **hyperborea** (Hulst, 1900)
6810 **nigroseriata** (Pack., 1874)
    fautaria (Hulst, 1888)

    HOMOCHLODES Hulst, 1896
6811 **lactispargaria** (Wlk., 1861), rev. stat.
    immersata (Wlk., 1862), rev.
      syn.
6812 **fritillaria** (Gn., 1857)
6813 **disconventa** (Wlk., 1860), rev. stat.

    NYCTIPHANTA Hulst, 1896
6814 **laetula** Hulst, 1896

    GUENERIA Pack., 1876
6815 **similaria** (Wlk., 1860)
    basiaria (Wlk., 1863)
    congrua (Wlk., 1866)
    basiata Pack., 1876

Anagogini

    SLOSSONIA Hulst, 1898
6816 **rubrotincta** Hulst, 1898

    SELENIA Hbn., 1823
6817 **alciphearia** Wlk., 1860
    perangulata Hulst, 1896
    ornata B. & McD., 1914, sum. f.
6818 **kentaria** (G. & R., 1867)
    glaucata B. & McD., 1917, sum. f.

    METANEMA Gn., 1857
6819 **inatomaria** Gn., 1857

6820 **determinata** Wlk., 1866
    carnaria (Pack., 1873)

    METARRANTHIS Warr., 1894
6821 **warnerae** (Harv., 1874), emend.
    warneri (Harv., 1874), incorr. gender
    a. **cappsaria** Rupert, 1943
6822 **duaria** (Gn., 1857)
    adustaria (Wlk., 1866), rev. syn.
    septentrionaria B. & McD., 1917, n.
      syn.
    a. **hamaria** (Gn., 1857), n. stat.
    panisaria (Wlk., 1860), rev. syn.
    agreasaria (Wlk., 1860), rev. syn.
6823 **angularia** B. & McD., 1917
6824 **amyrisaria** (Wlk., 1860), n. stat.
    franclemonti Rupert, 1943, n. syn.
6825 **indeclinata** (Wlk., 1861), rev. syn.
6826 **hypocharia** (H.-S., 1854)
    mestusata (Wlk., 1860)
    broweri Rupert, 1943, rev. syn.
6827 **refractaria** (Gn., 1857), rev. stat.
    foedaria (Wlk., 1866), rev. syn.
6828 **homuraria** (G. & R., 1868)
    amethystaria (Stkr., 1899)
6829 **lateritiaria** (Gn., 1857)
6830 **pilosaria** (Pack., 1876)
6831 **apiciaria** (Pack., 1876)
6832 **obfirmaria** (Hbn., 1823)
6833 **mollicularia** (Zell., 1872)

    CEPPHIS Hbn., 1823
    PRIOCYCLA Gn., 1857
    HETRIONE Poppius, 1887
6834 **decoloraria** (Hulst, 1886)
    jucundaria (Stkr., 1899)
6835 **armataria** (H. & S., 1855)

    ANAGOGA Hbn., 1823
    AZINEPHORA Steph., 1827
    NUMERIA Dup., 1829
6836 **occiduaria** (Wlk., 1861)
    plagifasciata (Wlk., 1863)
    californiaria (Pack., 1871)

    PROBOLE H.-S., 1855
    HYPERETIS Gn., 1857, rev. syn.
6837 **alienaria** H.-S., 1855
    nyssaria (Gn., 1857), n. comb.
    insinuaria (Gn., 1857), n. comb.
    persinuaria (Gn., 1857), n. comb.
    subsinuaria (Gn., 1857), n. comb.
6838 **amicaria** (H.-S., 1855), n. comb.
    exsinuaria (Gn., 1857), n. comb.
    aesionaria (Wlk., 1860), n. comb.
    neonaria (Wlk., 1860), n. comb.
    laticincta (Wlk., 1861), n. comb.
6839 **nepiasaria** (Wlk., 1860), n. comb.

    PLAGODIS Hbn., 1823
    EURYMENE Dup., 1829
6840 **serinaria** H.-S., 1855
    subprivata (Wlk., 1863)
    rosaria (Grt., 1876)
    floscularia Grt., 1881
6841 **kuetzingi** (Grt., 1876)
    nigrescaria Hulst, 1887
    keutzingi auth., missp.
6842 **phlogosaria** (Gn., 1857)
    altruaria Pears., 1907, spr. f.

intermediaria B. & McD., 1917
  a. **iris** Rupert, 1949
  b. **bowmanaria** Mun., 1959
  c. **approximaria** Dyar, 1899
  d. **keutzingaria** Pack., 1876
  e. **purpuraria** Pears., 1907
    schuylkillensis Grossb., 1908, sum. f.
  f. **illinoiaria** Mun., 1959
6843 **fervidaria** (H. S., 1854)
  emargataria (Gn., 1857)
  excavaria (Morr., 1873)
  arrogaria (Hulst, 1886), sum. f.
6844 **alcoolaria** (Gn., 1857)
  kempii Hulst, 1900, sum. f.

Ourapterygini

PHILTRAEA Hulst, 1896
6845 **elegantaria** (Hy. Edw., 1881)
6846 **utahensis** Buckett, 1971
6847 **surcaliforniae** Buckett, 1971
6848 **latifoliae** Buckett, 1971
6849 **albimaxima** Buckett, 1971
6850 **paucimacula** B. & McD., 1918
6851 **monillata** Buckett, 1971

ERIPLATYMETRA Grt., 1873
  EPIPLATYMETRA Hulst, 1896,
  missp.
6852 **coloradaria** (G. & R., 1867)
6853 **lentifluata** B. & McD., 1917
6854 **grotearia** (Pack., 1876)
  vidularia Grt., 1882
  angularia (Grossb., 1908)

MELEMAEA Hulst, 1896
6855 **magdalena** Hulst, 1896
6856 **virgata** Tayl., 1906

LYCHNOSEA Grt., 1883
6857 **helveolaria** (Hulst, 1881)
  aulularia Grt., 1883
  runcinaria Stkr., 1899
6858 **intermicata** (Wlk., 1862)
  pervaria (Pack., 1873)
  interminaria (Grt., 1877)

NEOTERPES Hulst, 1896
6859 **ephelidaria** (Hulst, 1886)
  kunzei Hulst, 1898, form
6860 **trianguliferata** (Pack., 1871)
  notataria (Hulst, 1886)
  a. **costimacula** (Warr., 1901)
6861 **edwardsata** (Pack., 1871)
6862 **graefiaria** (Hulst, 1887)

CARIPETA Wlk., 1863
6863 **divisata** Wlk., 1863
  albopunctata (Morr., 1874)
  nigraria Fbs., 1924, form
6864 **piniata** (Pack., 1870)
  seductaria Stkr., 1899
6865 **aequaliaria** Grt., 1883
6865.1 **suffusata** Guedet, 1939
6866 **interalbicans** Warr., 1904, rev. stat.
6867 **angustiorata** Wlk., 1863
  criminosa Swett, 1906
6868 **latiorata** Wlk., 1863
6869 **aretaria** (Wlk., 1860)
  subochrearia Grt., 1883

6870 **pulcherrima** (Guedet, 1941), n. comb.
  waltonaria (Sperry, 1949)
6871 **hilumaria** (Hulst, 1886)
6872 **canidiaria** (Stkr., 1899)
6873 **ocellaria** (Grossb., 1907)
6874 **macularia** (B. & McD., 1916)

SNOWIA Neum., 1884
6875 **montanaria** Neum., 1884

NEMERIS Rindge, 1981
6876 **speciosa** (Hulst, 1896)
  albocrenulata Cass., 1927
6876.1 **percne** Rindge, 1981
6876.2 **sternitzkyi** Rindge, 1981

MERIS Hulst, 1896
6877 **paradoxa** Rindge, 1981
6878 **suffusaria** McD., 1940
6879 **patula** Rindge, 1981
6879.1 **alticola** Hulst, 1896
6879.2 **cultrata** Rindge, 1981

DESTUTIA Grossb., 1908
6880 **flumenata** (Pears., 1906), n. comb.
6881 **novata** Grossb., 1908
6882 **oblentaria** (Grt., 1883)
6883 **excelsa** (Stkr., 1878)
  a. **simpliciaria** (Grt., 1883)
  b. **olivata** (B. & McD., 1917)
    cochizeata (C. & S., 1923)

BESMA Capps, 1943
6884 **endropiaria** (G. & R., 1867)
  fatuaria (Stkr., 1899)
6885 **quercivoraria** (Gn., 1857)
  aeliaria (Wlk., 1860)
  textrinaria (G. & R., 1867)
  trilinearia (Pack., 1876)
  incongruaria (Hulst, 1887)
6886 **rubritincta** (C. & S., 1925)
  nigripuncta (C. & S., 1925)
6887 **sesquilinearia** (Grt., 1883)
  cavillaria (Hulst, 1886), n. syn., sum. f.
  novellata (Hulst, 1886)

LAMBDINA Capps, 1943
6888 **fiscellaria** (Gn., 1857)
  flagitiaria (Gn., 1857)
  peccataria (Swett, 1909)
  johnsoni (Swett, 1913)
  turbataria (B. & McD., 1917)
  a. **lugubrosa** (Hulst, 1903)
  b. **somniaria** (Hulst, 1886)
6889 **pultaria** (Gn., 1857), rev. stat.
  scitata (Wlk., 1863)
  invexata (Wlk., 1863)
6890 **flavilinearia** (B. & McD., 1913), rev.
  stat.
6891 **laeta** (Hulst, 1900), rev. stat.
6892 **pellucidaria** (G. & R., 1867)
6893 **canitiaria** Rupert, 1944
6894 **fervidaria** (Hbn., 1827)
  a. **athasaria** (Wlk., 1860)
  aequaliaria (Wlk., 1860)
  seminudata (Wlk., 1863)
  siccaria (Wlk., 1866)
  bibularia (G. & R., 1867)
  semiundaria (Pack., 1867), missp.
6895 **vitraria** (Grt., 1883)

punctata (Hulst, 1898), n. syn.
jacularia (B. & McD., 1917), n. syn.,
  spr. f.
6896 **phantoma** (B. & McD., 1916), rev. stat.

EVITA Capps, 1943
6897 **hyalinaria** (Grossb., 1908)

CINGILIA Wlk., 1862
  CATERVA Grt., 1876
6898 **catenaria** (Drury, 1773)
  devinctaria (Gn., 1857)
  humeralis Wlk., 1862
  rubiferaria Swett, 1910
  immacularia Swett, 1914

NEPYTIA Hulst, 1896
6899 **umbrosaria** (Pack., 1873)
  a. **nigrovenaria** (Pack., 1876)
6900 **regulata** B. & McD., 1916
6901 **disputata** McD., 1940
6902 **janetae** Rindge, 1967
6903 **juabata** C. & S., 1922
6904 **lagunata** C. & S., 1923
6905 **swetti** B. & B., 1923
6906 **canosaria** (Wlk., 1863)
  pulchraria (Minot, 1869)
  piniaria (Pack., 1870)
  fuscaria B. & Benj., 1923
6907 **phantasmaria** (Stkr., 1899)
6908 **semiclusaria** (Wlk., 1863)
  fumosaria (Stkr., 1899)
6909 **pellucidaria** (Pack., 1873)
6910 **freemani** Mun., 1963

SICYA Gn., 1857
6911 **crocearia** Pack., 1873, n. stat.
6912 **macularia** (Harr., 1850)
  solfataria Gn., 1857, form
  sublimaria Gn., 1857
  truncataria Gn., 1857
  agyllaria (Wlk., 1860), n. syn.
  calipusaria (Wlk., 1860)
  faustinaria Stkr., 1899
  a. **cruzensis** Dyar, 1922
  b. **lewisi** Swett, 1914
6913 **laetula** B. & McD., n. stat.
6914 **pergilvaria** B. & McD., 1917
6915 **morsicaria** (Hulst, 1886)
  snoviaria (Hulst, 1888), n. syn.
6916 **olivata** B. & McD., 1916
  subviridis C. & S., 1922

EUASPILATES Pack., 1874
6917 **spinataria** Pack., 1874

EUCATERVA Grt., 1882
6918 **variaria** Grt., 1882
  lymax (Druce, 1898)
  labesaria Grt., 1882
6919 **bonniwelli** C. & S., 1922

ACANTHOTOCA Fletcher, 1979
  ACANTHOPHORA Hulst, 1896,
  preocc. by Sollas, 1873
6920 **graefi** (Hulst, 1896)

PLATAEA H.-S., 1855
  GORYTODES Gn., 1857
  APICRENA Pears., 1911

6921 **calcaria** (Pears., 1911)
    triangularia B. & McD., 1916
    dulcinia Dyar, 1923
6922 **personaria** (Hy. Edw., 1881)
    lessaria Pears., 1907
    pasadenaria W. S. Wright, 1917
6923 **ursaria** C. & S., 1922
6924 **californiaria** H.-S., 1855
    uncanaria (Gn., 1857)
6925 **diva** Hulst, 1896
6926 **trilinearia** (Pack., 1873)
    dulciaria (Grt., 1880)
    astrigaria B. & McD., 1918

EUSARCA Hbn., 1813
    APICIA Gn., 1857, n. syn.
6927 **falcata** (Pack., 1873), n. comb.
6928 **terraria** (McD., 1940), n. comb.
6929 **argillaria** (Hulst, 1886), n. comb.
6930 **galbanaria** (Hulst, 1886), n. comb.
6931 **detractaria** (B. & McD., 1916), n. comb.
6932 **lutzi** (W. S. Wright, 1920), n. comb.
6933 **fundaria** (Gn., 1857), n. comb.
    juncturaria (Gn., 1857), n. comb.
    impexaria (Gn., 1857), n. comb.
    incopularia (Gn., 1857), n. comb.
    thasusaria (Wlk., 1860), n. comb., ident. uncert.
    arbuaria (Wlk., 1860), n. comb., ident. uncert.
    eldanaria (Wlk., 1860), n. comb., ident. uncert.
    carcearia (Wlk., 1860), n. comb., ident. uncert.
    basifusata (Wlk., 1863), n. comb.
    effascinaria (Hulst, 1886), n. comb.
6934 **subflavaria** (Pears., 1906), n. comb.
6935 **distycharia** (Gn., 1857), n. comb.
6936 **packardaria** (McD., 1940), n. comb.
6937 **venosaria** (McD., 1940), n. comb.
6938 **subcineraria** (Grossb., 1908), n. comb.
6939 **geniculata** (Hulst, 1886), n. comb.
6940 **tibiaria** (McD., 1940), n. comb.
6941 **confusaria** Hbn., 1813
    remissaria (Gn., 1857), n. comb.
    imbraria (Gn., 1857), n. comb.
    superaria (Gn., 1857), n. comb.
    majoraria (Gn., 1857), n. comb.
    ineffusaria (Gn., 1857), n. comb.
    floridaria (Gn., 1857), n. comb.
    phasianaria (Gn., 1857), n. comb.
    interlinearia (Gn., 1857), n. comb.
    metrocamparia (Gn., 1857), n. comb.
    pandaria (Wlk., 1860), n. comb.
    mimaria (Hulst, 1886), n. comb., n. syn.
    subochrearia (Hulst, 1900), n. comb.
6942 **graceiaria** (Kirkwood, 1951), n. comb.

SOMATOLOPHIA Hulst, 1896
6943 **desolata** Rindge, 1980
6944 **montana** Rindge, 1980
6945 **ectrapelaria** (Grossb., 1908)
6946 **pallescens** McD., 1940
    obliterata Warr., 1904, infrasubsp.
6947 **haydenata** (Pack., 1876)
    umbripennis Hulst, 1896
6948 **incana** Rindge, 1980

6948.1 **vatia** Rindge, 1980
6948.2 **petila** Rindge, 1980
6948.3 **simplicia** (B. & McD., 1918)
6948.4 **cuyama** Comst., 1940

PHERNE Hulst, 1896
    CTENOTETRACIS Warr., 1894, unused sr. syn.
6949 **placeraria** (Gn., 1857)
    mellitularia (Hulst, 1886)
6950 **parallelia** (Pack., 1873)
6951 **sperryi** McD., 1935
6952 **subpunctata** (Hulst, 1898)
    a. *vernalaria* W. S. Wright, 1917

SYNAXIS Hulst, 1896
6953 **fuscata** Hulst, 1898
6954 **jubararia** (Hulst, 1886)
    a. *sericeata* B. & McD., 1917
6955 **pallulata** (Hulst, 1887)
6956 **cervinaria** (Pack., 1871)
    aurantiacaria (Pack., 1873)
6957 **triangulata** (B. & McD., 1916)
6958 **formosa** (Hulst, 1896)
6959 **barnesii** (Hulst, 1896)
6960 **hirsutaria** (B. & McD., 1913)
6961 **mosesiani** Sala, 1971
6962 **brunneilinearia** (Grossb., 1907)

TETRACIS Gn., 1857
6963 **crocallata** Gn., 1857
    allediusaria Wlk., 1860
    a. *aspilatata* Gn., 1857
6964 **cachexiata** Gn., 1857
    lorata Grt., 1864

EUGONOBAPTA Warr., 1894
6965 **nivosaria** (Gn., 1857)

EUTRAPELA Hbn., 1809
    SYNDALIMIA Hbn., 1821
    ABBOTTANA Hulst, 1896
6966 **clemataria** (J. E. Smith, 1797)
    transducens (Wlk., 1860), n. comb.
    transferens (Wlk., 1860), n. comb.
    transfingens (Wlk., 1860), n. comb.

OXYDIA Gn., 1857
6967 **vesulia** (Cram., 1779), extralim.
    a. **transponens** (Wlk., 1860)
    antilleana (Prout, 1932)
6968 **cubana** (Warr., 1906)
6969 **gueneei** (Warr., 1904)
6970 **nimbata** Gn., 1857
    noctuitaria Wlk., 1860
    vitiligata Felder & Rogenhofer, 1875
6971 **mundata** Gn., 1857
    zonulata Hulst, 1886
6972 **masthala** Druce, 1892

PHYLLODONTA Warr., 1894
    LYCIMNA; B. & McD., 1916, not Wlk., 1860
6973 **peccataria** (B. & McD., 1916), n. comb.

PATALENE H.-S., 1854
    DREPANODES Gn., 1857
    HALESA Wlk., 1860
    SYSSAURA; auth., not Hbn., 1819
6974 **olyzonaria** (Wlk., 1860)

    syzygiaria (Hulst, 1886), n. comb.
    sesquilinea (Grt., 1870), n. comb.
    a. **puber** (G. & R., 1867), n. stat.
    varus (G. & R., 1867)
    aquosus (G. & R., 1867)
    juniperaria (Pack., 1871)
    hortularia (Hulst, 1886), n. comb., ident. uncert.
    perizomaria (Hulst, 1886)
6975 **epionata** (Gn., 1857), n. comb.
    pionaria (Wlk., 1860), n. comb.
    tellesaria (Wlk., 1860), n. comb.
    pappiaria (Wlk., 1860), n. comb.
    bicesaria (Wlk., 1860), n. comb.
    oemearia (Wlk., 1860), n. comb.
    amytisaria (Wlk., 1860), n. comb.
    ochrea (Butl., 1878), n. comb.
    spadicearia (Mösch., 1888), n. comb.
6976 **nicoaria** (Wlk., 1860), n. comb., ident. uncert.

PROCHOERODES Grt., 1883
    CHOERODES Gn., 1857, preocc. by Leidy, 1852
    AESCHROPTERYX Butler, 1883
6977 **truxaliata** (Gn., 1857)
6978 **amplicineraria** (Pears., 1906)
6979 **accentuata** (B. & McD., 1918)
6980 **nonangulata** (Stkr., 1899)
6981 **forficaria** (Gn., 1857)
    nubilata (Pack., 1871)
    a. **catenulata** Grt., 1883
    anfracta (Hulst, 1886)
    b. **combinata** McD., 1940
6982 **transversata** (Drury, 1770)
    goniata (Gn., 1857)
    contingens (Wlk., 1860)
    transposita (Wlk., 1860)
    transfindens (Wlk., 1860)
    a. **incurvata** (Gn., 1857)
    transvertens (Wlk., 1860)
    transmutans (Wlk., 1860)
6983 **olivata** Warr., 1904

PITYEJA Wlk., 1861
6984 **ornata** Rindge, 1958
    picta; B. & Benj., 1924, not Schaus, 1898

NEPHELOLEUCA Butler, 1883
6985 **politia** (Cram., 1777)
6986 **floridata** (Grt., 1883)

ANTEPIONE Pack., 1876
6987 **thisoaria** (Gn., 1857)
    arcasaria (Wlk., 1860)
    depontanata (Grt., 1864)
    sulphuraria (Pack., 1873)
    furciferata (Pack., 1876), spr. f.
6988 **comstocki** Sperry, 1939
6989 **indiscretata** (Hy. Edw., 1884)
    vanusaria (Stkr., 1899)
6990 **imitata** Hy. Edw., 1884
    costinotata (Tayl., 1906)
6991 **constans** (Hulst, 1898)
6992 **ochreata** (Hulst, 1898)
    arizonata (Tayl., 1906), form.
    dyari (Grossb., 1908)
    ligata (Grossb., 1908)
6993 **hewesata** Sperry, 1948

SERICOPTERA H.-S., 1856
   RIPULA Gn., 1857
6994 **virginaria** (Hulst, 1886)
   vestalis (Hulst, 1898)
   insularis Warr., 1909, n. syn.

SABULODES Gn., 1857
   PHENGOMMATAEA Hulst, 1896
6995 **aegrotata** (Gn., 1857)
   caberata; auth., not Gn., 1857
   arsesaria (Wlk., 1860)
   cottlei B. & Benj., 1926
6996 **dissimilis** (Hulst, 1898)
6997 **sericeata** B. & McD., 1917
6998 **huachuca** Rindge, 1978
6999 **mabelata** (Sperry, 1948)
7000 **niveostriata** (Ckll., 1894)
   gertruda (Hulst, 1896)
7001 **olifata** (Guedet, 1939)
7002 **duoangulata** (C. & S., 1923)
7003 **spoliata** (Grossb., 1908)
   a. **berkleyata** (W. S. Wright, 1917)
   b. **lagunata** (C. & S., 1923)
7004 **edwardsata** (Hulst, 1886)

ENYPIA Hulst, 1896
7005 **venata** (Grt., 1883)
   perangulata Hulst, 1896
   elaborata C. & S., 1923
   eddyi B. & Benj., 1929
7006 **griseata** Grossb., 1908
   moillieti Blkmre., 1926
7007 **packardata** Tayl., 1906
7008 **coolidgi** C. & S., 1923

NEMATOCAMPA Gn., 1857
7009 **limbata** (Haw., 1809)
   filamentaria Gn., 1857
   chagnoni (Swett, 1913)
   a. **orfordensis** C. & S., 1922
7010 **expunctaria** Grt., 1872
   brunneolineata (Hulst, 1900), n. syn.
7011 **brehmeata** (Grossb., 1907)

GEOMETRINAE

Nemoriini

CHLOROSEA Pack., 1873
7012 **nevadaria** Pack., 1873
   proutaria Pears., 1911
7013 **banksaria** Sperry, 1944
   a. **gracearia** Sperry, 1946
7014 **margaretaria** Sperry, 1944
7015 **roseitacta** Prout, 1912

NEMORIA Hbn., 1818
   RACHEOSPILA Gn., 1857
   APLODES Gn., 1857
   HIPPARCHISCUS Walsh, 1864
   EUNEMORIA Pack., 1873
    (November), preocc. by Pack.,
    1873 (July)
   ANAPLODES Pack., 1876
   ANNEMORIA Pack., 1876
7016 **pulcherrima** (B. & McD., 1916)
   naidaria (Swett, 1916)
7017 **mutaticolor** Prout, 1912
7018 **unitaria** (Pack., 1873)
   junctolinearia (Graef, 1881)

   hudsonaria (Tayl., 1906)
   unilinearia (Tayl., 1908)
7019 **latirosaria** (Pears., 1906)
7020 **aemularia** B. & McD., 1918
7021 **arizonaria** (Grt., 1883)
   olivearia Cass., 1927
7022 **daedalea** Fgn., 1969
7023 **viridicaria** (Hulst, 1880)
7024 **subsequens** Fgn., 1969
7025 **diamesa** Fgn., 1969
7026 **albaria** (Grt., 1883)
7027 **pistaciaria** (Pack., 1876)
   unistrigata (Gump., 1895)
7028 **extremaria** (Wlk., 1861)
7029 **elfa** Fgn., 1969
7030 **tuscarora** Fgn., 1969
7031 **catachloa** (Hulst, 1898)
7032 **outina** Fgn., 1969
7033 **lixaria** (Gn., 1857)
   inclusaria (Wlk., 1861)
   texana (Hulst, 1898)
   associaria (B. & McD., 1917)
   knobelaria (Cass., 1927)
7034 **saturiba** Fgn., 1969
7035 **darwiniata** (Dyar, 1904)
   oregonensis Cass., 1927
   mentastii Guedet, 1941
   a. **punctularia** B. & McD., 1918
   californica Prout, 1932
7036 **zelotes** Fgn., 1969
7037 **obliqua** (Hulst, 1898)
   bellonaria (Stkr., 1899)
   a. **hennei** (Sperry, 1953)
7038 **splendidaria** (Grossb., 1910)
7039 **strigataria** (Grossb., 1910)
7040 **zygotaria** (Hulst, 1886)
7041 **leptalea** Fgn., 1969
   delicataria (Dyar, 1908), preocc. by
    Mösch., 1881
7042 **caerulescens** Prout, 1912
7043 **intensaria** (Pears., 1911)
7044 **festaria** (Hulst, 1886)
7044.1 **albilineata** Cass., 1927
7045 **bifilata** (Wlk., 1863)
   abdominaria (B. & McD., 1917)
   a. **planuscula** Fgn., 1969
7046 **bistriaria** Hbn., 1818
   rubrolinearia (Pack., 1873)
   brunnearia (Pack., 1876)
   a. **rubromarginaria** (Pack., 1876)
7047 **rubrifrontaria** (Pack., 1873)
   packardaria (Grt., 1882)
7048 **mimosaria** (Gn., 1857)
   tractaria (Wlk., 1861)
   venustus (Walsh, 1864)
   latiaria (Pack., 1873)
   approximaria (Pack., 1873), ident.
   uncert.
   coniferaria (Pack., 1884)
7049 **glaucomarginaria** (B. & McD., 1917)
7050 **rindgei** Fgn., 1969

PHRUDOCENTRA Warr., 1895
   MELOCHLORA Warr., 1901
   NESIPOLA Warr., 1909
7051 **centrifugaria** (H.-S., 1870)
   protractaria (H.-S., 1870)
   hollandaria (Hulst, 1886)
   jaspidiaria (Hulst, 1886)
   anomalaria (Mösch., 1890)

   viridipurpurea (Hulst, 1898)
   heterospila (Hamp., 1904)
7052 **neis** (Druce, 1892)

DICHORDA Warr., 1900
7053 **iridaria** (Gn., 1857)
   remotaria (Wlk., 1861)
   a. **latipennis** (Hulst, 1898)
7054 **consequaria** (Hy. Edw., 1884)
   perpendiculata Warr., 1904
7055 **illustraria** (Hulst, 1886)
7056 **rectaria** (Grt., 1877)
   a. **cockerelli** Sperry, 1939

Dichordophorini

DICHORDOPHORA Prout, 1913
7057 **phoenix** (Prout, 1912)

Synchlorini

SYNCHLORA Gn., 1857
   EUNEMORIA Pack., 1873
7058 **aerata** (F., 1798)
   glaucaria (Gn., 1857)
   mimicata (Wlk., 1866)
   rubivora (Riley, 1869)
   rubrifrontaria Pack., 1873
   gracilaria (Pack., 1873)
   rubivoraria Pack., 1876, emend.
   rufofrontaria Gump., 1895, missp.
   a. **albolineata** Pack., 1873, rev. stat.
   albolinearia Pack., 1876. emend.
   b. **liquoraria** Gn., 1857, rev. stat.
   tricoloraria (Pack., 1874)
7059 **frondaria** Gn., 1858
   denticularia (Wlk., 1861), n. syn.
   minuata (Wlk., 1866)
   albicostaria (H.-S., 1870), n. syn.
   excurvaria Pack., 1873, rev. syn.
   pallida (Warr., 1900), n. syn.
   a. **avidaria** Pears., 1917
7060 **xysteraria** (Hulst, 1886), rev. stat.
   gerularia; auth., not Hbn., 1823
7060.1 **gerularia** (Hbn., 1823)
   ocellata (Stoll, 1791), preocc. by L.,
    1758
   stollaria (Gn., 1857)
   marginiplaga (Wlk., 1861)
   rufidorsaria (Snell., 1874)
   jucunda (Felder, 1875)
7061 **herbaria** (F., 1794)
   sitellaria (Gn., 1857)
   congruata (Wlk., 1861)
   hulstiana Dyar, 1901, n. syn.
7062 **irregularia** (B. & McD., 1918)
7063 **noel** (Sperry, 1949)
7064 **cupedinaria** (Grt., 1880)
   louisa Hulst, 1898

CHETEOSCELIS Prout, 1912
   MEROCHLORA Prout, 1912
7065 **bistriaria** (Pack., 1876)
   undinaria (Stkr., 1878)
   clarkei Sperry, 1939
7066 **pectinaria** (Grossb., 1910)
7067 **faseolaria** (Gn., 1857)
   perviridaria (Pack., 1876)
7068 **graefiaria** (Hulst, 1886)
   eutraphes Prout, 1912

Lophochoristini

LOPHOCHORISTA Warr., 1904
7069 **lesteraria** (Grossb., 1910)

EUEANA Prout, 1912
7070 **niveociliaria** (H.-S., 1870)
    saltusaria (Hulst, 1886)

Hemitheini

CHLOROCHLAMYS Hulst, 1896
7071 **chloroleucaria** (Gn., 1857)
    indiscriminata (Wlk., 1863)
    densaria (Wlk., 1863)
    deprivata (Wlk., 1863)
    desolataria (H.-S., 1870)
    flavilineata (Riley, 1870)
    rectilinea (Zell., 1872)
7072 **triangularis** Prout, 1912
7073 **appellaria** Pears., 1911
    rubromediaria C. & S., 1925
    hesperia Sperry, 1951
7074 **phyllinaria** (Zell., 1872)
    zelleraria (Pack., 1876)
    vertaria Pears., 1908
    curvifera Prout, 1912
    fletcheraria Sperry, 1949

CHLOROPTERYX Hulst, 1896
7075 **tepperaria** (Hulst, 1886)
7076 **nordicaria** (Schaus, 1901)
7077 **paularia** (Mösch., 1886)
    punctata (Warr., 1904)

XEROCHLORA Fgn., 1969
7078 **viridipallens** (Hulst, 1896)
    volantaria (Pears., 1906)
7079 **inveterascaria** (Swett, 1907)
7080 **martinaria** (Sperry, 1948)
7081 **masonaria** (Schaus, 1897)
    hyperalla (Prout, 1933)
7082 **mesotheides** Fgn., 1969

HEMITHEA Dup., 1829
    CHLOROCHROMA Dup., 1845
7083 **aestivaria** (Hbn., 1799)
    strigata (O. F. Müller, 1764), preocc.
      by Scop., 1763

HETHEMIA Fgn., 1969
7084 **pistasciaria** (Gn., 1857)
    euchloraria (Gn., 1857)
    subcroceata (Wlk., 1862)
    superata (Wlk., 1866)
    gratata (Pack., 1876)
    dyarii (Hulst, 1900)
    a. **insecutata** (Wlk., 1862), n. stat.
    auranticolorata (Stkr., 1899)

MESOTHEA Warr., 1901
    CHLOROCHROMA Gump., 1895,
      preocc. by Dup., 1845
7085 **incertata** (Wlk., 1862)
    oporaria (Zell., 1872)
    a. **viridipennata** (Hulst, 1896)
    marinaria (Stkr., 1899)

STERRHINAE
by CHARLES V. COVELL, Jr

Sterrhini

EUMACRODES Warr., 1905
    EUPTYCHOPODA Prout, 1910
7086 **yponomeutaria** (Gn., 1857)
    semidecora (Wlk., 1863)

EUACIDALIA Pack., 1873
    OBELOPTERYX Warr., 1906
7087 **sericearia** Pack., 1873
    sericeata auth., missp.
    albescens (Cass., 1931), n. syn.
7088 **puerta** Cass., 1931, rev. stat.
7089 **quakerata** Cass., 1927
7090 **brownsvillea** Cass., 1931
7091 **nigridaria** Cass., 1931, ident. uncert.

PROTOPROUTIA McD., 1939
7092 **rusticaria** McD., 1939
7093 **laredoata** (Cass., 1931)

LOBOCLETA Warr., 1906
    METASIOPSIS Prout, 1910, rev.
      syn.
7094 **ossularia** (Gey., 1837)
    ossulata auth., missp.
    sublataria (Gn., 1857)
    temnaria (Gn., 1857)
    magniferaria (Wlk., 1861)
    favillifera (Wlk., 1866)
    repletaria (Wlk., 1866)
7095 **granitaria** (Pack., 1871)
    granitata auth., missp.
7096 **quaesitata** (Hulst, 1880)
    ferruginata (Cass., 1931)
    chockolatella (Cass., 1931), n. syn.
7097 **plemyraria** (Gn., 1857), rev. stat.
    balistaria (Gn., 1857), rev. syn.
    balistraria auth., missp.
    perirrorata (Pack., 1873), n. syn.
    rufescens (Hulst, 1896)
    moricaria Dyar, 1923, rev. syn.
    helena (Cass., 1931), n. syn.
7098 **lanceolata** (Hulst, 1896)
7099 **griseata** (Cass., 1931)
7100 **peralbata** (Pack., 1873)
    longipennata (Pack., 1873)

IDAEA Tr., 1825
    ACIDALIA; Tr., 1825, not Hbn.,
      1819
    STERRHA Hbn., 1825
    ARRHOSTIA Hbn., 1825
    PYCTIS Hbn., 1825
    PTYCHOPODA Curt., 1826
    HYRIA Steph., 1829
    ANIA Steph., 1831
    GONIACIDALIA Pack., 1873, n.
      syn.
    PYTHODORA Meyr., 1886
    JANARDA Moore, 1888
    MNESTERODES Meyr., 1889
    XENOCENTRIS Meyr., 1889
    ARGIA Gump., 1890
    GNIDIA Gump., 1890, preocc. by
      Koch, 1839
    PELAGIA Gump., 1890

    APHROGENEIA Gump., 1890
    ANDRAGRUPOS Hamp., 1891
    CARPHOXERA Riley, 1891
    LOPHOPHLEPS Hamp., 1891
    CNIDIA Gump., 1892, missp.
    LIMERIA Stgr., 1892
    SYNOMILA Hulst, 1896, rev. syn.
    PTENOPODA Hulst, 1896
    HEMIPOGON Warr., 1897
    LEPTACME Warr., 1897
    STROPHOPTILA Warr., 1897
    BRACHYPROTA Warr., 1897
    POLYGRAPHODES Warr., 1897
    CACORISTA Warr., 1899
    ANTEOIS Warr., 1900
    HYRIOGONA Warr., 1900
    NEOCHRYSA Warr., 1900
    POGONOGYA Warr., 1900
    PROSPASTA Warr., 1900
    CYSTEOPHORA Hulst, 1900, rev.
      syn.
    THYSANOTRICHA Warr., 1903
    DEINOPYGIA Warr., 1904
    SCHEMATORHAGES Warr., 1905
    LOBURA Warr., 1906
    OMOPERA Warr., 1906
    ARGYROSCELIA Warr., 1907
    PAREUPITHEX Warr., 1907
    EUPHENOLIA Grossb., 1907, rev.
      syn.
    HIRTHESTES Dognin, 1914
7101 **minuta** (Schaus, 1901), n. comb.
7102 **bonifata** (Hulst, 1887), n. comb.
    ptelearia (Riley, 1891), n. comb.
    delicata (Hulst, 1896), n. comb.
7103 **nibseata** (Cass., 1931), n. comb.
7104 **microphysa** (Hulst, 1896), n. comb.
7105 **scintillularia** (Hulst, 1888), n. comb.
    scintillaria auth., missp.
7106 **insulensis** (Rindge, 1958), n. comb.
7107 **pervertipennis** (Hulst, 1900), n.
    comb.
7108 **furciferata** (Pack., 1873), n. comb.
    parvularia (Hulst, 1888), n. syn.
7109 **celtima** (Schaus, 1901), n comb.
7110 **basinta** (Schaus, 1901), n. comb.
    pallimedia (Grossb., 1908), n. comb.,
      rev. syn.
7111 **skinnerata** (Grossb., 1907), n. comb.
7112 **productata** (Pack., 1876), n. comb.
    subochreata (Hulst, 1887), n. comb.,
      n. syn.
    australis (Hulst, 1896), n. comb., n.
      syn.
7113 **miranda** (Hulst, 1896), n. comb.
7114 **demissaria** (Hbn., 1831)
    inclusaria (Wlk., 1863)
    ferrugata (Pack., 1876)
    russata (Hulst, 1887)
    a. **columbia** (McD., 1927)
7115 **eremiata** (Hulst, 1887), n. comb.
    flavescens (Hulst, 1896), n. comb.
7116 **gemmata** (Pack., 1876), n. comb.
7117 **occidentaria** (Pack., 1874), n. comb.
    occidentata auth., missp.
7118 **hilliata** (Hulst, 1887), n. comb.
7119 **micropterata** (Hulst, 1900), n. comb.
7120 **violacearia** (Wlk., 1861), n. comb.
    lacteola (Lintner, 1878), n. syn.
    lacteolata auth., missp.

floridaria (Hulst, 1888)
nimbicolor (Hulst, 1896)
benubia (Cass., 1931), n. syn.
7121 **ostentaria** (Wlk., 1861), n. comb.
marceata (Cass., 1931), n. syn.
7122 **tacturata** (Wlk., 1861), n. comb.
eburneata; auth., not Gn., 1857
albidula (Hulst, 1896), n. comb.
7123 **obfusaria** (Wlk., 1861)
obfustaria auth., missp.
punctofimbriata (Pack., 1873), rev. syn.
pimacoata (Cass., 1931), n. syn.
7124 **retractaria** (Wlk., 1861), n. comb.
pallida (Hulst, 1896), n. comb., n. syn.
7125 **rotundopennata** (Pack., 1876), n. comb.
hanhami (Hulst, 1898), n. comb.
7126 **dimidiata** (Hufn., 1767)
scutulata (D. & S., 1775)
scutata (F., 1787), n. comb.
scutularia (Hbn., 1798)
delictata (Prout, 1913), n. comb.
roseata (Turati, 1913), n. comb.
antiaurica (Wehrli, 1931), n. comb.
tauricola (Wehrli, 1934), n. comb.

PAOTA Hulst, 1896
7127 **fultaria** (Grt., 1882)

PIGIA Gn., 1857
ARCOBARA Wlk., 1863, rev. syn.
7128 **multilineata** Hulst, 1887
7129 **perlineata** (Schaus, 1913), rev. stat.

ODONTOPTILA Warr., 1897
7130 **obrimo** (Druce, 1892)
siculodaria Schaus, 1901, rev. syn.
albiserpentata (Pears., 1906), rev. syn.

PTYCHAMALIA Prout, 1932
7131 **dorneraria** (B. & McD., 1913)

Cosymbiini

PLEUROPRUCHA Mösch., 1890
APALLACTA Mösch., 1890
DEPTALIA Hulst, 1896
7132 **insulsaria** (Gn., 1857)
insularia auth., missp.
placidaria (Gn., 1857)
invariata (Wlk., 1863)
persimilata (Grt., 1863)
7133 **asthenaria** (Wlk., 1861)
imparata (Wlk., 1863)

CYCLOPHORA Hbn., 1822
COSYMBIA Hbn., 1823
LEUCOPHTHALMIA Hbn., 1823
CODONIA Hbn., 1823
EPHYRA Dup., 1829, preocc. by Peron & Lesueur, 1810
MATELLA Gistl, 1848
ZONOSOMA Led., 1853
EUEPHYRA Pack., 1873
7134 **culicaria** (Gn., 1857)
7135 **dataria** (Hulst, 1887)
piazzaria (W. S. Wright, 1924), rev. syn.
microps (Prout, 1936), n. syn.

7136 **packardi** (Prout, 1936)
7137 **myrtaria** (Gn., 1857)
ignotaria (Wlk., 1863)
triseriata (Prout, 1936)
7138 **benjamini** (Prout, 1936)
7139 **pendulinaria** (Gn., 1857)
quadriannulata (Wlk., 1863)
dilucidaria (Rothke, 1920)
nigricaria (Rothke, 1920), melanic f.
griseor (McD., 1927), rev. syn.
7140 **nanaria** (Wlk., 1861)
nanularia (H.-S., 1870)
serrulata (Pack., 1873), rev. syn.
obscura (Druce, 1898)

SEMAEOPUS H.-S., 1855
CYPHOPTERYX Gn., 1857
CNEMODES Gn., 1857
ISSA Wlk., 1867
LIGONIA Mösch., 1882
DICHROMATOPODIA Warr., 1895
DYSEPHYRA Warr., 1895, rev. syn.
HETEREPHYRA Warr., 1895
CHAROMMATAEA Hulst, 1896
XENOSTIGMA Warr., 1900
SCHISTOCOLPIA Warr., 1906
PARADMETA Warr., 1907
PARAZEUXIS Warr., 1907
DASYCOSYMBIA Grossb., 1912, rev. syn.
7141 **ella** (Hulst, 1896)
ellatina (Hulst, 1896), form
boweri (Sperry, 1940), n. syn.
7142 **gracilata** (Grossb., 1912)
7143 **cantona** (Schaus, 1901)
7144 **caecaria** (Hbn., 1823)
punctata (Stoll, 1790), preocc. by Cl., 1759
fartaria (Gn., 1857)
occipitaria (H.-S., 1870)
distinctata (Warr., 1900)
grisea (Warr., 1906)
rubella (Warr., 1907)
7145 **marginata** (Schaus, 1901)

Timandrini

HAEMATOPIS Hbn., 1823
HAEMATOPSIS auth., missp.
7146 **grataria** (F., 1798)
saniaria Hbn., 1823
successaria (Wlk., 1860)
annettearia Haim., 1907, form

CALOTHYSANIS Hbn., 1823
TIMANDRA Dup., 1829
BRADYEPETES Steph., 1831
7147 **amaturaria** (Wlk., 1866)
effusaria Prout, 1936

Scopulini

ACRATODES Gn., 1857
7148 **suavata** (Hulst, 1900)
fusaria (Hamp., 1903)
roseicosta (B. & McD., 1913), n. syn.
davisi (Grossb., 1917), n. syn.

SCOPULA Schrank, 1802
ACIDALIA; Tr., 1823, not Hbn., 1819
LEPTOMERIS Hbn., 1825
CRASPEDIA Hbn., 1825
DOSITHEA Dup., 1829
CYMATIDA Sodoffsky, 1837
CYMATOIDES Agassiz, 1847
PYLARGE H.-S., 1855
PHYLETIS Gn., 1857
LYCAUGES Butler, 1879
TRICHOCLADA Meyr., 1886
RUNECA Moore, 1888
LONGULA Stgr., 1892
SYNELYS Hulst, 1896
INDUNA Warr., 1897
ACIDALINA Stgr., 1898
PLEIONOCENTRA Warr., 1898
CHLOROCRASPEDIA Warr., 1899
LIPOCENTRIS Warr., 1905
PSILEPHYRA Bastelberger, 1909
HOLARCTIAS Prout, 1913, n. syn.
7149 **lautaria** (Hbn., 1831)
myrmidonata (Gn., 1857)
minutularia (Hulst, 1880)
7150 **eburneata** (Gn., 1858)
subsignaria (Wlk., 1861)
chionaeata (H.-S., 1870)
blandula (Warr., 1906)
7151 **aemulata** (Hulst, 1896)
tawneata (Cass., 1931)
7152 **compensata** (Wlk., 1861)
obluridata (Hulst, 1887)
7153 **apparitaria** (Wlk., 1861)
responsaria (Wlk., 1861)
atomaria (Warr., 1897)
approbata (Warr., 1900)
trias (Warr., 1904)
tincta (Warr., 1904)
7154 **plantagenaria** (Hulst, 1887)
canthema (Schaus, 1901)
hieronyma Prout, 1922
7155 **benitaria** (B. & McD., 1913)
7156 **umbilicata** (F., 1794)
indoctaria (Wlk., 1861)
nigroapicata (Thierry-Mieg, 1892)
crenatilinea (Warr., 1901)
cugia (Schaus, 1901)
peruviana Prout, 1922
7157 **cacuminaria** (Morr., 1874)
7158 **purata** (Gn., 1857)
nigrocandida (Hulst, 1898)
7159 **limboundata** (Haw., 1809)
alabastaria; auth., not **alabastraria** Hbn., 1816
enucleata (Gn., 1857), form
restrictata (Wlk., 1861)
reconditaria (Wlk., 1861)
mensurata (Wlk., 1866)
continuaria (Wlk., 1866)
nigrodiscalis (Hulst, 1898), n. syn.
relevata (Swett, 1907), form
adornata (Prout, 1907)
7160 **timandrata** (Wlk., 1861)
rufilinearia (Wlk., 1861)
7161 **ordinata** (Wlk., 1861)
puraria (Wlk., 1861)
candidaria (Pack., 1873)
7162 **ancellata** (Hulst, 1887)

7163 **fuscata** (Hulst, 1887)
7164 **junctaria** (Wlk., 1861)
    a. **quinquelinearia** (Pack., 1871)
      impunctata (Warr., 1904)
      vestalialis (B. & McD., 1913)
    a. **johnstonaria** McD., 1941, emend.
      johnsonaria McD., 1941, incorr.
      orig. spell.
7165 **quadrilineata** (Pack., 1876)
    quadrilinearia (Hulst, 1896), missp.
    persimilis (Hulst, 1898)
7166 **frigidaria** (Mösch., 1860)
    defixaria (Wlk., 1861)
    impauperata (Wlk., 1861)
    arcticaria (Wlk., 1863)
    okakaria (Pack., 1867)
7167 **siccata** McD., 1939
7168 **septentrionicola** McD., 1939
7169 **inductata** (Gn., 1857)
    anticaria (Wlk., 1860)
    suppressaria (Wlk., 1862), rev. syn.,
      ident. uncert.
    consecutaria (Wlk., 1866)
    sobria (Wlk., 1866), dark form
    delicata (Cass., 1931)
    oliveata (Cass., 1931)
7170 **luteolata** (Hulst, 1880)
    subfuscata (Tayl., 1906)
7171 **sideraria** (Gn., 1857)
    magnetaria (Gn., 1857)
    californiaria (Pack., 1871), form
    pacificaria (Pack., 1871)
    rubrolinearia (Pack., 1873)
    bucephalaria (B. & McD., 1918),
      preocc. by Chrétien, 1909
    chretieni (B. & Benj., 1923), form
7172 **sentinaria** (Gey., 1837), n. comb.
    spuriaria (Christoph, 1858)
    gracilior (Butler, 1893)

LEPTOSTALES Mösch., 1890, rev.
    stat.
    XYSTROTA Hulst, 1896, n. syn.
    SCELOLOPHIA Hulst, 1896, n.
      syn.
    CALYPTOCOME Warr., 1900, n.
      syn.
    WAUCHULA Hulst, 1900, n. syn.
7173 **pannaria** (Gn., 1857), n. comb.
    tremularia (Wlk., 1863), n. comb.
    purpurissata (Grt., 1871), n. comb.
    formosa (Hulst, 1896), n. comb.
    borrigaria (Warr., 1906), n. comb.
7174 **crossii** (Hulst, 1900), n. comb.
    crossi, missp.
7175 **hepaticaria** (Gn., 1857), n. comb.
    purpurascens (Hulst, 1900), n. comb.,
      n. syn.
7176 **rubrotincta** (Hulst, 1900), ident.
    uncert.
7177 **laevitaria** (Gey., 1837), n. comb.
    floridata (Pack., 1873), n. comb.
7178 **oblinataria** Mösch., 1890
    scintillans (Warr., 1904)
7179 **rubromarginaria** (Pack., 1871), n.
    comb.
    rubromarginata, missp.
    erythrata (Hulst, 1880), n. comb.
    volucrata (Hulst, 1887), n. comb.
7180 **ferruminaria** (Zell., 1872)

LOPHOSIS Hulst, 1896
7181 **labeculata** (Hulst, 1887)
    roseotincta (Hulst, 1896)

LARENTIINAE

Hydriomenini

DYSSTROMA Hbn., 1825
    CHLOROCLYSTA Hbn., 1825
    POLYPHASIA Steph., 1831
    EUTHALIA Steph., 1831, preocc.
      by Hbn., 1819
7182 **citrata** (L., 1761)
    marmorata (F., 1794)
    immanata (Haw., 1809)
    omicronata (Don., 1811), ident.
      uncert.
    amaenata (Steph., 1831)
    immanaria (Doubleday, 1849)
    mulleolata (Hulst, 1881), n. syn.
    a. **glacialis** (Hulst, 1898), n. stat.
      longula (Hulst, 1898), n. syn.
7183 **hewlettaria** W. S. Wright, 1927, n.
    stat.
7184 **sobria** Swett, 1917
    subumbrata Swett, 1917
    swetti Blkmre., 1921
7185 **suspectata** (Mösch., 1874), rev. stat.
    infuscata; McD., 1946, not
      Tengström, 1869
    a. **mackieata** C. & S., 1923, n.
      stat.
7186 **ochrofuscaria** Swett, 1917
7187 **truncata** (Hufn., 1767), extralim.
    centumnotata (Schulze, 1775)
    russata (D. & S., 1775)
    perfuscata (Haw., 1809)
    commanotata (Haw., 1809)
    saturata (Steph., 1831)
    russaria (Bdv., 1840)
    a. **traversata** (Kellicott, 1886)
7188 **walkerata** (Pears., 1909)
7189 **hersiliata** (Gn., 1857)
    flammifera (Wlk., 1862)
    mirandata (Tayl., 1910), rev. syn.
    a. **cervinifascia** (Wlk., 1862)
    b. **manitoba** McD., 1939
      infumata McD., 1939, n. syn.
7190 **rutlandia** McD., 1943
    perfuscata McD., 1943, preocc. by
      Haw., 1809
7191 **formosa** (Hulst, 1896)
    boreata (Tayl., 1910), preocc. by
      Curt., 1836
    admiranda McD., 1927, n. syn.
    gilvifasciata McD., 1939, n. syn.
    a. **occidentata** (Tayl., 1910), rev.
      stat.
    mutata (Tayl., 1910), n. syn.
7192 **colvillei** Blkmre., 1926
7193 **rectiflavata** McD., 1941
7194 **brunneata** (Pack., 1867)
    casloata (Tayl., 1910), n. syn.
    a. **ethela** (Hulst, 1896), n. stat.
7195 **mancipata** (Gn., 1857)
    subochreata (Pack., 1871)
    leoninata (Pack., 1871), n. syn.
    a. **hulstata** (Tayl., 1907), n. stat.
    b. **decorata** (Tayl., 1910), n. stat.

EULITHIS Hbn., 1821
    LYGRIS Hbn., 1825
    EUPHIA Hbn., 1825
    STEGANOLOPHIA Steph., 1829
    PHYLACE Hulst, 1896, n. syn.
    NEOLEXIA Hulst, 1896, n. syn.
7196 **diversilineata** (Hbn., 1813)
7197 **gracilineata** (Gn., 1857)
7198 **luteolata** (Hulst, 1896), n. comb.
7199 **propulsata** (Wlk., 1862), n. comb.
    packardata (Lint., 1878), n. comb.
    ximena (Ellsworth, 1918), n. comb.
7200 **populata** (L., 1758)
    dotata (L., 1758)
    popularia (Bdv., 1840)
7201 **testata** (L., 1761)
    colorata (F., 1794)
    achatinata (Hbn., 1796–99)
    achatinaria (Bdv., 1840)
    insulicola Stgr., 1901
7202 **mellinata** (F., 1787)
    associata (Bkh., 1794)
    marmorata (Hbn., 1796–99)
    spinaciata (Haw., 1809)
    marmoraria (Bdv., 1840)
7203 **molliculata** (Wlk., 1862)
    remotata (Wlk., 1862), n. comb., rev.
      syn.
7204 **destinata** (Mösch., 1860)
    lugubrata (Mösch., 1862), n. comb.
    similis (Wlk., 1862), n. comb.
    nubilata (Pack., 1867), n. comb.
    a. **triangulata** (Pack., 1873), n. comb.
      montanata (Pack., 1873), n. comb.
    b. **schistacea** (Warr., 1901), n. comb.
      griseata (C. & S., 1922), n. comb.,
      n. syn.
      bowmani (C. & S., 1922), n. comb.,
      n. syn.
    c. **harveyata** (Tayl., 1906), n. comb.
7205 **flavibrunneata** (McD., 1943), n.
    comb.
    unicolorata (McD., 1943), n. comb.,
      n. syn.
7206 **explanata** (Wlk., 1862)
    cunigerata (Wlk., 1863), n. comb.,
      form
    disjunctaria (Pack., 1873), n. comb.
    brunneomaculata (Bates, 1886), n.
      comb.
7207 **xylina** (Hulst, 1896)
    a. **speciosa** (Hulst, 1896), n. comb.
7208 **serrataria** (B. & McD., 1917), n.
    comb.

EURHINOSEA Pack., 1873
7209 **flavaria** Pack., 1873

EUSTROMA Hbn., 1825
    ANTEPIRRHOE Warr., 1905
7210 **semiatrata** (Hulst, 1881)
    nubilata; Pack., 1871, not Pack.,
      1867
    macdunnoughi Blkmre., 1920, form
7211 **fasciata** B. & McD., 1918
7212 **atrifasciata** (Hulst, 1888)
    delimitata (Warr., 1895)

ECLIPTOPERA Warr., 1894
    DIACTINIA Warr., 1898

7213 **silaceata** (D. & S., 1875), extralim.
  insulata (Haw., 1809)
  cuneata (Don., 1810)
  silacearia (Bdv., 1840)
  a. **albolineata** (Pack., 1873), n. comb.
    deflavata; auth., part, not Stgr., 1871
7214 **atricolorata** (G. & R., 1867), n. comb.
  atrocolorata auth., missp.

COLOSTYGIA Hbn., 1825
  CALOSTIGIA auth., missp.
  AMOEBE Hbn., 1825
  AMAEBE auth., missp.
  ERINOBIA Steph., 1850
7215 **turbata** Hbn., 1826, extralim.
  a. **circumvallaria** (Tayl., 1906)

PLEMYRIA Hbn., 1825
7216 **georgii** Hulst, 1896
  a. **benesignata** (B. & McD., 1917)

THERA Steph., 1831
7217 **juniperata** (L., 1758)
  juniperaria (Bdv., 1840)
  procteri Brower, 1940
7218 **contractata** (Pack., 1873)
7219 **otisi** (Dyar, 1904)
7220 **latens** B. & McD., 1917

CERATODALIA Pack., 1876
7221 **gueneata** Pack., 1876
  excurvata (Grt., 1883)

HYDRIOMENA Hbn., 1825
  YPSIPETES Steph., 1829
  HYPSIPETES Steph., 1831, emend. preocc. by Vigors, 1831
  KARACIDARIA Matsumura, 1925
7222 **tuolumne** B. & McD., 1917
7223 **exculpata** B. & McD., 1917
  tribulata B. & McD., 1917, form
  a. **josepha** McD., 1954
  b. **nanata** McD., 1954
7224 **expurgata** B. & McD., 1918
  a. **franclemonti** McD., 1952
  b. **nicolensis** McD., 1954
  c. **alticola** McD., 1954
7225 **shasta** B. & McD., 1917
7226 **borussata** B. & McD., 1918
  brunneata B. & McD., 1918, *lapsus calami*
7227 **henshawi** Swett, 1912
7228 **irata** Swett, 1910
  niveifascia Swett, 1916, form
  niveifasciata Swett, 1918
  a. **lolata** McD., 1954
  b. **quaesitata** B. & McD., 1918
7229 **perfracta** Swett, 1910
  a. **marmorata** B. & McD., 1917
  b. **centralis** McD., 1952
  c. **monoensis** McD., 1952
7230 **charlestonia** McD., 1954
7231 **marinata** B. & McD., 1917
  a. **exasperata** B. & McD., 1917
7232 **edenata** Swett, 1909
  a. **prasinata** McD., 1954
  b. **baueri** McD., 1954
  c. **grandis** B. & McD., 1917

  d. **olivata** W. S. Wright, 1916
  e. **pallidata** W. S. Wright, 1916
  f. **indistincta** McD., 1952
7233 **furtivata** McD., 1939
7234 **johnstoni** McD., 1954
7235 **divisaria** (Wlk., 1860)
  a. **brunnescens** McD., 1954
  b. **frigidata** (Wlk., 1863)
7236 **renunciata** (Wlk., 1862)
  a. **columbiata** Tayl., 1906
  b. **pernigrata** B. & McD., 1917
  c. **viridescens** McD., 1954
7237 **transfigurata** Swett, 1912
  a. **manitoba** B. & McD., 1917
7238 **bistriolata** (Zell., 1872)
7239 **pluviata** (Gn., 1857)
  a. **meridianata** McD., 1954
7240 **obliquilinea** B. & McD., 1917
7241 **rita** McD., 1954
7242 **arizonata** B. & McD., 1917
7243 **albimontanata** McD., 1939
7244 **sierrae** B. & McD., 1917
7245 **nevadae** B. & McD., 1917
7246 **californiata** Pack., 1871
7247 **crokeri** Swett, 1910
  a. **comstocki** McD., 1944
    waltoni McD., 1944, form
7248 **glaucata** (Pack., 1874)
7249 **muscata** B. & McD., 1917
7250 **chiricahuata** Swett, 1909
7251 **modestata** B. & McD., 1917
7252 **mississippiensis** McD., 1952
7253 **feminata** McD., 1944
7254 **ruberata** (Freyer, 1831)
  variegata (Prout, 1914), form
  a. **pallula** McD., 1944
7255 **macdunnoughi** Swett, 1918
7256 **septemberata** McD., 1952
7257 **furcata** (Thunb., 1784)
  sordidata (F., 1794)
  elutata (Hbn., 1796–99)
  fuscoundata (Don., 1806), form
  elutaria (Bdv., 1840)
  a. **fergusoni** McD., 1954
7258 **quinquefasciata** (Pack., 1871)
  viridata Hulst, 1903, form
  periclata Swett, 1910, form
7259 **catalinata** McD., 1943
7260 **costipunctata** B. & McD., 1912
7261 **albifasciata** (Pack., 1874)
  resecta Swett, 1910, form
  puncticaudata B. & McD., 1917, form
  beldenae Guedet, 1941
  a. **victoria** B. & McD., 1917
  b. **reflata** Grt., 1882
    abacta (Hulst, 1898)
7262 **cochiseata** Swett, 1909
  cochizeata B. & McD., 1917, missp.
  swetti B. & McD., 1917, form
7263 **speciosata** (Pack., 1874)
  agassizi Swett, 1910, form
  taylori Swett, 1910, form
  ameliata Swett, 1915, form
7264 **morosata** B. & McD., 1917
  gravis McD., 1954, form
7265 **barnesata** Swett, 1909
7266 **cyriadoides** McD., 1954
  cyriades; B. & McD., 1916, not Druce, 1896

7267 **sperryi** McD., 1952
7268 **bryanti** McD., 1943
7269 **clarki** W. S. Wright, 1920
7270 **magnificata** Tayl., 1906
7271 **gracillima** McD., 1944
7272 **regulata** Pears., 1909
7273 **furculoides** B. & McD., 1917
7274 **peratica** Rindge, 1956
7275 **similaris** Hulst, 1896
  glenwoodata Swett, 1909
  a. **terminipunctata** B. & McD., 1916
7276 **nubilofasciata** (Pack., 1871)
  sparsimacula Hulst, 1896, form
  banavahrata Stkr., 1899
  scalata Warr., 1904
  raptata Swett, 1910, form
  cumulata Swett, 1910, form
  cupidata Swett, 1910, form
  vulnerata Swett, 1910, form
7277 **manzanita** Tayl., 1906

HYMENODRIA McD., 1954
7278 **mediodentata** (B. & McD., 1911)

ERSEPHILA Hulst, 1896
7279 **indistincta** Hulst, 1898
7280 **grandipennis** Hulst, 1896

CARPTIMA Pears., 1906
7281 **hydriomenata** Pears., 1906

CYCLICA Grt., 1882
  CATOCLOTHIS Hulst, 1896, n. syn.
7282 **frondaria** Grt., 1882

GROSSBECKIA B. & McD., 1912
7283 **semimaculata** B. & McD., 1912

EUTREPSIA H.-S., 1855
7284 **inconstans** (Gey., 1837)
  cephisaria (Grt., 1881)

TRIPHOSA Steph., 1829
  UMBROSINA Bruand, 1847
7285 **haesitata** (Gn., 1857)
  pustularia Hy. Edw., 1885
  a. **affirmaria** (Wlk., 1860)
    indubitata (Grt., 1882)
7286 **affirmata** (Gn., 1857)
7287 **californiata** (Pack., 1871)
  packardata (Grossb., 1907)
7288 **bipectinata** B. & McD., 1917

MONOSTOECHA Fletcher, 1978
  MONOTAXIS Hulst, 1898, preocc. by Bennet, 1830
7289 **semipectinata** (Hulst, 1898)

CORYPHISTA Hulst, 1896
7290 **meadii** (Pack., 1874)
  badiaria (Hy. Edw., 1885), form
  optimata (Stkr., 1899)
  a. **atlantica** Mun., 1954
  b. **fumosa** Comst., 1967

HYDRIA Hbn., 1822
  CALOCALPE Hbn., 1825
  EUCOSMIA Steph., 1831

7291 **undulata** (L., 1758)
undularia (Bdv., 1840)
7292 **prunivorata** (Fgn., 1955)

RHEUMAPTERA Hbn., 1822
EULYPE Hbn., 1825
MELANIPPE Dup., 1829
7293 **hastata** (L. 1758)
hastaria (Bdv., 1840)
a. **gothicata** (Gn., 1857)
furcifascia (Wlk., 1862)
7294 **subhastata** (Nolcken, 1870)
moestata (Nolcken, 1870)
a. **albodecorata** (Blkmre., 1920)
b. **confusa** (McD., 1937)
c. **stygiata** (McD., 1937)

ARCHIRHOE Herbulot, 1951
CAMPTOGRAMMA; auth., not
Steph., 1831
7295 **neomexicana** (Hulst, 1896)
elisata (Stkr., 1899)
7296 **indefinata** (Grossb., 1907)
7297 **associata** (McD., 1941)
7298 **multipunctata** (Tayl., 1906)

PTEROCYPHA H.-S., 1855
7299 **floridata** (Wlk., 1863), n. comb.
defensata Wlk., 1862, indent. uncert.
decertaria (H.-S., 1870), ident.
uncert.
australata (Hulst, 1886), n. comb., n.
syn.

ENTEPHRIA Hbn., 1825
GLAUCOPTERYX Hbn., 1825
PHAESYLOIDES Bruand, 1847
TRICHOCHLAMYS Hulst, 1896,
n. syn.
DASYURIS; auth., part, not Gn.,
1857
7300 **aurata** (Pack., 1867)
7301 **multivagata** (Hulst, 1881)
7302 **takuata** (Tayl., 1908)
7303 **lagganata** (Tayl., 1908)
7304 **nigrescens** (Hulst, 1900)
7305 **inventaraia** (Grt., 1882)
7306 **polata** (Dup., 1830), extralim.
a. **brullei** (Lefebvre, 1836)
fumidotata (Wlk., 1863)
b. **punctipes** (Curt., 1835)
c. **bradorata** (Mun., 1951)
d. **ursata** (Mun., 1951)
e. **kidluitata** (Mun., 1951)
f. **aleutiata** (Mun., 1951)

MESOLEUCA Hbn., 1825
7307 **ruficillata** (Gn., 1857)
7308 **gratulata** (Wlk., 1862)
brunneicillata (Pack., 1874)
bakeri Guedet, 1941, n. syn.
a. **latialbata** B. & McD., 1917

SPARGANIA Gn., 1857
7309 **viridescens** (Grossb., 1910)
7310 **aurata** (Grt., 1882)
daira (Druce, 1893)
7311 **bellipicta** Warr., 1901
illustrata B. & McD., 1917
7312 **magnoliata** Gn., 1857

ziczacata (Wlk., 1862)
placidata (Wlk., 1862)
incommodata (Wlk., 1862)
cumatilis (G. & R., 1867)
a. **pernotata** (Hulst, 1898)
b. **quadripunctata** (Pack., 1871)
c. **ruptata** B. & McD., 1917
7313 **luctuata** (D. & S., 1775)
lugubrata (Stgr., 1871)
kodiakata (Pack, 1874)
a. **obductata** (Mösch., 1860)
concordata (Wlk., 1862)

HAMMAPTERA H.-S., 1855
7314 **parinotata** (Zell., 1872)
a. **densata** (Grossb., 1909)

PSALIODES Gn., 1857
7315 **fervescens** Dyar, 1920

PERIZOMA Hbn., 1825
EMMELESIA Steph., 1831
7316 **basaliata** (Wlk., 1862)
explagiata (Wlk., 1863)
7317 **grandis** (Hulst, 1896)
saanichata (Swett, 1915), form
saawichata (Swett, 1915), missp.
7318 **alaskae** (Hulst, 1896)
7319 **actuata** (Pears., 1909)
7320 **alchemillata** (L., 1758)
rivulata (D. & S., 1775)
nassata (F., 1787)
alchemillaria (Bkh., 1794)
7321 **interrupta** (Grossb., 1910)
7322 **ochreata** (Grossb., 1910)
7323 **oxygramma** (Hulst, 1896)
tahoensis B. & McD., 1916
7324 **curvilinea** (Hulst, 1896)
occidens (Hulst, 1898)
a. **foxi** (W. S. Wright, 1924)
7325 **costiguttata** (Hulst, 1896)
7326 **epictata** B. & McD., 1916
7327 **ablata** (Hulst, 1896)
7328 **custodiata** (Gn., 1857)
gueneeata (Pack., 1876)
carnata (Pack., 1874), form
polygrammata (Hulst, 1896), form

ANTICLEA Steph., 1831
EAROPHILA Gump., 1887
LAURENTIA W. S. Wright, 1916,
preocc. by Rag., 1888 (missp. of
Larentia?)
7329 **vasiliata** Gn., 1857
vasaliata auth., missp.
spilosaria (Wlk., 1860), n. comb.
rigidata (Wlk., 1863), n. comb.
niveifasciata (Hulst, 1900), n. comb.,
form
7330 **multiferata** (Wlk., 1863), n. comb.
7331 **switzeraria** (W. S. Wright, 1916), n.
comb.
swettaria (W. S. Wright, 1916), n.
comb., form
7332 **pectinata** (Rindge, 1967), n. comb.

Stamnodini

STAMNODES Gn., 1857
7333 **gibbicostata** (Wlk., 1862)

costinotata (Wlk., 1863)
strigularia (Minot, 1869)
oeneiformis (Harv., 1874)
7334 **blackmorei** Swett, 1915
7335 **albiapicata** Grossb., 1910
7336 **reckseckeri** Pears., 1910
rickseckeri auth., emend.
7337 **affiliata** Pears., 1911
7338 **delicata** (Grossb., 1908)
7339 **mendocinoensis** Dyar, 1923
7340 **annellata** (Hulst, 1887)
catastrophata Dyar, 1923
gratificata Dyar, 1923, form
7341 **coenonymphata** (Hulst, 1900)
prunata W. S. Wright, 1924, form
pallidata W. S. Wright, 1924, form
brunneata W. S. Wright, 1924, form
wrightii C. & S., 1925, form
7342 **eldridgensis** Swett, 1917
7343 **marinata** W. S. Wright, 1920
7344 **splendorata** Pears., 1909
7345 **apollo** Cass., 1920
7346 **artemis** Rindge, 1958
7347 **formosata** (Stkr., 1878)
dryadata (Hulst, 1880)
sponsata (Grt., 1882)
7348 **lampra** Rindge, 1958
7349 **topazata** (Stkr., 1899)
a. **albida** B. & McD., 1912
b. **apicata** B. & McD., 1917
7350 **franckata** (Pears., 1909)
7351 **modocata** W. S. Wright, 1920
7352 **cassinoi** Swett, 1917
7353 **seiferti** (Neum., 1882)
7354 **fervefactaria** (Grt., 1881)
7355 **deceptiva** B. & McD., 1918

STAMNOCTENIS Warr., 1901
7356 **morrisata** (Hulst, 1887)
volucer (Hulst, 1896)
7357 **pearsalli** (Swett, 1914)
7358 **rubrosuffusa** (Grossb., 1912)
7359 **ululata** (Pears., 1912)
7360 **costimacula** (Grossb., 1912)
kelseyi (W. S. Wright, 1927)
a. **pallula** McD., 1941
7361 **similis** (W. S. Wright, 1927)
7362 **vernon** Guedet, 1939

MARMOPTERYX Pack., 1874
7363 **marmorata** (Pack., 1871)
a. **odontata** Hulst, 1896
7364 **tessellata** Pack., 1874
7365 **watsoni** Cass., 1920
7366 **animata** Pears., 1906

HETERUSIA Hbn., 1827–31
SCORDYLIA Gn., 1857
7367 **atalantata** (Gn., 1857)

Xanthorhoini

XANTHORHOE Hbn., 1825
MALENYDRIS Hbn., 1825
OCHYRIA Hbn., 1825
COREMIA Gn., 1845, preocc. by
Audinet-Serville, 1834
7368 **labradorensis** (Pack., 1867)
emendata (Pears., 1914), n. syn.
designata; auth., not Hufn., 1767

7369 **packardata** McD., 1945
    gynandrata (Pears., 1914),
      infrasubsp.
7370 **abrasaria** (H.-S., 1856), extralim.
    ligularia (Gn., 1857)
    baicalata (Bremer, 1864)
    a. **congregata** (Wlk., 1862)
      nigrofasciata (Pack., 1867)
    b. **trilineata** (Warr., 1904)
      salvata (Pears., 1913)
    c. **aquilonaria** C. & S., 1922
7371 **iduata** (Gn., 1857)
    planata Tayl., 1908
7372 **macdunnoughi** Swett, 1918
7373 **ramaria** S. & C., 1920
    a. **delectaria** C. & S., 1922
7374 **incursata** (Hbn., 1809–13), extralim.
    a. **lagganata** S. & C., 1920
    b. **harveyata** C. & S., 1922
7375 **reclivisata** S. & C., 1920
7376 **baffinensis** McD., 1939
7377 **dodata** S. & C., 1920
7378 **algidata** (Mösch., 1874)
7379 **pontiaria** Tayl., 1906
7380 **fossaria** Tayl., 1906
    a. **blackmorei** Swett, 1918
    b. **atlinensis** Swett, 1918
7381 **montanata** (D. & S., 1775), extralim.
    implicata (Villers, 1789)
    implicaria (Haw., 1809)
    montanaria (Tr., 1828)
    fuscomarginata (Stgr., 1871)
    a. **lapponica** (Stgr., 1871)
7382 **marinensis** McD., 1945
7383 **spaldingaria** (Grossb., 1907)
7384 **munitata** (Hbn., 1800–9)
    fulvata (F., 1787), preocc. by
      Forster, 1771
    munitaria (Bdv., 1840)
    strigata (Pack., 1867)
    immediata (Grt., 1882), n. syn.
    anticostiata (Stkr., 1899)
    a. **convallaria** (Gn., 1857)
    b. **nemorella** Hulst, 1896
    illocata Hulst, 1896
7385 **alticolata** B. & McD., 1916
7386 **defensaria** (Gn., 1857)
    californiata (Pack., 1871)
    amorata (Hulst, 1900)
    thanataria Swett, 1916
    mephistaria (Swett, 1915), form
    suppuraria Swett, 1916
    conciliaria Swett, 1916, form
    gigantaria Swett, 1916, form
7387 **offensaria** McD., 1941
7388 **ferrugata** (Cl., 1759)
    unidentaria (Haw., 1809), form
    inclinataria (Wlk., 1863)
    a. **infumata** B. & McD., 1917
    b. **alaskae** C. & S., 1925
7389 **borealis** Hulst, 1896
7390 **lacustrata** (Gn., 1857)
7391 **dentilinea** B. & McD., 1913
7392 **mirabilata** (Grt., 1883)
7393 **columelloides** B. & McD., 1913

EPIRRHOE Hbn., 1825
7394 **alternata** (Müller, 1764)
    sociata (Bkh., 1794)
    subtristata (Haw., 1809)

    degenerata (Haw., 1809)
    contristata (Don., 1811)
7395 **plebeculata** (Gn., 1857)
    rubrosuffusata (Pack. 1871)
    a. **vivida** B. & McD., 1917
7396 **sperryi** Herbulot, 1951
    tristata; auth., not L., 1758
7397 **medeifascia** (Grossb., 1908)

EUPHYIA Hbn., 1825
7398 **swetti** Cass., 1927
7399 **unangulata** (Haw., 1809), extralim.
    amniculata (Hbn., 1809–13)
    amnicularia (Bdv., 1840)
    a. **intermediata** (Gn., 1857)
    ideata (Wlk., 1862)
7400 **implicata** (Gn., 1857)
    a. **multilineata** (Pack., 1871)
    williamsi (Swett, 1910)
    b. **grandiosa** (Hulst, 1898)
7401 **minima** C. & S., 1922

ENCHORIA Hulst, 1896
7402 **osculata** Hulst, 1896
7403 **lacteata** (Pack., 1876), n. comb.
7404 **herbicolata** (Hulst, 1896), n. comb.

LOXOFIDONIA Pack., 1876
7405 **acidaliata** (Pack., 1874)

ZENOPHLEPS Hulst, 1896
7406 **lignicolorata** (Pack., 1874)
    a. **victoria** Tayl., 1906
7407 **pallescens** McD., 1938
7408 **alpinata** Cass., 1927
    rosa Cass., 1927, form
7409 **obscurata** Hulst, 1896
    a. **infumata** B. & McD., 1917

PSYCHOPHORA Kby., 1824
7410 **sabini** Kby., 1824
    sabiniaria (Pack., 1876)
    a. **polaris** Hulst, 1903
7411 **phocata** (Mösch., 1862)
7412 **suttoni** Heinr., 1942
7413 **immaculata** (Skin. & Mengel, 1892)

ORTHONAMA Hbn., 1825
    PLEMYRIA Hbn., 1825 (p. 334),
      preocc. by Hbn., 1825 (p. 327)
    NYCTEROSEA Hulst, 1896
    NYCTOSIA auth., missp.
    PERCNOPTILOTA Hulst, 1896
7414 **obstipata** (F., 1794)
    fluviata (Hbn., 1796–99)
    gemmata (Hbn., 1796–99)
    angustata (Haw., 1809)
    albicinctata (Haw., 1809)
    gemmaria (Bdv., 1840)
    lapillata (Gn., 1857), n. comb.
    baccata (Gn., 1857), n. comb.
    exagitata (Wlk., 1862), n. comb.
    intrusata (Wlk., 1862), n. comb.
    peracutata (Wlk., 1862), n. comb.
    obruptata (Wlk., 1863), n. comb.
    signataria (Wlk., 1863), n. comb.
    alternata (Wlk., 1866), n. comb.
    pigrata (Wlk., 1866), n. comb.
    brunneipennis (Hulst, 1896), n. comb.
    mortuaria (Schaus, 1908), n. comb.

    marginata (Mathew, 1906)
    obsoleta (Mathew, 1906)
    olivacea (Mathew, 1906)
7415 **evansi** (McD., 1920), n. comb.
7416 **centrostrigaria** (Woll., 1858), n. comb.
    mediata (Wlk., 1862), n. comb.
    latirupta (Wlk., 1866), n. comb.
    luscinata (Zell., 1873), n. comb.
    interruptata (Rebel, 1894), n. comb.
    paranensis (Schaus, 1901), n. comb.
    custodiata; Warr., 1905, not Gn.,
      1857

DISCLISIOPROCTA Wallgr., 1861
    CAMPTOGRAMMA; auth., not
      Steph., 1831
    CAMPTOLINA Schaus, 1940, n.
      syn.
7417 **stellata** (Gn., 1857), n. comb.
    fuscovariata (Wallgr., 1860), n. syn.
    impauperata (Wlk., 1862), n. comb.
    balteolata (H.-S., 1865), n. comb.,
      ident. uncert.
    albosignata (Pack., 1873), n. comb.
    foedata (Warr., 1900), n. comb.

HERRESHOFFIA Sperry, 1949
7418 **gracea** Sperry, 1949

Asthenini

HYDRELIA Hbn., 1825
7419 **lucata** (Gn., 1857)
7420 **condensata** (Wlk., 1862)
7421 **terraenovae** Krogerus, 1954
7422 **inornata** (Hulst, 1896)
    exhumata (Pears., 1906)
7423 **albifera** (Wlk., 1866)
    albogilvaria (Morr., 1874)
    triseriata (Pack., 1874)
7424 **brunneifasciata** (Pack., 1876)

VENUSIA Curt., 1839
    NOMENIA Pears., 1905
7425 **cambrica** Curt., 1839
    erutaria (Bdv., 1840)
    nebulosaria (Freyer, 1852)
    cambricaria Gn., 1857
    scitularia (Wlk., 1860)
7426 **duodecemlineata** (Pack., 1873)
    unipecta (Pears., 1906)
    a. **secunda** (Pears., 1906)
7427 **obsoleta** (Swett, 1916)
7428 **comptaria** (Wlk., 1860)
    perlineata (Pack., 1874)
    salienta (Pears., 1905)
    palumbes Franc., 1938, melanic f.
7429 **pearsalli** (Dyar, 1906)

TRICHODEZIA Warr., 1895
    NEODEZIA Warr., 1904
7430 **albovittata** (Gn., 1857)
    propriaria (Wlk., 1862)
    reciprocata (Wlk., 1862)
    a. **tenuifasciata** B. & McD., 1917
7431 **albofasciata** (Grt., 1863)
7432 **californiata** (Pack., 1871)

Operophterini

EPIRRITA Hbn., 1822
  OPORINIA Hbn., 1825
  OPORABIA Steph., 1831
7433 **autumnata** (Bkh., 1794), extralim.
  autumnaria (Doubleday, 1849)
  approximaria (Weaver, 1852)
  addendaria (White, 1878)
  a. **henshawi** (Swett, 1917)
  b. **omissa** (Harrison, 1932)
  oblita (Harrison, 1932), form
  dayi (Harrison, 1932), form
  coarctata (Harrison, 1932), form
  suffusa (Harrison, 1932), form
  latifasciata (Harrison, 1942), form
7434 **undulata** (Harrison, 1942)
  extensa (Harrison, 1942)
7435 **pulchraria** (Tayl., 1907)

OPEROPHTERA Hbn., 1825
  OPOROPHTERA McD., 1938, missp.
  RACHELA Hulst, 1896
  PARAPTERA Hulst, 1896, n. syn.
7436 **brumata** (L., 1758)
  grisearia (de Villers, 1789)
  brumaria (Esp., 1800)
  vulgaris (Steph., 1829)
  myricaria (Cooke, 1882)
7437 **bruceata** (Hulst, 1886)
  groenlandica de Lesse, 1951
  a. **hyperborea** (Hulst, 1896), n. stat.
7438 **occidentalis** (Hulst, 1896)
  latipennis (Hulst, 1896), form
7439 **danbyi** (Hulst, 1896), n. comb.

Eudulini

EUBAPHE Hbn., 1823, p. 20
  EUDULE; auth., not Hbn., 1823, p. 14
  AMERIA Wlk., 1854
  EUPHANESSA Pack., 1864
  LEPTIDULE Butler, 1877
7440 **mendica** (Wlk., 1854)
  biseriata (H.-S., 1855)
7441 **meridiana** (Slosson, 1889)
7442 **helveta** (Barnes, 1907)
7443 **rotundata** (C. & S., 1922)
7444 **unicolor** (Rob., 1869)
  hyalina (Hulst, 1898)

Eupitheciini

HORISME Hbn., 1825
7445 **intestinata** (Gn., 1857)
  indoctrinata (Wlk., 1863)
  impleta (Wlk., 1866)
7446 **incana** Swett, 1918
  a. **columbia** McD., 1927
7447 **rectilineata** (Tayl., 1907)
7448 **gillettei** (Hulst, 1898)

EUPITHECIA Curt., 1825
  TEPHROCLYSTIA Hbn., 1825
  DYSCYMATOGE Hbn., 1825
  TARACHIA Hbn., 1825
  LEUCOCORA Hbn., 1825
  ARCYONIA Hbn., 1825
  EUCYMATOGE Hbn., 1825
  HYPEPIRRITIS Hbn., 1825
  DIETZEA Schütze, 1956
  PETERSENIA Schütze, 1958
  BOHATSCHIA Schütze., 1960
7449 **palpata** Pack., 1873
7450 **transcanadata** MacKay, 1951
7451 **slossonata** McD., 1949
7452 **albimontanata** McD., 1940
7453 **peckorum** Heitzman and Enns, 1977
7454 **longidens** (Hulst, 1896)
  a. **kerrvillaria** C. & S., 1924
7455 **ornata** (Hulst, 1896)
  exornata B. & McD., 1912
7456 **monacheata** C. & S., 1922
  carolata McD., 1944
7457 **terrestrata** McD., 1944
7458 **karenae** Leuschner, 1966
7459 **columbiata** (Dyar, 1904)
  a. **erpata** Pears., 1908
  b. **holbergata** MacKay, 1951
7460 **maestosa** (Hulst, 1896)
  dyarata Tayl., 1906
  a. **harlequinaria** (Dyar, 1905)
7461 **subvirens** Dietze, 1875
  laisata Stkr., 1899
  diegata McD., 1940
7462 **castellata** McD., 1944
7463 **chiricahuata** McD., 1944
7464 **insolabilis** (Hulst, 1900)
7465 **catalinata** McD., 1944
7466 **edna** (Hulst, 1896)
7467 **owenata** McD., 1944
7468 **longipalpata** Pack., 1876
7469 **sabulosata** McD., 1944
7470 **macrocarpata** McD., 1944
7471 **placidata** Tayl., 1908
7472 **unicolor** (Hulst, 1896)
  cenataria C. & S., 1922
7473 **pseudotsugata** MacKay, 1951
7474 **miserulata** Grt., 1863
  nebulosa (Hulst, 1896)
  plumbaria (Hulst, 1900)
  grossbeckiata Swett, 1907
  a. **zela** S. & C., 1919
7475 **chlorofasciata** Dietze, 1872, rev. stat.
7476 **misturata** (Hulst, 1896)
  subfoveata (Dyar, 1904)
  insignificata Tayl., 1906
  sublineata Tayl., 1906
  minorata Tayl., 1907
  scelestata Tayl., 1907
  a. **delzurata** C. & S., 1922
  b. **frostiata** Swett, 1907
  conformata Pears., 1908
7477 **harveyata** Tayl., 1906
7478 **bivittata** (Hulst, 1896)
7479 **pygmaeata** (Hbn., 1796–99), extralim.
  pygmaearia Bdv., 1840
  palustraria Doubleday, 1850
  a. **obumbrata** Tayl., 1906
  fortunata Pears., 1909
7480 **bryanti** Tayl., 1906
  modesta Tayl., 1906
7481 **coloradensis** (Hulst, 1896)
  carolinensis Grossb., 1907
  spenceata Cass., 1927
7482 **cretata** (Hulst, 1896)
7483 **regina** Tayl., 1906
7484 **undata** Freyer, 1842
  scriptaria H.-S., 1847
7485 **borealis** (Hulst, 1898)
7486 **jejunata** McD., 1949
7487 **subfuscata** (Haw., 1809)
  castigata (Hbn., 1809–13)
  castigaria Bdv., 1840
  implicata Wlk., 1862
  blancheata Cooke, 1881
  latipennis (Hulst, 1898)
7488 **tripunctaria** H.-S., 1852
  albipunctata; Haw., 1809 not Hufn., 1767
7489 **luteata** Pack., 1867
  catskillata Pears., 1908
  fasciata Tayl., 1910
  a. **bifasciata** (Dyar, 1904)
7490 **harrisonata** MacKay, 1951
7491 **fletcherata** Tayl., 1907
  promulgata Pears., 1909
  dolorosata Pears., 1910
7492 **casloata** (Dyar, 1904)
  kasloata auth., emend.
7493 **bradorata** McD., 1930
7494 **sheppardata** McD., 1938
7495 **affinata** Pears., 1908
7496 **rotundopuncta** Pack., 1871
  californiata Gump., 1888
7497 **sierrae** (Hulst, 1896)
  conceptata Pears., 1909
  joymaketa Cass., 1925
7498 **litoris** McD., 1946
7499 **quakerata** Pears., 1909
  apacheata Cass., 1927
  conglomerata McD., 1946
7500 **bolterii** (Hulst, 1900)
7501 **palmata** C. & S., 1922
7502 **piccata** Pears., 1910
7503 **pretansata** Grossb., 1908
7504 **neomexicana** McD., 1946
7505 **alpinata** Cass., 1927
7506 **prostrata** McD., 1938
7507 **persimulata** McD., 1938
7508 **exudata** Pears., 1909
7509 **herefordaria** C. & S., 1923
  herfordaria C. & S., 1923, incorr. orig. spell.
7510 **cazieri** Kirkwood, 1961
7511 *number omitted*
7512 **macdunnoughi** Rindge, 1952
  suspiciosata; McD., 1949, not Deitz, 1875
7513 **nabokovi** McD., 1946
7514 **biedermanata** C. & S., 1922
7515 **cupressata** Pears., 1910
7516 **albigrisata** Pears., 1909
7517 **gibsonata** Tayl., 1910
  chagnoni Swett, 1911
7518 **intricata** (Zett., 1939), extralim.
  a. **taylorata** Swett, 1907, rev. stat., *arceuthata*; auth., not Freyer, 1842
7519 **uinta** Rindge, 1956
7520 **satyrata** (Hbn., 1809–13), extralim.
  satyraria Bdv., 1840
  fagicolaria Robson & Gardner, 1886
  a. **fumata** Tayl., 1910
  b. **intimata** Pears., 1908
  c. **dodata** Tayl., 1906
  divinula C. & S., 1924
  mackieata C. & S., 1925
7521 **terminata** Tayl., 1908
  slocanata Tayl., 1908

7522 **nimbicolor** (Hulst, 1896)
    obscurior (Hulst, 1896)
    adornata Tayl., 1906
    incresata Pears., 1910
    inclarata C. & S., 1924
7523 **strattonata** Pack., 1873
7524 **cimicifugata** Pears., 1908
7525 **grata** Tayl., 1910
7526 **russeliata** Swett, 1908
    a. **brauneata** Swett, 1908
7527 **ammonata** McD., 1929
7528 **fumosa** (Hulst, 1896)
7529 **coagulata** Gn., 1857
    geminata Pack., 1873
    packardata Tayl., 1907
    meritata Pears., 1908
7530 **swettii** Grossb., 1907
    ritaria Cass., 1927
7531 **indistincta** Tayl., 1910
7531 **zygadeniata** Pack., 1876
    tenebrescens (Hulst, 1900)
7533 **cretaceata** (Pack., 1874)
    fenestrata Millière, 1874
7534 **nimbosa** (Hulst, 1896)
    plenoscripta (Hulst, 1900)
    a. **bindata** Pears., 1910
7535 **behrensata** Pack., 1876
    perillata Pears., 1912
    monterata C. & S., 1922
7536 **multiscripta** (Hulst, 1896)
7537 **sewardata** Bolte, 1977
7538 **gelidata** Mösch., 1860
    hyperboreata Stgr., 1861
    flebilis (Hulst, 1900)
    lagganata Tayl., 1910
    compactata Tayl., 1910
    nordeggensis C. & S., 1922
7539 **multistrigata** (Hulst, 1896)
    spaldingi Tayl., 1910
7540 **perfusca** (Hulst, 1898)
    cootenaiata (Dyar, 1904)
    kootenaiata B. & McD., 1912
    alberta Tayl., 1906
    a. **youngata** Tayl., 1906
      winnata Tayl., 1910
7541 **hanhami** Tayl., 1906
7542 **filmata** Pears., 1908
7543 **annulata** (Hulst, 1896)
    limnata Pears., 1909
    orfordata Cass., 1927
7544 **vinsullata** MacKay, 1951
7545 **usurpata** Pears., 1909
7546 **olivacea** Tayl., 1906
7547 **cognizata** Pears., 1910
7548 **lachrymosa** (Hulst, 1900)
7549 **georgii** McD., 1929
7550 **kananaskata** MacKay, 1951
7551 **pusillata** (D. & S., 1775),
    extralim.
    sobrinata (Hbn., 1814–17)
    sobrinaria Bdv., 1840
    scotica Dietze, 1910
    a. **interruptofasciata** Pack., 1873,
      rev. stat.
      quebecata Tayl., 1910
7552 **niphadophilata** (Dyar, 1904)
7553 **subcolorata** (Hulst, 1898)
7554 **appendiculata** McD., 1946
7555 **zelmira** S. & C., 1920
7556 **vitreotata** Cass., 1927

7557 **segregata** Pears., 1910
    bonita C. & S., 1925
7558 **pinata** Cass., 1925
7559 **tenuata** Hulst, 1880
7560 **phyllisae** Rindge, 1963
7561 **agnesata** Tayl., 1908
    barnesi C. & S., 1922
7562 **huachuca** Grossb., 1908
7563 **woodgatata** (C. & S., 1923)
7564 **stellata** (Hulst, 1896)
7565 **bowmani** C. & S., 1923
7566 **niveifascia** (Hulst, 1898)
    perbrunneata Tayl., 1906
7567 **joanata** C. & S., 1922
    balboata C. & S., 1925
7568 **flavigutta** (Hulst, 1896)
7569 **sperryi** McD., 1939
7570 **johnstoni** McD., 1946
7571 **dichroma** McD., 1946
7572 **rindgei** McD., 1949
7573 **cocoata** Pears., 1908
7574 **albicapitata** Pack., 1876
7575 **mutata** Pears., 1908
7576 **helena** Tayl., 1906
7577 **columbrata** McD., 1940
7578 **spermaphaga** (Dyar, 1917)
7579 **purpurissata** Grossb., 1908
    a. **valariata** Pears., 1910
      muriflua (Dyar, 1923)
7580 **mystiata** Cass., 1925
7581 **gilvipennata** C. & S., 1922
7582 **miamata** Cass., 1925
7583 **scabrogata** Pears., 1912
    vistata C. & S., 1922
7584 **hohokamae** Rindge, 1963
7585 **adequata** Pears., 1910
7586 **acutipennis** (Hulst, 1898)
7587 **subapicata** Gn., 1857
    occidentaliata (Pack., 1871)
7588 **shirleyata** C. & S., 1922
7589 **sinuata** McD., 1946
7590 **redingtonia** McD., 1949
7591 **gilata** Cass., 1925
7592 **plumasata** McD., 1946
7593 **deserticola** McD., 1946
7594 **anticaria** Wlk., 1862
    explanata Wlk., 1862
7595 **pertusata** McD., 1938
7596 **tricolorata** Cass., 1927
7597 **carneata** McD., 1946
7598 **classicata** Pears., 1909
7599 **penumbrata** (Pears., 1912)
7600 **graefii** (Hulst, 1896)
    a. **vancouverata** (Tayl., 1906)
      stikineata C. & S., 1922
    b. **tulareata** C. & S., 1922
7601 **nevadata** Pack., 1871
    a. **geneura** S. & C., 1919
    b. **morensa** C. & S., 1922
    c. **moirata** S. & C., 1919
    d. **probata** S. & C., 1919
7602 **implorata** (Hulst, 1896)
7603 **cestata** (Hulst, 1896)
7604 **cestatoides** McD., 1949
7605 **ravocostaliata** Pack., 1876

      NASUSINA Pears., 1908
7606 **inferior** (Hulst, 1896)
7607 **vaporata** (Pears., 1912)
7608 **mendicata** (B. & McD., 1918)

7609 **minuta** (Hulst, 1896)

      PRORELLA B. & McD., 1918
7610 **emmedonia** (Grossb., 1908)
7611 **gypsata** (Grt., 1882)
7612 **discoidalis** (Grossb., 1908)
7613 **leucata** (Hulst, 1896)
7614 **albida** (C. & S., 1923)
    ruthiata (Cass., 1927)
7615 **ochrocarneata** McD., 1949
7616 **irremorata** (Dyar, 1923)
7617 **tremorata** McD., 1949
7618 **remorata** (Grossb., 1907)
7619 **desperata** (Hulst, 1896)
7620 **artestata** (Grossb., 1908)
7621 **mellisa** (Grossb., 1908)
7622 **insipidata** (Pears., 1910)
7623 **opinata** (Pears., 1909)
7624 **protoptata** (McD., 1938)

      CHLOROCLYSTIS Hbn., 1825
      DYSERGA Petersen, 1909
      CALLICLYSTIS Dietze, 1910
7625 **rectangulata** (L., 1758)
    nigrosericeata (Haw., 1809), form
    sericeata (Haw., 1809)
    subaerata (Hbn., 1814–17)
    rectangularia (Bdv., 1840)

Lobophorini

      CARSIA Hbn., 1825
      CELMA Steph., 1831
7626 **sororiata** (Hbn., 1809–13), extralim
    paludata (Thunb., 1788), preocc. by
      L. 1767
    imbutata (Hbn., 1809–13)
    imbutaria (Bdv., 1840)
    a. **alpinata** Pack., 1873, rev. stat.
      boreata Pack., 1873
    b. **labradoriensis** (Sommer, 1897),
      rev. stat.
    c. **thaxteri** Swett, 1917, rev. stat.
    d. **columbia** McD., 1939, rev. stat.

      APLOCERA Steph., 1827
      ANAITIS Dup., 1829
      LARISSA Curt., 1830
7627 **plagiata** (L., 1758)
    duplicata (F., 1775)
    plagiaria (Bdv., 1840)

      LITHOSTEGE Hbn., 1825
7628 **rotundata** Pack., 1874
    arizonata Grt., 1883
7629 **fuscata** (Grossb., 1906)
7630 **marcata** B. & McD., 1916
7631 **elegans** (Grossb., 1909)
7632 **angelicata** Dyar, 1923
7633 **deserticola** B. & McD., 1916

      SCELIDACANTHA Hulst, 1896
7634 **triseriata** (Pack., 1874)
    virginata (Graef, 1881)

      ACASIS Dup., 1844
      CYSTEOPTERYX Hulst, 1896
      AGIA Hulst, 1896
7635 **viridata** (Pack., 1873)
    eborata (Hulst, 1896)

TRICHOPTERYX Hbn., 1825
   NOTHOPTERYX Prout, 1909
7636 veritata Pears., 1907, n. comb.
   februalis (Dyar, 1923), n. comb., n.
     syn.

CLADARA Hulst, 1896
   NYCTOBIA Hulst, 1896, preocc.
     by Thorell, 1869
7637 limitaria (Wlk., 1860)
   fusifasciata (Wlk., 1862), n. comb.
   lobophorata (Wlk., 1862), n. comb.
   longipennis (Wlk., 1866), n. comb.
   reiffi (Swett, 1910), n. comb., form
    a. **nigroangulata** (Stkr., 1899), n.
     comb.
7638 anguilineata (G. & R., 1867), n.
   comb.
   vernata (Pack., 1873), part, n. comb.
   eastmani (Reiff, 1913), n. comb.,
    form
7639 atroliturata (Wlk., 1863)
   geminata (G. & R., 1866)
   vernata (Pack., 1873), part

LOBOPHORA Curt., 1825
   PHILOPSIA Hulst, 1896
   TALLEDEGA Hulst, 1896
7640 nivigerata Wlk., 1862
   a. **tabulata** (Hulst, 1896)
7641 montanata Pack., 1874
   nocticolata (Hulst, 1881)
7642 simsata Swett, 1920
7643 magnoliatoidata (Dyar, 1904)
7644 canavestita (Pears., 1906)

HETEROPHLEPS H.-S., 1854
   NANNIA Hulst, 1896
7645 refusaria (Wlk., 1861)
   harveiata Pack., 1876
   refusata auth., missp.
7646 morensata (Hulst, 1896)
7647 triguttaria H.-S., 1854
   quadrinotata (Wlk., 1863)
   hexaspilata (Wlk., 1866)

DYSPTERIS Hbn., 1818
7648 abortivaria (H.-S., 1855)

EPIPLEMIDAE
by DOUGLAS C. FERGUSON

EPIPLEMA H.-S., 1855
   EPIPLEURA Wlk., 1861, missp.
   EVERSMANNIA Stgr., 1871
7649 incolorata (Gn., 1857)

CALLIZZIA Pack., 1876
7650 amorata Pack., 1876
7651 certiorara Pears, 1906

ANTIPLECTA Warr., 1900
7652 triangularis Warr., 1906

CALLEDAPTERYX Grt., 1868
7653 dryopterata Grt., 1868
   erosiata Pack., 1876

PHILAGRAULA Hulst, 1896
7654 slossoniae Hulst, 1896

EROSIA Gn., 1857
7655 incendiata Gn., 1857

SCHIDAX Hbn., 1818
7656 coronaria Stkr., 1899

SEMATURIDAE
by DOUGLAS C. FERGUSON

ANURAPTERYX Hamp., 1918
7657 crenulata B. & L., 1919

URANIIDAE
by DOUGLAS C. FERGUSON

URANIA F., 1807
   CYDIMON Dalman, 1825
   DASYCEPHALUS Swainson, 1833
   LEILUS Wlk., 1854
   URANIDIA Westwood, 1879
7658 fulgens (Wlk., 1854)
   cacica (Gn., 1857)

Mimallonoidea

MIMALLONIDAE
by JOHN G. FRANCLEMONT

LACOSOMA Grt., 1864
7659 chiridota Grt., 1864
7660 arizonicum Dyar, 1898

NANITETA Franc., 1973
7661 elassa Franc., 1973

CICINNUS Blanchard, 1852
   SACCOPHORA Harr., 1841,
    preocc. by Brandt, 1837
   PEROPHORA Harr., 1841, preocc.
    by Lister, 1834, Weigmann, 1835
   PTOCHOPSYCHE Grt., 1896
7662 melsheimeri (Harr., 1841)
   egenaria (Wlk., 1866)

Bombycoidea

APATELODIDAE
by JOHN G. FRANCLEMONT

APATELODES Pack., 1864
   ASTASIA Harr., 1841, preocc. by
    Ehrenberg, 1830
7663 torrefacta (J. E. Smith, 1797)
   floridana Hy. Edw., 1886
7664 pudefacta Dyar, 1904
   uvada Barnes, 1904

OLCECLOSTERA Butler, 1878
7665 angelica (Grt., 1864)
   hyalinopuncta (Pack., 1864)
7666 indistincta (Hy. Edw., 1886)
7667 seraphica (Dyar, 1906)

BOMBYCIDAE
by JOHN G. FRANCLEMONT

BOMBYX L., 1758
7668 mori (L., 1758), only in domestication

LASIOCAMPIDAE
by JOHN G. FRANCLEMONT

MACROMPHALIINAE

HYPOPACHA N. & D., 1893
7669 grisea (Neum., 1882)

TOLYPE Hbn., 1820
   PLANOSA Fitch, 1856
   TOLYP Cass., 1928, missp.
7670 velleda (Stoll, 1791)
   candidatus Cass., 1928
7671 austella Franc., 1973
7672 mayelisae Franc., 1973
7673 laricis (Fitch, 1856)
   minuta (Grt., 1864)
7674 notialis Franc., 1973
7675 minta Dyar, 1906
7676 glenwoodii Barnes, 1900
7677 distincta French, 1890
7678 nigricaria Cass., 1928
7679 dayi Blkmre., 1921
7680 lowriei B. & McD., 1918

APOTOLYPE Franc., 1973
7681 brevicrista (Dyar, 1895)
   vemerila (Dyar, 1914)
   nigrocristata (Walter, 1928)
7682 blanchardi Franc., 1973

ARTACE Wlk., 1855
7683 cribraria (Ljungh, 1825)
   punctistriga Wlk., 1855
7684 colaria Franc., 1973

GASTROPACHINAE

HETEROPACHA Harv., 1874
7685 rileyana Harv., 1874

PHYLLODESMA Hbn., 1820
   EPICNAPTERA Rambur, 1866
   AMMATOCAMPA Wallgr., 1869
7686 occidentis (Wlk., 1855)
   ilicifolia; J. E. Smith, 1797, not L.,
    1758
   carpinifolia (Bdv., 1868)
   texana Lajonquière, 1969, part
7687 americana (Harr., 1841)
   canadensis Lajonquière, 1969
   sanctilaurentis Lajonquière, 1969
   ferruginea (Pack., 1864), form
    clarinervata Lajonquière, 1969
   a. **rockiesensis** Lajonquière, 1969
   b. **borealis** Lajonquière, 1976
    occidentis; Lajonquière, 1969, not
     Wlk., 1855
   c. **alascensis** (Pack., 1870)
    (available?)
   d. **californica** (Pack., 1872)
    carpinifolia; Lajonquière, 1969 &
     1976, not Bdv., 1868

angelorum Lajonquière, 1969
mildei (Stretch, 1872), form
roseata (Stretch, 1872)
franciscana Lajonquière, 1969,
form
e. **dyari** (Rivers, 1893)
f. **celsivolans** Lajonquière, 1969
transiens Lajonquière, 1969
arizonensis Lajonquière, 1969
texana Lajonquière, 1969
7688 **coturnix** Lajonquière, 1969
cinerascens Lajonquière, 1969, form
ardens Lajonquière, 1969, form
a. **nevadensis** Lajonquière, 1969
preciosa Lajonquière, 1969, form

LASIOCAMPINAE

CALOECIA B. & McD., 1911
7689 **juvenalis** (B. & McD., 1911)
7690 **entima** Franc., 1973
juvenalis B. & McD., 1911, part

QUADRINA Grt., 1881
7691 **diazoma** Grt., 1881
oweni (Barnes, 1906)

DICOGASTER B. & McD., 1911
7692 **coronada** (Barnes, 1904)
valens (Dyar, 1905), form

GLOVERIA Pack., 1872
7693 **howardi** (Dyar, 1896)
7694 **medusa** (Stkr., 1898)
7695 **gargamelle** (Stkr., 1884)
7696 **arizonensis** Pack., 1872
dentata Hy. Edw., 1884
dolores N. & D., 1893
7697 **sphingiformis** B. & McD., 1910

MALACOSOMA Hbn., 1820
CLISIOCAMPA Curt., 1828
7698 **disstria** Hbn., 1820
neustria; J. E. Smith, 1797, not L.,
1758
nubilis (Guér.-Méneville, 1832)
sylvatica (Harr., 1841)
drupacearum (Bdv., 1868)
erosa (Stretch, 1881)
thoracicoides (N. & D., 1893)
sylvaticoides (N. & D., 1893)
perversa (N. & D., 1893)
astriata Reiff, 1913
anita Reiff, 1913
7699 **constrictum** (Hy. Edw., 1874)
strigosa (Stretch, 1881)
a. **austrinum** Stehr, 1968
7700 **tigris** (Dyar, 1902)
disstria; N. & D., 1894, part, not
Hbn., 1820
inducta Dyar, 1906
onissa (Dyar, 1911)
texana Dyar, 1928
7701 **americanum** (F., 1793)
castrensis; J. E. Smith, 1797, not L.,
1758
pensylvanica (Guér.-Méneville, 1832)
decipiens (Wlk., 1855)
frutetorum (Bdv., 1868)
7702 **californicum** (Pack., 1864)

californica (Wlk., 1865)
pseudoneustria (Bdv., 1868)
thoracica (Stretch, 1881)
perlutea (N. & D., 1893)
a. **ambisimile** (Dyar, 1893)
b. **recenseo** Dyar, 1928
c. **pluviale** (Dyar, 1893)
d. **lutescens** (N. & D., 1893)
e. **fragile** (Stretch, 1881)
mus (Neum., 1893)
7703 **incurvum** (Hy. Edw., 1882)
constrictina (N. & D., 1893)
a. **discoloratum** (Neum., 1893)

SATURNIIDAE
by DOUGLAS C. FERGUSON

CITHERONIINAE

EACLES Hbn., 1819
CEROCAMPA Kirby & Spence, 1826
BASILONA Bdv., 1868
CRENUDIA Burmeister, 1879
7704 **imperialis** (Drury, 1773)
imperatoria (J. E. Smith, 1797)
didyma (Beauvois, 1805)
punctatissima Neum., 1891
a. **pini** Michener, 1950
b. **nobilis** Neum., 1891
7705 **oslari** Roths., 1907

CITHERONIA Hbn., 1819
CERATOCAMPA Harr., 1833
DRYOCAMPA; Duncan, 1841, not
Harr., 1833
7706 **regalis** (F., 1793)
regia (J. E. Smith, 1797)
infernalis Stkr., 1883
saengeri Neum., 1891
7707 **splendens** (Druce, 1886), extralim.
a. **sinaloensis** Hoffmann, 1942
7708 **sepulcralis** G. & R., 1865

SPHINGICAMPA Walsh, 1864
ADELOCAMPA Pack., 1905
BOUVIERINA Michener, 1949
7709 **bicolor** (Harr., 1841)
distigma Walsh, 1864
immaculata (Jewett, 1882)
suprema Neum., 1885
7710 **heiligbrodti** (Harv., 1877)
7711 **hubbardi** (Dyar, 1903)
7712 **bisecta** (Lint. 1879)
nebulosa Neum., 1890
7713 **blanchardi** Fgn., 1971
7714 **albolineata** (G. & R., 1866)
raspa (Bdv., 1872)

DRYOCAMPA Harr., 1833
7715 **rubicunda** (F., 1793)
pallida Bowles, 1875
semialba (Schüssler, 1936), nom.
nud.
sperryi (Bower, 1942)
a. **alba** Grt., 1874
roseilinea (Schaus, 1920)

ANISOTA Hbn., 1820
7716 **stigma** (F., 1775)

stygma (Bdv., 1872)
a. **fuscosa** Fgn., 1971
7717 **manitobensis** McD., 1921
7718 **consularis** Dyar, 1896
7719 **senatoria** (J. E. Smith, 1797)
7720 **peigleri** Riotte, 1975
7721 **finlaysoni** Riotte, 1969
7722 **oslari** Roths., 1907
skinneri Biederman, 1908
neomexicana Brehme, 1909
7723 **virginiensis** (Drury, 1773)
astynome (Olivier, 1790)
a. **pellucida** (J. E. Smith, 1797)
sinulis Riotte, 1970
b. **discolor** Fgn., 1971

HEMILEUCINAE

Hemileucini

COLORADIA Blake, 1863
7724 **pandora** Blake, 1863
loiperda Dyar, 1912
a. **lindseyi** B. & Benj., 1926
b. **davisi** B. & Benj., 1926
bonniwelli B. & Benj., 1926
chiricahua B. & Benj., 1926
7725 **doris** Barnes, 1900
lois Dyar, 1911
duffneri B. & Benj., 1926
7726 **luski** B. & Benj., 1926
7726.1 **velda** Johnson and Walter, 1981

HEMILEUCA Wlk., 1855
EUCHRONIA Pack., 1864
PSEUDOHAZIS G. & R., 1866
HERA Harr., 1869
EULEUCOPHAEUS Pack., 1872
ARGYRAUGES Grt., 1882
7727 **tricolor** (Pack., 1872)
7728 **hualapai** (Neum., 1882)
7729 **oliviae** Ckll., 1898
grisea Ckll., 1914
suffusa Ckll., 1914
7730 **maia** (Drury, 1773)
proserpina (F., 1775)
lintneri Ckll., 1914
7731 **nevadensis** Stretch, 1872
californica Wgt., 1888
artemis Pack., 1893
latifascia B. & McD., 1916
7732 **lucina** Hy. Edw., 1887
obsoleta Reiff, 1910
lutea Reiff, 1910
7733 **grotei** G. & R., 1868
7734 **diana** Pack., 1874
7735 **juno** Pack., 1872
yavapai Neum., 1881
7736 **electra** Wgt., 1884
rickseckeri J. H. Watson, 1912
watsoni Schüssler, 1934
a. **clio** B. & McD., 1918
7737 **burnsi** J. H. Watson, 1910
ilmae J. H. Watson, 1910
nigrovenosa J. H. Watson, 1910
conjuncta J. H. Watson, 1912
paradoxa J. H. Watson, 1913
bifasciata Schüssler, 1934
7738 **neumoegeni** Hy. Edw., 1881
7739 **chinatiensis** (Tinkham, 1943)

7740 **griffini** Tuskes, 1978
7741 **hera** (Harr., 1841)
    pica Wlk., 1855
    chrysocarena (Harr., 1869)
    a. *marcata* (Neum., 1891)
      gunderi (Hill, 1924)
7742 **magnifica** (Rotger, 1948), n. stat.
7743 **nuttalli** (Stkr., 1875)
    arizonensis (Stkr., 1878)
    washingtonensis (Medlar, 1944)
    a. *uniformis* (Ckll., 1914)
7744 **eglanterina** (Bdv., 1852)
    denudata (Neum., 1891)
    boisduvali (Oberth., 1912)
    harrisi (Oberth., 1914)
    a. *shastaensis* (Grt., 1880)
      normalis (Dyar & Ckll., 1914)
    b. **annulata** Fgn., 1971

AUTOMERIS Hbn., 1819
    PROTAUTOMERIS Pack., 1903
    PROTAUMERIS auth., missp.
    AGLIOPSIS Bouvier, 1929
7745 **randa** Druce, 1894
7746 **io** (F., 1775)
    corollaria (Perry, 1810)
    varia (Wlk., 1855)
    argus N. & D., 1893
    fuscus; Luther, 1907, not Wlk.,
      1855
    lutheri Ckll., 1914
    coloradensis Ckll., 1914
    texana B. & Benj., 1923
    caeca Igel, 1928
    mexicana Draudt, 1929
    packardi Schüssler, 1934
    a. *lilith* (Stkr., 1878)
    b. *neomexicana* B. & Benj., 1923
7747 **iris** (Wlk., 1865), extralim.
    a. *hesselorum* Fgn., 1972
7748 **cecrops** (Bdv., 1875), extralim.
    mexicana Bouvier, 1929
    herse Hoffmann, 1942
    a. *pamina* (Neum., 1882)
      aurosea (Neum., 1882)
      chiricahuana Schüssler, 1934
7749 **zephyria** Grt., 1882
    zephyriata B. & Benj., 1924

HYLESIA Hbn., 1820
    MICRATTACUS Wlk., 1855
    HYLOSIA H.-S., 1855
7750 **coinopus** Dyar, 1913

SATURNIINAE

Saturniini

SATURNIA Schr., 1802
    PAVONIA Hbn., 1819
    HERAEA Hbn., 1822
    CALOSATURNIA Sm., 1886
    EUDIA Jordan, 1911
7751 **mendocino** Behrens, 1876
7752 **walterorum** Hogue & Johnson,
    1958
    meridionalis; Johnson, 1940, not
      Calberla, 1887
7753 **albofasciata** (Johnson, 1938)

AGAPEMA N. & D., 1894
7754 **galbina** (Clem., 1860)
    dyari Ckll., 1914
    a. *anona* (Ottol., 1903)
      interrupta Draudt, 1929
7755 **solita** Fgn., 1972
7756 **homogena** Dyar, 1908

ANTHERAEA Hbn., 1819
    TELEA Hbn., 1819
    METOSAMIA Druce, 1892
7757 **polyphemus** (Cram., 1776)
    fenestra; Perry, 1811, not L., 1758
    polypheme (Hbn., 1819)
    flava (Grt., 1903)
    fumosus (Wurster, 1930)
    wilfriedi (Sageder, 1933)
    brunnea (Sageder, 1933)
    nigra (Schüssler, 1936)
    albida (Bouvier, 1936)
    intermedia (Bouvier, 1936)
    a. *olivacea* (Ckll., 1914)
      nigrescens (Schüssler, 1936)
      vinacea (Schüssler, 1936)
    b. *oculea* (Neum., 1883)
      aurelia (Druce, 1892)

ACTIAS Leach, 1815
    ECHIDNA Hbn., 1807, preocc. by
      Forster, 1788
    TROPAEA Hbn., 1819
    PLECTOPTERON Hutton, 1846
7758 **luna** (L., 1758)
    dictynna (Wlk., 1855)
    rossii Ross, 1872
    bolli (Wagner, 1876)
    maasseni Kby., 1892
    rubromarginata (Davis, 1912)
    rubrosuffusa (Ckll., 1914)
    mariae (Benj., 1922)
    lacrimens (Niepelt, 1932)

Attacini

SAMIA Hbn., 1819
    PHILOSAMIA Grt., 1874
7759 **cynthia** (Drury, 1773)
    advena (J. W. Watson, 1912)
    alboabdominalis (Schüssler, 1926)
    viridis Mezger, 1928
    bicolorata Mezger, 1928
    punctata Mezger, 1928
    fenestrella Mezger, 1928

ROTHSCHILDIA Grt., 1896
7760 **cincta** (Tepper, 1883)
    jorulla; auth., not Westwood, 1853
    guerreronis Draudt, 1929
    pseudoguerreronis Hoffmann, 1942
7761 **forbesi** Benj., 1934
    draudti Benj., 1934
7762 **orizaba** (Westwood, 1853)
    ochracea Draudt, 1929
    prionidia Draudt, 1929
    paradoxa Hoffmann, 1942

EUPACKARDIA Ckll., 1912
7763 **calleta** (Westwood, 1853)
    polyommata (Tepper, 1883)
    semicaeca Ckll., 1914

    caeca Draudt, 1929
    digueti Bouvier, 1936

CALLOSAMIA Pack., 1864
7764 **promethea** (Drury, 1773)
    caeca Ckll., 1914
7765 **angulifera** (Wlk., 1855)
    aurantiaca Igel, 1928
7766 **securifera** (Maassen, 1873)
    carolina (Jones, 1908)

HYALOPHORA Duncan, 1841
    PLATYSAMIA Grt., 1865
7767 **cecropia** (L., 1758)
    diana (Castiglioni, 1790), n. syn.
    macula (Reiff, 1911)
    uhlerii (Polacek, 1928)
    obscura (Sageder, 1933)
    albofasciata (Sageder, 1933)
7768 **columbia** (S. I. Smith, 1865)
7769 **gloveri** (Stkr., 1872)
    reducta (Neum., 1891)
    a. *nokomis* (Brodie, 1894)
      winonah (Brodie, 1894)
7770 **euryalus** (Bdv., 1855)
    californica (Grt., 1865)
    ceanothi (Bdv., 1869)
    rubra (N. & D., 1894)
    parvimacula (Grt., 1903)
    kasloensis (Ckll., 1914)
    cedrosensis (Ckll., 1914)

Sphingoidea

SPHINGIDAE

by RONALD W. HODGES

SPHINGINAE

Sphingini

AGRIUS Hbn., 1819
    HERSE; auth.
7771 **cingulata** (F., 1775)
    affinis (Goeze, 1780)
    pungens (Eschscholtz, 1821)
    druraei (Don., 1810)
    decolora (Hy. Edw., 1882)

COCYTIUS Hbn., 1819
    AMPHONYX Poey, 1832
    ANCISTROGNATHUS Wallgr., 1858
7772 **antaeus** (Drury, 1773)
    jatrophae (F., 1775)
    hydaspus (Cram., 1777)
    medor (Stoll, 1782)
7773 **duponchel** (Poey, 1832)
    godartii (Bdv., 1875)
    rivularis (Butler, 1875)
    affinis Roths., 1894

NEOCOCYTIUS Hodges, 1971
7774 **cluentius** (Cram., 1776)

MANDUCA Hbn., 1807
    PHLEGETHONTIUS Hbn., 1819
    PROTOPARCE Burmeister, 1856

MACROSILA Wlk., 1856
SYZYGIA G. & R., 1865
DILUDIA G. & R., 1865
CHLAENOGRAMMA Sm., 1887
7775 **sexta** (L., 1763)
   carolina (L., 1764)
   lycopersici (Bdv., 1875)
   nicotianae (Bdv., 1875)
7776 **quinquemaculata** (Haw., 1803)
   celeus (Hbn., 1821)
   wirti (Schaus, 1927), ab.
7777 **occulta** (R. & J., 1903)
7778 **rustica** (F., 1775)
   chionanthi (J. E. Smith, 1797)
7779 **albiplaga** (Wlk., 1856)
7780 **brontes** (Drury, 1773)
   pamphilius (Cram., 1782)
   collaris (Wlk., 1856)
   cubensis (Grt., 1865)
7781 **muscosa** (R. & J., 1903)
7782 **florestan** (Cram., 1782)
   brevimargo (Butler, 1875)
   cabnal (Schaus, 1932)
7783 **jasminearum** (Guér., 1829–1831)
   rotundata (Roths., 1894)

DOLBA Wlk., 1856
7784 **hyloeus** (Drury, 1773)
   prini (J. E. Smith, 1797), repl. name
   floridensis B. P. Clark, 1919
   hylaeus auth., missp.

DOLBOGENE R. & J., 1903
7785 **hartwegii** (Butler, 1875)
   manni B. P. Clark, 1917
   clarki (Hoffmann, 1924)

CERATOMIA Harr., 1839
DAREMMA Wlk., 1856
ISOGRAMMA R. & J., 1903,
   preocc. by Meek & Worthen, 1873
AUTOGRAMMA Jordan, 1946
7786 **amyntor** (Geyer, 1835)
   quadricornis Harr., 1839
   ulmi Bdv., 1875
7787 **undulosa** (Wlk., 1856)
   repentinus Clem., 1859
   polingi B. P. Clark, 1929
   borealis B. P. Clark, 1929
   engeli Chermock & Chermock, 1940,
   form
7788 **sonorensis** Hodges, 1971
7789 **catalpae** (Bdv., 1875)
   kanawahensis Sweadner, Chermock &
     Chermock, 1940, race
   kansensis Howe & Howe, 1950
7790 **hageni** Grt., 1874

ISOPARCE R. & J., 1903
7791 **cupressi** (Bdv., 1875)

SAGENOSOMA Jordan, 1946
DICTYOSOMA R. & J., 1903,
   preocc. by Temminck & Schlepel,
   1877
7792 **elsa** (Stkr., 1878)

PARATREA Grt., 1903
ATREUS Grt., 1886, preocc. by
   Koch, 1837
ATREIDES Holl., 1903

7793 **plebeja** (F., 1777)

SPHINX L., 1758
SPECTRUM Scop., 1777
HYLOICUS Hbn., 1819
LETHIA Hbn., 1819
HERSE Agassiz, 1846
LINTNERIA Butler, 1877
GARGANTUA Kby., 1892, preocc.
   by Jullien, 1880
MESOSPHINX Ckll., 1920
7794 **lugens** Wlk., 1856
   andromedae Bdv., 1870
7795 **geminus** R. & J., 1903
7796 **eremitus** (Hbn., 1823)
   sordida Harr., 1839
   mccrearyi B. P. Clark, 1929
7797 **eremitoides** Stkr., 1874
7798 **separata** Neum., 1885
7799 **istar** (R. & J., 1903)
7800 **chisoya** (Schaus, 1932)
7801 **leucophaeta** Clem., 1859
   lanceolata R. Felder, 1868
7802 **chersis** (Hbn., 1823)
   cinerea Harr., 1839
   oreodaphne Hy. Edw., 1873
   pallescens (R. & J., 1903)
7803 **vashti** Stkr., 1878
   albescens Tepper, 1881
   mordecai McD., 1923
   gerhardi (B. & Benj., 1924)
7804 **libocedrus** Hy. Edw., 1881
   insolita Lint., 1884
7805 **perelegans** Hy. Edw., 1874
   vancouverensis Hy. Edw., 1874
7806 **asella** (R. & J., 1903)
7807 **canadensis** Bdv., 1875
   plota Stkr., 1875
7808 **franckii** Neum., 1893
7809 **kalmiae** J. E. Smith, 1797
7810 **gordius** Cram., 1780
   poecila Stephens, 1828
   oslari (R. & J., 1903)
   borealis B. & P. Clark, 1920
   coxeyi Cadbury, 1931, form
   campestris McD., 1931
7811 **luscitiosa** Clem., 1859
   una Skin., 1903, var.
   bombax B. & Benj., 1927, race
   borealis B. P. Clark, 1930, preocc. by
     B. P. Clark, 1920
   benjamini B. P. Clark, 1932, repl.
   name
7812 **drupiferarum** J. E. Smith, 1797
   utahensis Hy. Edw., 1881
   marginalis B. P. Clark, 1936
7813 **dollii** Neum., 1881
   coloradus Sm., 1887
   australis B. P. Clark, 1922
7814 **sequoiae** Bdv., 1868
   engelhardti B. P. Clark, 1919
7815 **pinastri** L., 1758
   saniptri Stkr., 1876

LAPARA Wlk., 1856
ELLEMA Clem., 1859
EXEDRIUM Grt., 1882
7816 **coniferarum** (J. E. Smith, 1797)
   cana (Butler, 1877)
   halicarnie (Stkr., 1880)
7817 **bombycoides** Wlk., 1856

   harrisii (Clem., 1859)
   pineum (Lint., 1872)

Smerinthini

PROTAMBULYX R. & J., 1903
7818 **strigilis** (L., 1771)
   rubripennis (Butler, 1877), var.
7819 **carteri** R. & J., 1903

AMPLYPTERUS Hbn., 1819
AMBLYPTERUS Sm., 1888,
   missp.
7820 **donysa**; auth., not Druce, 1889

SMERINTHUS Latr., 1802
DILINA Dalm., 1816
EUSMERINTHUS Grt., 1877
COPISMERINTHUS Grt., 1886
7821 **jamaicensis** (Drury, 1773)
   geminatus Say, 1824
   tripartitus Grt., 1886, var.
   flavitincta Nixon, 1912, var.
   clarkii Franck, 1913, ab.
   gamma Ckll., race
7822 **cerisyi** Kby., 1837
   ophthalmica Bdv., 1855
   pallidulus Hy. Edw., 1875, var.
   vancouveriensis Butler, 1877
   astarte Stkr., 1885
   nigrescens B. P. Clark, 1919, ab.
   borealis B. P. Clark, 1929
7823 **saliceti** Bdv., 1875

PAONIAS Hbn., 1819
CALASYMBOLUS Grt., 1873
7824 **excaecatus** (J. E. Smith, 1797)
   pavonina Geyer, 1837
   pecosensis Ckll., 1905, var.
   borealis (B. P. Clark, 1929)
7825 **myops** (J. E. Smith, 1797)
   rosacearum (Bdv., 1836)
   cerasi (Bdv., 1875), var.
   sorbi (Bdv., 1875), var.
   tiliastri (Bdv., 1875)
   occidentalis (B. P. Clark, 1919)
   mccrearyi (B. P. Clark, 1929)
7826 **astylus** (Drury, 1773)
   io (Guér., 1829–31)
   integerrima (Harr., 1839)

LAOTHOE F., 1807
AMORPHA Hbn., 1809
CRESSONIA G. & R., 1865
7827 **juglandis** (J. E. Smith, 1797), n.
   comb.
   pallens (Stkr., 1873), n. comb.
   instibilis (Butler, 1877), n. comb.
   robinsonii (Butler, 1877), n. comb.
   hyperbola (Slosson, 1890), var., n.
   comb.
   alpina (B. P. Clark, 1927), n. comb.
   manitobae (B. P. Clark, 1930), n.
   comb.

PACHYSPHINX R. & J., 1903
7828 **modesta** (Harr., 1839)
   princeps (Wlk., 1856)
   populicola (Bdv., 1875)
   cablei (Reizenstein, 1881)
   borealis B. P. Clark, 1929

7829 **occidentalis** (Hy. Edw., 1875)
   imperator (Stkr., 1878)
   kunzei R. & J., 1903, sum. f.

MACROGLOSSINAE

Dilophonotini

PSEUDOSPHINX Burmeister, 1856
7830 **tetrio** (L., 1771)
   hasdrubal (Cram., 1780)
   obscura Butler, 1877

ISOGNATHUS Felder & Felder, 1862
   TATOGLOSSUM Butler, 1877
7831 **rimosus** (Grt., 1865)
   menechus (Grt., 1865)
   andae (G. & R., 1868)
   silenus (G. & R., 1868)
   excelsior (Bdv., 1875)
   inclitus Hy. Edw., 1887

ERINNYIS Hbn., 1819
   DILOPHONOTA Burmeister, 1856
7832 **alope** (Drury, 1770)
   flavicans (Goeze, 1780)
   fasciata (Swainson, 1823)
   edwardsii (Butler, 1881)
7833 **lassauxii** (Bdv., 1859)
   merianae Grt., 1865
   janiphae (Bdv., 1875)
7834 **ello** (L., 1758)
7835 **oenotrus** (Cram., 1782)
   penaeus (F., 1787)
   picta (Sepp, 1848)
   melancholica Grt., 1865
   piperis (G. & R., 1868)
   hippothoon (Burmeister, 1878)
7836 **crameri** (Schaus, 1898)
7837 **obscura** (F., 1775)
   stheno Geyer, 1829
   cinerosa G. & R., 1865, nom. nud.
   rhaebus (Bdv., 1870)
7838 **domingonis** (Butler, 1875)
   festa (Hy. Edw., 1882)
7839 **guttularis** (Wlk., 1856)

PHRYXUS Hbn., 1819
   GRAMMODIA R. & J., 1903
7840 **caicus** (Cram., 1777)

PACHYLIA Wlk., 1856
7841 **ficus** (L., 1758)
   crameri (Mén., 1857)
   lyncea Clem., 1859

PACHYLIOIDES Hodges, 1971
7842 **resumens** (Wlk., 1856)
   inconspicua (Wlk., 1856)
   versuta (Clem., 1859)
   tristis (G. & R., 1868)

MADORYX Bdv., 1875
7843 **pseudothyreus** (Grt., 1865)

CALLIONIMA Luc., 1857
   CALLIOMMA Wlk., 1856, preocc.
     by Agassiz, 1846
   EUCHERYX Bdv., 1875
7844 **parce** (F., 1775)
   licastus (Cram., 1782)

   galianna (Burmeister, 1856)
7845 **falcifera** (Gehlen, 1943)

PERIGONIA H.-S., 1854
7846 **lusca** (F., 1777)
   bahamensis B. P. Clark, 1919

AELLOPOS Hbn., 1819
7847 **tantalus** (L., 1758)
   ixion (L., 1758)
   zonata (Drury, 1773)
   terpunctata (Goeze, 1780)
   sisyphus (Burmeister, 1856)
7848 **clavipes** (R. & J., 1903)
7849 **titan** (Cram., 1777)
7850 **fadus** (Cram., 1776)
   annulosus (Swainson, 1823)
   balteata (Kirtland, 1851)

ENYO Hbn., 1819
   EPISTOR Bdv., 1875
7851 **lugubris** (L., 1771)
   fegeus (Cram., 1780)
   luctuosus (Bdv., 1875)
7852 **ocypete** (L., 1758)
   camertus (Cram., 1780)
   danus (Cram., 1780)

HEMARIS Dalm., 1816
   HAEMORRHAGIA G. & R., 1865
   CHAMAESESIA Grt., 1877
7853 **thysbe** (F., 1775)
   pelasgus (Cram., 1780)
   cimbiciformis (Steph., 1828)
   ruficaudis (Kby., 1837)
   fuscicaudis (Wlk., 1856)
   buffaloensis (G. & R., 1867)
   floridensis (G. & R., 1867)
   uniformis (G. & R., 1868)
   etolus (Bdv., 1875)
   pyramus (Bdv., 1875)
7854 **gracilis** (G. & R., 1865)
7855 **diffinis** (Bdv., 1836)
   thetis (Bdv., 1855)
   axillaris (G. & R., 1868)
   tenuis Grt., 1873
   marginalis Grt., 1873
   palpalis Grt., 1874
   grotei (Butler, 1874)
   fumosa (Stkr., 1874)
   aethra (Stkr., 1875)
   cynoglossum Hy. Edw., 1875
   rubens Hy. Edw., 1875
   metathetis Butler, 1877
   ariadne (B. & McD., 1910)
   mcdunnoughi B. P. Clark, 1927
   jordani B. & Benj., 1927
7856 **senta** (Stkr., 1878)
   brucei French, 1890

Philampelini

EUMORPHA Hbn., 1807
   PHOLUS Hbn., 1819
   DAPHNIS Hbn., 1819
   ARGEUS Hbn., 1819
   DUPO Hbn., 1819
   PHILAMPELUS Harr., 1839
7857 **anchemola** (Cram., 1780)
7858 **satellitia** (L., 1771)
   licaon (Cram., 1776)

7859 **pandorus** (Hbn., 1821)
   ampelophaga (Wlk., 1856)
7860 **intermedia** (B. P. Clark, 1917)
7861 **achemon** (Drury, 1773)
   crantor (Cram., 1777)
7862 **eacus** (Cram., 1780)
   megeacus (Hbn., 1819)
7863 **typhon** (Klug, 1836)
7864 **vitis** (L., 1758)
   hornbeckiana (Harr., 1839)
   linnei (G. & R., 1865)
7865 **fasciata** (Sulz., 1776)
   jussieuae (Hbn., 1819)
   strigilis (Vogel, 1822)
7866 **labruscae** (L., 1758)
   clotho (F., 1775)

Macroglossini

CAUTETHIA Grt., 1865
   OENOSANDA Wlk., 1856, preocc.
     by Wlk., 1856
7867 **grotei** Hy. Edw., 1882
7868 **spuria** (Bdv., 1875)
7869 **yucatana** (B. P. Clark, 1919)

SPHECODINA Blanchard, 1840
   BRACHYNOTA Bdv., 1870
   MAREDUS Kby., 1880
7870 **abbottii** (Swainson, 1821)

DEIDAMIA Clem., 1859
   TRICHOLON Bdv., 1875
7871 **inscripta** (Harr., 1839)

ARCTONOTUS Bdv., 1852
7872 **lucidus** Bdv., 1852
   clarki B. & Benj., 1923

AMPHION Hbn., 1819
7873 **floridensis** B. P. Clark, 1920, n. stat.
   nessus (Cram., 1777), preocc. by
     Drury, 1773

PROSERPINUS Hbn., 1819
   LEPISESIA Grt., 1865
   POGOCOLON Bdv., 1875
   DIENECES Butler, 1881
7874 **gaurae** (J. E. Smith, 1797)
   circae Hy. Edw., 1882
   deceptiva B. & Benj., 1924
7875 **juanita** (Stkr., 1877)
   oslari R. & J., 1903
7876 **clarkiae** (Bdv., 1852)
   victoria (Grt., 1874)
7877 **flavofasciata** (Wlk., 1856)
   ulalume (Stkr., 1878)
   rachel (Bruce, 1901)
7878 **vega** (Dyar, 1903)
7879 **terlooii** Hy. Edw., 1875
   terlooti auth., missp.

EUPROSERPINUS G. & R., 1865
7880 **phaeton** G. & R., 1865
   erato (Bdv., 1868)
   mojave Comst., 1938
7881 **euterpe** Hy. Edw., 1888
7882 **wiesti** Sperry, 1939

MACROGLOSSUM Scop., 1777
   PSITHYROS Hbn., 1822

RHAMPHOSCHISMA Wallgr., 1858
BOMBYLIA Kby., 1892
7883 **stellatarum** (L., 1758)
flavidum (Retzius, 1783)
nigrum Cosmovici, 1892

DARAPSA Wlk., 1856
AMPELOECA R. & J., 1903
7884 **versicolor** (Harr., 1839)
lutescens (B. P. Clark, 1920), ab.
7885 **myron** (Cram., 1780)
pampinatrix (J. E. Smith, 1797)
cnotus (Hbn., 1823)
lutescens (B. P. Clark, 1920), ab.
texana (B. P. Clark, 1920)
7886 **pholus** (Cram., 1776)
choerilus (Cram., 1780)
azaleae (J. E. Smith, 1797)
chlorinda (Butler, 1877)
brodiei B. P. Clark, 1929

XYLOPHANES Hbn., 1819
DEILONCHE Grt., 1886
GONENYO Roths., 1894
7887 **pluto** (F., 1777)
croesus (Dalm., 1823)
thorates (Hbn., 1827–31)
7888 **porcus** (Hbn., 1823)
7889 **falco** (Wlk., 1856)
fugax (Bdv., 1870)
mexicana (Erschoff, 1877)
7890 **tersa** (L., 1771)
7891 **libya** (Druce, 1878)

HYLES Hbn., 1819
CELERIO Agassiz, 1846
HAWAIINA Tutt, 1903
7892 **euphorbiae** (L., 1758)
7893 **gallii** (Rottemburg, 1775)
intermedia (Kby., 1837)
chamaenerii (Harr., 1839)
oxybaphi (Clem., 1859)
canadensis (Gn., 1868)
7894 **lineata** (F., 1775)
daucus (Cram., 1777)

Noctuoidea

NOTODONTIDAE

by JOHN G. FRANCLEMONT

CLOSTERA Samouelle, 1819
MELALOPHA Hbn., 1806, suppr.
(ICZN Op. 97)
ICHTHYURA Hbn., 1819
MELALOPHA Hbn., 1822
PYGAERA; auth., part, not Ochs.,
1810
7895 **albosigma** Fitch, 1856
specifica (Dyar, 1892), n. stat.
7896 **inclusa** (Hbn., 1829–31)
anastomosis; J. E. Smith, 1797, not
L., 1758
americana Harr., 1841
inversa (Pack., 1864)
palla (French, 1882)
a. **jocosa** (Hy. Edw., 1886)

7897 **inornata** (Neum., 1882)
7898 **strigosa** (Grt., 1882)
luculenta (Hy. Edw., 1886), n. stat.
7899 **paraphora** (Dyar, 1921)
7900 **brucei** (Hy. Edw., 1885)
a. **multnoma** (Dyar, 1892)
alethe (N. & D., 1893)
7901 **apicalis** (Wlk., 1855)
vau Fitch, 1859
indentata (Pack., 1864)
a. **ornata** (G. & R., 1868)
incarcerata Bdv., 1869
astoriae (Hy. Edw., 1886), n. stat.
bifira (Hy. Edw., 1886), n. stat.

DATANA Wlk., 1855
PETASIA; Westwood, 1837, not
Steph., 1828
PYGAERA; Harr., 1841, not Ochs.,
1810
EUMETOPONA Fitch, 1856
7902 **ministra** (Drury, 1773)
ruficollis Wlk., 1862
a. **californica** Dyar, 1890
7903 **angusii** G. & R., 1866
ministra; J. E. Smith, 1797, part, not
Drury, 1773
7904 **drexelii** Hy. Edw., 1884
7905 **major** G. & R., 1866
ministra; J. E. Smith, 1797, part, not
Drury, 1773
7906 **contracta** Wlk., 1855
7907 **integerrima** G. & R., 1866
a. **cochise** Dyar, 1905
7908 **perspicua** G. & R., 1865
a. **mesillae** Ckll., 1897, rev. stat.
discalis Dyar, 1923, rev. stat.
eileena Dyar, 1923
infusa Dyar, 1923, form
perfusa Dyar, 1923, n. stat.
b. **opposita** B. & Benj., 1927
7909 **robusta** Stkr., 1878
7910 **modesta** Beutenmüller, 1890
7911 **ranaeceps** (Guér.-Méneville, 1832)
floridana Graef, 1880
palmii Beutenmüller, 1890, n, stat.
7912 **diffidens** Dyar, 1917
7913 **neomexicana** Doll, 1911
7914 **chiriquensis** Dyar, 1895

NADATA Wlk., 1855
ALASTOR Bdv., 1869, preocc. by
Lepeletier, 1841
7915 **gibbosa** (J. E. Smith, 1797)
doubledayi Pack., 1864
rubripennis N. & D., 1893, n.
stat.
7916 **oregonensis** Butler, 1881, n. stat.
behrensii Hy. Edw., 1885, n. syn.

HYPERAESCHRA Butler, 1880
7917 **georgica** (H.-S., 1855)
7918 **tortuosa** Tepper, 1881

PERIDEA Steph., 1828
LOPHODONTA Pack., 1864
MESODONTA Matsumura, 1920
FUSADONTA Matsumura, 1920
7919 **basitriens** (Wlk., 1855)
7920 **angulosa** (J. E. Smith, 1797)

7921 **ferruginea** (Pack., 1864)
a. **reducta** (McD., 1932)

PHEOSIA Hbn., 1819
LEIOCAMPA Steph., 1828
7922 **rimosa** Pack., 1864
californica (Stretch, 1872)
7923 **portlandia** Hy. Edw., 1886
descherei (Neum., 1892)

ODONTOSIA Hbn., 1819
7924 **elegans** (Stkr., 1885)
7925 **grisea** (Stkr., 1885), rev. stat.
nototaria (Hy. Edw., 1885)

NOTODONTA Ochs., 1810
NOTODON Meigen, 1830
7926 **scitipennis** Wlk., 1862, rev. stat.
stragula Grt., 1864, rev. stat
a. **manitou** N. & D., 1893
b. **ochreata** B. & McD., 1912
7927 **pacifica** Behr, 1892
7928 **simplaria** Graef, 1881

NERICE Wlk., 1855
7929 **bidentata** Wlk., 1855

ELLIDA Grt., 1876
7930 **caniplaga** (Wlk., 1856)
transversata (Wlk., 1865)
gelida Grt., 1876

GLUPHISIA Bdv., 1828
GLYPHIDIA Agassiz, 1846, emend.
MELIA Neum., 1892, preocc. by
Bosc, 1813
EUMELIA Neum., 1893
CERURIDIA Dyar, 1893
PARAGLUPHISIA Djakonov,
1926, n. syn.
7931 **septentrionis** Wlk., 1855
septentrionalis auth.
clandestina (Wlk., 1861)
trilineata Pack., 1864
opaca B. & Benj., 1927, form
a. **quinquelinea** Dyar, 1892
b. **ridenda** Hy. Edw., 1886
rupta Hy. Edw., 1886
c. **albofascia** Hy. Edw., 1886
formosa Hy. Edw., 1886
7932 **wrightii** Hy. Edw., 1886, rev. stat.
7933 **avimacula** Hudson, 1891
slossoniae (Dyar, 1893), form
slossonii (Dyar, 1893) incorr. orig. spell.
7934 **lintneri** (Grt., 1877)
pretians Franc., 1941, form
7935 **severa** Hy. Edw., 1886
danbyi (Neum., 1892), form
normalis Dyar, 1904, form

FURCULA Lamarck, 1816
HARPYIAS Hbn., 1819, n. syn.
CERURA; auth., not Schr., 1802
HARPYIA; auth., not Ochs., 1810
DICRANURA; auth., not R. L., 1817
7936 **borealis** (Guér.-Méneville, 1832), n. comb.
furcula; J. E. Smith, 1797, not Cl.,
1759
7937 **cinerea** (Wlk., 1865), n. comb.
a. **cinereoides** (Dyar, 1890)

b. **paradoxa** (Dyar, 1892)
paradoxa (Behr, 1885), nom. nud.
c. **placida** (Dyar, 1892)
d. **wileyi** (Dyar, 1922)
7938 **nivea** (Neum., 1891), n. comb.
a. **meridionalis** (Dyar, 1892)
b. **niveata** (B. & Benj., 1924)
7939 **occidentalis** (Lint., 1878), n. comb.
a. **gigans** (McD., 1922)
deorum (Dyar, 1922)
7940 **scolopendrina** (Bdv., 1869), n. comb.
albicoma (Stkr., 1885), rev. stat.
pluvialis (Dyar, 1922), rev. stat.
a. **aquilonaris** (Lint., 1878)
7941 **modesta** (Hudson, 1891), n. comb.

CERURA Schr., 1802
ANDRIA Hbn., 1806, suppr.
(ICZN Op. 97)
HARPYIA; auth., not Ochs., 1810
DICRANURA R. L., 1817
ANDRIA Hbn., 1822
PANIA Dalman, 1823
DICRANOURA Berthold, 1827
7942 **scitiscripta** Wlk., 1865
a. **multiscripta** Riley, 1875, rev. stat.
b. **canadensis** McD., 1932
7943 **candida** Lint., 1878
7944 **rarata** Wlk., 1865

NYSTALEA Gn., 1852
CYRRHESTA Wlk., 1857
EUNYSTALEA Grt., 1895
CONGRUIA Dyar, 1908
7945 **indiana** Grt., 1884
7946 **collaris** Schaus, 1910
quaesita Draudt, 1932
7947 **eutalanta** Dyar, 1921
ebalea; auth., not Cram., 1784

ELYMIOTIS Wlk., 1857
CICYNNA Wlk., 1858
7948 **notodontoides** Wlk., 1857
sericea (Wlk., 1858)
phaleroides (Wlk., 1865)

HIPPIA Mösch., 1878
EDEMA; auth., not Wlk., 1855
HARMA Wlk., 1858, preocc. by
Doubleday, 1848
ELASMIA Mösch., 1886, preocc.
by Hbn., 1822
7949 **packardii** (Morr., 1875)
7950 **insularis** (Grt., 1866)
lignosa (Mösch., 1886)

SYMMERISTA Hbn., 1821
EDEMA Wlk., 1855
7951 **albifrons** (J. E. Smith, 1797)
albicosta (Hbn., 1808–09)
7952 **canicosta** Franc., 1946
7953 **leucitys** Franc., 1946
7954 **sauvis** (Barnes, 1901)
7955 **zacualpana** (Draudt, 1932)

PENTOBESA Schaus, 1901
7956 **valta** Schaus, 1901
roberta (Dyar, 1909)
placida Schaus, 1910

DASYLOPHIA Pack., 1864
7957 **anguina** (J. E. Smith, 1797)
cuculifera (H.-S., 1855)
punctata (Wlk., 1865)
signata (Wlk., 1865)
cana (Wlk., 1869)
puntagorda Slosson, 1892, rev. stat.
a. **saturata** Barnes, 1901
7958 **thyatiroides** (Wlk., 1862)
interna Pack., 1864
tripartita (Wlk., 1865)
7959 **seriata** (Druce, 1887)
melanopa Barnes, 1901
indecoris Schaus, 1911

NOTELA Schaus, 1901
7960 **jaliscana** Schaus, 1901
ramosa Draudt, 1932

DIDUGUA Druce, 1891
7961 **argentilinea** Druce, 1891

AFILIA Schaus, 1901
7962 **oslari** Dyar, 1904
moqui Barnes, 1904

SKEWESIA Dyar, 1916
SCEVESIA Dyar, 1916, incorr.
orig. spell.
7963 **angustiora** (B. & McD., 1910), n.
comb.

CARGIDA Schaus, 1901
7964 **pyrrha** (Druce, 1898), extralim.
a. **intensa** Roths., 1917

RIFARGIA Wlk., 1862
7965 **bichorda** (Hamp., 1901)
7966 **distinguenda** (Wlk., 1856), stray
dubia (Mösch., 1877)
7967 **lineata** Druce, 1887, stray

LITODONTA Harv., 1876
7968 **hydromeli** Harv., 1876
fusca Harv., 1876, form
7969 **contrasta** B. & McD., 1910, rev. stat.
7970 **wymola** (Barnes, 1905)
7971 **aonides** (Stkr., 1899), n. comb.
7972 **gigantea** B. & Benj., 1924
7973 **alpina** Benj., 1932

MISOGADA Wlk., 1865
7974 **unicolor** (Pack., 1864)
marina (Pack., 1864)
sobria Wlk., 1865

MACRUROCAMPA Dyar, 1893
FENTONIA; auth., not Butler, 1881
7975 **marthesia** (Cram., 1780)
tesella (Pack., 1864)
turbida (Wlk., 1865)
elongata (G. & R., 1867)
nigra (F. H. Chermock, 1929), form
a. **manitobensis** (McD., 1932)
7976 **dorothea** Dyar, 1896
miranda Dyar, 1905, rev. stat.
a. **frisoni** (B. & Benj., 1927)

HETEROCAMPA Doubleday, 1841
CERITA Wlk., 1855

TRICHOTIS Felder, 1868
BALIA Doubleday, 1869, n. syn.
7977 **astarte** Doubleday, 1841
picta (Felder, 1868)
menas (Doubleday, 1869)
chapmanii Grt., 1881
perolivata Pack., 1895
7978 **astartoides** Benj., 1932
7979 **simulans** B. & Benj., 1924
7980 **rufinans** (Dyar, 1921)
7981 **secessionis** Benj., 1932
7982 **varia** Wlk., 1855
georgiana (Dyar, 1921)
baryspus (Dyar, 1921)
7983 **obliqua** Pack., 1864
trouvelotii Pack., 1864
brunnea G. & R., 1867
7984 **cubana** Grt., 1866
7985 **subrotata** Harv., 1874
superba Hy. Edw., 1885
a. **celtiphaga** Harv., 1874
7986 **ditta** B. & McD., 1910
agapa Schaus, 1928, n. syn.
7987 **benitensis** A. Blanchard, 1971
7988 **belfragei** Grt., 1879
7989 **incongrua** B. & Benj., 1924
7990 **umbrata** Wlk., 1855
athereo (Doubleday, 1869)
a. **pulverea** G. & R., 1867
nigra F. H. Chermock, 1927, form
7991 **averna** B. & McD., 1910
pasathelys (Dyar, 1921)
7992 **amanda** B. & L., 1921
7993 **lunata** Hy. Edw., 1884
plumosa Hy. Edw., 1886
7994 **guttivitta** (Wlk., 1855)
albiplaga (Wlk., 1856)
mucorea (H.-S., 1856)
harrisii (Pack., 1864)
cinereus (Pack., 1864)
indeterminata (Wlk., 1865)
doubledayi Scudder, 1869
hugoi Chermock & Chermock, 1940,
form
7995 **biundata** Wlk., 1855
semiplaga Wlk., 1861
olivata (Pack., 1864)
viridescens (Wlk., 1865), n. syn.
mollis Wlk., 1865
7996 **ruficornis** Dyar, 1905
7997 **zayasi** (Torre & Alayo, 1959), n. record

LOCHMAEUS Doubleday, 1841, rev. stat.
TADANA Wlk., 1855
7998 **manteo** Doubleday, 1841
cinerascens (Wlk., 1855)
subalbicans (Grt., 1864)
7999 **bilineata** (Pack., 1864)
turbida (Wlk., 1865)
associata (Wlk., 1865)
ulmi (Harr., 1869)
boweri (Chermock & Chermock,
1940), form
a. **exsanguis** (Dyar, 1906)

THEROA Schaus, 1901
8000 **zethus** (Druce, 1898)

PRAESCHAUSIA Benj., 1932
8001 **zapata** (Schaus, 1920)

URSIA B. & McD., 1911
8002 **noctuiformis** B. & McD., 1911
8003 **furtiva** A. Blanchard, 1971

SCHIZURA Doubleday, 1841
 DICENTRIA H.-S., 1856
 OEDEMASIA Pack., 1864
 COELODASYS Pack., 1864
8004 **biedermani** B. & McD., 1911
 clammenhoa (Dyar, 1914)
8005 **ipomoeae** Doubleday, 1841
 ipomaeae Doubleday, 1841, incorr.
  orig. spell.
 biguttatus (Pack., 1864)
 confusa (Wlk., 1865)
 ducens (Wlk., 1865)
 corticea (Wlk., 1865)
 compta (Wlk., 1865)
 nigrosignata (Wlk., 1865)
 telifer (Grt., 1880), form
 cinereofrons (Pack., 1864), form
  ustipennis (Wlk., 1865)
8006 **badia** (Pack., 1864)
 significata (Wlk., 1865)
8007 **unicornis** (J. E. Smith, 1797)
 edmandsii (Pack., 1864)
 semirufescens (Wlk., 1865), n. syn.
 humilis (Wlk., 1865)
 a. **deserta** Barnes, 1929
 b. **conspecta** (Hy. Edw., 1874)
8008 **errucata** Dyar, 1906
8009 **apicalis** (G. & R., 1866)
8010 **concinna** (J. E. Smith, 1797)
 nitida (Pack., 1864)
 a. **salicis** (Hy. Edw., 1876)
  riversii (Behr, 1890)
8011 **leptinoides** (Grt., 1864)
 mustelina (Pack., 1864)

OLIGOCENTRIA H.-S., 1856
 IANASSA Wlk., 1855, preocc. by
  Muenster, 1839, Agassiz, 1846
 XYLINOIDES Pack., 1864
 HATIMA Wlk., 1865
 PHYA Druce, 1887
 DICENTRIA; auth., not H.-S., 1856
8012 **semirufescens** (Wlk., 1865), n. comb.
 eximia (Grt., 1881)
8013 **alpica** (Benj., 1932), n. comb.
8014 **pallida** (Stkr., 1899), n. comb.
8015 **coloradensis** (Hy. Edw., 1885), n.
 comb.
8016 **perangulata** (Hy. Edw., 1882), n. comb.
8017 **lignicolor** (Wlk., 1855)
 virgata (Pack., 1864)
 lignigera (Wlk., 1865)
8018 **pinalensis** (Benj., 1932), n. comb.
8019 **paradisus** (Benj., 1932), n. comb.
 delicatoides (Benj., 1932), n. comb.,
  n. syn.
8020 **delicata** (Dyar, 1905)

EUHYPARPAX Beutenmüller, 1893
 RIBALDIA Dyar, 1916, n. syn.
8021 **rosea** Beutenmüller, 1893

HYPARPAX Hbn., 1825
 SANGATA Wlk., 1860
8022 **aurora** (J. E. Smith, 1797)
 rosea (Wlk., 1860)

 venusta (Wlk., 1865)
8023 **venus** Neum., 1892
8024 **minor** (B. & Benj., 1924), n. stat.
8025 **aurostriata** Graef, 1888, rev. stat.
8026 **perophoroides** (Stkr., 1876)
 tyria Slosson, 1894

LIRIMIRIS Wlk., 1865
8027 **truncata** (H.-S., 1856)

ASTYLIS Bdv., 1872
 CRINO; auth., not Hbn., 1821
 CRINODES; auth., not H.-S., 1856
8028 **biedermani** (Skin., 1905), n. comb.

PSEUDHAPIGIA Schaus, 1901
8029 **brunnea** Schaus, 1901
 estrella (Barnes, 1904)

HEMICERAS Gn., 1852
 ECREGMA Wlk., 1858
 COMIDAVA Wlk., 1863
 EPICORIA Wlk., 1865
 GADIANA Wlk., 1865
 EULOPHOPTERYX Mösch., 1877
8030 **cadmia** Gn., 1852, extralim. (reported
 in error?)
 obliquilinea (Wlk., 1862)

DIOPTIDAE
by JOHN G. FRANCLEMONT

PHRYGANIDIA Pack., 1864
8031 **californica** Pack., 1864

ZUNACETHA Wlk., 1863
8032 **annulata** (Guér.-Méneville, 1844)
 bipartita Wlk., 1863
 nervosa (Felder & Rogenhofer, 1875)

ARCTIIDAE
by JOHN G. FRANCLEMONT

PERICOPINAE

Pericopini

GNOPHAELA Wlk., 1854
 OMOIALA Grt., 1864
 LAMPROSOMA Grt., 1864,
  preocc. by Kirby, 1819
 CALLALUCIA Grt., 1865
8033 **clappiana** Holl., 1891
 ruidosensis Ckll., 1904, rev. stat.
8034 **latipennis** (Bdv., 1852)
 hopfferi G. & R., 1868
8035 **aequinoctialis** (Wlk., 1854)
 disjuncta Hy. Edw. 1885, rev. stat.
8036 **discreta** Stretch, 1875, rev. stat.
 arizona French, 1884
 arizonae auth.
 morrisoni Druce, 1885
8037 **vermiculata** (Grt., 1864), rev. stat.
 continua Hy. Edw., 1881

COMPOSIA Hbn., 1820
 COCASTRA Bdv., 1870
8038 **fidelissima** H.-S., 1866

 olympia (Butler, 1871)
 a. **vagrans** Bates, 1934
  olympia; French, 1890, not Butler, 1871

PHALOESIA Wlk., 1854
8039 **saucia** Wlk., 1854
 gentilis (Bdv., 1870)
 fulvicollis Butler, 1875
 venezuelae Butler, 1875
 chalybea Butler, 1875
 chalybdea auth., missp.

DARITIS Wlk., 1855
 DORIMENA Bdv., 1870
8040 **howardi** Hy. Edw., 1886
 thetis; auth., not Klug, 1836

Doaini

DOA N. & D., 1894
 EMYDIA; auth.
8041 **ampla** (Grt., 1878)

LEUCULODES Dyar, 1903
8042 **lacteolaria** (Hulst, 1886)
 lacteolata (Hulst, 1896), missp.

LITHOSIINAE

Lithosiini

EILEMA Hbn., 1819
 LEXIS; auth.
 TRIGRIOIDES Fbs., 1960, missp.,
  not Tigrioides Butler, 1877
8043 **bicolor** (Grt., 1864), n. comb.
 argillacea (Pack., 1864)
8044 **complana** (L., 1758), extralim.
 allegheniensis (Holl., 1903)

CRAMBIDIA Pack., 1864
8045 **lithosioides** Dyar, 1898
8045.1 **pallida** Pack., 1864
8046 **uniformis** Dyar, 1898
8047 **dusca** B. & McD., 1913
8048 **myrlosea** Dyar, 1917
8049 **suffusa** B. & McD., 1912
8050 **impura** B. & McD., 1913
8051 **casta** (Pack., 1869)
 candida (Hy. Edw., 1873)
8052 **pura** B. & McD., 1913
8053 **cephalica** (Grt. & Rob., 1870)

AGYLLA Wlk., 1854
 SALAPOLA Wlk., 1863
 CRAMBOMORPHA Felder, 1868
 MACROCRAMBUS Kby., 1892
8054 **septentrionalis** B. & McD., 1911

INOPSIS Felder, 1868
 JNOPSIS Felder, 1868, variant spell.
8055 **modulata** (Hy. Edw., 1884)
8056 **funerea** (Grt., 1883), n. comb.

GNAMPTONYCHIA Hamp., 1900
8057 **ventralis** B. & L., 1921

GARDINIA Kby., 1892
 CHRYSOCALE; Butler, 1877, not
  Wlk., 1854

GARDINICA Strand, 1932, repl. name
8058 **anopla** Hering, 1925, n. record.
   anopia (Bryk., 1931), missp.

   CISTHENE Wlk., 1854
   ILLICE Wlk., 1859
   PYRALIDIA Felder, 1868
   BYSSOPHAGA Stretch, 1872
   OZODANIA Dyar, 1899
8059 **subrufa** (B. & McD., 1913)
   schwarziorum (Dyar, 1899), part
8060 **unifascia** G. & R., 1868
   ruptifascia (B. & McD., 1918)
   mexicana (Draudt, 1918)
8061 **kentuckiensis** (Dyar, 1904)
8062 **liberomacula** (Dyar, 1904)
   basijuncta (B. & McD., 1918)
8063 **deserta** (Felder, 1868)
   nexa (Bdv., 1869)
   grisea (Pack., 1872)
8064 **faustinula** (Bdv., 1869)
   fusca Stretch, 1872
8065 **dorsimacula** (Dyar, 1904)
8066 **tenuifascia** Harv., 1875
   schwarziorum (Dyar, 1899)
   interrupta (Draudt, 1918)
8067 **plumbea** Stretch, 1885
   injecta (Dyar, 1904)
   gamma (Dyar, 1904)
   flavicosta (Draudt, 1918)
8068 **striata** Ottol., 1898
   apicipicta (Strand, 1922)
8069 **perrosea** (Dyar, 1904)
8070 **angelus** (Dyar, 1904)
8071 **subjecta** Wlk., 1854
   bellicula (Dyar, 1921)
8072 **packardii** (Grt., 1863)
8073 **conjuncta** (B. & McD., 1913)
8074 **barnesii** (Dyar, 1904)
   flavula (B. & McD., 1918)
   flava (Draudt, 1919)
   costimacula (Draudt, 1918)
8075 **picta** (B. & McD., 1918)
   texensis (Draudt, 1918)
8076 **juanita** B. & Benj., 1925
8077 **coronado** C. Knowlton, 1967
8078 **martini** C. Knowlton, 1967

   PTYCHOGLENE Felder, 1868
8079 **coccinea** (Hy. Edw., 1886)
8080 **phrada** Druce, 1889
   flammans Dyar, 1898
8081 **sanguineola** (Bdv., 1870)

   PROPYRIA Hamp., 1898
8082 **schausi** (Dyar, 1898)
   fulgens; N. & D., 1893, not Hy.
     Edw., 1881

   LYCOMORPHA Harr., 1839
   ANATOLMIS Pack., 1864
8083 **grotei** (Pack., 1864)
   palmerii Pack., 1872
   a. **pulchra** Dyar, 1898
8084 **regulus** (Grinnell, 1903)
8085 **fulgens** (Hy. Edw., 1881)
   tenuimargo Holl., 1903
8086 **splendens** B. & McD., 1912
8087 **pholus** (Drury, 1773)
   a. **miniata** Pack., 1872

8088 **desertus** Hy. Edw., 1881

   HYPOPREPIA Hbn., 1827–31
8089 **miniata** (Kby., 1837)
   vittata (Harr., 1841)
   a. **mississippiensis** B. & Benj., 1926
8090 **fucosa** Hbn., 1827–31
   a. **subornata** N. & D., 1893
   inornata Ottol., 1895
   dollii Dyar, 1906, form
   b. **tricolor** (Fitch, 1857), n. stat.
   plumbea Hy. Edw., 1886, rev. stat.
8091 **cadaverosa** Stkr., 1880
8092 **inculta** Hy. Edw., 1882

   HAEMATOMIS Hamp., 1900
8093 **mexicana** (Druce, 1885)

   BRUCEIA Neum., 1893
8094 **pulverina** Neum., 1893
8095 **hubbardi** Dyar, 1898

   EUDESMIA Hbn.,1823
   RUSCINO Wlk., 1854
   GERBA Wlk., 1865
8096 **arida** (Skin., 1906)
8097 **laetifera** (Wlk., 1865), rev. stat., n.
   record
   latifasciatus (Butler, 1875)

   CLEMENSIA Pack., 1864
   UXIA Wlk., 1866
   REPA Wlk., 1866
8098 **albata** Pack., 1864
   albida (Wlk., 1866)
   cana (Wlk., 1866)
   a. **umbrata** Pack., 1872
   irrorata Hy. Edw., 1873

   PAGARA Wlk., 1856
   COMACLA Wlk., 1865
   VANESSODES G. & R., 1870
8099 **simplex** Wlk., 1856
   murina (Wlk., 1865)
   clarus (G. & R., 1870)
   texana (French, 1889)
8100 **fuscipes** (Grt., 1883)

   NEOPLYNES Hamp., 1900
8101 **eudora** (Dyar, 1894)

Afridini

   AFRIDA Mösch., 1886
   ARESIA B. & McD., 1913
8102 **ydatodes** Dyar, 1913
   parva (B. & McD., 1913)
8103 **minuta** (Druce, 1885)

   COMACHARA Franc., 1939
8104 **cadburyi** Franc., 1939

Acsalini

   ACSALA Benj., 1935
8104.1 **anomala** Benj., 1935

ARCTIINAE

Callimorphini

   UTETHEISA Hbn., 1819
   DEIOPEIA Curt., 1827
   EUCHELIA Bdv., 1828, part
   UTETHESIA Moore, 1859, missp.
   DIOPEA Burmeister, 1878, missp.
8105 **ornatrix** (L., 1758)
   stretchii (Butler, 1877)
   pura (Butler, 1877)
   daphoena Dyar, 1914
   butleri Dyar, 1914
8106 **bella** (L., 1758)
   venusta (Dalm., 1823)
   speciosa (Wlk., 1854)
   hybrida (Butler, 1877)
   intermedia (Butler, 1877)
   terminalis N. & D., 1893
   nova Sm., 1910
   grossbecki Strand, 1919

   HAPLOA Hbn., 1820
   CALLIMORPHA; auth.
   HYPERCOMPA; Wlk., 1855, not
     Steph., 1828
8107 **clymene** (Brown, 1776)
   interruptomarginata Palisot de
     Beauvois, 1821
   comma (Wlk., 1855)
8108 **colona** (Hbn., 1800–03)
   clymene (Esp., 1794), preocc. by
     Brown, 1776
   carolina (Harr., 1841)
   conscita (Wlk., 1865), form
   lactata (Sm., 1887)
   consita Dyar, 1902, missp.
   fulvicosta (Clem., 1860), form
   duplicata N. & D., 1893
8109 **reversa** (Stretch, 1885), rev. stat.
   suffusa (Sm., 1887)
8110 **contigua** (Wlk., 1855)
   lumbonigera Dyar, 1902, form
   lumbonigra Seitz, 1919, missp.
8111 **lecontei** (Guér.-Méneville, 1832)
   leucomelas H.-S., 1855
   lecontii (Lint., 1874), missp.
   militaris (Harr., 1841), form
   confinis (Wlk., 1855)
   harrisii Dyar, 1902
   harrisi Strand, 1919, missp.
   smithii Dyar, 1902, form
   smithi Strand, 1919, missp.
   dyarii Merrick, 1903, form
   vestalis (Pack., 1864), form
   ochroleuca Strand, 1919
   albanchlora (Grt., 1888), hyb.?
   suffusca Franc., 1938, rev. stat.,
     form
8112 **confusa** (Lyman, 1887)
   triangularis Sm., 1899, form
   lymani Dyar, 1902, form

   TYRIA Hbn., 1819
   HIPOCRITA Hbn., 1806, suppr.
     (ICZN Op. 97)
   EUCHELIA Bdv., 1828
   CALLIMORPHA; auth., part
8113 **jacobaeae** (L., 1758) introduced,

biological control, established
jacobeae auth., missp.
senecionis (Latr., 1809), nom. nud.
senecionis (Godt., 1823)

Arctiini

HOLOMELINA H.-S., 1856
CROCOTA; auth., not Hbn., 1823
CROCATA auth., missp.
EUBAPHE; auth., not Hbn., 1823
COTHOCIDA Wlk., 1865
CYTORUS Grt., 1866
CATHOCIDA Hamp., 1901, missp.
8114 **laeta** (Guér.-Méneville, 1832)
  a. **treatii** (Grt., 1865)
    rubropicta (Pack., 1887)
8115 **costata** (Stretch, 1885)
  opelloides (Graef, 1887)
  intermedia (Graef, 1887), ♀
  a. **parvula** (N. & D., 1893)
    cocciniceps (Schaus, 1901)
    pallipennis (B. & McD., 1918)
8116 **ostenta** (Hy. Edw., 1881)
  calera Barnes, 1907
8117 **cetes** (Druce, 1897), n. record
8818 **opella** (Grt., 1863)
  flava (B. & Benj., 1925), form
  obscura (Stretch, 1885), form
  belmaria (Ehrmann, 1895), form
  rubricosta (Ehrmann, 1895)
8119 **nigricans** (Reak., 1864), rev. stat.
  nigrifera (Wlk., 1865)
8120 **lamae** (Free., 1941)
8121 **aurantiaca** (Hbn., 1827–31)
  bimaculata (Saund., 1869)
8122 **rubicundaria** (Hbn., 1827–31), rev. stat.
  brevicornis (Wlk., 1854)
  marginata (Druce, 1885)
  belfragei (Stretch, 1885)
  diminutiva (Graef, 1887)
  rosa (French, 1890)
8123 **ferruginosa** (Wlk., 1854)
  quinaria (Grt., 1863), rev. syn.
  choriona (Reak., 1864)
  trimaculosa (Reak., 1864)
  buchholzi Wyatt, 1963
8124 **immaculata** (Reak., 1864), rev. stat.
8125 **fragilis** (Stkr., 1878)

LEPTARCTIA Stretch, 1872
8126 **californiae** (Wlk., 1855)
  lena (Bdv., 1869)
  adnata (Bdv., 1869)
  melanis (Bdv., 1869), mispl.?
  californica (Bdv., 1869), mispl.?
  fulvofasciata Butler, 1881
  wrightii French, 1889
  decia (Bdv., 1869), form
    boisduvalii Butler, 1881
    latifasciata Butler, 1881
    albifascia French, 1889
    occidentalis French, 1889
  dimidiata Stretch, 1872, form
    stretchii Butler, 1881

PARASEMIA Hbn., 1820
NEMEOPHILA Steph., 1828
EUPSYCHOMA Grt., 1865

8127 **plantaginis** (L., 1758)
  petrosa (Wlk., 1855)
  caespitis (G. & R., 1868)
  cichorii (G. & R., 1868)
  hospita (D. & S., 1775), form
  modesta (Pack., 1864), form
    alascensis (Stretch, 1906)
  geometrica (Grt., 1865), form
  geddesi (Neum., 1884), form
  scudderi (Pack., 1864), form
    selwynii (Hy. Edw., 1885)

DODIA Dyar, 1901
8128 **albertae** Dyar, 1901

PYRRHARCTIA Pack., 1864
ISIA; auth., part
8129 **isabella** (J. E. Smith, 1797)
  californica Pack., 1864

SEIRARCTIA Pack., 1864
SEIARCTIA Stretch, 1872, missp.
8130 **echo** (J. E. Smith, 1797)
  niobe (Stkr., 1885)

ESTIGMENE Hbn., 1820
LEUCARCTIA Pack., 1864
8131 **acrea** (Drury, 1773)
  caprotina (Drury, 1773)
  [*menthrastrina* (Martyn), not published]
  acria (F., 1793), missp.
  acraea auth., missp.
  pseuderminea (Harr., 1823)
  californica (Pack., 1864)
  packardii (Schaupp, 1882)
  rickseckeri (Behr, 1893)
  klagesii (Ehrmann, 1894)
  echo (Roths., 1910), part, ♀
  a. **arizonensis** Roths., 1910
8132 **albida** (Stretch, 1873)

SPILOSOMA Curt., 1825
RHAGONIS Wlk., 1862
SPILARCTIA Butler, 1875
ELPIS Dyar, 1893
8133 **latipennis** Stretch, 1872
8134 **congrua** Wlk., 1855
  a. **antigone** Stkr., 1878
    athena Stkr., 1899
8135 **vestalis** Pack., 1864
  echo (Roths., 1910), part, ♂
  amelaina (Dyar, 1893), form
8136 **dubia** (Wlk., 1855), rev. stat.
  prima Slosson, 1889, n. syn.
8137 **virginica** (F., 1798)
  congrua Wlk., 1855, part, ♀
  fumosa Stkr., 1900, form
8138 **vagans** (Bdv., 1852)
  rufula (Bdv., 1855)
  bicolor (Wlk., 1862)
  punctata (Pack., 1864)
  proba (Hy. Edw., 1881), form
    walsinghamii (Butler, 1881)
  a. **kasloa** (Dyar, 1904)
8139 **pteridis** Hy. Edw., 1874
  danbyi (N. & D., 1893)
  a. **rubra** (Neum., 1881)

HYPHANTRIA Harr., 1841
8140 **cunea** (Drury, 1773)
  liturata (Goeze, 1781), n. syn.
  punctatissima (J. E. Smith, 1797)
  budea (Hbn., 1823)
  textor (Harr., 1828), rev. syn.
  mutans (Wlk., 1856)
  punctata Fitch, 1857
  pallida (Pack., 1864)
  candida (Wlk., 1865), rev. syn.
  suffusa Stkr., 1900
  brunnea Stkr., 1900

EUERYTHRA Harv., 1876
8141 **phasma** Harv., 1876
8142 **trimaculata** Sm., 1887

ALEXICLES Grt., 1883
8143 **aspersa** Grt., 1883

TURUPTIANA Wlk., 1869
8144 **permaculata** (Pack., 1872)
  reducta (Grt., 1877)
  caeca (Stkr., 1885)
8145 **extrema** (Wlk., 1855), n. comb., n. record
  chilensis (Oberth., 1881), n. stat.

ECPANTHERIA Hbn., 1820
HYPERCOMPE Hbn., 1819, unused sr. syn.
8146 **scribonia** (Stoll, 1790)
  ocularia; auth., not F., 1775
  deflorata; auth., not F., 1775
  chryseis (Olivier, 1790)
  oculatissima; J. E. Smith, 1797, not F., 1775, emend.
  cunegonda; Palisot de Beauvois, 1811, not Cram., 1782, missp.
  confluens Oberth., 1881, rev. stat.
  a. **denudata** Slosson, 1888
8147 **muzina** Oberth., 1881
  albicollis Oberth., 1881
  abscondens Oberth., 1881
  depauperata Oberth., 1881
  sennettii Lint., 1885
8148 **oslari** Roths., 1910
8149 **suffusa** (Schaus, 1889)
  semiclara Stretch, 1906

ARACHNIS Gey., 1837
8150 **zuni** Neum., 1890
8151 **nedyma** Franc., 1966
8152 **picta** Pack., 1864
  a. **verna** B. & McD., 1918
  b. **maia** Ottol., 1896
  c. **citra** N. & D., 1893
  d. **hampsoni** Dyar, 1903
  e. **insularis** Clarke, 1941
  f. **meadowsi** Comst., 1942
8153 **apachea** Clarke, 1941
8154 **midas** B. & L., 1921
8155 **aulaea** Gey., 1837, extralim.
  incarnata (Wlk., 1855)
  a. **pompeia** Druce, 1894
    aulaea; auth., part

PHRAGMATOBIA Steph., 1828
8156 **fuliginosa** (L., 1758), extralim.
  a. **rubricosa** (Harr., 1841)

dallii Pack., 1870, n. syn.
(available?)
8157 **lineata** Newman & Donahue, 1966
8158 **assimilans** Wlk., 1855
franconia Slosson, 1891

NEOARCTIA N. & D., 1893
8159 **brucei** (Hy. Edw., 1888)
8160 **beanii** (Neum., 1891)
fuscosa (Neum., 1891), form
8161 **sordida** McD., 1921

PLATARCTIA Pack., 1864
HYPHORAIA; auth., not Hbn.,
1820
PARASEMIA; auth., not Hbn.,
1820
8162 **parthenos** (Harr., 1850)
borealis (Mösch., 1860)
parvimaculata (Brower, 1973), n. stat.
multimaculata (Brower, 1973), n. stat.
obsolescens (Brower, 1973), n. stat.

PLATYPREPIA Dyar, 1897
EPICALLIA; auth.
8163 **virginalis** (Bdv., 1852)
guttata (Bdv., 1852), form
ochracea (Stretch, 1872), form

ACERBIA Sotavalta, 1963
PLATYPREPIA; auth., part
8164 **alpina** (Quensel, 1802), extralim.
thulea (Dalman, 1823)
a. **johanseni** (Bang-Haas, 1927)

PARARCTIA Sotavalta, 1965
PARASEMIA; auth., part, not
Hbn., 1820
8165 **lapponica** (Thunb., 1791), extralim.
festiva (Bkh., 1790), preocc. by
Hufn., 1766
avia (Hbn., 1803–08)
a. **hyperborea** (Curt., 1835)
b. **gibsoni** (Bang-Haas, 1927)
c. **yarrowii** (Stretch, 1873)
remissa (Hy. Edw., 1888)
8165.1 **subnebulosa** (Dyar, 1899)

ARCTIA Schrank, 1802
HYPERCOMPE Hbn., 1806, suppr.
(ICZN Op. 97)
EYPREPIA Ochs., 1810
ZOOTE Hbn., 1820
HYPERCOMPE Hbn., 1822,
preocc. by Hbn., 1819
CHELONIA Godt., 1823, preocc.
by Brongniart, 1800
CALLARCTIA Pack., 1864
HYPERCOMPA Kby., 1892, missp.
EUPREPIA auth., missp.
8166 **caja** (L., 1758), extralim.
erinacea (Retzius, 1783)
a. **americana** Harr., 1841
b. **opulenta** (Hy. Edw., 1881)
virginivir Dyar, 1923, rev. stat.
c. **utahensis** (Hy. Edw., 1886)
transmontana (N. & D., 1893), rev.
stat.
d. **parva** Roths., 1910
e. **waroi** B. & Benj., 1927

KODIOSOMA Stretch, 1872
8167 **fulvum** Stretch, 1872
nigrum Stretch, 1872, form
eavesii Stretch, 1872, form
tricolor Stretch, 1872, form
8168 **otero** Barnes, 1907

APANTESIS Harr., 1841
ORODEMNIAS Wallgr., 1886
MIMARCTIA N. & D., 1894
ARCTIA; auth.
CALLARCTIA; auth.
8169 **phalerata** (Harr., 1841)
incompleta (Butler, 1881), n. syn.
rhoda (Butler, 1881)
incarnata (Stretch, 1906)
pulcherrima (Stretch, 1906)
naidella Strand, 1919, n. syn.
vittatula Strand, 1919, n. syn.
rhodana Strand, 1919, n. syn.
8170 **vittata** (F., 1787)
radians Wlk., 1855, n. syn.
colorata (Wlk., 1865)
floridana Cassino, 1918, n. syn.
ochracea Cassino, 1918, n. syn.
subterminalis Strand, 1919, n. syn.
8171 **nais** (Drury, 1773)
cuneata (Goeze, 1781)
[*defloriana* (Martyn), not published]
decorata (Saunders, 1863)
ochreata (Butler, 1881)
phaleratula Strand, 1919, n. syn.
8172 **quenseli** (Paykull, 1793)
strigosa (F., 1793)
a. **gelida** (Mösch., 1848)
8173 **cervinoides** (Stkr., 1876)
8174 **turbans** (Christoph, 1892)
8175 **virguncula** (W. Kby., 1837)
otiosa (N. & D., 1893), form
a. **speciosa** (Mösch., 1864)
8176 **anna** (Grt., 1864)
persephone (Grt., 1864), form
8177 **ornata** (Pack., 1864)
simplicior (Butler, 1881)
shastaensis (French, 1881)
perpicta (Dyar, 1893)
sulphuricella Strand, 1919, form
sulphurica; Hamp., 1901, not
Neum., 1885
achaia (G. & R., 1868), form
maculosa (Stretch, 1906)
rivulosa (Stretch, 1906)
californica (Cass., 1917)
barda (Hy. Edw., 1881), form, rev. stat.
ochracea-rivulosa (Stretch, 1906),
form
ochracea (Stretch, 1872), form
edwardsi (Stretch, 1872), form
obliterata (Stretch, 1885), form
a. **complicata** (Wlk., 1865)
8178 **hewletti** B. & McD., 1918, rev. stat.
8179 **nevadensis** (G. & R., 1866)
behrii (Stretch, 1872)
8180 **geneura** (Stkr., 1880), rev. stat.
a. **incorrupta** (Hy. Edw., 1881), rev.
stat.
sulphurica (Neum., 1885), form
ochracea (Neum., 1883), preocc.
by Stretch, 1872
b. **superba** (Stretch, 1873)

8181 **proxima** (Guér.-Méneville, 1844)
docta (Wlk., 1855)
mexicana (G. & R., 1866)
autholea (Bdv., 1869), rev. stat.
arizonensis (Stretch, 1873)
a. **mormonica** (Neum., 1885)
8182 **gibsoni** McD., 1937
8183 **bolanderi** (Stretch, 1872)
a. **confluentis** Strand, 1919
8184 **elongata** (Stretch, 1885)
dieckii (Neum., 1890)
8185 **blakei** (Grt., 1865)
8186 **williamsii** (Dodge, 1871)
determinata (Neum., 1881), form
a. **tooele** B. & McD., 1910
ophir B. & McD., 1910
8187 **celia** (Saund., 1863)
franconia (Hy. Edw., 1888), form
8188 **figurata** (Drury, 1773)
ceramica (Hbn., 1820)
snowi (Grt., 1875), form
excelsa (Neum., 1883), form
lugubris (Hulst, 1886), form
preciosa Nixon, 1911
8189 **f-pallida** (Stkr., 1878)
8190 **quadranotata** (Stkr., 1880)
moierra Dyar, 1914
8191 **placentia** (J. E. Smith, 1797)
flammea (Neum., 1881)
8192 **sociata** B. & McD., 1910, rev. stat.
8193 **favorita** (Neum., 1890)
favoritella Strand, 1919, form
8194 **phyllira** (Drury, 1773)
b-ata (Goeze, 1781)
dodgei (Butler, 1881)
8195 **oithona** (Stkr., 1878)
rectilinea (French, 1879)
conspicua (Stretch, 1906)
8196 **parthenice** (W. Kirby, 1837)
virguncula; Saunders, not W. Kirby,
1837
saundersii (Grt., 1864)
circa (Stretch, 1906)
approximata (Stretch, 1885), form
a. **intermedia** (Stretch, 1873), rev. stat.
stretchii (Grt., 1875), form
8197 **virgo** (L., 1758)
citrinaria (N. & D., 1893), form
simplex (Stretch, 1906), form
8198 **doris** (Bdv., 1869)
nerea (Bdv., 1869), form
michabo (Grt., 1875)
a. **minea** (Slosson, 1892)
8199 **arge** (Drury, 1773)
dione (F., 1775)
incarnatorubra (Goeze, 1781)
nervosa (N. & D., 1893), form
strigosa (Stretch, 1906), preocc. by
F., 1793

HYPERBOREA Grum-Grshimaïlo, 1900
8200 **czekanowskii** Grum-Grshimaïlo, 1900
czecanousci Hamp., 1920, emend.

Phaegopterini

HYPOCRISIAS Hamp., 1901
8201 **minima** (Neum., 1884)
armillata (Hy. Edw., 1884)
agelia (Druce, 1890)

HALYSIDOTA Hbn., 1819
    HALISIDOTA; auth.
    HALESIDOTA; auth.
8202 **cinctipes** Grt., 1866
8203 **tessellaris** (J. E. Smith, 1797)
    tessellata (Guér.-Méneville, 1832),
      missp.
    antiphola Walsh, 1864
    oslari Roths., 1909
    antipholella Strand, 1919
    tesselaroides Strand, 1919
8204 **harrisii** Walsh, 1864
    tessellaris; Harris, 1841
8205 **davisii** Hy. Edw., 1874
8205.1 **schausi** Roths., 1909
    pallida Roths., 1909, preocc. by
      Schaus, 1901
    mexiconis Strand, 1919, repl. name
8205.2 **underwoodi** Roths., 1909, extralim.
    bricenoi Roths., 1909
    meridensis Roths., 1909
    lucia Strand, 1919
8205.3 **fuliginosa** Roths., 1909, extralim.
    carinator Dyar, 1912
    meta Strand, 1919

LOPHOCAMPA Harr., 1841
    PHOEGOPTERA; Boisd., 1869, not
      H.-S., 1853, missp.
    THALESA Schaus, 1896
    SCHAUSIA Dyar, 1897, preocc. by
      Karsch, 1895
    EUSCHAUSIA Dyar, 1898
    LEPIDONEWTONIA Rego-Barros,
      1956
    AEMILIA; auth., part
8206 **roseata** (Wlk., 1866), n. comb.
    cinnamomea (Bdv., 1869)
    sanguivenosa (Neum., 1892)
    a. **occidentalis** (French, 1890)
8207 **significans** (Hy. Edw., 1888), n.
    comb., rev. stat.
8208 **ingens** (Hy. Edw., 1881)
    scapularis (Stretch, 1885)
8209 **argentata** (Pack., 1864)
    a. **subalpina** (French, 1890)
8210 **sobrina** (Stretch, 1872), rev. stat.
8211 **caryae** Harr., 1841
    annulifascia (Wlk., 1855)
    porphyrea (H.-S., 1855)
8212 **mixta** (Neum., 1882)
    pseudocarye (Roths., 1909)
8213 **pura** (Neum., 1882)
    flavescens (Roths., 1909)
8214 **maculata** Harr., 1841
    fulvoflava (Wlk., 1855)
    guttifera (H.-S., 1855)
    a. **agassizii** (Pack., 1864)
    californica (Wlk., 1865)
    angulifera (Wlk., 1866)
    salicis (Bdv., 1869)
    alni (Hy. Edw., 1876)
    eureka (Dyar, 1904)
    b. **texana** (Roths., 1909)
8215 **indistincta** (B. & McD., 1910)
8216 **annulosa** (Wlk., 1855)
    niveigutta (Wlk., 1856)

LEUCANOPSIS Rego–Barros, 1956
8217 **longa** (Grt., 1880), n. comb.

8217.1 **perdentata** (Schaus, 1901), n. comb.
8217.2 **lurida** (Hy. Edw., 1887), n. comb.
    otho (Barnes, 1901), n. comb., n.
      stat.

AEMILIA Kby., 1892
    AMELES Wlk., 1855, preocc. by
      Burmeister, 1838
8218 **ambigua** (Stkr., 1880)
    bolteri (Hy. Edw., 1885)
    syracosia (Druce, 1889)

APOCRISIAS Franc., 1966
8219 **thaumasta** Franc., 1966

HEMIHYALEA Hamp., 1901
8220 **splendens** B. & McD., 1910
    mansueta; Druce, 1884
    a. **griseiventris** Roths., 1935
8221 **labecula** (Grt., 1881)
8222 **edwardsii** (Pack., 1864)
    translucida (Wlk., 1865)
    quercus (Bdv., 1869)
    argillacea Roths., 1909
    ochreous [*sic*] Meadows, 1939, form
8223 **mansueta** (Hy. Edw., 1884),
    reinstated as species in US

CALIDOTA Dyar, 1900
8224 **laqueata** (Hy. Edw., 1886), rev. stat.
    strigosa; auth., part

OPHARUS Wlk., 1855
8225 **muricolor** (Dyar, 1898), n. comb.

CARALES Wlk., 1855, rev. stat.
    SENIA Mösch., 1877
8226 **arizonensis** (Roths., 1909), n. stat.
    fumata (B. & McD., 1910)
    astur; auth., part

PAREUCHAETES Grt., 1866
8227 **insulata** (Wlk., 1855)
    cadaverosa Grt., 1866
    affinis Grt., 1866

CYCNIA Hbn., 1818
    TANADA Wlk., 1856
8228 **inopinatus** (Hy. Edw., 1882)
    nivalis (Stretch, 1906), rev. stat.
8229 **pudens** (Hy. Edw., 1882), n. comb.
8230 **tenera** Hbn., 1818
    tenera (Hbn., 1808), unavail. (ICZN
     Op. 789)
    antica (Wlk., 1856)
    collaris (Fitch, 1857)
    a. **sciurus** (Bdv., 1869)
    yosemite (Hy. Edw., 1884)
8231 **oregonensis** (Stretch, 1873), n. comb.

EUCHAETES Harr., 1841, not
    preocc. by *Euchaetes* Dejean, 1835, a
    nom. nud.
    STENOPHAEA Hamp., 1901, n.
     syn.
    EUCHAETIAS Lyman, 1902, rev.
     stat.
    EUCHOETES auth., missp.
8232 **zella** (Dyar, 1903)
8233 **perlevis** Grt., 1882

8234 **fusca** (Roths., 1910)
8235 **helena** (Cass., 1928)
8236 **castalla** (B. & McD., 1910)
    griseopunctata (B. & McD., 1910), form
8237 **elegans** Stretch, 1873
8238 **egle** (Drury, 1773)
    cyclica Hy. Edw., 1884
8239 **gigantea** (B. & McD., 1910)
8240 **polingi** (Cass., 1928)
8241 **bolteri** Stretch, 1885
    scepsiformis Graef, 1887
8242 **antica** (Wlk., 1856)
    zonalis Grt., 1882
8243 **albicosta** (Wlk., 1855)
    fumida Hy. Edw., 1884

PYGOCTENUCHA Grt., 1883
    PROTOSIA Hamps., 1900
8244 **terminalis** (Wlk., 1854)
    harrisii (Bdv., 1869)
    votiva (Hy. Edw., 1884)
8245 **pyrrhoura** (Hulst, 1881)

LERINA Wlk., 1854
8246 **incarnata** Wlk., 1854
    robinsonii (Bdv., 1869)

ECTYPIA Clem., 1860
    ECTYPA Wlk., 1864, missp.
    EUVERNA N. & D., 1893
8247 **bivittata** Clem., 1860
    nigroflava (Graef, 1887)
8248 **mexicana** (Dognin, 1911), n. record
8249 **clio** (Pack., 1864)
    thona (Stkr., 1899)
    a. **jessica** (Barnes, 1900)

PYGARCTIA Grt., 1871
8250 **murina** (Stretch, 1885)
    oslari Roths., 1910, rev. stat.
    poliochroa Hamp., 1916
    polyochroa McD., 1938, missp.
    a. **albistrigata** B. & McD., 1913
8251 **neomexicana** Barnes, 1904
8252 **lorula** Dyar, 1914
8253 **roseicapitis** (N. & D., 1893)
8253.1 **flavidorsalis** B. & McD., 1913, n. stat.
8254 **spraguei** (Grt., 1875)
    conspicua (Neum., 1890)
8255 **abdominalis** Grt., 1871
    grossbecki Davis, 1913, n. syn.
8256 **eglenensis** (Clem., 1860)
    vivida (Grt., 1882), n. syn.

EUPSEUDOSOMA Grt., 1866
8257 **involutum** (Sepp, 1849)
    niveum (H.-S., 1855)
    niveum Grt., 1866, preocc. by H.-S.,
     1855
    bicolor (Roths., 1909), part, ♂
    a. **floridum** Grt., 1882
    immaculata (Graef, 1887)

BERTHOLDIA Schaus, 1896
8258 **trigona** (Grt., 1879)

TRICHROMIA Hbn., 1819
    NERITOS Wlk., 1855
    ANTILOBA Weymer, 1895
8259 **prophaea** (Schaus, 1905), stray?

CTENUCHINAE

CTENUCHA W. Kirby, 1837
   EUCTENUCHA Grt., 1873
8260 **venosa** Wlk., 1854
   tigrina Stkr., 1899
8261 **cressonana** Grt., 1863
   venosa; Hamp., 1898, part
   sanguinaria Stkr., 1878, form
   lutea Grt., 1902, form
8262 **virginica** (Esp., 1794)
   latreillana W. Kirby, 1837
8263 **multifaria** (Wlk., 1854)
   luteoscapus N. & D., 1893, form
8264 **rubroscapus** (Mén., 1857)
   walsinghamii Hy. Edw., 1873
   ochroscapus G. & R., 1868, form
   corvina Bdv., 1869
8265 **brunnea** Stretch, 1872

DAHANA Grt., 1875
8266 **atripennis** Grt., 1875

CISSEPS Franc., 1936
   SCEPSIS Wlk., 1854, preocc. by
     Wlk., 1850
8267 **fulvicollis** (Hbn., 1818)
   semidiaphana (Harr., 1839)
   a. **pallens** (Hy. Edw., 1886)
8268 **packardii** (Grt., 1865)
   matthewi (Hy. Edw., 1873)
   a. **cocklei** (Dyar, 1904)
8269 **wrightii** (Stretch, 1885)
   a. **gravis** (Hy. Edw., 1886)

LYMIRE Wlk., 1854
8270 **edwardsii** (Grt., 1881)

EUCEREON Hbn., 1819
   ERITHALES Poey, 1832
   ILIPA Wlk., 1854
   EUCEREEON Wlk., 1854, emend.
   THEAGES Wlk., 1855
   ACRIDOPSIS Butler, 1876
   GALETHALEA Butler, 1876
   EUCEREA auth.
8271 **carolina** (Hy. Edw., 1886)
   confine; Hamp., 1898, part
   cubensis Schaus, 1904
   ?confusum Roths., 1912

EMPYREUMA Hbn., 1818
8272 **affinis** Roths., 1912
   pugione; auth., part, not L., 1767
   lichas; auth., part, not Cram., 1775

MACROCNEME Hbn., 1819
8273 **chrysitis** (Guér.-Méneville., 1844)
   iole Druce, 1884

ANTICHLORIS Hbn., 1818
   ILLIPULA Wlk., 1854
   COPAENA H.-S., 1855
   ERIPHIA H.-S., 1856
   CERAMIDIA auth., part
8274 **viridis** Druce, 1884, extralim., often
   introduced
   musicola (Ckll., 1910)
   cyanopasta (Dognin, 1911)
   scintillocollaris (Roths., 1912)

caerulescens (Draudt, 1917)
importata (Strand, 1920)
butleri; auth.

EPISCEPSIS Butler, 1877
8275 **inornata** (Wlk., 1856)
   lamia; Butler, 1878, part.

PSILOPLEURA Hamp., 1898
8276 **vittatum** (Wlk., 1865)

PSEUDOSPHEX Hbn., 1818
   SPHECOMORPHA Hbn., 1806,
     nom. nud.
   MYRMECOPSIS Newman, 1850
   AMYCLES auth., not H.-S., 1855
     (emend. of *ANYCLES* Wlk.,
     1854)
8277 **strigosa** (Druce, 1884)

COSMOSOMA Hbn., 1823
   LAGARIA Wlk., 1854
   ERRUCA Wlk., 1854
   ILIPA Wlk., 1854
   ARISTODAEMA Wallgr., 1858
   REZIA Kby., 1892
8278 **festivum** (Wlk., 1854)
   aleus Schaus, 1889
8279 **teuthras** (Wlk., 1854), extralim.
   a. **cingulatum** Butler, 1876
     rubrigutta Skin., 1906
8280 **myrodora** Dyar, 1907
   auge; auth., not L.
   omphale; auth., not Hbn.

DIDASYS Grt., 1875
8281 **belae** Grt., 1875

SYNTOMEIDA Harr., 1839
   HIPPOLA Wlk., 1854
8282 **ipomoeae** (Harr., 1839)
   ferox (Wlk., 1854)
   euterpe (H.-S., 1855)
8283 **melanthus** (Cram., 1780), extralim.
   nycteus (Cram., 1784)
   apricans (Wlk., 1854)
   a. **albifasciata** Butler, 1876, n. stat.
8284 **epilais** (Wlk., 1854), extralim.
   a. **jucundissima** Dyar, 1907
8285 **hampsonii** Barnes, 1904
   befana Skin., 1906

PSEUDOCHARIS Druce, 1884
8286 **minima** (Grt., 1867)

HORAMA Hbn., 1819
   MASTIGOCERA Harr., 1839
   PHYLLOECIA Guér.-Méneville, 1844
   HORAMIA Wlk., 1854, missp.
   CALLICARUS Grt., 1866
   CALLICORUS Kby., 1892, missp.
   DRUCEA Kby., 1892
8287 **panthalon** (F., 1793), extralim.
   pretus; Wlk., 1856, not Cram., 1777
   tibialis (Butler, 1876)
   serena Schaus, 1924
   stoneri Lindsey, 1926
   a. **texana** (Grt., 1866)
     plumipes; Clem., 1860, not Drury,
     1773

pretus; Clem., 1862, not Cram.,
   1777
8288 **plumipes** (Drury, 1773)
   punctata (Guér.-Méneville, 1844)
   jalapensis Neum., 1890

LYMANTRIIDAE
by DOUGLAS C. FERGUSON

ORGYIINAE

Orgyiini

8289 see 8104.1

GYNAEPHORA Hbn., 1822
   CLADOPHORA Gey., 1832
   DASORGYIA Stgr., 1881
   KONOKAREHA Matsumura, 1928
8290 **rossii** Curt., 1835
8291 **groenlandica** (Wocke, 1874)

DASYCHIRA Hbn., 1809
   PARORGYIA Pack., 1864
8292 **tephra** Hbn., 1809
8293 **dorsipennata** (B. & McD., 1919)
   atomaria Wlk., 1856, preocc. by
     Wlk., 1855
8294 **vagans** (B. & McD., 1913)
   a. **grisea** (B. & McD., 1913)
     willingi (B. & McD., 1913)
8295 **mescalera** Fgn., 1977
8296 **basiflava** (Pack., 1864)
   clintoni (G. & R., 1866)
8297 **matheri** Fgn., 1977
8298 **meridionalis** (B. & McD., 1913)
   a. **memorata** Fgn., 1977
     achatina (J. E. Smith, 1797),
     preocc. by Sulz., 1776
   b. **kerrvillei** (B. & McD., 1913)
   c. **pallorosa** Fgn., 1977
8299 **atrivenosa** (Palm, 1873)
8300 **cinnamomea** (G. & R., 1866)
   aridensis (Benj., 1922)
8301 **leucophaea** (J. E. Smith, 1797)
8302 **obliquata** (G. & R., 1866)
   parallela (G. & R., 1866)
   lemmeri (B. & Benj., 1927)
8303 **dominickaria** Fgn., 1977
8304 **plagiata** (Wlk., 1865)
   montana (Beutenmüller, 1903)
   interposita (Dyar, 1911)
   pini (Dyar, 1911)
8305 **pinicola** (Dyar, 1911)
8306 **grisefacta** (Dyar, 1911)
   bonniwelli (B. & Benj., 1924)
   a. **ella** Bryk, 1934
     styx (B. & McD., 1911), preocc.
     by Bethune-Baker, 1911
8307 **manto** (Stkr., 1900)

ORGYIA Ochs., 1810
   NOTOLOPHUS Germ., 1812
   TRICHIOSOMA Luc., 1849
   ACYPHAS Wlk., 1855
   TEIA Wlk., 1855
   CLETHROGYNA Rambur, 1866
   MICROPTEROGYNA Rambur,
     1866

THYLACIGYNA Rambur, 1866
APTEROGYNIS Gn., 1875
HEMEROCAMPA Dyar, 1897
8308 **antiqua** (L., 1758), extralim.
paradoxa (Retzius, 1783)
confinis Grum-Grschimaïlo, 1891
modesta Heyne, 1899
infernalis Rebel, 1909
bukovina Strand, 1910
dilutior Schultz, 1910
grisea Denso, 1912
manchurica Matsumura, 1933
ovomaculata Schnaider, 1950
bicolor Lempke, 1959
approximata Lempke, 1959
delineata Lempke, 1959
a. **nova** Fitch, 1863
b. **badia** Hy. Edw., 1874
c. **argillacea** Fgn., 1977
8309 **vetusta** Bdv., 1852
gulosa Hy. Edw., 1881
8310 **magna** Fgn., 1977
8311 **cana** Hy. Edw., 1881
8312 **pseudotsugata** (McD., 1921)
a. **morosa** Fgn., 1977
b. **benigna** Fgn., 1977
8313 **detrita** Guér., 1831
inornata Beutenmüller, 1890
kendalli Riotte, 1972
8314 **definita** Pack., 1864
8315 **leuschneri** Riotte, 1972
a. **rindgei** Riotte, 1972
8316 **leucostigma** (J. E. Smith, 1797)
leucographa (Gey., 1832)
meridionalis Riotte, 1974
a. **intermedia** Fitch, 1856
borealis Fitch, 1856
obliviosa Hy. Edw., 1886
b. **plagiata** (Wlk., 1855)
wardi Riotte, 1971
c. **sablensis** Neil, 1979
d. **oslari** Barnes, 1900
libera Stkr., 1900
8317 **falcata** Schaus, 1896

Lymantriini

LYMANTRIA Hbn., 1819
LIPARIS Ochs., 1810, preocc. by
Scop., 1777
PORTHETRIA Hbn., 1819
HYPOGYMNA Billberg, 1820
SERICARIA Berthold, 1827
PSILURA Steph., 1828
ERASTA Gistl, 1848
MORASA Wlk., 1855
ENOME Wlk., 1855
PALASEA Wallgr., 1863
PEGELLA Wlk., 1866
SAROTHROPYGA Felder, 1874
NAGUNDA Moore, 1879
BARHONA Moore, 1879
PYRAMOCERA Butler, 1880
8318 **dispar** (L., 1758)
disparina (Müller, 1802)
nigra Selys Longchamps, 1857
burdigalensis (Mabille, 1876)
disparoides (Gaschet, 1876)
erebus (Thierry-Mieg, 1886)
semiobscura (Thierry-Mieg, 1886)

mediofusca Lambillion, 1898
major (Fuchs, 1899)
fasciata Lambillion, 1907, ab.
fasciata; Rebel, 1909, not
Lambillion, 1907, ab.
fraguarius Ribbe, 1910
insignata Schultz, 1910
angulifera Schultz, 1910
unifascia Schultz, 1910
submarginalis Schultz, 1910
brunnea Schulze, 1910
wladiwostockensis Strand, 1911
destrigata Klemensiewicz, 1912
suffusa Schulze, 1912
unicolor Lambillion, 1913
variegata Lambillion, 1913
alba Stauder, 1914
atra Heinr., 1916
lactea (Manon, 1926)
dealbata (Manon, 1926)
radiata (Manon, 1926)
asiatica Wnukowsky, 1926
obsoleta Wnukowsky, 1926
unicolor Wnukowsky, 1926 preocc.
by Lambillion, 1913
ochracea Wnukowsky, 1926
spectrum Klatt, 1928
praeterea Kardakoff, 1928
examinata Kardakoff, 1928
fasciatella Strand, 1934
andalusiaca Reinig, 1938
mediterranea Goldschmidt, 1940
bocharae Goldschmidt, 1940
rara Klatt, 1944
albofasciata Schnaider, 1950
obscurata Schnaider, 1950
fasciata Schnaider, 1950, preocc. by
Lambillion, 1907
marmorea Schnaider, 1950
grisea Schnaider, 1950
subbrunnea Schnaider, 1950
albicans Schnaider, 1950

LEUCOMA Steph., 1828
LARIA Schrank, 1802, preocc. by
Scop., 1763
STILPNOTIA Westwood, 1841
LEUCOSIA Rambur, 1866, preocc.
by Weber, 1795
NYMPHYXIS Grt., 1895
8319 **salicis** (L., 1758)
sohesti Capronnier, 1878
rubicunda Strand, 1901
nigrociliata (Fuchs, 1903)
doii (Matsumura, 1927)
neumanni (Bandermann, 1929)
radiosa S. G. Smith, 1954
infranigricosta Lempke, 1959

EUPROCTIS Hbn., 1819
NYGMIA Hbn., 1822
PORTHESIA Steph., 1828
ARTAXA Wlk., 1855
LACIDA Wlk., 1855
ANTIPHA Wlk., 1855
DULICHIA Wlk., 1855
CISPIA Wlk., 1855
LOPERA Wlk., 1855
ARNA Wlk., 1855
UROCOMA H.-S., 1855

SOMENA Wlk., 1856
UTIDAVA Wlk., 1862
COZOLA Wlk., 1865
ADLULLIA Wlk., 1865
THEMACA Wlk., 1865
ORVASCA Wlk., 1865
BEMBINA Wlk., 1865
GOGANA Wlk., 1866
CHOEROTRICHA Felder, 1874
CATAPHRACTES Felder, 1874
TEAROSOMA Felder, 1874
CHIONOPHASMA Butler, 1886
8320 **chrysorrhoea** (L., 1758)
auriflua (Esp., 1785)
phaeorrhoea (Don., 1813)
punctigera (Teich, 1889)
nigrosignata Bandermann, 1906
flavescens Rebel, 1909
abdominata Strand, 1910
punctella Strand, 1910
plumbociliata Heinr., 1916
xanthorrhoea Oberth., 1916
fumosa Chalmers-Hunt, 1951
8321 **similis** (Fuessly, 1775)
auriflua; F. 1787, not (Esp., 1785)
nyctea (Grum-Grschimaïlo, 1891)
xanthocampa (Dyar, 1905)
trimaculata (Strand, 1910)
quadrimaculata (Strand, 1910)
coreacola (Matsumura, 1933)
wilczynskia (Wize, 1934)
punctellata (Lempke, 1937)
varabilina (Bryk, 1948)
marginalis Cockayne, 1951
nigrostriata Cockayne, 1951
immaculata Lemke, 1959
fuscabdominata Lempke, 1959
nigricosta Lemke, 1959

NOCTUIDAE

by JOHN G. FRANCLEMONT & E. L. TODD
(except *Papaipema* and *Euxoa*)

HERMINIINAE

IDIA Hbn., 1813
Epizeuxis Hbn., 1818, n. syn.
Camptylochila Steph., 1834
Helia Dup., 1845, preocc. by Hbn.,
1818
Campylochila Agassiz, 1846, emend.
Helia Gn., 1854, preocc. by Hbn., 1818
Pseudaglossa Grt., 1874, n. syn.
8322 **americalis** (Gn., 1854), n. comb.
pulverosa (Sm., 1893), n. comb.
scriptipennis (Wlk., 1858), n. comb.
8323 **aemula** Hbn., 1813
aemulalis Hbn., 1821, emend., n.
comb.
undulalis (Steph., 1834), n. comb.
mollifera (Wlk., 1858), n. comb.
herminioides (Wlk., 1860), n. comb.
concisa (Wlk., 1860), n. comb.
8324 **majoralis** (Sm., 1895), n. comb.
8325 **suffusalis** (Sm., 1899), n. comb.
8326 **rotundalis** (Wlk., 1866), n. comb.
borealis (Sm., 1884), n. comb.
8327 **forbesi** (French, 1894), n. comb.
merricki (Sm., 1905), n. comb.

8328 **julia** (B. & McD., 1918), n. comb.
8329 **diminuendis** (B. & McD., 1918), n. comb.
8330 **scobialis** (Grt., 1880), n. comb.
8331 **laurenti** (Sm., 1893), n. comb.
8332 **terrebralis** (B. & McD.), n. comb.
8333 **denticulalis** (Harv., 1875), n. comb.
8334 **lubricalis** (Gey., 1832), n. comb.
    bistrigalis (Steph., 1834), n. comb.
    phaealis (Gn., 1854), n. comb.
    surrectalis (Wlk., 1859), n. comb.
    a. **occidentalis** (Sm., 1884), n. comb.
      intensalis (Sm., 1908), form, n. comb.
    b. **partitalis** (Sm., 1908), n. comb.
    c. **cobeta** (Sm., 1908), n. comb.
8335 **parvulalis** (B. & McD., 1911), n. comb.
8336 **gopheri** (Sm., 1899), n. comb.

REABOTIS Sm., 1903
8337 **immaculalis** (Hulst, 1886)

PHALAENOPHANA Grt., 1873
    NEOHERMINIA Druce, 1891
8338 **pyramusalis** (Wlk., 1859)
    gyasalis (Wlk., 1859)
    rurigena Grt., 1873
8339 **extremalis** (B. & McD., 1912)

ZANCLOGNATHA Led., 1857
    CLEPTOMITA Grt., 1873
    MEGACHYTA Grt., 1873
    PITYOLITA Grt., 1873
    EPIZEUXIS; McD., 1938, not Hbn., 1818
8340 **lituralis** (Hbn., 1818)
8341 **theralis** (Wlk., 1859)
    deceptricalis Zell., 1873
8342 **gypsalis** (Grt., 1880)
8343 **minoralis** Sm., 1895
    theralis; auth., not Wlk., 1859
8344 **inconspicualis** (Grt., 1883)
8345 **laevigata** (Grt., 1872)
    modestalis Dyar, 1903, form
    reversata Dyar, 1903, form
    obsoleta Sm., 1884, form
8346 **atrilineella** (Grt., 1873)
8347 **obscuripennis** (Grt., 1872)
8348 **pedipilalis** (Gn., 1854)
8349 **protumnusalis** (Wlk., 1859)
    minimalis Grt., 1878
8350 **martha** Barnes, 1928
8351 **cruralis** (Gn., 1854)
8352 **jacchusalis** (Wlk., 1859)
    marcidilinea (Grt., 1872)
8353 **ochreipennis** (Grt., 1872)
    a. **bryanti** Barnes, 1928, n. stat.
8354 **lutalba** (Sm., 1906)

CHYTOLITA Grt., 1873
8355 **morbidalis** (Gn., 1854)
8356 **petrealis** Grt., 1880
    punctiformis (Sm., 1895)
    fulicalis Sm., 1907

HORMISA Wlk., 1859
    LITOGNATHA Grt., 1873
    SISYRHYPENA Grt., 1873
    PALLACHIRA Grt., 1877
    XYLORMISA Fbs., 1922

8357 **absorptalis** Wlk., 1859
    nubilifascia (Grt., 1873)
8358 **litophora** (Grt., 1873)
8359 **bivittata** (Grt., 1877)
8360 **orciferalis** Wlk., 1859
    a. **pupillaris** (Grt., 1873)
    harti (French, 1894)
8361 **louisiana** (Fbs., 1922)

PHALAENOSTOLA Grt., 1873
    PHILOMETRA Grt., 1872, preocc. by Costa, 1846
    EPIDELTA Nye, 1975, n. syn.
8362 **metonalis** (Wlk., 1859)
    gaosalis (Wlk., 1859)
    inceptaria (Wlk., 1860)
    effusalis (Wlk., 1860)
    longilabris (Grt., 1872)
8363 **eumelusalis** (Wlk., 1859)
    serraticornis (Grt., 1872)
8364 **larentioides** Grt., 1873
    citima Grt., 1873, n. stat.
8365 **hanhami** (Sm., 1899)

TETANOLITA Grt., 1873
8366 **mynesalis** (Wlk., 1859)
    lixalis Grt., 1873
8367 **palligera** (Sm., 1884)
    greta Sm., 1909
8368 **floridana** (Sm., 1895)
    a. **fulata** Sm., 1909
8369 **negalis** B. & McD., 1912

BLEPTINA Gn., 1854
8370 **caradrinalis** Gn., 1854
    cloniasalis (Wlk., 1859)
8371 **inferior** Grt., 1872
    medialis Sm., 1895
8372 **sangamonia** B. & McD., 1912
8373 **flaviguttalis** B. & McD., 1912
8374 **minimalis** B. & McD., 1912
8375 **hydrillalis** Gn., 1854
    ulricusalis Wlk., 1859
    philetesalis Wlk., 1859
    atymnusalis (Wlk., 1859)
    phanasgalis Wlk., 1859

HYPENULA Grt., 1876
8376 **cacuminalis** (Wlk., 1859)
    biferalis (Wlk., 1859)
    opacalis Grt., 1876
8377 **caminalis** Sm., 1905

RENIA Gn., 1854
8378 **salusalis** (Wlk., 1859)
    brevirostralis Grt., 1872
8379 **factiosalis** (Wlk., 1859)
    clitosalis (Wlk., 1859)
    centralis Grt., 1872
    plenilinealis Grt., 1872
    alutalis Grt., 1872
    tilosalis Sm., 1909
8380 **nemoralis** B. & McD., 1918
8381 **discoloralis** Gn., 1854
    fallacialis (Wlk., 1859)
    generalis (Wlk., 1859)
    thraxalis (Wlk., 1859)
8382 **pulverosalis** Sm., 1895
8383 **hutsoni** Sm., 1906
8384 **rigida** Sm., 1905

8384.1 **flavipunctalis** (Gey., 1832)
    phalerosalis (Wlk., 1859)
    heliusalis (Wlk., 1859)
    pastorialis Grt., 1872
    belfragei Grt., 1872
    exserta Sm., 1909, n. syn.
    atrimacula Sm., 1910, n. syn., form
8385 **fraternalis** Sm., 1895
8386 **adspergillus** (Bosc, 1800)
    larvalis Grt., 1872, n. syn.
    restrictalis Grt., 1872, n. syn.
8387 **sobrialis** (Wlk., 1859)
8388 **subterminalis** B. & McD., 1912
8389 **mortualis** B. & McD., 1912

ARISTARIA Gn., 1854
8390 **theroalis** (Wlk., 1859)

CARTERIS Dognin, 1914
8391 **oculatalis** (Mösch., 1890)
    anser (Druce, 1891)

PHLYCTAINA Mösch., 1890
    APHLYCTAENA Kaye, 1927
8392 **irrigualis** Mösch., 1890
    griseirena (Hamp., 1898)

LASCORIA Wlk., 1859
    TORTRICODES Gn., 1854, preocc. by Gn., 1845
    GABERASA Wlk., 1866
    EULINTNERIA Grt., 1882
    EPITOMIPTERA Kaye, 1923
    ALBERTICODES Biezanko & Ruffinelli, 1963, n. syn.
8393 **ambigualis** Wlk., 1866
    bifidalis (Grt., 1872)
    indivisalis (Grt., 1872)
8394 **aon** Druce, 1891
8395 **alucitalis** Gn., 1854
8396 **orneodalis** Gn., 1854

PALTHIS Hbn., 1825
    CLANYMA Gn., 1854
8397 **angulalis** (Hbn., 1796)
    aracinthusalis (Wlk., 1859)
8398 **asopialis** (Gn., 1854)
    insignalis (Wlk., 1859)

REJECTARIA Gn., 1854
8399 **albisinuata** (Sm., 1905)

REDECTIS Nye, 1975
    DERCETIS Grt., 1878, preocc. by Muenster & Agassiz, 1834
    DIRCETIS Dyar, 1903, missp.
8400 **pygmaea** (Grt., 1878)
8401 **vitrea** (Grt., 1878)

MACRISTIS Schaus, 1916
8402 **bilinealis** (B. & McD., 1912)
8403 **schausi** B. & Benj., 1924
    flavipennis B. & Benj., 1924, form

RIVULINAE

RIVULA Gn., 1845
    PASIRA Moore, 1882, preocc. by Stål, 1859
8404 **propinqualis** Gn., 1854

OXYCILLA Grt., 1896
8405 **tripla** Grt., 1896
    mecyanalis (Druce, 1898)
8406 **basipallida** B. & McD., 1916
8407 **malaca** (Grt., 1873)
8408 **mitographa** (Grt., 1873)
8409 **ondo** (Barnes, 1907)

ZELICODES Grt., 1896
8410 **linearis** (Grt., 1883)

COLOBOCHYLA Hbn., 1825
    SALIA Hbn., 1806, suppr. (ICZN
      Op. 97 & 278)
    CHOLOBOCHYLA Hbn., 1826,
      missp.
    MADOPA Steph., 1829
    COLOBOCHILA Agassiz, 1846,
      emend.
    CALOBOCHYLA Wlk., 1859,
      missp.
    SALIA; auth., not Hbn., 1818
8411 **interpuncta** (Grt., 1872)
    saligna Zell., 1872
    a. *rufa* (Grt., 1883)

MELANOMMA Grt., 1875
8412 **auricinctaria** Grt., 1875

MYCTEROPHORA Hulst, 1896
8413 **inexplicata** (Wlk., 1862)
    slossoniae Hulst, 1898
8414 **monticola** Hulst, 1896
8415 **longipalpata** Hulst, 1896
8416 **geometriformis** Hill, 1924
8417 **rubricans** B. & McD., 1918

PARASCOTIA Hbn., 1825
    BOLETOBIA Bdv., 1840
    BOLITOBIA Agassiz, 1846, missp.
8418 **fuliginaria** (L., 1761), n. record
    carbonaria (F., 1787)

PROSOPARIA Grt., 1883
    FRIESIA B. & McD., 1912
8419 **perfuscaria** Grt., 1883
    anormalis (B. & McD., 1912)

HYPENODINAE

HYPENODES Doubleday, 1850
    SCHRANKIA H.-S., 1851, preocc.
      by Hbn., 1825
    SCHRANCKIA Gn., 1854, missp.
    THOLOMIGES Led., 1857
    MENOPSIMUS Dyar, 1907
8420 **caducus** (Dyar, 1907)
8421 **fractilinea** (Sm., 1908)
8422 **palustris** Fgn., 1954
    turfosalis; Krogerus, 1954, not
      Wocke, 1850?
8423 **sombrus** Fgn., 1954
8424 **franclemonti** Fgn., 1954

DASYBLEMMA Dyar, 1923
8425 **straminea** Dyar, 1923

DYSPYRALIS Warr., 1891
8426 **illocata** Warr., 1891
    humerata (Sm., 1908)

8427 **puncticosta** (Sm., 1908)
8428 **nigella** (Stkr., 1900)
    immuna (Sm., 1908)
8429 **noloides** B. & McD., 1916

PARAHYPENODES B. & McD.,
    1918
8430 **quadralis** B. & McD., 1918

SCHRANKIA Hbn., 1825
    HYPENODES Gn., 1854, preocc.
      by Doubleday, 1850
    HYPENOPSIS Dyar, 1913
8431 **macula** (Druce, 1891)

QUANDARA Nye, 1975
    LYCAUGESIA Hamp., 1912,
      preocc. by Dognin, 1910
8432 **brauneata** (Swett, 1913), n. comb.,
      from Geometridae

SIGELA Hulst, 1896
    PSEUDCRASPEDIA Hamp., 1898,
      n. syn.
    PSEUDOCRASPEDIA auth., missp.
8433 **penumbrata** Hulst, 1896
8434 **basipunctaria** (Wlk., 1861)
    melanosticta (Hamp., 1898)
8435 **eoides** (B. & McD., 1913), n. comb.

ABABLEMMA Nye, 1975
    MICROBLEMMA Hamp., 1910,
      preocc. by Semenow, 1889
8436 **grandimacula** (Schaus, 1911), n.
      comb.
8437 **brimleyana** (Dyar, 1914), n. comb.
8438 **bilineata** (B. & McD., 1916), n. comb.

PHOBOLOSIA Dyar, 1908
8439 **anfracta** (Hy. Edw., 1881)
    reincarnata Dyar, 1908

NIGETIA Wlk., 1866
8440 **formosalis** Wlk., 1866
    melanopa (Zell., 1872)

HYPENINAE

BOMOLOCHA Hbn., 1825
    MACRHYPENA Grt., 1873
    EUHYPENA Grt., 1873
    MEGAHYPENA Grt., 1873
    HYPENA; auth.
8441 **manalis** (Wlk., 1859)
8442 **baltimoralis** (Gn., 1854)
    benignalis (Wlk., 1859)
    laciniosa (Zell., 1872)
8442.1 **ramstadtii** (Wyatt, 1967)
8443 **bijugalis** (Wlk., 1859)
    toreuta (Grt., 1872)
    internalis (Rob., 1870), preocc.
    albisignalis (Zell., 1872)
    pallialis (Zell., 1872)
    fecialis Grt., 1881
8444 **palparia** (Wlk., 1861)
    scutellaris Grt., 1873
8445 **abalienalis** (Wlk., 1859)
8446 **deceptalis** (Wlk., 1859)
    perangulalis Harv., 1875
8447 **madefactalis** (Gn., 1854)

    caducalis (Wlk., 1858)
    damnosalis (Wlk., 1859)
    achatinalis (Zell., 1872)
    profecta (Grt., 1872)
8448 **sordidula** (Grt., 1872)
8449 **henloa** Sm., 1905
8450 **atomaria** Sm., 1903
    chicagonis Dyar, 1904
    perpallida Dyar, 1904
8451 **vega** Sm., 1900
8452 **edictalis** (Wlk., 1859)
    vellifera (Grt., 1873)
    lentiginosa (Grt., 1873)
8453 **umbralis** Sm., 1884
8454 **variabilis** (Druce, 1890), n. comb., n.
      record
    cachialis (Schaus, 1912)

LOMANALTES Grt., 1873
8455 **eductalis** (Wlk., 1859)
    laetulus Grt., 1873

OPHIUCHE Hbn., 1825
    ANEPISCHETOS Sm., 1900
    ANEPHISCHETOS auth., missp.
8456 **abjuralis** (Wlk., 1859)
    abzinalis (Wlk., 1859), missp.
    scissalis (Wlk., 1866)
    bipartita (Sm., 1900)
    lividalis; auth., not Hbn., 1790
8457 **minualis** (Gn., 1854)
    citata (Grt., 1872)
    trituberalis (Zell., 1872)
    obtectalis (Mösch., 1886)
8458 **annulalis** (Grt., 1876), n. comb.
8459 **degasalis** (Wlk., 1859)
    ancara (Druce, 1881)
8460 **porrectalis** (F., 1794)

HYPENA Schr., 1802
    ERICHILA Billberg, 1820
    HERPYZON Hbn., 1822
8461 **humuli** Harr., 1841
    germanalis Wlk., 1859
    evanidalis Rob., 1870
    olivacea Grt., 1873, form
    albopunctata Tepper, 1882, form
8462 **californica** Behr, 1870
8463 **decorata** Sm., 1884
8464 **modesta** Sm., 1895

PLATHYPENA Grt., 1873
8465 **scabra** (F., 1798)
    obesalis; Steph., 1834, not Tr., 1828
    erectalis (Gn., 1854)
    revoluta (Wlk., 1858, 14: 1472), n.
      syn.
    revoluta (Wlk., 1858, 15; 1835),
      preocc. by Wlk., 1858
    subrufalis (Grt., 1872)

TATHORHYNCHUS Hamp., 1894
    TATHORHYNCUS
      Bethune-Baker, 1911, missp.
8466 **exsiccatus** (Led., 1855)
    vinctalis (Wlk., 1866)
    angustiorata (Grt., 1882)

HEMEROPLANIS Hbn., 1818
    SCOPELOPUS Steph., 1829

PLEONECTYPTERA Grt., 1872
   COPTOCNEMIA Zell., 1872
8467 scopulepes (Haw., 1809)
   pyralis Hbn., 1818
   pyraloides Hbn., 1823
   scopulaepes (Steph., 1829), missp.
   inops (Steph., 1829)
   denticulata (Wlk., 1865)
   irrecta (Wlk., 1865)
   floccalis (Zell., 1872)
   scopelopes Seitz, 1940–46, missp.
   geometralis (Grt., 1872), form
8468 punitalis (Sm., 1907)
8469 secundalis (Sm., 1907)
8470 reversalis (Sm., 1907)
8471 habitalis (Wlk., 1859)
   phalaenalis (Grt., 1872)
8472 historialis (Grt., 1882)
   rectalis (Sm., 1907), form
8473 finitima (Sm., 1893)
   cumulalis (Dyar, 1910)
   a. tenalis (Sm., 1907)
     serena (Sm., 1911)
   a. concoloralis B. & Benj., 1924
8474 incusalis (Grt., 1881)
   subflavidalis (Grt., 1881)
8475 parallela (Sm., 1907)
8476 immaculalis (Harv., 1875)
8477 obliqualis (Hy. Edw., 1886)

   GLYMPIS Wlk., 1859
   ECREGMA Wlk., 1859, preocc. by
     Wlk., 1858
8478 concors (Hbn., 1823)
   agilaria (Druce, 1890)
   flavicapilla (Mösch., 1890)
   reniformis (Sm., 1903)

   SPARGALOMA Grt., 1873
8479 sexpunctata Grt., 1873

   PHYTOMETRA Haw., 1809
   PROTHYMIA Hbn., 1823
   NANTHILDA Blanchard, 1840
8480 ernestinana (Blanchard, 1840)
   coccineifascia (Grt., 1873)
8481 rhodarialis (Wlk., 1859)
   semipurpurea (Wlk., 1865), form
   confinisalis (Wlk., 1866)
   rosalba (Grt., 1873)
8482 apicata B. & McD., 1916
8483 obliqualis (Dyar, 1912)
   oma (Dyar, 1912)
   curvata B. & McD., 1916
8484 orgiae (Grt., 1875)
   subolivacea (Harv., 1875)

   NYCHIOPTERA Franc., 1966
8485 noctuidalis (Dyar, 1907)
8486 accola Franc., 1966
8487 opada Franc., 1966

   HORMOSCHISTA Mösch., 1890
8488 latipalpis (Wlk., 1858)
   pagenstecheri Mösch., 1890

   OMMATOCHILA Butler, 1894
8489 mundula (Zell., 1872)
   orba (Grt., 1877), n. syn.
   lagore (Druce, 1890)

CATOCALINAE

   PANGRAPTA Hbn., 1818
   MARMORINIA Gn., 1852
   PARGRAPTA Grt., 1872, missp.
8490 decoralis Hbn., 1818
   epionoides (Gn., 1852)
   geometroides (Gn., 1852)
   elegantalis (Fitch, 1856)
   recusans (Wlk., 1866)

   LEDAEA Druce, 1891
   LEGNA Wlk., 1865, preocc. by
     Wlk., 1858
8491 perditalis (Wlk., 1859)
   semilineata (Wlk., 1865)
   umbrifascia (Grt., 1873)

   ISOGONA Gn., 1852
   EUTOREUMA Grt., 1872
   PARORA Sm., 1900
   DIAPERA Hamp., 1926
8492 natatrix Gn., 1852
8493 tenuis (Grt., 1872)
8494 texana (Sm., 1900)
8495 punctipennis (Grt., 1883)
   acuna Barnes, 1907
8496 segura Barnes, 1907
8497 snowi (Sm., 1908)
8498 scindens (Wlk., 1858)

   METALECTRA Hbn., 1823
   STIMMIA Gn., 1852
   HOMOPYRALIS Grt., 1875
   METALESTRA B. & Benj., 1924,
     missp.
8499 discalis (Grt., 1876)
8500 quadrisignata (Wlk., 1858)
   contracta (Wlk., 1860)
   zonata (Wlk., 1865)
   tactus (Grt., 1875)
8501 bigallis (Sm., 1908)
8502 tantillus (Grt., 1875)
   monodia (Dyar, 1903)
8503 diabolica B. & Benj., 1924
8504 albilinea Richards, 1941
8505 richardsi Brower, 1941
8506 miserulata (Grt., 1882)
   irentis (Sm., 1905)
8507 edilis (Sm., 1906)
8508 cincta (Sm., 1905)

   ARUGISA Wlk., 1865
   DIALLAGMA Sm., 1900
8509 latiorella (Wlk., 1863)
   lutea (Sm., 1900)
8510 watsoni Richards, 1941
8511 punctalis (B. & McD., 1916)

   PSEUDORGYIA Harv., 1875
   ARETYPA Sm., 1903
8512 versuta Harv., 1875
8513 russula Grt., 1883
   pectinicoris (Sm., 1899)

   SCOLECOCAMPA Gn., 1852
   SCOLICOCAMPA Wlk., 1856,
     missp.
8514 liburna (Gey., 1837)
   ligni Gn., 1852

8515 atriluna Sm., 1903

   PHARGA Wlk., 1863
8516 pallens (B. & McD., 1911)

   PALPIDIA Dyar, 1898
   ECHINOCAMPA Franc., 1949, n.
     syn.
8517 pallidior Dyar, 1898
   cocophaga (Franc., 1949), n. comb.,
     n. syn.

   GABARA Wlk., 1866
   EUCALYPTRA Morr., 1875
   CILLA Grt., 1880
8518 obscura (Grt., 1883)
8519 gigantea (Sm., 1905)
8520 stygialis (Sm., 1903)
8521 infumata (Hamp., 1926)
8522 subnivosella Wlk., 1866
   nivealis (Sm., 1903)
   a. bipuncta (Morr., 1875)
8523 distema (Grt., 1880)
   strigata (Sm., 1902)
   depunctata (Strand, 1916)
   a. humeralis (Sm., 1903)
     umbonata (Sm., 1903)
     apicalis (Sm., 1903)
8524 pulverosalis (Wlk., 1866)
   minorata (Sm., 1903)

   PHYPROSOPUS Grt., 1872
   [September]
   SUDARIOPHORA Zell., 1872
   [December]
   PHIPROSOPUS Dyar, 1893,
     emend.
8525 callitrichoides Grt., 1872
   acutalis; Wlk., 1859, not H.-S., 1852,
     emend., & misident.
   nasutaria (Zell., 1872)
8526 calligrapha (Hamp., 1926)

   HYPSOROPHA Hbn., 1818
   GLOEE Hbn., 1808, suppr. (ICZN
     Op. 789)
   MONOGONA Gn., 1852
   TIAUSPA Wlk., 1858
   GLOCE Neave, 1939, missp.
8527 monilis (F., 1777)
8528 hormos Hbn., 1818

   PSAMMATHODOXA Dyar,
    1921
8529 cochlidioides Dyar, 1921

   CECHARISMENA Mösch., 1890
8530 cara Mösch., 1890
   cubalis (Schaus, 1916)
8531 abarusalis (Wlk., 1859)
8532 nectarea Mösch., 1890
   debora (Druce, 1898)

   MATILOXIS Schaus, 1916
8533 anartoides (Wlk., 1865)
   darconis (Schaus, 1913)

   PLUSIODONTA Gn., 1852
   PLUCIODONTA Desmarest, 1857,
     missp.

8534 **compressipalpis** Gn., 1852
  insignis Wlk., 1865
  suffusa Hill, 1924
8535 **amado** Barnes, 1907

CALYPTRA Ochs., 1816
  CALPE Tr., 1825
  CALPA Pagenstecher, 1909, missp.
  PERCALPE Berio, 1956
8536 **canadensis** (Bethune, 1865)
  purpurascens (Wlk., 1865)
  sobria (Wlk., 1865)

GONODONTA Hbn., 1818
  ATHYSANIA Hbn., 1823
  DOSA Wlk., 1865
8537 **sicheas** (Cram., 1777)
  hesione (Drury, 1782)
  uncina Hbn., 1818
8538 **sinaldus** Gn., 1852
  ginaldus Dyar, 1914, missp.
8539 **pyrgo** (Cram., 1777)
  serix Gn., 1852
8540 **nutrix** (Cram., 1782)
  acmeptera (Sepp, 1832–48)
8541 **unica** Neum., 1891
8542 **incurva** (Sepp, 1832–40)
  teretimacula Gn., 1852
  velata Wlk., 1858
  temperata Wlk., 1858
  dentata Felder & Rogenhofer, 1874
  soror; Stahl, 1883, not Cram.,
    1780
  elaborans Dyar, 1914

EUDOCIMA Billberg, 1820
  ELYGEA Billberg, 1820, n. syn.
  LEPTOPHARA Billberg, 1820,
    nom. nud., available Nye, 1975
  TRISSOPHAES Hbn., 1823, n.
    syn.
  OTHREIS Hbn., 1823, n. syn.
  MAENAS Hbn., 1823, preocc. by
    Walbaum, 1792
  RHYTIA Hbn., 1823, n. syn.
  ACACALLIS Hbn., 1823, n. syn.
  CORYCIA Hbn., 1823, n. syn.
  OPHIDERES Bdv., 1832, n. syn.
  ACACALIS Agassiz, 1846, emend.
  OPHIODERES Agassiz, 1846,
    emend.
  OTHRYIS Agassiz, 1846, emend.
  MOENAS Wlk., 1858, missp.
  KHADIRA Moore, 1881, n. syn.
  ADRIS Moore, 1881, n. syn.
  PURBIA Moore, 1881, n. syn.
  VANDANA Moore, 1881, n. syn.
  ARGADESA Moore, 1881, n. syn.
  HALASTUS Butler, 1892, n. syn.
  EUMAENAS Hamp., 1924
8543 **materna** (L., 1767)
  hybrida (F., 1775)
  apta (Wlk., 1858)
  serpentifera; Draudt, 1940–46, not
    Wlk., 1858, mislabelled figure

GONIAPTERYX Perty, 1833
  GONIOPTERYX Agassiz, 1846,
    emend.
  RHESCIPHA Wlk., 1866

  RESCIPHA Weymer, 1895, missp.
8544 **servia** (Cram., 1782)
  tullia Perty, 1833
  obtusa (Wlk., 1866)
  snowi (Skin., 1906)

ANOMIS Hbn., 1821
  COSMOPHILA Bdv., 1833
  ANOMUS Agassiz, 1846, emend.
  GONITIS Gn., 1852
  RUSICADA Wlk., 1858
  SCOEDISA Wlk., 1858
  GONOTIS Moore, 1882, missp.
8545 **erosa** Hbn., 1821
8546 **flava** (F., 1775), extralim.
  a. *fimbriago* (Steph., 1829)
    serrata B. & McD., 1913
8547 **commoda** (Butler, 1878)
  fulvida; auth., not Gn., 1852
8548 **impasta** Gn., 1852
  doctorium Dyar, 1913
8549 **luridula** Gn., 1852
8550 **texana** Riley, 1885
  fuscostigma Ckll., 1889
  albostigma Ckll., 1889, form
8551 **illita** Gn., 1852
8552 **exacta** Hbn., 1822
  conducta Wlk., 1858
  hostia (Harv., 1876)
8553 **editrix** (Gn., 1852)

ALABAMA Grt., 1895
  EUALABAMA Grt., 1896
8554 **argillacea** (Hbn., 1823)
  xylina (Say, 1828)
  grandipuncta (Gn., 1852)
  bipunctina (Gn., 1852)

SCOLIOPTERYX Germ., 1810
  PTERODONTA R.L., 1817
  EPHEMIAS Hbn., 1821
  EUPHAIS Hbn., 1822
  GONOPTERA Berthold, 1827
8555 **libatrix** (L., 1758)
  libatricus (Haw., 1809), missp.

LITOPROSOPUS Grt., 1869
8556 **futilis** (G. & R., 1868)
8557 **bahamensis** Hamp., 1926
8558 **coachella** Hill, 1921
8559 **confligens** (Wlk., 1857)

DIPHTHERA Hbn., 1809
  EUGLYPHIA Hbn., 1820
  DIPHTERA Steph., 1850,
    emend.
  NOROPSIS Gn., 1852
8560 **festiva** (F., 1775)
  hieroglyphica (Cram., 1777), preocc.
    by Drury, 1773
  elegans Hbn., 1809
  fastuosa (Gn., 1852)

RAPARNA Moore, 1882
  RAPARA Pagenstecher, 1909,
    missp.
8561 **melanospila** (Gn., 1852)

CONCANA Wlk., 1858
  THELIDORA Mösch., 1880

8562 **mundissima** Wlk., 1858
  splendens (Mösch., 1880)
  hoshea (Druce, 1889)

ASTICTA Hbn., 1823
  TOXOCAMPA Gn., 1841
8563 **victoria** (Grt., 1874)

RHOSOLOGIA Wlk., 1865
8564 **stigmaphiles** Dyar, 1914

OBRIMA Wlk., 1856
8565 **rinconada** Schaus, 1894
  pimaensis B. & Benj., 1925

BANDELIA Linds., 1923
  NOTHOPHILA B. & L., 1922,
    preocc. by Alexander, 1922
8566 **angulata** (B. & L., 1922)

EUANOTIA B. & McD., 1910
8567 **semirufa** B. & McD., 1910
8568 **clarki** B. & McD., 1916

EULEPIDOTIS Hbn., 1823
  PALINDIA Gn., 1852
  EULEPIDOTUS H.-S., 1856,
    missp.
8569 **dominicata** (Gn., 1852)
8570 **metamorpha** Dyar, 1914
8571 **micca** (Druce, 1889)
8572 **merricki** (Holl., 1902), quest. occur.

METALLATA Mösch., 1890
  SASERNA Druce, 1891
  EUTHERMESIA Butler, 1896
8573 **absumens** (Wlk., 1862)
  scissilinea (Wlk., 1862)
  inexacta (Wlk., 1865)
  canalis (Grt., 1874)
  guttula (Hy. Edw., 1882)
  variabilis Mösch., 1890
  minorata (Sm., 1900)
  fasciata (Sm., 1900)

ANTICARSIA Hbn., 1818
8574 **gemmatalis** Hbn., 1818
  gemmatilis auth., missp.

AZETA Gn., 1852
  MILYAS Wlk., 1858
8575 **repugnalis** (Hbn., 1825)
  vampoa Gn., 1852
  mirzah Gn., 1852
  fusilinea (Wlk., 1858)
  undulifera (Wlk., 1865)
  suffusa (Wlk., 1866)
  ferruginea (Sm., 1900)
8576 **schausi** (B. & Benj., 1924)

ANTIBLEMMA Hbn., 1823
  CAPNODES; auth.,? Gn., 1852
  EGRYRLON Sm., 1900
8577 **rufinans** (Gn., 1852)
  torrida (Wlk., 1865)
  lara (Druce, 1890)
  astyla (Mösch., 1890)
  punctivena (Sm., 1900)
  discerpta (Wlk., 1858), form
  marita (Schaus, 1901), form

8578 **filaria** (Sm., 1900)
8579 **concinnula** (Wlk., 1865)
    priscilla (Mösch., 1890)
    distacta (Hamp., 1898)

GONIOCARSIA Hamp., 1926
8580 **electrica** (Schaus, 1894)

EPITAUSA Wlk., 1857
    ORTHOGRAMMA Gn., 1852,
      preocc. by R.L., 1817
    ARCHANA Wlk., 1865
8581 **prona** (Mösch., 1880)

EPHYRODES Gn., 1852
8582 **cacata** Gn., 1852

ATHYRMA Hbn., 1823
    ATHYMA Pagenstecher, 1888, missp.
8583 **adjutrix** (Cram., 1780)
    ganglio Hbn., 1827–31, n. syn.

SYLLECTRA Hbn., 1819
    SYLECTRA Hbn., 1825, missp.
8584 **erycata** (Cram., 1780)
    ericata (Gn., 1852), missp.
    mirandalis Hbn., 1819

EPIDROMIA Gn., 1852
8585 **poaphilodes** (Gn., 1852)
    pannosa Gn., 1852
    delinquens (Wlk., 1858)
    profana (Wlk., 1858)

MASSALA Wlk., 1865
8586 **obvertens** (Wlk., 1858)
    turtur (Felder & Rogenhofer, 1874)
    larina Druce, 1890
    insularis (Mösch., 1890)

PANOPODA Gn., 1852
    SIAVANA Wlk., 1857
    HARVEYA Grt., 1873
8587 **rufimargo** (Hbn., 1818)
    cressoni Grt., 1863
    roseicosta Gn., 1852, form
    rubricosta Gn., 1852, form
8588 **carneicosta** Gn., 1852
    scissa (Wlk., 1865)
    combinata (Wlk., 1857)
8589 **repanda** (Wlk., 1858)
    auripennis (Grt., 1873)
8590 **rigida** (Sm., 1903)

PHOBERIA Hbn., 1818
    LYSSIA Gn., 1852
8591 **atomaris** Hbn., 1818
    orthosioides (Gn., 1852)
    porrigens (Wlk., 1858)
    ingenua (Wlk., 1858)
    orthosiodes Richards, 1939, missp.

CISSUSA Wlk., 1856
    ULOSYNEDA Sm., 1903
8592 **spadix** (Cram., 1780)
    remigipila (Gn., 1852)
    vegeta (Morr., 1876)
8593 **mucronata** (Grt., 1883)
8594 **indiscreta** (Hy. Edw., 1886)
8595 **subtermina** (Sm., 1900)

8596 **valens** (Hy. Edw., 1881)
    insperata (Grt., 1882)
    cervina (Hy. Edw., 1882), n. syn.

LITOCALA Harv., 1875
    LITA Harv., 1875, preocc. by Tr.,
      1833
8597 **sexsignata** (Harv., 1875)
    a. **deserta** Hy. Edw., 1881

MELIPOTIS Hbn., 1818
    HELIOTHIS Hbn., 1808 suppr.
      (ICZN Op. 789)
    LYNCESTIS Wlk., 1857
    GERESPA Wlk., 1858
    CORONTA Wlk., 1858
    JARASANA Moore, 1882
    MELIOPTIS Swinhoe, 1900, missp.
    BOLINA; auth., not Dup., 1845
8598 **perpendicularis** (Gn., 1852)
    marmoraris (Gn., 1852)
    disturbans (Wlk., 1858)
    excepta (Wlk., 1858)
    glaucipennis (Wlk., 1858)
    stolida (Wlk., 1858)
    stygialis Grt., 1878
    indistincta (Butler, 1878)
    limitata (Mösch., 1886)
8599 **fasciolaris** (Hbn., 1823)
    fascicularis (Gn., 1852), missp.
    cunearis (Gn., 1852)
    fuscaris (Gn., 1852)
    limitaris (Gn., 1852)
    illuminans (Wlk., 1858)
    lunearis McD., 1938, missp.
8600 **indomita** (Wlk., 1858)
    nigrescens (G. & R., 1866)
    ochreipennis (Harv., 1875)
    ochreifascia (Grt., 1875)
8601 **cellaris** (Gn., 1852)
    turbata (Wlk., 1858)
    insipida (Felder & Rogenhofer, 1874)
8602 **nigrobasis** (Gn., 1852)
8603 **januaris** (Gn., 1852)
    russaris (Gn., 1852)
    excavans (Wlk., 1858)
    subtilis (Wlk., 1858)
    parcicolor (H.-S., 1868)
    rectifascia (H.-S., 1868)
    surinamensis (Mösch., 1880)
    nebulosa (Maassen, 1890)
    argos Druce, 1900
    rectifasciata Wolcott, 1936, missp.
8604 **famelica** (Gn., 1852)
    bivittata (Wlk., 1858)
    leucomelana; auth., not H.-S., 1868
8605 **contorta** (Gn., 1852)
    bistriga (Wlk., 1858)
    striolaris (H.-S., 1868)
8606 **prolata** (Wlk., 1858)
8607 **jucunda** Hbn., 1818
    cinis (Gn., 1852)
    versabilis Harv., 1877
    a. **hadeniformis** (Behr, 1870)
      tetrica (Hy. Edw., 1878)
8608 **agrotoides** (Wlk., 1858)
    agrotipennis (Harv., 1875)
8609 **novanda** (Gn., 1852)
8610 **acontioides** (Gn., 1852)
    sinualis Harv., 1877

IANIUS Richards, 1939
8611 **mosca** (Dyar, 1910)

PANULA Gn., 1852
8612 **inconstans** Gn., 1852

FORSEBIA Richards, 1936
    ASYNEDA Richards, 1936
8613 **perlaeta** (Hy. Edw., 1882)
    aegrota (Hy. Edw., 1884), ♀ form
    aegrotata Sm., 1893, missp.
    flavofasciata (Stkr., 1898)

BULIA Wlk., 1858
    BIULA Wlk., 1858, preocc. by
      Wlk., 1857
    ARSISACA Wlk., 1866
    CIRRHOBOLINA Grt., 1875
    CIRRHBOLINA Dyar, 1903, missp.
    BULLIA Kimball, 1965, missp.
8614 **deducta** (Morr., 1875)
    pavitensis (Morr., 1875)
    incandescens (Grt., 1875)
    vulpina (Hy. Edw., 1882)
    albina Stkr., 1900
8615 **similaris** Richards, 1936
    deducta; auth. not Morr., 1875
    a. **californica** Richards, 1939

DRASTERIA Hbn., 1818
    SYNEDA Gn., 1852
8616 **mirifica** (Hy. Edw., 1878)
    a. **hastingsi** (Hy. Edw., 1878)
      perpallida (Hy. Edw., 1881)
    b. **klotsi** Richards, 1939
8617 **eubapta** Hamp., 1926
8618 **graphica** Hbn., 1818
    capiticola (Wlk., 1858)
    media (Morr., 1875)
    faceta (Hy. Edw., 1881)
    capticola McD., 1938, missp.
    a. **atlantica** B. & McD., 1918
8619 **occulta** (Hy. Edw., 1881)
8620 **ingeniculata** (Morr., 1876)

SYNEDOIDA Hy. Edw., 1878
8621 **scrupulosa** Hy. Edw., 1878
8622 **inepta** Hy. Edw., 1881
    morbosa Hy. Edw., 1881
    violescens (Hamp., 1926)
8623 **sabulosa** Hy. Edw., 1881
    a. **abrupta** (B. & McD., 1918)
8624 **nichollae** (Hamp., 1926)
    a. **garthi** (Richards, 1939)
8625 **biformata** Hy. Edw., 1878
8626 **ochracea** (Behr, 1870)
8627 **edwardsi** (Behr, 1870)
8628 **pallescens** (G. & R., 1866)
    tenella (Hy. Edw., 1881)
8629 **fumosa** (Stkr., 1898)
    a. **brunneifasciata** (B. & McD.,
      1916)
8630 **divergens** (Behr, 1870)
    socia (Behr, 1870), form
8631 **petricola** (Wlk., 1858)
    a. **athabasca** (Neum., 1884)
      crokeri (B. & Benj., 1924),
        form
      crockeri Richards, 1936, missp.
8632 **hudsonica** (G. & R., 1865)

    a. **heathi** (B. & McD., 1918)
      pedionis (Hamp., 1926)
    b. **seposita** (Hy. Edw., 1881)
8633 **nubicola** (Behr, 1870)
8634 **maculosa** (Behr, 1870)
8635 **perplexa** (Hy. Edw., 1884)
8636 **adumbrata** (Behr, 1870)
    a. **alleni** (Grt., 1877)
    b. **saxea** (Hy. Edw., 1881)
8637 **stretchii** (Behr, 1870)
8638 **pulchra** (B. & McD., 1918)
8639 **howlandi** (Grt., 1865)
    exquisita (Hamp., 1926)
8640 **tejonica** (Behr, 1870)
    perfecta (Hy. Edw., 1884)
    decepta (Stkr., 1898)
    nigromarginata (Stkr., 1898)
8641 **grandirena** (Haw., 1809)
    limbolaris (Gey., 1832)

HYPOCALA Gn., 1852
8642 **andremona** (Cram., 1784)
    filicornis Gn., 1852
    pierreti Gn., 1852
    hilli Lint., 1878

BORYZOPS Richards, 1936
8643 **purissima** (Dyar, 1910)

ORODESMA H.-S., 1868
    POLIOTHRIPA Hamp., 1912,
      preocc. by Hamp., 1902
    MEGASTOPOLIA Hamp., 1918, n.
      syn.
    ORODESMIA Wolcott, 1923,
      missp.
    LOIS Dyar, 1924
    BORYZOLA Hamp., 1926
8644 **apicina** H.-S., 1868
    lorina (Druce, 1890)
    juanita (Schaus, 1894)

HEMEROBLEMMA Hbn., 1818
    BLOSYRIS Hbn., 1822
    THERMESIA Hbn., 1823
    PEOSINA Gn., 1852
    BRUJAS Gn., 1852
    OBUCOLA Wlk., 1858
    BRUNJIA Pagenstecher, 1909,
      missp.
8645 **opigena** (Drury, 1773)
    pandrosa (Cram., 1776)

LATEBRARIA Gn., 1852
8646 **amphipyroides** Gn., 1852

THYSANIA Dalm., 1824
8647 **zenobia** (Cram., 1777)
8648 **agrippina** (Cram., 1776)

ASCALAPHA Hbn., 1809
    IDECHTHIS Hbn., 1821
    OTOSEMA Hbn., 1823
    EREBUS; auth., not Latr., 1810
8649 **odorata** (L., 1758)
    odora (L., 1764), missp.
    agarista (Cram., 1777)

TYRISSA Wlk., 1866
8650 **multilinea** B. & McD., 1913

LESMONE Hbn., 1818
    BENDIS Hbn., 1823
    AZATHA Wlk., 1858
    TRAMA Harv., 1875, preocc. by
      Heyden, 1837
    LEPIDOTRAMA Ckll., 1903
8651 **detrahens** (Wlk., 1858), n. comb.
    arrosa (Harv., 1875), n. comb.
8652 **fufius** (Schaus, 1894), n. comb.
8653 **hinna** (Gey., 1837), n. comb.
    pulverosa (Wlk., 1865) n. comb.
8654 **griseipennis** (Grt., 1882), n. comb.
8655 **formularis** (Gey., 1837), n. comb.
    impar (Gn., 1852) n. comb.
    fusifascia (Wlk., 1858) n. comb.
    postica (Wlk., 1865) n. comb.
    umbrata (Wlk., 1865) n. comb.
    irregularis; Felder & Rogenhofer,
      1874, not Hbn., 1808

BENDISODES Hamp., 1924
8656 **aeolia** (Druce, 1890)

HELIA Hbn., 1818
8657 **agna** (Druce, 1890), n. comb.

SELENISA Hayward, 1967
    SELENIS Gn., 1852, preocc. by
      Hope, 1839
8658 **sueroides** (Gn., 1852)
    monotropa (Grt., 1876), n. syn.

HETERANASSA Sm., 1899
8659 **mima** (Harv., 1876)
8660 **minor** (Sm., 1899)
8661 **fraterna** (Sm., 1899)

COXINA Gn., 1852
8662 **cinctipalpis** (Sm., 1899)

EUCLYSTIS Hbn., 1823
    EUCLISTIS Hbn., 1826, missp.
    FOCILLA Gn., 1852
    JOCILLA H.-S., 1858, missp.
    GALAPHA Wlk., 1858, List, 15:
      1544, not List, 15: 1850
8663 **sytis** (Gn., 1852)
8664 **guerini** (Gn., 1852)

COENIPETA Hbn., 1818
    ACOLASIS Hbn., 1821
    ACOLASIA Hbn., 1826, missp.
    COENOPETA Agassiz, 1846, missp.
    CAENIPETA Gn., 1852, missp.
8665 **bibitrix** (Hbn., 1823)
    glaucescens Wlk., 1866
    meskei (Hy. Edw., 1882)

METRIA Hbn., 1823
    SAFIA Gn., 1852
    CAMPOMETRA Gn., 1852
    YRIAS Gn., 1852
    RHUBUNA Wlk., 1858
    PLACONIA Mösch., 1880
    RHUBINA Hamp., 1913, missp.
8666 **amella** (Gn., 1852)
    integerrima (Wlk., 1858)
    stylobata (Harv., 1876)
8667 **bilineata** (Sm., 1899)
8668 **celia** (Cram., 1782)

KAKOPODA Sm., 1900
    MERIDYRIAS Hamp., 1926, n.
      syn.
8669 **cincta** Sm., 1900

TOXONPRUCHA Mösch., 1890
    YRIAS; auth., not Gn., 1852
    SYNYRIAS Hamp., 1926
8670 **pardalis** (Sm., 1908)
8671 **strigalis** (Sm., 1903)
8672 **volucris** (Grt., 1883)
    a. **terminalis** (Sm., 1907)
8673 **repentis** (Grt., 1881)
8674 **crudelis** (Grt., 1882)
8675 **clientis** (Grt., 1882)
8676 **psegmapteryx** (Dyar, 1913)

ZALEOPS Hamp., 1926
8677 **umbrina** (Grt., 1883)
    protea (Sm., 1906)
8678 **paressa** (Sm., 1906)

MATIGRAMMA Grt., 1872
8679 **pulverilinea** Grt., 1872
8680 **rubrosuffusa** Grt., 1882
    laena Grt., 1882, n. syn.
8681 **obscurior** Strand, n. stat.
8682 **metaleuca** Hamp., 1913

PSEUDANTHRACIA Grt., 1874
8683 **coracias** (Gn., 1852)
    cornix (Gn., 1852)
    cinerea (Morr., 1875)

ZALE Hbn., 1818
    LEMUR Hbn., 1808, suppr. (ICZN
      Op. 789)
    PHAEOCYMA Hbn., 1818
    OMOPTERA Guér.-Méneville,
      1832
    OMOPTERUS Guér.-Méneville,
      1844, missp.
    XYLIS Gn., 1852
    HOMOPTERA Gn., 1852, missp.
    YPSIA Gn., 1852
    NEPHELINA Kby., 1897
8684 **exhausta** (Gn., 1852)
    vernifera (Wlk., 1858)
    privata (Wlk., 1865)
8685 **viridans** (Gn., 1852)
8686 **strigimacula** (Gn., 1852)
    erilda Schaus, 1940, n. syn.
8687 **fictilis** (Gn., 1852)
    guadulpensis (Gn., 1852)
    terrosa (Gn., 1852)
    posterior (Wlk., 1858)
8688 **sabena** (Schaus, 1901)
8689 **lunata** (Drury, 1773)
    edusa (Drury, 1773), ♂ form
    putrescens (Guér.-Méneville, 1832)
    saundersi (Bethune, 1865), form
    lunatoides Strand, 1917, form
    a. **salicis** (Behr, 1870)
    rosae (Behr, 1870)
8690 **smithi** Haimbach, 1928
    sexplagiata; auth., not Wlk., 1858
8691 **declarans** (Wlk., 1858)
    uniformis (Morr., 1875), ♀ form
8692 **galbanata** (Morr., 1876)
    lunifera; auth., not Hbn., 1818

penna (Morr., 1876)

lineosa; Franc., 1950, not Wlk., 1858

8693 **edusina** (Harv., 1875)

atritincta (Harv., 1875), ♀ form

edusinoides Strand, 1917

8694 **aeruginosa** (Gn., 1852)

plenipennis (Wlk., 1858)

nigrior Strand, 1917

8695 **undularis** (Drury, 1773)

nigricans (Bethune, 1865)

umbripennis Grt., 1876, form

albosquamulata Strand, 1917

8696 **insuda** (Sm., 1908)

8697 **minerea** (Gn., 1852)

lineosa (Wlk., 1858)

involuta (Wlk., 1858)

albofasciata (Bethune, 1865)

minereoides Strand, 1917, ♀ form

a. **norda** (Sm., 1908)

minereana Strand, 1917, ♂ form

minereella Strand, 1917, ♀ form

mexicana Haimbach, 1928, n. syn.

8698 **phaeocapna** Franc., 1950

8699 **obliqua** (Gn., 1952)

8700 **squamularis** (Drury, 1773)

lapidaria Haimbach, 1928, ♀ form

8701 **confusa** McD., 1940

8702 **submediana** Strand, 1917

lemmeri McD., 1943, form

8703 **duplicata** (Bethune, 1865)

benesignata (Harv., 1875), form

franclemonti McD., 1943, form

a. **largera** (Sm., 1908)

8704 **helata** (Sm., 1908)

ruperti McD., 1943, form

8705 **bethunei** (Sm., 1908)

8706 **buchholzi** McD., 1943

8707 **metatoides** McD., 1943

8708 **metata** (Sm., 1908)

8709 **curema** (Sm., 1908)

8710 **rubiata** (Sm., 1908)

8711 **rubi** (Hy. Edw., 1881)

8712 **termina** (Grt., 1883)

yavapai (Sm., 1908)

8713 **lunifera** (Hbn., 1818)

calycanthata; auth., not J. E. Smith, 1797

intenta (Wlk., 1858)

cingulifera (Wlk., 1858)

woodi (Grt., 1877)

8714 **calycanthata** (J. E. Smith, 1797)

dealbata Strand, 1917, form

8715 **colorado** (Sm., 1908)

8716 **unilineata** (Grt., 1876)

purpureobrunnea Strand, 1917

8717 **horrida** Hbn., 1818

8718 **perculta** Franc., 1964

EUPARTHENOS Grt., 1876

PARTHENOS Hbn., 1823, preocc. by Hbn., 1819

CATOCALIRRHUS Andrews, 1877

8719 **nubilis** (Hbn., 1823)

fasciata (Beutenmüller, 1907)

unilineata Chermock & Chermock, 1940, form

a. **apache** (Poling, 1901)

b. **osiris** B. & Benj., 1926

EUBOLINA Harv., 1875

8720 **impartialis** Harv., 1875

ALLOTRIA Hbn., 1823

8721 **elonympha** (Hbn., 1818)

OPHISMA Gn., 1852

8722 **tropicalis** Gn., 1852

crocimacula Gn., 1852

detrahens Wlk., 1858

luteiplaga Wlk., 1858

confundens Wlk., 1858

stigmatifera Wlk., 1858

fugiens Wlk., 1858

morbillosa Felder & Rogenhofer, 1874

MIMOPHISMA Hamp., 1926

8723 **delunaris** (Gn., 1852)

ablunaris; auth., not Gn., 1852

DYSGONIA Hbn., 1823

PARALLELIA; auth., not Hbn., 1818

8724 **consobrina** (Gn., 1852), n. comb.

redditura (Wlk., 1858), n. comb.

8725 **similis** (Gn., 1852), n. comb.

apicalis (Gn., 1852), form, n. comb.

concolor (Grt., 1893), n. comb.

8726 **smithii** (Gn., 1852), n. comb.

PARALLELIA Hbn., 1818

8727 **bistriaris** Hbn., 1818

amplissima (Wlk., 1858)

CUTINA Wlk., 1866

8728 **albopunctella** Wlk., 1866

strigulataria (Sm., 1900)

8729 **distincta** (Grt., 1883)

inquieticolor (Dyar, 1922)

FOCILLIDIA Hamp., 1913

8730 **texana** Hamp., 1913

brunnior Strand, 1916, form

EUCLIDIA Ochs., 1816

ECTYPA Billberg, 1820

EUCLIDINA McD., 1937

8731 **cuspidea** (Hbn., 1818)

8732 **ardita** Franc., 1957

CAENURGIA Wlk., 1858

LITOSEA Grt., 1875

8733 **chloropha** (Hbn., 1818)

convalescens (Gn., 1852)

purgata Wlk., 1858

socors Wlk., 1858

8734 **togataria** (Wlk., 1862)

adversa (Grt., 1875)

CAENURGINA McD., 1937

8735 **annexa** (Hy. Edw., 1890)

conspicua (Sm., 1900)

8736 **caerulea** (Grt., 1873)

aquamarina (Felder & Rogenhofer, 1874)

livida (Letcher, 1896)

8737 **distincta** (Neum., 1884)

8738 **crassiuscula** (Haw., 1809)

erichto (Gn., 1852)

sobria (Wlk., 1858)

ochrea (Grt., 1873), form

erictho (Hy. Edw., 1876), missp.

8739 **erechtea** (Cram., 1780)

patibilis (Wlk., 1858)

narrata (Wlk., 1858)

agricola (G. & R., 1867)

mundula (G. & R., 1867)

parva Blkmre., 1920, form

CALLISTEGE Hbn., 1823

EUCLIDIMERA Hamp., 1913

8740 **intercalaris** (Grt., 1882), n. comb.

dyari (Sm., 1903), n. comb.

8741 **diagonalis** (Dyar, 1898), n. comb.

8742 **triangula** (B. & McD., 1918), n. comb.

MOCIS Hbn., 1823

REMIGIA Gn., 1852

PELAMIA Gn., 1852

BARATHA Wlk., 1865

CAUNINDA Moore, 1885

PELOMIA Warr., 1913, missp.

8743 **latipes** (Gn., 1852)

repanda; auth., not F., 1794

indentata (Harv., 1875)

8744 **marcida** (Gn., 1852)

perlata (Wlk., 1858)

8745 **texana** (Morr., 1875)

hexastylus (Harv., 1875)

8746 **disseverans** (Wlk., 1858)

acuta (Wlk., 1865)

munda; auth., not (Wlk., 1865)

CELIPTERA Gn., 1852

LITOMITUS Grt., 1864

PARACELIPTERA Draudt, 1940

8747 **frustulum** Gn., 1852

elongatus (Grt., 1864)

discissa (Wlk., 1865)

8748 **valina** (Schaus, 1901)

PTICHODIS Hbn., 1818

PHURYS; auth., not Gn., 1852

8749 **vinculum** (Gn., 1852)

8750 **herbarum** (Gn., 1852)

lima (Gn., 1852)

obversa (Wlk., 1858)

dissocians (Wlk., 1858)

bifasciata (Bates, 1886)

8751 **bistrigata** Hbn., 1818

8752 **pacalis** (Wlk., 1858), rev. comb.

irrorata (Grt., 1878)

8753 **ovalis** (Grt., 1883)

8754 **bucetum** (Grt., 1883)

campanilis (Sm., 1905)

ARGYROSTROTIS Hbn., 1821

AGNOMONIA Hbn., 1831

CROCHIPHORA Hbn., 1831, n. syn.

ARGYROSTROTUS Agassiz, 1846, emend.

POAPHILA Gn., 1852

8755 **herbicola** (Gn., 1852)

8756 **contempta** (Gn., 1852)

8757 **diffundens** (Wlk., 1858)

8758 **carolina** (Sm., 1905), n. comb.

8759 **flavistriaria** (Hbn., 1831), n. comb.

perplexa (Gn., 1852)

perspicua (Wlk., 1858)
    glans (Grt., 1877)
8760 **sylvarum** (Gn., 1852)
8761 **erasa** (Gn., 1852)
8762 **quadrifilaris** (Hbn., 1831)
    obsoleta (Grt., 1876), form
8763 **deleta** (Gn., 1852)
    placata (Grt., 1878)
8764 **anilis** (Drury, 1773)
    sesquistriaris (Hbn., 1831)

DORYODES Gn., 1857
    THEMMA Wlk., 1863
    TUNZA Wlk., 1863
8765 **bistrialis** (Gey., 1832)
    acutaria (H.-S., 1852)
8766 **grandipennis** B. & McD., 1918
8767 **spadaria** Gn., 1857
    divisa (Wlk., 1863)
    promptella (Wlk., 1863)
8768 **tenuistriga** B. & McD., 1918

SPILOLOMA Grt., 1873
    STRENOLOMA Grt., 1880
8769 **lunilinea** Grt., 1873

CATOCALA Schr., 1802
    HEMIGEOMETRA Haw., 1809
    EPHESIA Hbn., 1818
    LAMPROSIA Hbn., 1821
    BLEPHARIDIA Hbn., 1822
    EUNETIS Hbn., 1823
    CORISCE Hbn., 1823
    ASTIOTES Hbn., 1823
    MORMONIA Hbn., 1823
    EUCORA Hbn., 1823
    CATOCALLA Hbn., 1823, missp.
    BLEPHARONIA Hbn., 1825
    CORISEE Wlk., 1858, missp.
    MORMOSIA Wlk., 1858, missp.
    ANDREWSIA Grt., 1882
    CATABAPTA Hulst, 1884
    ANDREUSIA Hamp., 1913, missp.
    ASTIODES Forster & Wohlfahrt,
      1970, missp.
8770 **innubens** Gn., 1852
    hinda French, 1881
    innubenta (Strand, 1914)
    flavidalis Grt., 1874
    scintillans G. & R., 1866, form
8771 **piatrix** Grt., 1864
    a. **dionyza** Hy. Edw., 1885
8772 **consors** (J. E. Smith, 1797)
    pensacola Reiff, 1919
    a. **sorsconi** B. & Benj., 1924
8773 **epione** (Drury, 1773)
    marginata F., 1775
8774 **muliercula** Gn., 1852
    peramans Hulst, 1884
8775 **antinympha** (Hbn., 1823)
    paranympha; Drury, 1773, not L.,
      1767
    affinis Westwood, 1837
    melanympha Gn., 1852
    multoconspicua Reiff, 1919, form
8776 **coelebs** Grt., 1874
    phoebe Hulst, 1885, form
    phoebe Hy. Edw., 1885, preocc. by
      Hulst, 1885
8777 **badia** G. & R., 1866

8778 **habilis** Grt., 1872
    basalis Grt., 1876
8779 **serena** Edw., 1864
8780 **robinsoni** Grt., 1872
    curvata French, 1882, form
    missouriensis Schwarz, 1915, form
8781 **judith** Stkr., 1874
    levettei Grt., 1874
8782 **flebilis** Grt., 1872
    carolina Holl., 1903
8783 **angusi** Grt., 1876
    edna Beutenmüller, 1907
    lucetta Hy. Edw., 1881, form
8784 **obscura** Stkr., 1873
    simulatilis Grt., 1874
    obvia Schwarz, 1919, form
8785 **residua** Grt., 1874
8786 **sappho** Stkr., 1874
    cleis Cass., 1918, form
8787 **agrippina** Stkr., 1874
    subviridis Harv., 1877, form
    barnesi French, 1900
8788 **retecta** Grt., 1872
    a. **luctuosa** Hulst, 1884
8789 **ulalume** Stkr., 1878
8790 **dejecta** Stkr., 1880
8791 **insolabilis** Gn., 1852
    insolabilella (Strand, 1914)
8792 **vidua** (J. E. Smith, 1797)
    viduata Gn., 1852, missp.
    desperata Gn., 1852
8793 **maestosa** Hulst, 1884
    viduata; auth., not Gn., 1852
    guenei Grt., 1887
    moderna Grt., 1900
8794 **lacrymosa** Gn., 1852
    evelina French, 1881, form
      emilia Hy. Edw., 1881
    zelica French, 1881, form
    paulina Hy. Edw., 1880, form
    albomarginata Cass., 1917, form
8795 **palaeogama** Gn., 1852
    annida Fager, 1882, form
    phalanga Grt., 1864, form
    denussa Ehrman, 1893, form
8796 **nebulosa** Edw., 1864
    ponderosa Grt., 1866
8797 **subnata** Grt., 1864
    subnatana (Strand, 1914)
8798 **neogama** (J. E. Smith, 1797)
    communis Grt., 1872
    snowiana Grt., 1876
    mildredae Franc., 1938, form
    a. **loretta** B. & McD., 1918
8799 **euphemia** Beutenmüller, 1907
    arizonae (Strand, 1914), preocc. by
      Grt., 1873
8800 **aholibah** Stkr., 1874
    ellenensis Reiff, 1920
    coloradensis Beutenmüller, 1903,
      form
8801 **ilia** (Cram., 1776)
    duplicata Worthington, 1883
    decorata Worthington, 1883, form
    conspicua Worthington, 1883, form
    uxor Gn., 1852, preocc. by Hbn.,
      1800–03
    albomacula Butler, 1892
    iliana Strand, 1914
    obsoleta Worthington, 1883, form

    umbrosa Worthington, 1883, form
    confusa Worthington, 1883
    normani Bartsch, 1916, form
    santanas Reiff, 1920, form
    hulsti Reiff, 1920
    a. **zoe** Behr, 1870
    osculata Hulst, 1884
    reiffi Cass., 1917
8802 **cerogama** Gn., 1852
    aurella Fischer, 1885
    bunkeri Grt., 1876, form
    eliza Fischer, 1885, form
    ruperti Franc., 1938, form
8803 **relicta** Wlk., 1858
    bianca Hy. Edw., 1880
    phrynia Hy. Edw., 1880, form
    clara Beutenmüller, 1903, form
    fischeri Meyer, 1958, ab.
    a. **elda** Behr, 1887
8804 **marmorata** Edw., 1864
8805 **unijuga** Wlk., 1858
    lucilla Worthington, 1883
    fletcheri Beutenmüller, 1903
    agatha Beutenmüller, 1907, form
    a. **patricia** Cass., 1917
      helena Cass., 1917
    cassinoi Beutenmüller, 1918, form
8806 **parta** Gn., 1852
    petulans Hulst, 1884
    perplexa Stkr., 1873, form
    forbesi Franc., 1938, form
8807 **irene** Behr, 1870
    volumnia Hy. Edw., 1880, form
    virgilia Hy. Edw., 1880, form
    a. **valeria** Hy. Edw., 1880
8808 **luciana** Stkr., 1874
    luciana Hy. Edw., 1875, preocc. by
      Stkr., 1874
    nebraskae Dodge, 1875, form
    somnus Dodge, 1881, form
8809 **allusa** Hulst, 1884
    frenchi Poling, 1901
8810 **cleopatra** Stkr., 1874
    perdita Stkr., 1874
    cleopatra Hy. Edw., 1875, preocc. by
      Stkr., 1874
    perdita Hy. Edw., 1875, preocc. by
      Stkr., 1874
    barbara Cass., 1918
    a. **caerulea** Beutenmüller, 1907
8811 **faustina** Stkr., 1873
    rubra Cass., 1918, form
    carlota Beutenmüller, 1897, form
    zillah Stkr., 1878, form
    lydia Beutenmüller, 1907, form
8812 **hermia** Hy. Edw., 1880
    vesta B. & McD., 1918, form
    a. **verecunda** Hulst, 1884
      rosa Beutenmüller, 1918, form
    diantha Beutenmüller, 1907, form
    ritana Beutenmüller, 1918, form
8813 **sheba** Cass., 1919
8814 **californica** Edw., 1864
    mariana Stkr., 1874
    mariana Hy. Edw., 1875, preocc. by
      Stkr., 1874
    a. **edwardsi** Kusnezov, 1903
    eldoradensis Beutenmüller, 1907
    b. **elizabeth** Cass., 1918
8815 **francisca** Hy. Edw., 1880

8816 **erichi** Brower, 1976
8817 **briseis** Edw., 1864
    briseana Strand, 1914, form
    clarissima Beutenmüller, 1918, form
    albida Beutenmüller, 1907, form
    a. **minerva** Cass., 1917
8818 **grotiana** Bailey, 1879
    georgeana Beutenmüller, 1918
8819 **pura** Hulst, 1880
    nigra Eastman, 1916, form
8820 **nevadensis** Beutenmüller, 1907
    montana Beutenmüller, 1907, form
8821 **semirelicta** Grt., 1874
    atala Cass., 1918, form
8822 **meskei** Grt., 1873
    beaniana Grt., 1878
    rosalinda Hy. Edw., 1880
    mescei Hamp., 1912, emend.
    krombeini Franc., 1938, form
    a. **orion** McD., 1922
        concolorata McD., 1922, form
8823 **hippolyta** Stkr., 1874
    hippolyta Hy. Edw., 1875, preocc. by
        Stkr., 1874
    walteri Schwarz, 1923, form
8824 **babayaga** Stkr., 1884
8825 **jessica** Hy. Edw., 1877
8826 **stretchi** Behr, 1870
    portia Hy. Edw., 1880
    sierrae Beutenmüller, 1897, form
    a. **margherita** Beutenmüller, 1918
8827 **elsa** Beutenmüller, 1918
8828 **arizonae** Grt., 1873
    aspasia; auth., not Stkr., 1874
    roseata Cass., 1919, form
    sara French, 1883, form
    arizonensis Strand, 1914, form
    huachuca Beutenmüller, 1918
    ?juncturana Strand, 1914
    ?juncturella Strand, 1914
    ?juncturelloides Strand, 1914
    a. **augusta** Hy. Edw., 1875
8829 **junctura** Wlk., 1858
    walshii Edw., 1864
    walshi auth., missp.
    julietta French, 1916, form
8830 **texanae** French, 1902
8831 **electilis** Wlk., 1858
    aspasia Stkr., 1874
    cassandra Hy. Edw., 1875
    electilella Strand, 1914
8832 **cara** Gn., 1852
    a. **carissima** Hulst, 1880
    sylvia Hy. Edw., 1880
8833 **concumbens** Wlk., 1858
    diana Hy. Edw., 1880, ab.
    hilli Grt., 1883, ab.
8834 **amatrix** (Hbn., 1809–13)
    nurus Wlk., 1858
    selecta Wlk., 1858, form
    a. **editha** Edw., 1875
    pallida B. & McD., 1918, form
8835 **delilah** Stkr., 1874
    adoptiva Grt., 1874
    calphurnia Hy. Edw., 1880
    umbella B. & Benj., 1927, form
    a. **desdemona** Hy. Edw., 1882
        umbra B. & Benj., 1927, form
    b. **utahensis** Cass., 1918
        swetti B. & Benj., 1927, form

8836 **andromache** Hy. Edw., 1885
    a. **benjamini** Brower, 1937
8837 **frederici** Grt., 1872
8838 **chelidonia** Grt., 1881
8839 **mcdunnoughi** Brower, 1937
8840 **illecta** Wlk., 1858
    magdalena Stkr., 1874
8841 **abbreviatella** Grt., 1872
8842 **nuptialis** Wlk., 1858
    myrrha Stkr., 1874
8843 **whitneyi** Dodge, 1874
    vhitneyi Hamp., 1912, missp.
    obscura Draudt, 1939, form
8844 **amestris** Stkr., 1874
    anna Grt., 1874
    westcottii Grt., 1878, form
    westcotti auth., missp.
8845 **messalina** Gn., 1852
    belfragiana Harv., 1875
    jocasta Stkr., 1875
8846 **sordida** Grt., 1877
    metalomus Mayfield, 1922, form
    engelhardti Lemmer, 1937, form
8847 **gracilis** Edw., 1864
    tela Strand, 1914
    cinerea Mayfield, 1922, form
    lemmeri Mayfield, 1923, form
8848 **louiseae** J. Bauer, 1965
8849 **andromedae** Gn., 1852
    a. **tristis** Edw., 1864
8850 **herodias** Stkr., 1876
    a. **gerhardi** B. & Benj., 1927
8851 **coccinata** Grt., 1872
    chiquita Bartsch, 1916
    circe Stkr., 1876, form
    a. **sinuosa** Grt., 1879
8852 **verrilliana** Grt., 1875
    votiva Hulst, 1884
    werneri Biederman, 1909, form
    verneri Hamp., 1912, missp.
    a. **beutenmulleri** B. & McD.,
        1910
8853 **violenta** Hy. Edw., 1880
    chiricahua Poling, 1901
    chiracahua McD., 1938, missp.
8854 **ophelia** Hy. Edw., 1880
    a. **dolli** Beutenmüller, 1907
8855 **miranda** Hy. Edw., 1881
8856 **orba** Kusnezov, 1903
8857 **ultronia** (Hbn., 1823)
    mopsa Hy. Edw., 1880
    lucinda Beutenmüller, 1907, form
    adriana Hy. Edw., 1880, form
    celia Hy. Edw., 1880, form
    nigrescens Cass., 1917, form
8858 **crataegi** Saund., 1876
    pretiosa Lint., 1876, form
8859 **texarkana** Brower, 1976
    bridwelli Brower, 1976, form
8860 **lincolnana** Brower, 1976
8861 **johnsoniana** Brower, 1976
8862 **californiensis** Brower, 1976
8863 **mira** Grt., 1876
    a. **dana** Cass., 1918
8864 **grynea** (Cram., 1780)
    polygama Gn., 1852
    nuptula Wlk., 1858
    constans Hulst, 1884
8865 **praeclara** G. & R., 1866
8866 **manitoba** Beutenmüller, 1908

8867 **blandula** Hulst, 1884
    a. **manitobensis** Cass., 1918
8868 **titania** Dodge, 1900
    distincta Schwarz, 1919
8869 **alabamae** Grt., 1875
8870 **olivia** Hy. Edw., 1880
8871 **dulciola** Grt., 1881
8872 **clintoni** Grt., 1864
8873 **similis** Edw., 1864
    amasia (J. E. Smith, 1797), part
    formula G. & R., 1866
    aholah Stkr., 1874, form
    isabella Hy. Edw., 1880, form
8874 **minuta** Edw., 1864
    hiseri Cass., 1918, form
    parvula Edw., 1864, form
    mellitula Hulst, 1884, form
    eureka Schwarz, 1919, form
    obliterata Schwarz, 1919, form
8875 **grisatra** Brower, 1936
8876 **micronympha** Gn., 1852
    timandra Hy. Edw., 1880, form
    hero Hulst, 1884, form
    hero Hy. Edw., 1885, preocc. by
        Hulst, 1884
    atarah Stkr., 1874, form
    gisela Meyer, 1880, form
    a. **fratercula** G. & R., 1866
        jacquenetta Hy. Edw., 1880, form
        jaquenetta McD., 1938, missp.
        helene Pilate, 1882, form
        ouwah Poling, 1901, form
8877 **connubialis** Gn., 1852
    sancta Hulst, 1884
    virens French, 1886
    cordelia Hy. Edw., 1880, form
    pulverulenta Brower, 1940, form
8878 **amica** (Hbn., 1818)
    androphila Gn., 1852
    aurantiaca Reiff, 1916
    nerissa Hy. Edw., 1880, form
    suffusa Beutenmüller, 1903, form
    a. **lineella** Grt., 1872
        novangliae Reiff, 1916
        curvifascia Brower, 1936, form
        melanotica Reiff, 1916, form
8879 **jair** Stkr., 1897

PLUSIINAE

Abrostolini

ABROSTOLA Ochs., 1816
    UNCA Oken, 1815, suppr. (ICZN
        Op. 417)
    HABROSTOLA Sodoffsky, 1837
    INGURIDIA Butler, 1879
    UNCA Lhomme, 1929
8880 **ovalis** Gn., 1852
8881 **urentis** Gn., 1852
8882 **parvula** B. & McD., 1916
    mariana Walter, 1928
8883 **microvalis** Ottol., 1919

MOURALIA Wlk., 1858
8884 **tinctoides** (Gn., 1852)
    annulifera Wlk., 1858
    cossoides (Roths., 1917)

Plusiini

ARGYROGRAMMA Hbn., 1823
  PLUSIA; Fbs., 1954, part, not
    Ochs., 1816
8885 **verruca** (F., 1794)
  omicron (Hbn., 1821)
  omega Hbn., 1823
  rutila (Wlk., 1865)
8886 **basigera** (Wlk., 1865)
  laticlavia (Morr., 1872)

TRICHOPLUSIA McD., 1944
  PLUSIA; Fbs., 1954, part, not
    Ochs., 1816
  DIACHRYSIA; auth., part, not
    Hbn., 1821
8887 **ni** (Hbn., 1800–03)
  humilis (Wlk., 1858)
  extrahens (Wlk., 1858)
  significans (Wlk., 1858)
  innata (H.-S., 1868)
  brassicae (Riley, 1870)
  echinocystidis (Stkr., 1874)
  comma (Schultz, 1907)
  deserticola (Oberth., 1913)
  u-notata (Strand, 1917)
8888 **abrota** (Druce, 1889)

AGRAPHA Hbn., 1821
  PLUSIA; Fbs., 1954, part, not
    Ochs., 1816
  CTENOPLUSIA Dufay, 1970, n. syn.
  ACANTHOPLUSIA Dufay, 1970,
    n. syn.
8889 **oxygramma** (Gey., 1832), n. comb.
  indigna (Wlk., 1858)
  parallela (Wlk., 1858)
  collateralis (H.-S., 1868)

PSEUDOPLUSIA McD., 1944
  PLUSIA; Fbs., 1954, part, not
    Ochs., 1816
8890 **includens** (Wlk., 1858)
  oo (Cram., 1782), preocc. by L.,
    1758
  rogationis; auth., not (Gn., 1852)
  dyaus (Grt., 1875)
  culta (Lint., 1885)
  ooana (Strand, 1917)

AUTOPLUSIA McD., 1944
  PLUSIA; Fbs., 1954, part, not
    Ochs., 1816
8891 **egena** (Gn., 1852)
8892 **olivacea** (Skin., 1917)
8893 **egenoides** (Strand, 1917)
8894 **illustrata** (Gn., 1852)
  egenella (H.-S., 1868)
  abeona (Druce, 1889)
  roxana (Druce, 1894)

RACHIPLUSIA Hamp., 1913
  PLUSIA; Fbs., 1954, part, not
    Ochs., 1816
8895 **ou** (Gn., 1852)
  fratella (Grt., 1874)
  pedalis (Grt., 1875)
  ouella (Strand, 1917)
  ouana (Strand, 1917)

DIACHRYSIA Hbn., 1821
  DYACHRYSIA Gey., 1832, missp.
  PLUSIA; auth., not Ochs., 1816
8896 **aeroides** (Grt., 1864)
8897 **balluca** Gey., 1832

ALLAGRAPHA Franc., 1964
  AGRAPHA; auth., not Hbn., 1821
  PLUSIA; auth., part, not Ochs.,
    1816
8898 **aerea** (Hbn., 1802–03)

PSEUDEVA Hamp., 1913
  PLUSIA; Fbs., 1954, part, not
    Ochs., 1816
8899 **purpurigera** (Wlk., 1858)
8900 **palligera** (Grt., 1881)
  rubigera Hamp., 1913

POLYCHRYSIA Hbn., 1821
  POLYCHRISIA Bethune-Baker,
    1906, missp.
  PLUSIA; Fbs., 1954, part, not
    Ochs., 1816
8901 **moneta** (F., 1787), extralim.
  a. **trabea** (Sm., 1895)
  esmeralda; auth., not Oberth., 1880
8902 **morigera** (Hy. Edw., 1886)

EUCHALCIA Hbn., 1821
  ADEVA McD., 1944
  PLUSIA; Fbs., 1954, part, not
    Ochs., 1816
8903 **albavitta** (Ottol., 1902)
  a. **hutsoni** (Sm., 1904)

CHRYSANYMPHA Grt., 1896
  PLUSIA; Fbs., 1954, part, not
    Ochs., 1816
8904 **formosa** (Grt., 1865)

EOSPHOROPTERYX Dyar, 1902
  PLUSIA; Fbs., 1954, part, not
    Ochs., 1816
8905 **thyatyroides** (Gn., 1852)

AUTOGRAPHA Hbn., 1821
  PHYTOMETRA; Hamp., 1913,
    part, not Haw., 1809
  PLUSIA; auth., not Ochs., 1816
8906 **bonaerensis** (Berg, 1882)
  solida Ottol., 1902
8907 **biloba** (Steph., 1830)
8908 **precationis** (Gn., 1852)
  tana (Strand, 1917)
8909 **rubida** Ottol., 1902
8910 **sansoni** Dod, 1910
8911 **bimaculata** (Steph., 1830)
  u-brevis (Gn., 1852)
  adapta (Strand, 1917)
8912 **mappa** (G. & R., 1868)
8913 **pseudogamma** (Grt., 1875)
  freya (Strand, 1917)
8914 **californica** (Speyer, 1875)
  russea (Hy. Edw., 1886), form
8915 **pasiphaeia** (Grt., 1873)
8916 **flagellum** (Wlk., 1858)
  monodon (Grt., 1875)
  insolita (Sm., 1895)
8917 **metallica** (Grt., 1875)

  bractea; Grt., 1874, not D. & S.,
    1775
  scapularis (Hy. Edw., 1882)
  lenzi (French, 1889)
  kasloensis (Strand, 1917)
8918 **corusca** (Stkr., 1885)
8919 **speciosa** Ottol., 1902
8920 **labrosa** (Grt., 1875)
8921 **v-alba** Ottol., 1902
8922 **ottolenguii** Dyar, 1903
  arctica Ottol., 1902, preocc. by
    Mösch, 1884
8923 **ampla** (Wlk., 1858)
  alterna (Stkr., 1885)

ANAGRAPHA McD., 1944
  PLUSIA; Fbs., 1954, part, not
    Ochs., 1816
8924 **falcifera** (Kby., 1837)
  falcigera (Wlk., 1858), missp.
  simplex (Gn., 1852), form
  simplicima (Ottol., 1902)

SYNGRAPHA Hbn., 1821
  CALOPLUSIA Sm., 1884
  AUTOGRAPHA; Dyar, 1903, part
  PLUSIA; Fbs., 1954, part, not
    Ochs., 1816
8925 **altera** (Ottol., 1902)
  a. **variana** (Ottol., 1902)
8926 **octoscripta** (Grt., 1874)
  a. **pallida** (Ottol., 1902)
  b. **zeta** (Ottol., 1902)
  c. **magnifica** (Ottol., 1919)
  d. **epsilon** (Ottol., 1902)
8927 **epigaea** (Grt., 1875)
  epigaella (Strand, 1917)
8928 **selecta** (Wlk., 1858)
  viridisigma; Sm., 1893, not Grt.,
    1874
8929 **viridisigma** (Grt., 1874)
  viridisignata (Grt., 1875), emend.
  selecta; auth., not Walk., 1858
8930 **orophila** Hamp., 1908
8931 **snowii** (Hy. Edw., 1884)
8932 **sackenii** (Grt., 1877)
  sacceni Hamp., 1913, emend.
8933 **lula** Strand, 1917
  snovi; Hamp., 1913, emend., not Hy.
    Edw., 1884
  diversigna (Ottol., 1919)
8934 **borea** (Auriv., 1890)
8935 **diasema** (Bdv., 1828)
8936 **u-aureum** (Gn., 1852)
  groenlandica (Stgr., 1857)
  arctica (Mösch., 1884)
  a. **vaccinii** (Hy. Edw., 1886)
8937 **interrogationis** (L., 1758)
  aemula (F., 1787)
  aurosignata (Don., 1808)
  borealis (Reuter, 1893)
  a. **herschelensis** (Benj., 1933)
  altera; Gibson, 1920, not Ottol.,
    1902
8938 **surena** (Grt., 1882)
8939 **alias** (Ottol., 1902)
  a. **interalia** (Ottol. 1919)
8940 **abstrusa** Eichlin, 1978
8941 **cryptica** Eichlin, 1979
8942 **rectangula** (W. Kby., 1837)

mortuorum (Gn., 1852)
    demaculata Strand, 1917
    a. **nargenta** (Ottol., 1919)
8943 **angulidens** (Sm., 1891)
    plusioides Strand, 1917
    a. **excelsa** (Ottol., 1902)
      excelsana Strand, 1917
      alta (Ottol., 1919)
8944 **celsa** (Hy. Edw., 1881)
    altera; Hamp., 1913, not Ottol., 1902
    alterana Strand, 1917
    a. **sierra** (Ottol., 1919)
8945 **montana** (Pack., 1869)
8946 **microgamma** (Hbn., 1823), extralim.
    a. **nearctica** Fgn., 1955
8947 **devergens** (Hbn., 1809–13), extralim.
    a. **alticola** (Wlk., 1858)
8948 **parilis** (Hbn., 1808–09)
    quadriplaga (Wlk., 1857)
8949 **ignea** (Grt., 1863)
    a. **simulans** McD., 1944
      hochenwarthi; auth., not Hochen.,
        1785
      hochenvarthi Hamp., 1913, part,
        emend.

PLUSIA Ochs., 1816
    CHRYSASPIDIA Hbn., 1821
    EUCHALCIA; Dyar, 1902, not
      Hbn., 1821
    PALAEOPLUSIA Hamp., 1913
    PHYTOMETRA; Hamp., 1913, not
      Haw., 1809
8950 **putnami** Grt., 1873
    punctistigma (Strand, 1917)
    a. **mendocinensis** (Strand, 1917)
8951 **nichollae** (Hamp., 1913), n. comb.
8952 **contexta** Grt., 1873
8953 **venusta** Wlk., 1865
    striatella Grt., 1873

EUTELIINAE

AON Neum., 1892
8954 **noctuiformis** Neum., 1892

MARATHYSSA Wlk., 1865
    MARASMALUS Grt., 1872
    SCHAZAMA Schaus, 1906
8955 **inficita** (Wlk., 1865)
    histrio (Grt., 1873)
    a. **minus** Dyar, 1921
8956 **basalis** Wlk., 1865
    ventilator (Grt., 1872)

PAECTES Hbn., 1818
    INGURA Gn., 1852
    ORTHOCLOSTERA Butler, 1878
    CALLINGURA Butler, 1894
8957 **oculatrix** (Gn., 1852)
8958 **flabella** (Grt., 1879)
8959 **pygmaea** Hbn., 1818
    abrostoloides; Wlk., 1866, not Gn.,
      1852
    praepilata (Grt., 1875)
8960 **burserae** (Dyar, 1901)
    lunodes; Grossb., 1917, not Gn., 1852
8961 **declinata** (Grt., 1879)
8962 **abrostoloides** (Gn., 1852)
    producta (Wlk., 1855)

8963 **delineata** (Gn., 1852)
8964 **acutangula** Hamp., 1912
8965 **nubifera** Hamp., 1912
    devincta; Grossb., 1917, not Wlk., 1858
8966 **obrotunda** (Gn., 1852)
8967 **arcigera** (Gn., 1852)
    nana (Wlk., 1865)
    murina (Druce, 1889), part

EUTELIA Hbn., 1823
    EUTESIA Hbn., 1826, missp.
    EURHIPIA Bdv., 1828
    PENICILLARIA Gn., 1852
    ELEALE Wlk., 1862, preocc. by
      Newman, 1841
    RIPOGENUS Grt., 1865
    PHALGA Moore, 1881
    ZOBIA Saalmüller, 1891
    TARGALLODES Holl., 1894
    ALOTSA Swinhoe, 1900
    ATACIRA Swinhoe, 1900
    SILACIDA Swinhoe, 1900
    ENTELIA Lower, 1901, missp.
    TAMSEALE Nye, 1975, n. syn.
8968 **pulcherrima** (Grt., 1865)
    dentifera Wlk., 1865
8968.1 **furcata** (Wlk., 1865)
    distracta Wlk., 1865
    nattereri Druce, 1898
    pyrastis Hamp., 1905, n. syn.

SARROTHRIPINAE

Risobini

BAILEYA Grt., 1895
    LEPTINA Gn., 1852, preocc. by
      Meigen, 1830
    BAYLEYA Lucas, 1900, missp.
8969 **doubledayi** (Gn., 1852)
8970 **ophthalmica** (Gn., 1852)
8971 **dormitans** (Gn., 1852)
    latebricola (Grt., 1863)
8972 **levitans** (Sm., 1906)
8973 **australis** (Grt., 1881)

Sarrothripini

CHARACOMA Wlk., 1863
    CORTICATA Wlk., 1864
    PARAXIA Mösch., 1890
    HYPOTHRIPA Hamp., 1894
8974 **nilotica** (Rogenhofer, 1881)
    chamaeleon (Mösch., 1890)
    littora (Bethune-Baker, 1894)
    proteella (Walsh, 1898)
    laurea (Druce, 1898)
    nigrinotata Warr., 1913
    nigrimacula Warr., 1913
    basibrunnea Warr., 1913
    submediana Strand, 1917
    albifascia Draudt, 1939

NYCTEOLA Hbn., 1822
    SARROTHRIPUS Curt., 1824
    AXIA Hbn., 1825
    SARROTHRIPA Dup., 1834,
      missp.
    SARROTHRIPUS Agassiz, 1846,
      emend.

    SUBRITA Wlk., 1866
    SAROTHRIPA Rogenhofer, 1882,
      missp.
    SAROTRICHA Meyr., 1888,
      emend.
    ICASMA Turner, 1902
8975 **frigidana** (Wlk., 1863)
    favillana (Wlk., 1863)
    latifasciella (Wlk., 1866)
    lintnerana (Speyer, 1875)
    a. **britana** McD., 1943
8976 **columbiana** (Hy. Edw., 1873)
8977 **cinereana** N. & D., 1893
8978 **scriptana** (Wlk., 1863)
    metaspilella (Wlk., 1866)
8979 **fletcheri** Rindge, 1961

ISCADIA Wlk., 1857
    ENCALYPTA Mösch., 1890
    EUCALYPTA Hamp., 1912, missp.
8980 **aperta** Wlk., 1857
    schildei (Mösch., 1890)

Collomenini

MOTYA Wlk., 1859
    LUSSA Grt., 1883
    PLEURASYMPIEZA Mösch.,
      1890
    CASANDRIA; auth., part, not Wlk.,
      1857
8981 **abseuzalis** Wlk., 1859
    nigroguttata (Grt., 1883)
    smithii (Mösch., 1890)
    tumidicosta (Hamp., 1898)

COLLOMENA Mösch., 1890
    CASANDRIA; auth., part, not
      Wlk., 1857
8982 **filifera** (Wlk., 1857)
    inflexa (Morr., 1875)
    elota Mösch., 1890, n. syn.

NOLINAE

MEGANOLA Dyar, 1898
    SARBENA; auth., not Wlk., 1862
8983 **minuscula** (Zell., 1872)
    a. **phylla** (Dyar, 1898)
    b. **eucalyptula** (Dyar, 1923)
8984 **minor** Dyar, 1899
8985 **fuscula** (Grt., 1881)
8986 **dentata** Dyar, 1899
8987 **varia** (B. & L., 1921)
    extusata (Dyar, 1923)
8988 **conspicua** Dyar, 1898
    bicrenuscula (Dyar, 1923)

NOLA Leach, 1815
    ROESELIA Hbn., 1825
    CELAMA Wlk., 1865
    LEBENA Wlk., 1866
    ARGYROPHYES Grt., 1873
8989 **pustulata** (Wlk., 1865)
    nigrofasciata Zell., 1872
    obaurata (Morr., 1874)
8990 **cilicoides** (Grt., 1873)
    a. **eurypennis** (Dyar, 1923)
8991 **sorghiella** Riley, 1882
    portoricensis Mösch., 1890

8992 **triquetrana** (Fitch, 1856)
  trinotata (Wlk., 1866)
  sexmaculata Grt., 1877
8993 **minna** Butler, 1881
  hyemalis Stretch, 1885
8994 **aphyla** (Hamp., 1900)
8995 **ovilla** Grt., 1875
8996 **clethrae** Dyar, 1899
8997 **lagunculariae** Dyar, 1901
  obliquata (B. & McD., 1913)
8998 **apera** Druce, 1897
  involuta Dyar, 1898
  a. *exposita* Dyar, 1898

ACONTIINAE

Cydosiini

CYDOSIA [Westwood] *in* Duncan
  1841
  EGGYNA Wlk., 1866
  PENTHETRIA Hy. Edw., 1881,
    preocc. by Meigen, 1803
  TANTURA Kby., 1892
8999 **aurivitta** G. & R., 1868
  imitella Stretch, 1873, form
  majuscula Hy. Edw., 1881, form
9000 **nobilitella** (Cram., 1780)
  histrio (F., 1781)
  nobilis (Hbn., 1819), emend.
  submutata (Wlk., 1866), n. syn.
  hilarella (Snell., 1878), n. comb., n.
    syn.
  cyanella Gn., 1879, n. syn.
  jamaicensis Ckll., 1896
  westwoodi Druce, 1897
9001 **phaedra** Druce, 1897

Eustrotiini

TRIPUDIA Grt., 1877
  COBUBATHA; auth., not Wlk.,
    1863
9002 **inquaesita** (B. & Benj., 1924), n.
    comb.
9003 **quadrifera** (Zell., 1874), n. comb.
9004 **grapholithoides** (Mösch., 1890), n.
    comb., rev. stat., n. record
9005 **balteata** Sm., 1900
9006 **luda** (Druce, 1898), n. comb.
9007 **dimidata** (Sm., 1905), n. comb.
9008 **luxuriosa** Sm., 1900
9009 **flavofasciata** Grt., 1877
9010 **versuta** (Hy. Edw., 1881), n. comb.,
    rev. stat.
  goyanensis (Hamp., 1910), n. syn.
  olivacea (Grossb., 1917), n. syn.
9011 **limbata** (Hy. Edw., 1881), n. comb.
  anaea (Druce, 1898)

COBUBATHA Wlk., 1863
  NERASTRIA McD., 1937, n. syn.
9012 **metaspilaris** Wlk., 1863
  signiferana (Wlk., 1865)
  numa; Kimball, 1965, not Druce,
    1889
9013 **numa** (Druce, 1889)
9014 **lixiva** (Grt., 1882), n. comb.
  numa; auth., part
9015 **basicinerea** (Grt., 1882), n. comb.

9016 **antonita** (Dyar, 1911), n. comb.
9017 **orthozona** (Hamp., 1910), n. comb.
  santarita (Dyar, 1911)
9018 **dividua** (Grt., 1879), n. comb.
  opipara (Hy. Edw., 1881), form
9019 **hippotes** (Druce, 1889), n. record
9020 **albiciliata** (Sm., 1903), n. comb.,
    missp.?
  bifasciata (B. & McD., 1912), n.
    comb.

EXYRA Grt., 1875
9021 **fax** (Grt., 1873)
9022 **rolandiana** Grt., 1877
9023 **nigrocaput** (Morr., 1874), rev. stat.
  ridingsii (Riley, 1875), rev. stat.
9024 **semicrocea** (Gn., 1852)
  hubbardiana Dyar, 1904, form
  immaculata Benj., 1922, form

ORUZA Wlk., 1862
  CURVATULA Stgr., 1892
  VITTAPPRESSA Bethune-Baker,
    1906
  VITTAPRESSA Neave, 1940
    missp.
9025 **albocostaliata** (Pack., 1876)
9026 **albocostata** (Druce, 1899)

HOMOCERYNEA B. & McD., 1913
9027 **cleoriformis** B. & McD., 1913

PHOENICOPHANTA Hamp., 1910
  CALLOSTOLIS Dyar, 1928
9028 **bicolor** B. & McD., 1916
  polyrrhoda (Walter, 1928)
9029 **modestula** Dyar, 1924

OZARBA Wlk., 1865
  ACONTIOLA Stgr., 1900
9030 **aeria** (Grt., 1881)
  fannia; B. & McD., 1913, not Druce,
    1889
9031 **propera** (Grt., 1882), n. comb.
9032 **catilina** (Druce, 1889), n. comb.
9033 **nebula** B. & McD., 1918

AMIANA Dyar, 1904
9034 **niama** Dyar, 1904

HYPERSTROTIA Hamp., 1910
  PROTOCRYPHIA B. & McD.,
    1918
9035 **nana** (Hbn., 1818)
9036 **aetheria** (Grt., 1879)
9037 **pervertens** (B. & McD., 1918)
9038 **villificans** (B. & McD., 1918)
9039 **flaviguttata** (Grt., 1882)
9040 **secta** (Grt., 1879)

SEXSERRATA B. & Benj., 1922
9041 **hampsoni** B. & Benj., 1922

GROTELLAFORMA B. & Benj.,
  1922
9042 **lactea** (Stretch, 1885)
  calora (Barnes, 1907)

HOMOLAGOA B. & McD., 1912
9043 **grotelliformis** B. & McD., 1912

THIOPTERA Franc., 1950
  XANTHOPTERA Gn., 1852,
    preocc. by Sodoffsky, 1837
  FLAVALA Berio, 1966
9044 **nigrofimbria** (Gn., 1852)
9045 **aurifera** (Wlk., 1858)
  nigrofimbria; Druce, 1889, not Gn.,
    1852
  tripuncta (Mösch., 1890)

LITHACODIA Hbn., 1818
9046 **bellicula** Hbn., 1818
  semichalcea (Wlk., 1865)
9047 **muscosula** (Gn., 1852)
9048 **albidula** (Gn., 1852)
  intractabilis (Wlk., 1860)
  cretiferana (Wlk., 1863)
9049 **synochitis** (G. & R., 1868)
9050 **concinnimacula** (Gn., 1852)
  parvimacula (Grt., 1880)
9051 **musta** (G. & R., 1868)
9052 **costaricana** Strand, 1917, n. stat., n.
    record
9053 **carneola** (Gn., 1852)
  biplaga (Wlk., 1858)
9054 **indeterminata** B. & McD., 1918
9055 **penthis** (Schaus, 1904), n. record

HOMOPHOBERIA Morr., 1875
  NEOERASTRIA McD., 1937, n.
    syn.
9056 **cristata** Morr., 1875
  caduca (Grt., 1876), n. syn.
  retis (Grt., 1879)
9057 **apicosa** (Haw., 1809)
  nigritula (Gn., 1852)
  undulifera (Wlk., 1856)

NEOTARACHE B. & Benj., 1922
  NEOTARCHE B. & Benj., 1922,
    missp.
9058 **deserticola** B. & Benj., 1922

CAPIS Grt., 1882
9059 **curvata** Grt., 1882

ARGILLOPHORA Grt., 1873
9060 **furcilla** Grt., 1873

CERMA Hbn., 1818
  ACHATIA Hbn., 1808, suppr.
    (ICZN Op. 789)
  CHAMYRIS Gn., 1852
  PACHYCERMA Van Duzee, 1897
9061 **cora** Hbn., 1818
  festa (Gn., 1852)
9062 **cerintha** (Tr., 1826)
  obscura (Dyar, 1923)
9063 **sirius** (B. & McD., 1918)

CERATHOSIA Sm., 1887
9064 **tricolor** Sm., 1887

LEUCONYCTA Hamp., 1909
  BRYOCODIA Hamp., 1910
9065 **diphteroides** (Gn., 1852)
  obliterata (Grt., 1864), form
9066 **lepidula** (Grt., 1874)
  a. *avirida* (Sm., 1906)

DIASTEMA Gn., 1852
NIPISTA Wlk., 1858
9067 **tigris** Gn., 1852
lineata (Wlk., 1858)
9068 **cnossia** (Druce, 1889), n. comb.

AMYNA Gn., 1852
ILATTIA Wlk., 1859
BERRESA Wlk., 1859
LOCHIA Wlk., 1865
STRIDOVA Wlk., 1873
PTERAETHOLIX Grt., 1873
CHYTORYZA Grt., 1876
HESPERIMORPHA Saalmüller, 1880
CHYTORHIZA Luc., 1902, missp.
ILLATIA Hamp., 1910, missp.
9069 **bullula** (Grt., 1873)
mexicana Strand, 1917
concolorata B. & Benj., 1924, form
9070 **octo** (Gn., 1852)
axis Gn., 1852
stricta (Wlk., 1858)
flavigutta (Wlk., 1858)
perfundens (Wlk., 1858)
cephusalis (Wlk., 1859)
colon Gn., 1862
vexabilis (Wallgr., 1863)
inornata (Wlk., 1865)
obstructa (Wlk., 1865)
leucospila (Wlk., 1865)
stigmatula (Snell., 1872)
albigutta (Wlk., 1873)
bavia (Felder & Rogenhofer, 1874)
orbica (Morr., 1874)
undulifera Butler, 1875
tecta (Grt., 1876)
monotretalis (Mabille, 1879)
supplex (Swinhoe, 1885)
rufa (Bethune-Baker, 1906)

ALEPTINA Dyar, 1902
9071 **inca** Dyar, 1913
texana B. & McD., 1913

PARACRETONA Dyar, 1912
9072 **aleptivoides** (B. & McD., 1912)
xithon Dyar, 1912

COPIBRYOPHILA Sm., 1900
9073 **angelica** Sm., 1900

METAPONPNEUMATA Mösch., 1890
PRORACHIA Hamp., 1908
9074 **rogenhoferi** Mösch., 1890
daria (Druce, 1898)
darioides (Strand, 1916)
dariella (Strand, 1916)

AIRAMIA B. & Benj., 1926
9075 **albiocula** (B. & McD., 1918)

Eublemmini

EUMICREMMA Berio, 1954
EUBLEMMA; auth., part, not Hbn., 1821
9076 **minima** (Gn., 1852)
pennula (Felder & Rogenhofer, 1874)

carmelita (Morr., 1876)
pallida (Schaus, 1904)

EUMESTLETA Butler, 1892
EUBLEMMA; auth., part, not Hbn., 1821
9077 **cinnamomea** (H.-S., 1868)
margaritae (Berg, 1882)
laphyra (Druce, 1890), part
rosescens (Hamp., 1898)
subcinnamomea (Strand, 1917)
9078 **recta** (Gn., 1852)
obliqualis (F., 1794), preocc. by Gmel., 1790
flammicincta (Wlk., 1865)
pallescens (H.-S., 1868)
stalii (Wallgr., 1871)
patula (Morr., 1875)
patruelis (Grt., 1876)
laphyra (Druce, 1898), part
brunneoochracea (Strand, 1917)
luteipennis (Strand, 1917)
9079 **irresoluta** (Dyar, 1919), n. comb., n. record

PROROBLEMMA Hamp., 1910
9080 **testa** B. & McD., 1913

ARAEOPTERON Hamp., 1893
ARAEOPTERUM Hamp., 1895, emend.
ARAEOPTERA Hamp., 1910, emend.
9081 **vilhelmina** Dyar, 1916

Acontiini

TARACHIDIA Hamp., 1898
9082 **albitermen** B. & McD., 1916
9083 **parvula** (Wlk., 1865)
georgica (Grt., 1881)
9084 **bicolorata** (B. & McD., 1912)
9085 **semiflava** (Gn., 1852)
9086 **clausula** (Grt., 1883)
9087 **venustula** (Wlk., 1865)
discoidalis (Wlk., 1866)
fortunata (Grt., 1882)
perita (Grt., 1882)
subcitrinalis (Hulst, 1886)
9088 **virginalis** (Grt., 1881)
tenuescens (Sm., 1902)
9089 **binocula** (Grt., 1875)
9090 **candefacta** (Hbn., 1831)
minuta; Haw., 1809, not Hbn., 1808–09
haworthana (Westwood, 1851)
debilis (Wlk., 1857)
neomexicana (Sm., 1900)
candefactella Strand, 1916
9091 **dorneri** (B. & McD., 1913), n. comb.
9092 **huita** (Sm., 1903)
9093 **heonyx** Dyar, 1913
9094 **cuta** (Sm., 1905)
9095 **erastrioides** (Gn., 1852)
9096 **libedis** (Sm., 1900)
9097 **nannodes** Hamp., 1910
9098 **phecolisca** (Druce, 1889)
tenuicula; auth., not Morr., 1875
9099 **alata** (Sm., 1905)
9100 **albimargo** B. & McD., 1916

9101 **tortricina** (Zell., 1872)
obsoleta (Grt., 1877)
deleta (Hy. Edw., 1884)
modesta (Hy. Edw., 1884)
a. **fumata** (Sm., 1905), n. syn. & n. stat.

FRUVA Grt., 1877
NEPTUNIA B. & McD., 1911, preocc. by Renier, 1847
UNIPTENA Nye, 1975, n. syn.
9102 **fasciatella** (Grt., 1975)
9103 **hutsoni** (Sm., 1906) n. comb.
9104 **pulchra** (B. & McD., 1910), n. comb.

CONOCHARES Sm., 1905
CONOCHARIS McD., 1943, missp.
9105 **acutus** Sm., 1905
9106 **rectangulus** McD., 1943
9107 **alter** (Sm., 1903), corrected gender
altera (Sm., 1903), orig. spell.
conocharodes (Hamp., 1910)
9108 **catalina** (Sm., 1906)
9109 **elegantulus** (Harv., 1876)
semiopaca (Grt., 1878)
9110 **arizonae** (Hy. Edw., 1878)
interrupta Sm., 1905
seminivealis (Hulst, 1886)

THERASEA Grt., 1895
CONACONTIA Sm., 1900
9111 **augustipennis** (Grt., 1875)
angustipennis (Grt., 1877), missp.
9112 **flavicosta** (Sm., 1900)
9113 **huachuca** (Sm., 1903)
9114 **orba** (Sm., 1903)

PONOMETIA H.-S., 1868
HELIODORA Neum., 1891
GRAEPERIA Grt., 1895
TORNACONTIA Sm., 1900
9115 **exigua** (F., 1793)
indubitans (Wlk., 1857), n. syn.
costalis (Wlk., 1858), n. syn., ♀
dimidiata (Wlk., 1865)
ochricosta H.-S., 1868, n. syn., ♀
citrina (Druce, 1889)
magnifica (Neum., 1891), n. syn., ♀
9116 **macdunnoughi** (B. & Benj., 1923)
albitermen (B. & McD., 1916), part
9117 **megocula** (Sm., 1900)
9118 **tripartita** (Sm., 1903)
mediatrix (Dyar, 1904)
9119 **sutrix** (Grt., 1880)

HEMISPRAGUEIA B. & Benj., 1923
9120 **idella** (Barnes, 1905)

SPRAGUEIA Grt., 1875
HELIOCONTIA Hamp., 1910, n. syn.
MNESIPYRGA Meyr., 1913, n. syn.
SPRAGUERIA Kohler, 1979, missp.
9121 **magnifica** Grt., 1883
9122 **dama** (Gn., 1852)
transmutata (Wlk., 1865)
trifariana (Wlk., 1865)

pardalis Grt., 1881
   concolor (Mösch., 1890)
9123 **cleta** (Druce, 1889)
9124 **perstructana** (Wlk., 1865)
   felina (H.-S., 1868)
   tigridula (H.-S., 1868)
   phaenna (Druce, 1889)
   mata (Druce, 1898)
9125 **guttata** Grt., 1875
9126 **onagrus** (Gn., 1852)
9127 **leo** (Gn., 1852)
   onagrus; H.-S., 1858, not Gn., 1852
9128 **jaguaralis** Hamp., 1910
   rudisana; Druce, 1889, not Wlk.,
     1865
9129 **funeralis** Grt., 1881
9130 **obatra** (Morr., 1875)
   plumbifimbriata Grt., 1877
   velata (Stkr., 1898)
9131 **apicalis** (H.-S., 1868), n. comb.,
   extralim.
   obliquella (Strand, 1912)
   trichostrota (Meyr., 1913), n. syn.
   a. *apicella* (Grt., 1872), n. stat.
     accepta (Hy. Edw., 1881)
     truncatula (Zell., 1873)
     marmorea; Druce, 1889, part, not
       Butler, 1879
9132 **margana** (F., 1794), n. comb.
   subapicana (Wlk., 1863)
   rudisana (Wlk., 1865)
   sordida Grt., 1882
   inorata Grt., 1883
   ochracea (Mösch., 1890)
   variegata (Mösch., 1890)
   canofusa Hamp., 1898
   tarasca Schaus, 1904

ACONTIA Ochs., 1816
   TARACHE Hbn., 1823
   DESMOPHORA Steph., 1829
   EUPHASIA Steph., 1830
   HELIOTHERA Sodoffsky, 1837
   PORROTHA Gistl, 1848
   TIMA Wlk., 1858, preocc. by
     Eschscholtz, 1829
   TRICHOTARACHE Grt., 1875, n.
     syn.
9133 **apela** Druce, 1889
   philomela Druce, 1889
9134 **sutor** (Hamp., 1910)
9135 **tenuicula** (Morr., 1875)
   nuicola Sm., 1900, n. syn.
   meskei Sm., 1900, n. syn.
   mescei (Hamp., 1910), emend.
   carcharodonta (Hamp., 1910), n. syn.
9136 **aprica** (Hbn., 1802)
   alboater (Haw., 1809)
   biplaga Gn., 1852
   unocula Frery, 1852
   redita Felder & Rogenhofer, 1874, n.
     syn.
   apricana (Strand, 1916)
   apricanoides (Strand, 1916)
   apricella (Strand, 1916)
9137 **lactipennis** (Harv., 1875)
9138 **abdominalis** (Grt., 1877)
   luta (Strand, 1917)
   mala (Strand, 1916)
9139 **knowltoni** McD., 1940

9140 **flavipennis** (Grt., 1873)
   delutea (Strand, 1916)
   discolutea (Strand, 1916)
9141 **assimilis** (Grt., 1875)
9142 **quadriplaga** Sm., 1900, rev., stat.
9143 **tetragona** Wlk., 1858
   redita; Druce, 1889, part, not Felder
     & Rogenhofer, 1874
   alessandra Sm., 1903
   ceyvestensis Dyar, 1904, n. syn.
   redota; Draudt, 1936
   gonoides (McD., 1943), n. syn.
9144 **dacia** Druce, 1889
   curvilinea (B. & McD., 1913)
9145 **terminimaculata** (Grt., 1873)
   pulchella (Grt., 1874), ♀
9146 **delecta** Wlk., 1858
   metallica Grt., 1865
9147 **bella** (B. & Benj., 1922)
9148 **lucasi** Sm., 1900
   aniluna Sm., 1905, ♀
   pima Sm., 1905
9149 **expolita** (Grt., 1882)
   embolima Druce, 1889
9150 **arida** Sm., 1900
9151 **cora** (B. & McD., 1918)
9152 **major** Sm., 1900
9153 **lanceolata** (Grt., 1879)
   lanceolatana (Strand, 1916)
9154 **sedata** (Hy. Edw., 1881)
   gonella Stkr., 1898
   niveicollis Sm., 1902
   gonellana (Strand, 1916)
   a. *cacola* Sm., 1907
9155 **acerba** (Hy. Edw., 1881)
9156 **disconnecta** Sm., 1903
9157 **bilimeki** Felder & Rogenhofer, 1874
   axendra (Schaus, 1898), n. syn., ♀
   bilimeci (Hamp., 1910), missp.
9158 **areloides** (B. & McD., 1912)
9159 **areli** Stkr., 1898
   monstrosa (Strand, 1916)
9160 **chea** Druce, 1889
9161 **cretata** (G. & R., 1870)
   neocula Sm., 1900
   schwarzii Sm., 1900
   schwarzi auth., missp.
   schvarzi (Hamp., 1910), missp.
9162 **eudryada** Sm., 1905
9163 **coquillettii** Sm., 1900
   coquilletti auth., missp.
9164 **behrii** Sm., 1900
   behri auth., missp.
9165 **semiatra** Sm., 1902

EUSCEPTIS Hbn., 1823
   EUGRAPHIA Gn., 1852
9166 **flavifrimbriata** Todd, 1971
   effusa; auth., not Druce, 1889

PSEUDALYPIA Hy. Edw., 1874
9167 **crotchii** Hy. Edw., 1874
   atrata Hy. Edw., 1884, form

Bagisarini

BAGISARA Wlk., 1858
   ATETHMIA; auth., not Hbn., 1821
9168 **repanda** (F., 1793)
   subusta (Hbn., 1821)

inusta (Gn., 1852)
   erecta (Wlk., 1858)
   dispartita (Wlk., 1858)
   incidens Wlk., 1858
   congesta (Wlk., 1858)
   trilinea (Wlk., 1865)
   venusta (Berg, 1882)
   unipunctata (Mösch., 1890)
9169 **rectifascia** (Grt., 1874)
9170 **pacifica** Schaus, 1911, n. record
9171 **demura** Dyar, 1913
   anotla Dyar, 1914, n. syn.
9172 **buxea** (Grt., 1881)
   delicia (Dyar, 1907)
   gustata Dyar, 1923
9173 **oula** Dyar, 1913
9174 **albicosta** Schaus, 1911, n. record
9175 **gulnare** (Stkr., 1880)
9176 **tristicta** (Hamp., 1898)
   amorata (Barnes, 1907)

PANTHEINAE

PANTHEA Hbn., 1820
   ELATINA Dup., 1845
   AUDELA Wlk., 1861
   PLATYCERURA Pack., 1864
   DIPHTHERA; Hamp., 1913, not
     Hbn., 1809
9177 **acronyctoides** (Wlk., 1861)
   leucomelana Morr., 1876
   a. **albosuffusa** McD., 1937
9178 **virginaria** (Grt., 1880)
9179 **portlandia** Grt., 1896
   a. **suffusa** McD., 1942
9180 **angelica** (Dyar, 1921)
9181 **gigantea** (French, 1890)
9182 **furcilla** (Pack., 1864)
9183 **pallescens** McD., 1937
   a. **centralis** McD., 1942
   atrescens McD., 1942, form

COLOCASIA Ochs., 1816
   LEPTOSTOLA Billberg, 1820
   DEMAS Steph., 1829
   PHINECA Wlk., 1856
   CALOCASIA Hamp., 1913, missp.
9184 **flavicornis** (Sm., 1884)
   infanta (Sm., 1911)
   electa (Sm., 1911), n. stat.
9185 **propinquilinea** (Grt., 1873)

PSEUDOPANTHEA McD., 1942
9186 **palata** (Grt., 1880)
   a. **egua** (Dyar, 1922)
   b. **utahensis** McD., 1942

LICHNOPTERA H.-S., 1856
9187 **decora** (Morr., 1875)

CHARADRA Wlk., 1865
9188 **pata** (Druce, 1894)
   patens Sm., 1908, missp.
   basiflava Sm., 1908
9189 **deridens** (Gn., 1952)
   circulifera (Wlk., 1857)
   contigua Wlk., 1865
   sudena Sm., 1908
   nigrosuffusana Strand, 1917, form
   fumosa Draudt, 1924

9189.1 **ingenua** Sm., 1906
9190 **dispulsa** Morr., 1875

    MELENETA Sm., 1908
     ZAZUNGA Dyar, 1911
9191 **antennata** Sm., 1908
    moes (Dyar, 1920)

    RAPHIA Hbn., 1821
     RHAPHIA Agassiz, 1846, missp.
     ANODONTA Rambur, 1858,
      preocc. by Lamarck, 1799
     PAPHIA Hill, 1927, missp.
9192 **abrupta** Grt., 1864
    flexuosa (Wlk., 1865), n. syn.
9193 **frater** Grt., 1864
    personata (Wlk., 1865)
9194 **piazzi** Hill, 1927
9195 **coloradensis** Putnam-Cramer, 1886
9196 **elbea** Sm., 1908
9197 **pallula** Hy. Edw., 1886
9198 **cinderella** Sm., 1903

ACRONICTINAE

Acronictini

ACRONICTA Ochs., 1816
    APATELE Hbn., 1806, suppr.
     (ICZN Op. 97 & 278)
    TRIAENA Hbn., 1818
    HYBOMA Hbn., 1820
    JOCHEAERA Hbn., 1820
    PHARETRA Hbn., 1820, preocc. by
     Bolten, 1798
    ARCTOMYSCIS Hbn., 1820
    APATELE Hbn., 1822
    ACRONYCTA Tr., 1825, emend.
    APATELA Steph., 1829, emend.
    ACRONYCTIA Meigen, 1831,
     missp.
    COMETA Sodoffsky, 1837
    SEMAPHORA Gn., 1841
    SEMATOPHORA Agassiz, 1846,
     emend.
    MICROCOELIA Gn., 1852
    MEGACRONYCTA Grt., 1873
    LEPITOREUMA Grt., 1873
    EULONCHE Grt., 1873
    PLATAPLECTA Butler, 1878
    ARTOMYSCIS Butler, 1879,
     missp.
    MASTIPHANES Grt., 1882
    VIMINIA Chapman, 1890
    CUSPIDIA Chapman, 1890
    PSEUDEPUNDA Butler, 1890
    TRICHOLONCHE Grt., 1896
    PHILORGYIA Grt., 1896
    CHAMAEPORA Warr., 1909
    PSEUDOPUNDA Hamp., 1909,
     missp.
    SUBACRONICTA Kozhanchikov,
     1950
9199 **rubricoma** Gn., 1852
9200 **americana** (Harr., 1841)
    aceris; J. E. Smith, 1797, not L., 1758
    acericola Gn., 1852
    a. **obscura** (Hy. Edw., 1886)
    b. **eldora** Sm., 1905
9201 **hastulifera** (J. E. Smith, 1797)

9202 **insita** Wlk., 1856
    a. **denvera** Sm., 1905
9203 **dactylina** Grt., 1874
9204 **hesperida** Sm., 1897
9205 **lepusculina** Gn., 1852
    populi Riley, 1870
    cretata Sm., 1897
    cinderella Sm., 1897
    chionochroa Hamp., 1909
    transversata Sm., 1897, form
    canadensis Sm., 1898, form
    similana Sm., 1905, form
    tonitra Sm., 1908
    a. **felina** (Grt., 1879)
     frigida Sm., 1897
     pacifica Sm., 1897
    b. **cyanescens** Hamp., 1909
     metra Sm., 1911
     turpis Sm., 1911
     amicora Sm., 1911
9206 **leporina** (L., 1758), extralim.
    a. **vulpina** (Grt., 1883)
     sancta (Hy. Edw., 1888)
    b. **cretatoides** (Benj., 1936)
     cretata; auth., not Sm., 1897
    c. **moesta** (Dyar, 1904)
    d. **cassinoi** (B. & Benj., 1927)
9207 **innotata** Gn., 1852
    graefii (Grt., 1863)
    griseor (Dyar, 1904)
9208 **betulae** Riley, 1884
9209 **radcliffei** (Harv., 1875)
    a. **vancouverensis** Strand, 1916
9210 **tota** (Grt., 1879)
9211 **tritona** (Hbn., 1818)
9212 **grisea** Wlk., 1856
    pudorata Morr., 1875
    a. **revellata** Sm., 1897
9213 **tartarea** Sm., 1903
9214 **falcula** (Grt., 1877)
9215 **parallela** (Grt., 1879)
9216 **albarufa** Grt., 1874
    walkeri Andrews, 1877
9217 **exempta** Dyar, 1922
9218 **mansueta** Sm., 1897
9219 **connecta** Grt., 1873
    a. **albina** McD., 1940
9220 **rapidan** (Dyar, 1912)
9221 **funeralis** G. & R., 1866
    fuscalis B. & Benj., 1924, form
9222 **paupercula** Grt., 1874
9223 **lepetita** Sm., 1908
9224 **quadrata** Grt., 1874
9225 **vinnula** (Grt., 1864)
    percolens Franc., 1938, form
9226 **superans** Gn., 1852
    superba Franc., 1938, form
9227 **laetifica** Sm., 1897
9228 **furcifera** Gn., 1852
    a. **manitoba** Sm., 1897
9229 **hasta** Gn., 1852
    a. **telum** Gn., 1852
     wanda Buchholz, 1917, form
9230 **thoracica** (Grt., 1880)
9231 **strigulata** Sm., 1897
9232 **atristrigata** (Sm., 1900)
9233 **theodora** Schaus, 1894
9234 **beameri** Todd, 1958
9235 **spinigera** Gn., 1852
    harveyana Grt., 1875

9236 **morula** G. & R., 1868
    ulmi Harr., 1869
    columboides Franc., 1938, form
9237 **interrupta** Gn., 1852
    occidentalis G. & R., 1866
    sagittaria Harr., 1869
    elisabeta Sm., 1907, form
    elizabeta B. & McD., 1917, missp.
9238 **lobeliae** Gn., 1852
    grotei Butler, 1893
9239 **valliscola** A. Blanchard, 1968
9240 **pruni** Harr., 1869
    smithii Butler, 1893
    prunata (B. & Benj., 1924), form
9241 **fragilis** (Gn., 1852)
    spectans (Wlk., 1861)
    atrior Franc., 1938, form
    a. **minella** (Dyar, 1898)
    b. **fragiloides** (B. & Benj., 1924)
9242 **exilis** Grt., 1874
9243 **ovata** Grt., 1873
9244 **modica** Wlk., 1856
9245 **haesitata** (Grt., 1882)
9246 **clarescens** Gn., 1852
    centriferruginea Strand, 1916
9247 **tristis** Sm., 1911
9248 **hamamelis** Gn., 1852
    guyasuta (Chermock & Chermock,
     1940), form
9249 **increta** Morr., 1974
9250 **inclara** Sm., 1900
    a. **inconstans** Sm., 1911
9251 **retardata** (Wlk., 1861)
    dissecta G. & R., 1870
9252 **caesarea** Sm., 1905
9253 **subochrea** Grt., 1874
9254 **afflicta** Grt., 1864
    dolens Druce, 1889
    schmalzriedi Lemmer, 1937, form
9255 **brumosa** Gn., 1852
    a. **persuasa** (Harv., 1875)
    b. **liturata** Sm., 1897
9256 **marmorata** Sm., 1897
9257 **impleta** Wlk., 1856
    luteicoma G. & R., 1870
    krautwormi (Chermock, 1927), form
    a. **illita** Sm., 1897
9258 **sperata** Grt., 1873
    a. **speratina** Sm., 1905
9259 **noctivaga** Grt., 1864
9260 **auricoma** (F., 1787) introduced,
    established Nfld.
    auricoma (D. & S., 1775), nom. nud.
    similis (Haw., 1809)
    pepli (Hbn., 1809–13)
    auricomma Freyer, 1858, missp.
    alpina Freyer, 1858
    pyhaevaarae Hoffmann, 1893
9261 **impressa** Wlk., 1856
    fasciata Wlk., 1856
    verrilli G. & R., 1870
    lemmeri (Chermock & Chermock,
     1940), form
    a. **emaculata** Sm., 1897
9262 **acla** (Benj., 1933)
9263 **distans** (Grt., 1879)
    scintillans Franc., 1938, form
    a. **dolorosa** (Dyar, 1904)
9264 **longa** Gn., 1852
    xylinoides Gn., 1852

xyliniformis Gn., 1852
pallidicoma (Grt., 1878)
set (B. & Benj., 1927), form
9265 **extricata** (Grt., 1882)
9266 **lithospila** Grt., 1874
9267 **barnesii** Sm., 1897
9268 **perdita** Grt., 1874
9269 **edolata** (Grt., 1881)
9270 **othello** Sm., 1908
9271 **arioch** Stkr., 1898
9272 **oblinita** (J. E. Smith, 1797)
salicis Harr., 1869
insolita Grt., 1873, form
9273 **sagittata** McD., 1940
9274 **lanceolaria** (Grt., 1875)

MEROLONCHE Grt., 1882
9275 **lupini** (Grt., 1873)
lupina McD., 1938, missp.
9276 **spinea** (Grt., 1876)
9277 **dolli** B. & McD., 1918
9278 **atlinensis** B. & Benj., 1927
9279 **ursina** Sm., 1898
ursini Sm., 1911, missp.

SIMYRA Ochs., 1816
CNEPHATA Billberg, 1820
SYMIRA Hbn., 1821, missp.
ASEMA Sodoffsky, 1837
NIMYRA Gn., 1841, missp.
ABLEPHARON Grt., 1873
SIYMRA Warr., 1912, missp.
OMMATOSTOLIDEA Benj., 1933,
n. syn.
9280 **henrici** (Grt., 1873)
evanida (Grt., 1873)
fumosa (Morr., 1874), form
a. **julitae** (Benj., 1933), n. syn., n.
stat.

AGRIOPODES Hamp., 1908
9281 **fallax** (H.-S., 1854)
9282 **geminata** (Sm., 1903)
9283 **tybo** (Barnes, 1904)
9284 **teratophora** (H.-S., 1854)
inscripta (Wlk., 1858)

POLYGRAMMATE Hbn., 1818
GRAMMOPHORA Gn., 1852
9285 **hebraeicum** Hbn., 1818
hebraea Gn., 1852, missp.

HARRISIMEMNA Grt., 1873
9286 **trisignata** (Wlk., 1856)
sexguttata (Harr., 1869)

Bryophilini

CRYPHIA Hbn., 1818
CERMA; auth., not Hbn., 1818
POECILIA Schrank, 1802, preocc.
by Schneider, 1801
JASPIDIA Hbn., 1806 suppr.
(ICZN Op. 97 & 278)
EUTHALES Hbn., 1820
JASPIDIA Hbn., 1822
BRYOPHILA Tr., 1825
BRYOLEUCA Hamp., 1908
BRYONYCTA Boursin, 1955, n.
syn.

SCYTHOBRYA Boursin, 1960
BRYOPSIS Boursin, 1970
9287 **olivacea** (Sm., 1891), n. comb.
9288 **fascia** (Sm., 1903), n. comb.
9289 **flavidior** (B. & McD., 1911), n. comb.
9290 **oaklandiae** (B. & McD., 1911), n.
comb.
9291 **nanoides** Franc. & Todd, new name
nana (B. & McD., 1911), preocc. by
Hbn., 1818
9292 **cuerva** (Barnes, 1907), n. comb.
9293 **pallida** (B. & L., 1922), n. comb.
9294 **galva** (Stkr., 1898), n. comb.
9295 **sarepta** (Barnes, 1907), n. comb.
9296 **viridata** (Harv., 1876), n. comb.
9297 **albipuncta** (B. & McD., 1910), n. comb.

AGARISTINAE

XEROCIRIS Ckll., 1904
CIRIS Grt., 1863, preocc. by Koch,
1846
9298 **wilsonii** (Grt., 1863)
vilsoni Hamp., 1910, emend.

EUDRYAS Bdv., 1836
EUTHISANOTIA; auth., not Hbn.,
1831
CYPHOCAMPA Harr., 1869
PARATHISANOTIA Kiriakoff,
1977, n. syn.
9299 **unio** (Hbn., 1827–31)
9300 **brevipennis** Stretch, 1872
9301 **grata** (F., 1793)
assimilis Bdv., 1874

GERRA Wlk., 1865
FENARIA Grt., 1882
9302 **radicalis** Wlk., 1865
adrasta (Druce, 1889)
9303 **sevorsa** (Grt., 1882)
aedessa (Druce, 1889)
meridionalis Draudt, 1919
luteomacula Strand, 1916

GERRODES Hamp., 1908
9304 **minatea** Dyar, 1912
longipes; auth., not Druce, 1889

NEOTUERTA Kiriakoff, 1977
MISA; auth., not Karsch, 1895
TUERTA; auth., not Wlk., 1869
9305 **hemicycla** (Hamp., 1904)

CAULARIS Wlk., 1858
9306 **lunata** Hamp., 1904

EUSCIRRHOPTERUS Grt., 1866
(July)
HETERANDRA H.-S., 1866 (Sept.)
COPIDRYAS Grt., 1876, n. syn.
EUSCHIRRHOPTERUS Kby.,
1892, missp.
LAQUEA Jordan, 1896
EUSCHIRROPTERUS Hamp.,
1901, missp.
LAGUEA Strand, 1912, missp.
HORTONIUS Dyar, 1926, n. syn.
9307 **gloveri** G. & R., 1868
euenemus (Dyar, 1926), n. syn.

9308 **cosyra** (Druce, 1896)

PSYCHOMORPHA Harr., 1839
9309 **epimenis** (Drury, 1782)
glaucopis Harr., 1839
9310 **euryrhoda** Hamp., 1910

EUPSEUDOMORPHA Dyar, 1893
EDWARDSIA Neum., 1880,
preocc. by Costa, 1834
EUEDWARDSIA Kby., 1892,
preocc. by Grt., 1882
9311 **brillians** (Neum., 1880)

ALYPIODES Grt., 1883
ALYPIOIDES auth., missp.
ALYPIODES Dyar, Neave, 1939,
erroneous author
9312 **bimaculata** (H.-S., 1853)
crescens (Wlk., 1856)
grotei (Bdv., 1869)
trimaculata (Bdv., 1874)
flavilinguis Grt., 1883
dugesii Ckll., 1895
9313 **geronimo** (Barnes, 1900)

ALYPIA Hbn., 1818
LEUCOSEMIA Mén., 1857
9314 **octomaculata** (F., 1775)
albomaculata (Cram., 1784)
bimaculata (Gmel., 1790)
octomaculalis Hbn., 1818
quadriguttalis Hbn., 1818
9315 **matuta** Hy. Edw., 1883
octomaculata; Hamp., 1901, part, not
F., 1775
9316 **wittfeldii** Hy. Edw., 1883
octomaculata; Hamp., 1901, part, not
F., 1775
9317 **dipsaci** G. & R., 1868
dipsaci (Bdv., 1869), preocc. by G. &
R., 1868
9318 **langtoni** Couper, 1865
sacramenti G. & R., 1868
sacramenti (Bdv., 1869), preocc. by
G. & R. 1868
hudsonia Hy. Edw., 1884
hudsonica B. & McD., 1917, missp.
sacramento McD., 1938, missp.
9319 **ridingsii** Grt., 1865
9320 **mariposa** G. & R., 1868
lunata Stretch, 1872

ANDROLOMA Grt., 1873
9321 **maccullochii** (Kby., 1837)
a. **lorquini** (G. & R., 1868)
b. **similis** (Stretch, 1872)
edwardsi Bdv., 1874
conjuncta (Hy. Edw., 1883)
9322 **disparata** (Hy. Edw., 1884)
gracilenta (Graef, 1887)
desperata (Kby., 1892), missp.
9323 **brannani** (Stretch, 1872)
9324 *number omitted*

AMPHIPYRINAE

Apameini

APAMEA Ochs., 1816
 ABROMIAS Billberg, 1820
 SEPTIS Hbn., 1821
 XYLOPHASIA Steph., 1829
 HAMA Steph., 1829
 AGROSTOBIA Boie, 1835
 SYMA Steph., 1850, preocc. by
  Lesson, 1827
 DIMYA Moore, 1882, preocc. by
  Rouault, 1850
9325 **cuculliformis** (Grt., 1875)
9326 **verbascoides** (Gn., 1852)
9327 **inebriata** Fgn., 1977
9328 **nigrior** (Sm., 1891)
9329 **cariosa** (Gn., 1852)
 idonea (Grt., 1882)
 cluna (Stkr., 1898)
9330 **dionea** (Sm., 1899)
9331 **cristata** (Grt., 1878)
9332 **vulgaris** (G. & R., 1866)
9333 **lignicolora** (Gn., 1852)
 a. **quaesita** (Grt., 1876)
 b. **atriclava** (B. & McD., 1913)
9334 **antennata** (Sm., 1891)
 a. **purpurissata** (B. & McD., 1913)
 washingtonensis (Strand, 1916)
9335 **grotei** (B. & McD., 1914)
 auranticolor; auth., not Grt., 1873
9336 **atrosuffusa** (B. & McD., 1913)
9337 **maxima** (Dyar, 1904)
9338 **acera** (Sm., 1900)
9339 **auranticolor** (Grt., 1873)
 a. **barnesii** (Sm., 1899)
 b. **sora** (Sm., 1903)
9340 **genialis** (Grt., 1874)
9341 **vultuosa** (Grt., 1875)
9342 **multicolor** (Dyar, 1904)
9343 **apamiformis** (Gn., 1852)
 contenta (Wlk., 1857)
9344 **plutonia** (Grt., 1883)
9345 **perpensa** (Grt., 1881)
 perpenoa (Grt., 1881), incorr. orig.
 spell.
9346 **occidens** (Grt., 1878)
 coloradensis (Strand, 1916), form
9347 **albina** (Grt., 1874)
9348 **amputatrix** (Fitch, 1857)
 arctica (Freyer, 1842), preocc.
 amica (Harr., 1862), preocc.
 pluviosa (Wlk., 1865)
 formosus (Ellsworth, 1918)
9349 **castanea** (Grt., 1874)
 cymosa (Grt., 1874)
 cymosana (Strand, 1916), form
9350 **smythi** Franc., 1952
9351 **alia** (Gn., 1852)
 suffusca (Morr., 1875)
 rorulenta (Sm., 1904), form
9352 **semilunata** (Grt., 1881)
9353 **inordinata** (Morr., 1875)
 a. **montana** (Sm., 1890)
 b. **columbiae** (Strand, 1916)
9354 **centralis** (Sm., 1891)
9355 **unita** (Sm., 1904)
9356 **spaldingi** (Sm., 1909)
 umbrifacta (Hamp., 1910)

9357 **cinefacta** (Grt., 1881)
 a. **albertae** (Strand, 1916)
9358 **parcata** (Sm., 1903)
9359 **commoda** (Wlk., 1857)
 satina (Stkr., 1898)
 a. **alberta** (Sm., 1903)
 illustra (Sm., 1908)
9360 **impulsa** (Gn., 1852)
9361 **mixta** (Grt., 1881)
9362 **indocilis** (Wlk., 1856)
 lona (Stkr., 1898)
 a. **runata** (Sm., 1899)
 ferens (Sm., 1903)
 enigra (Sm., 1904), form
 separans (Grt., 1881), form
9363 **ampliata** (McD., 1940)
9364 **finitima** Gn., 1852
 a. **cerivana** (Sm., 1900)

AGROPERINA Hamp., 1908
9365 **lateritia** (Hufn., 1766)
 assimilis (Doubleday, 1847)
 expallescens (Stgr., 1882)
 amanda (Swinhoe, 1901)
 borealis (Strand, 1903), form
9366 **obliviosa** (Wlk., 1858)
9367 **dubitans** (Wlk., 1856)
 insignata (Wlk., 1857)
 sputator (Grt., 1873)
 sputatrix (Grt., 1875)
 a. **cogitata** (Sm., 1891)
9368 **geminimacula** (Dyar, 1904), n. comb.
9369 **inficita** (Wlk., 1857)
 belangeri (Morr., 1874)
9370 **conradi** (Grt., 1879)
 citima (Grt., 1883)
9371 **popofensis** (Sm., 1900)
 a. **indela** Sm., 1910
 b. **lineosa** Sm., 1910
 palliderufa Strand, 1916
 pendina Sm., 1910, form
 nada Strand, 1916
 saturatior Strand, 1916
9372 **lutosa** (Andrews, 1877)
9373 **helva** (Grt., 1875)

PROTAGROTIS Hamp., 1903
9374 **niveivenosa** (Grt., 1879)
 viralis (Grt., 1881)
9375 **extensa** Sm., 1905
 flavistriga (Sm., 1905)
9376 **obscura** B. & McD., 1911
9377 **nichollae** Hamp., 1908

CRYMODES Gn., 1841
9378 **burgessi** (Morr., 1874)
9379 **ona** (Sm., 1909)
 stygia (Dyar, 1914)
9380 **relicina** (Morr., 1875)
9381 **migrata** (Sm., 1903)
9382 **devastator** (Brace, 1819)
 abjecta (Gn., 1852)
 ordinaria (Wlk., 1856)
 contenta (Wlk., 1856)
 speciosa (Morr., 1874), form
9383 **longula** (Grt., 1879)
9384 **murrayi** (Gibs., 1920)
 johanseni (McD., 1933)
9385 **exulis** (Dup., 1836)
 gelata (Dup., 1836)

 maillardi; auth., not Gey., 1834
 diffua (Gey., 1837)
 groenlandica (Dup., 1838)
 gelida Gn., 1852
 borea Gn., 1852
 borea (H.-S., 1852), preocc. by Gn.,
 1852
 a. **alticola** (Sm., 1891)

LUPERINA Bdv., 1828
 LUPERNIA Doubleday, 1850,
 missp.
 LYPERINA Spuler, 1908, emend.
 PALLUPERINA Hamp., 1920
9386 **trigona** Sm., 1902
9387 **innota** Sm., 1908
9388 **venosa** (Sm., 1903)
9389 **enargia** B. & Benj., 1926
9390 **virguncula** (Sm., 1899), n. comb.
9391 **passer** (Gn., 1852)
 incallida (Wlk., 1857)
 loculata (Morr., 1874)
 conspicua (Morr., 1874), form
 a. **birnata** (Sm., 1908)
9392 **morna** (Stkr., 1878), n. comb.
 hulstii (Grt., 1880)
9393 **stipata** (Morr., 1875)

TRICHOPLEXIA Hamp., 1908
9394 **exornata** (Mösch., 1860)
 a. **contradicta** (Sm., 1895)

OMMATOSTOLA Grt., 1873
9395 **lintneri** Grt., 1873

EREMOBINA McD., 1937
9396 **claudens** (Wlk., 1857)
 leucoscelis (Grt., 1874)
 fibulata (Morr., 1874)
 a. **albertina** (Hamp., 1908)
9397 **hilli** (Grt., 1876)
9398 **jocasta** (Sm., 1900)
9399 **unicincta** (Sm., 1902)
9400 **hanhami** (B. & Benj., 1924)

OLIGIA Hbn., 1821
 PROCUS Oken, 1815, suppr.
 (ICZN Op. 417)
 MIANA Steph., 1829
 PROCUS Agassiz, 1846
 PHOTEDES Led., 1857
 PHOTHEDES Led., 1857, orig.
 missp.
 MESOLIGIA Boursin, 1965
9401 **indirecta** (Grt., 1875)
9402 **chlorostigma** (Harv., 1876)
 viridimusca (Sm., 1899)
9403 **marina** (Grt., 1874)
9404 **modica** (Gn., 1852)
 subcedens (Wlk., 1857)
9405 **tusa** (Grt., 1878)
9406 **fractilinea** (Grt., 1874)
 vulgivaga (Morr., 1874), form
 modiola (Grt., 1879), form
 mactatoides (B. & McD., 1912)
 una (Stkr., 1900), form
 misera (Grt., 1881), form
 a. **albescens** B. & McD., 1910
9407 **arbora** (B. & McD., 1912)
9408 **exhausta** (Sm., 1903)

9409 **hausta** (Grt., 1882)
9410 **crytora** (Franc., 1950)
9411 **semicana** (Wlk., 1865)
    latireptana (Wlk., 1865)
9412 **laevigata** (Sm., 1898)
9413 **tonsa** (Grt., 1880)
    fasciata B. & McD., 1912
    a. **subjuncta** (Sm., 1898)
9414 **violacea** (Grt., 1881)
    a. **rampartensis** B. & Benj., 1923
    b. **columbia** McD., 1927
9415 **bridghami** (G. & R., 1866)
    a. **iridis** Dyar, 1923
9416 **minuscula** (Morr., 1874)
    a. **grahami** Benj., 1933
9417 **egens** (Wlk., 1857)
    a. **ferrealis** (Grt., 1883)
    transfrons (Neum., 1884)
9418 **obtusa** (Sm., 1902)
9419 **mactata** (Gn., 1852)
    a. **allecto** (Sm., 1899)
9420 **illocata** (Wlk., 1857)
    stigmata (Grt., 1876)

EUROS Hy. Edw., 1881
    PROTOPHANA Hamp., 1906
9421 **cervina** (Hy. Edw., 1890)
9422 **proprius** Hy. Edw., 1881

COBALOS Sm., 1899
    COBALIODES Dyar, 1903
9423 **angelicus** Sm., 1899
9424 **franciscanus** Sm., 1899

MEROPLEON Dyar, 1924 .
9425 **cosmion** Dyar, 1924
9426 **titan** Todd, 1958
9427 **diversicolor** (Morr., 1874)
9428 **ambifusca** (Newman, 1948)

LEMMERIA B. & Benj., 1926
    BRACHYCOSMIA Hamp., 1906,
    preocc. by Butler, 1890
9429 **digitalis** (Grt., 1882)

SELICANIS Sm., 1900
9430 **cinereola** Sm., 1900

PARASTICHTIS Hbn., 1821
    PARASTICTUS Agassiz, 1846,
    missp.
    DYSCHORISTA Led., 1857
    TAENIOSEA Grt., 1874
9431 **discivaria** (Wlk., 1856)
    perbellis (Grt., 1874)
    gentilis (Grt., 1874), form

XYLOMOIA Stgr., 1892
    XYLOMOEA Hamp., 1909, missp.
9432 **didonea** (Sm., 1894)
    draudti B. & Benj., 1926, n. stat.
9433 **chagnoni** B. & McD., 1917

SPARTINIPHAGA McD., 1937
9434 **includens** (Wlk., 1858)
    norma (Morr., 1875)
    penita (Morr., 1875)
    mariae (Grt., 1877)
    lunaris (Strand, 1916), form
9435 **inops** (Grt., 1881)

    insipidella (Strand, 1916)
    insipida (Stkr., 1900), form
9436 **panatela** (Sm., 1904)

HYPOCOENA Hamp., 1908
9437 **inquinata** (Gn., 1852)
    orientalis (Grt., 1882)
9438 **variana** (Morr., 1876)
9439 **basistriga** (McD., 1933)
9440 **rufostrigata** (Pack., 1867)
    punctivena (Sm., 1894)
    rufostriga Hamp., 1910, missp.
9441 **enervata** (Gn., 1852)
    fodiens (Gn., 1852)
9442 **orphnina** (Dyar, 1912)
9443 **defecta** (Grt., 1874)

PHRAGMATIPHILA Hamp., 1908
9444 **interrogans** (Wlk., 1856)

BENJAMINIOLA Strand, 1928
    BUCHHOLZIA B. & Benj., 1926,
    preocc. by Michaelsen, 1886
    EUBUCHHOLZIA B. & Benj.,
    1929
9445 **colorada** (Sm., 1900)
    cirphidia (Hamp., 1910)
    leucanidia (Hamp., 1910)
    colorado McD., 1938, missp.

MAMMIFRONTIA B. & L., 1922
9446 **leucania** B. & L., 1922
9447 **rileyi** Benj., 1936

ARCHANARA Wlk., 1866
    ARCHARANA Viette, 1947, missp.
9448 **alameda** (Sm., 1903)
9449 **oblonga** (Grt., 1882)
    permagna (Grt., 1883)
    subcarnea (Kellicott, 1883)
9450 **subflava** (Grt., 1882)
9451 **laeta** (Morr., 1875)

MACRONOCTUA Grt., 1874
9452 **onusta** Grt., 1874

HELOTROPHA Led., 1857, rev. stat.
9453 **reniformis** (Grt., 1874)
    atra (Grt., 1874), form
    insignata (Strand, 1916)

AMPHIPOEA Billberg, 1820
9454 **velata** (Wlk., 1865)
    sera (G. & R., 1868)
9455 **lunata** (Sm., 1891)
    albilunata (Sm., 1899)
9456 **interoceanica** (Sm., 1899)
9457 **americana** (Speyer, 1875)
    lusca (Sm., 1891)
    atlantica (Sm., 1899)
    a. **pacifica** (Sm., 1899)
9458 **cottlei** (McD., 1948)
9459 **senilis** (Sm., 1892)
9460 **keiferi** (Benj., 1935)
9461 **erepta** (Grt., 1881)
    a. **ryensis** (Bird, 1913)
9462 **flavostigma** (B. & Benj., 1924)

PARAPAMEA Bird, 1927
9463 **buffaloensis** (Grt., 1877)

    latia (Stkr., 1899)
    simplicissima (Bird, 1916), form

PAPAIPEMA Sm., 1899
by ERIC L. QUINTER [nos. 9464–9509]

    OCHRIA; Grt., 1874, not Hbn.,
    1821
    APAMEA; Grt., 1874, not Ochs.,
    1816
    GORTYNA; auth., not Ochs., 1816
    HYDROECIA; auth., missp., not
    Gn., 1841
    EMBOLOECIA Hamp., 1908, n.
    syn.
    PAIPEMA Rockburne & Hdwk.,
    1970, missp.
9464 **cerina** (Grt., 1874)
9465 **duovata** (Bird, 1902)
    rutila; (Bird, 1900), not Gn., 1852
9466 **cataphracta** (Grt., 1864)
9467 **sulphurata** Bird, 1926, n. stat.
9468 **aerata** (Lyman, 1901)
    imperspicua; auth., not Bird, 1908
    limpida; auth., not Gn., 1852
    fluxa Bird, 1911, form, n. stat.
9469 **polymniae** Bird, 1917
9470 **araliae** Bird & Jones, 1921
    araliae McD., 1938, attributed
    incorrectly to Bird
9471 **arctivorens** Hamp., 1910
    rutila; auth., not Gn., 1852
    artivorens Muller, 1965, missp.
9472 **harrisii** (Grt., 1881)
    harrisi (Grt., 1890), and auth.,
    missp.
    rubiginosa Bird, 1911, form
    mulieris Strand, 1916
9473 **impecuniosa** (Grt., 1881)
    inquaesita; Holl., 1903, not G. & R.,
    1868
9474 **sauzalitae** (Grt., 1875), n. comb.
    purpurifascia; Grt., 1873, not G. &
    R., 1868
    sanzalitae anon., 1882, missp.
    sauzaelita Sm., 1891, missp.
    sauzaelitae auth., missp.
    erubescens Bird, 1911
    papaipemoides (B. & Benj., 1929), form
9475 **angelica** (Sm., 1899)
9476 **verona** (Sm., 1899)
    anargyrea (Dyar, 1908), n. syn.
    impecuniosa; Hamp., 1910, part, not
    Grt., 1881
    anargyria Bird, 1944, missp.
9477 **astuta** Bird, 1907
9478 **leucostigma** (Harr., 1841)
    purpurifascia (G. & R., 1868)
    pterisii; Hamp., 1910, part, not Bird,
    1907
    luteipicta Strand, 1916
9479 **lysimachiae** Bird, 1914
    purpurifascia; G. & R., 1868, part, ♂
9480 **pterisii** Bird, 1907
    triorthia (Dyar, 1908)
    pterissi Moore, 1955, missp.
9481 **stenocelis** (Dyar, 1907)
    stenoscelis auth., missp.
9482 **speciosissima** (G. & R., 1868)
    regalis Wyatt & Beer, 1927, form

9483 **inquaesita** (G. & R., 1868)
  quaesita Grt., 1881, missp.
  wyatti B. & Benj., 1929, form
9484 **rutila** (Gn., 1852)
  merriccata Bird, 1907, incorr. orig.
    spell., n. syn.
  placida Bird, 1921, n. syn.
  merricata Draudt, 1926, Tietz, 1936,
    missp.
  merrickata McD., 1938, emend.
  baptisiae; Forbes, 1954, part, not
    Bird, 1902
9485 **baptisiae** (Bird, 1902)
  circumlucens; Bird, 1900, not Sm.,
    1899
  a. **limata** Bird, 1908, n. stat.
    ochroptena; B. & McD., 1912,
      part, not Dyar, 1908
    vaha Benj., 1933, n. syn.
    depictata Benj., 1933, n. syn., n.
      stat.
    ochroptenoides Bird, 1944, form,
      nom. nud.
9486 **birdi** (Dyar, 1908). rev. stat.
  marginidens; Sm., 1899, part, not
    Gn., 1852
9487 **pertincta** Dyar, 1920
9488 **insulidens** (Bird, 1902)
9489 **dribi** B. & Benj., 1926
9490 **nepheleptena** (Dyar, 1908)
  appasionata; Sm., 1899, part, missp.,
    not Harv., 1876
  moeseri Bird, 1911
9491 **circumlucens** (Sm., 1899)
  ochroptena (Dyar, 1908), n. stat.
  marginidens; Dyar, 1908, not Gn.,
    1852
  humuli Bird, 1915, n. stat.
  ochropenta Bird, 1916, missp.
9492 **marginidens** (Gn., 1852)
  nephrasyntheta (Dyar, 1908), n. syn.
  nephasyntheta (Lyman, 1908), missp.
9493 **appassionata** (Harv., 1876)
  appasionata auth., missp.
  horni Strand, 1915
9494 **eryngii** Bird, 1917
  nephrasyntheta; auth., not Dyar, 1908
9495 **furcata** (Sm., 1899)
9496 **nebris** (Gn., 1852)
  nitela; Holl., 1903, not Gn., 1852
  nitela (Gn., 1852), form
  nitella (Morris, 1860), missp.
  nebris; Holl., 1903, not Gn., 1852
9497 **necopina** (Grt., 1876)
  imperturbata Bird, 1907
  mulleri Buchholz, 1957, n. syn.
9498 **silphii** Bird, 1915
9499 **duplicata** Bird, 1908
  duplicatus Bird, 1908, incorr. orig.
    spell.
  obsolescens Strand, 1916
9500 **maritima** Bird, 1909
  necopina; auth., not Grt., 1876
9501 **eupatorii** (Lyman, 1905)
9502 **nelita** (Stkr., 1898)
  errans B. & McD., 1912, n. syn.
  linda Bird, 1916, form
  orbicularis Strand, 1916
  melita Tietz, 1936, missp.
  engelhardti Bird, 1939, n. syn.

  unimoda; Fbs., 1954, part, not Sm.,
    1894
9503 **rigida** (Grt., 1877)
9504 **aweme** (Lyman, 1908)
  aveme Hamp., 1910, missp.
9505 **cerussata** (Grt., 1864)
  cerrusata auth., missp.
  cerrussata (Sm., 1899), missp.
  cerussata; Dyar, 1903, auth.,
    attributed incorrectly to G. & R.
9506 **sciata** Bird, 1908
  limpida; auth., not Gn., 1852
  limpida; Hamp., 1910, part, not Gn.,
    1852
9507 **limpida** (Gn., 1852)
9508 **beeriana** Bird, 1923
  laciniariae Bird, 1923, form
  laciniaria Hessel, 1954, missp.
9509 **unimoda** (Sm., 1894)
  frigida (Sm., 1899), n. stat.
  thalictri (Lyman, 1905), form
    cerrussata; Sm., 1899, part, missp.,
      not Grt., 1864
    frigida; Hamp., 1910, part, Draudt,
      1926, part, not Sm., 1899
    terminalis Strand, 1916
  perobsoleta (Lyman, 1905)
  imperspicua Bird, 1908, n. syn.

HYDRAECIA Gn., 1841
  HYDROECIA auth. missp.
  HYDROOECIA H.-S., 1849, missp.
  HADROECIA Schaus, 1894, missp.
9510 **pallescens** Sm., 1899
9511 **medialis** Sm., 1892
9512 **obliqua** Harv., 1876
9513 **immanis** Gn., 1852
9514 **micacea** (Esp., 1789)
  cypriaca (Hbn., 1800–03)
9515 **perobliqua** Hamp., 1910
9516 **stramentosa** Gn., 1852
  diplocyma Hamp., 1910
9517 **intermedia** B. & Benj., 1924
9518 **ximena** (B. & Benj., 1924)
9519 **columbia** (B. & Benj., 1924)

ACHATODES Gn., 1852
9520 **zeae** (Harr., 1841)
  sandix Gn., 1852

PERANIANA Strand, 1942
  PERANIA B. & McD., 1910,
    preocc. by Thorell, 1890
9521 **dissociata** (B. & McD., 1910)

IODOPEPLA Franc., 1964
9522 **u-album** (Gn., 1852)
  v-album (Wlk., 1856), missp.
  w-album (Grt., 1874), missp.
  baliola (Morr., 1874)
  purpuripennis (Grt., 1875)

BELLURA Wlk., 1865
  ARZAMA Wlk., 1865
  SPHIDA Grt., 1878
  ARZAMOPSIS Dyar, 1922
9523 **gortynoides** Wlk., 1865
  vulnifica (Grt., 1873)
  melanopyga (Grt., 1881), form
  diffusa (Grt., 1878), form

9524 **brehmei** (B. & McD., 1916)
9525 **obliqua** (Wlk., 1865)
  obliquata (G. & R., 1868)
  a. **pallida** (B. & Benj., 1924)
  b. **gargantua** (Dyar, 1913)
  c. **anoa** (Dyar, 1913)
9526 **densa** (Wlk., 1865)
  oecogenes (Dyar, 1913)

ASEPTIS McD., 1937
9527 **fumosa** (Grt., 1879)
9528 **catalina** (Sm., 1899)
9529 **perfumosa** (Hamp., 1918)
9530 **fumeola** (Hamp., 1908)
  probata (B. & McD., 1910)
9531 **ethnica** (Sm., 1899)
9532 **binotata** (Wlk., 1865)
  extersa (Wlk., 1865)
  rubiginosa (Wlk., 1865)
  a. **curvata** (Grt., 1874)
9533 **adnixa** (Grt., 1880)
9534 **paviae** (Stkr., 1874)
  inconspicua (Sm., 1893)
9535 **pausis** (Sm., 1899)
9536 **genetrix** (Grt., 1878)
9537 **dilara** (Stkr., 1899)
9538 **cara** (B. & McD., 1912)
9539 **susquesa** (Sm., 1908)
9540 **monica** (B. & McD., 1918)
9541 **serrula** (B. & McD., 1918)
9542 **bultata** (Sm., 1906)
9543 **characta** (Grt., 1880)
  a. **erica** (Sm., 1905)
  b. **luteocinerea** (Sm., 1900)
  c. **pluraloides** (McD., 1922)

EUPLEXIA Steph., 1829
9544 **brillians** B. & McD., 1911
9545 **benesimilis** McD., 1922

PHLOGOPHORA Tr., 1825
  SOLENOPTERA Dup., 1845,
    preocc. by Audinet-Serville, 1832
  BROTOLOMIA Led., 1857
  RACOPTERA Scott, 1858
  MESOLOMIA Sm., 1893
9546 **iris** Gn., 1852
9547 **periculosa** Gn., 1852
  v-brunneum (Grt., 1875), form

CONSERVULA Grt., 1874
  APPANA Moore, 1881
9548 **anodonta** (Gn., 1852)

ENARGIA Hbn., 1821
  EUPERIA Gn., 1839
9549 **decolor** (Wlk., 1858)
  discolor Sm., 1893, missp.
  sia Strand, 1916
  mia Strand, 1916
9550 **infumata** (Grt., 1874)
  punctirena (Sm., 1900)
9551 **mephisto** Franc., 1939

IPIMORPHA Hbn., 1821
  ZENOBIA Oken, 1815, suppr.
    (ICZN Op. 417)
  PLASTENIS Bdv., 1840
  ZENOBIA Agassiz, 1846, preocc. by
    Gray, 1821

IPHIMORPHA Mösch., 1886,
missp.
9552 nanaimo Barnes, 1905
9553 viridipallida B. & McD., 1916
9554 subvexa Grt., 1876
9555 pleonectusa Grt., 1873
    aequilinea (Sm., 1883)
    a. manitobae Strand, 1916

CHYTONIX Grt., 1874
    CHYTONYX Neave, 1939, missp.
9556 palliatricula (Gn., 1852)
    iaspis (Gn., 1852), form
9557 sensilis Grt., 1881
    submediana Strand, 1916, form
    macdonaldi Benj., 1922
9558 ruperti Franc., 1941
9559 divesta (Grt., 1874), n. comb.
    laticlava Sm., 1903

DYPTERYGIA Steph., 1829
    DIPTERYGIA Agassiz, 1846,
    missp.
9560 rozmani Berio, 1974
    scabriuscula; auth., not L., 1758
9561 patina (Harv., 1875)
    minorata Barnes, 1907
9562 dolens (Druce, 1909), n. record
    daemonassa (Dyar, 1910), n.
    syn.

ANDROPOLIA Grt., 1895
9563 diversilineata (Grt., 1877)
    illepida (Grt., 1879)
    resoluta (Sm., 1894)
    submissa Sm., 1911, form
9564 contacta (Wlk., 1856)
    aspera (Morr., 1874)
    diffusilis (Harv., 1878)
    a. pulverulenta (Sm., 1891)
9565 pallifera (Grt., 1877)
9566 sansar (Stkr., 1898)
9567 ochracea (Sm., 1900)
9568 dispar (Sm., 1900)
9569 olorina (Grt., 1876)
    a. australiae Grt., 1895
9570 aedon (Grt., 1880)
9571 theodori (Grt., 1878)
    a. epichysis (Grt., 1880)
    b. vancouvera Strand, 1916
9572 olga Sm., 1911
9573 extincta (Sm., 1900)
9574 lichena B. & McD., 1912

RHIZAGROTIS Sm., 1890
9575 cloanthoides (Grt., 1881)
9576 albalis (Grt., 1878)
    defectipes (Dyar, 1921)
    a. actona Sm., 1910

PSEUDOHADENA Alphéraky, 1889
9577 vulnerea (Grt., 1883)

HYPPA Dup., 1845
9578 xylinoides (Gn., 1852)
    contraria (Wlk., 1857)
    ancocisconensis (Morr., 1875)
9579 contrasta McD., 1946
9580 brunneicrista Sm., 1902
9581 indistincta Sm., 1894

NEDRA Clarke, 1940
    DELTA; auth., not Saalmüller, 1891
9582 ramosula (Gn., 1852)
9583 stewarti (Grt., 1875)
9584 dora Clarke, 1940
9585 hoeffleri Clarke, 1940

AGROTISIA Hamp., 1908
    HARRISONIA Schaus, 1923,
    preocc. by Ferris, 1922
9586 evelinae Benj., 1933

PROPERIGEA B. & Benj., 1926
    NEPERIGEA McD., 1937, n. syn.
9587 loculosa (Grt., 1881)
    continentis (Dyar, 1916)
    oziphona (Dyar, 1916)
9588 albimacula (B. & McD., 1912), n.
    comb.
9589 costa (B. & Benj., 1923), n. comb.
9590 continens (Hy. Edw., 1885), n. comb.
9591 mephisto (A. Blanchard, 1968), n.
    comb.
9592 tapeta (Sm., 1900), n. comb.
9593 seitzi (B. & Benj., 1926), n. comb.
9594 suffusa (B. & McD., 1912), n. comb.
    a. rubida (B. & McD., 1912), n.
    comb.
9595 perolivalis (B. & McD., 1912), n.
    comb.
9596 niveirena (Harv., 1876), n. comb.
    pohono (Sm., 1899), n. comb.

HEMIBRYOMIMA B. & Benj., 1927
9597 chryselectra (Grt., 1880)
    benigna (Hy. Edw., 1885)

PSEUDOBRYOMIMA B. & Benj.,
1927
9598 distans (B. & McD., 1912)
9599 muscosa (Hamp., 1906)
9600 fallax (Hamp., 1906)
    falsa; Grt., 1880, ♀, not Grt., 1880, ♂
    uintara (Sm., 1911), form

PSEUDANARTA Grt., 1878
9601 flavidens (Grt., 1879)
9602 actura Sm., 1908
9603 basivirida (B. & McD., 1911), n.
    comb.
9604 caeca Dod, 1913
9605 crocea (Hy. Edw., 1875)
9606 flava (Grt., 1874)
    dupla Sm., 1908
9607 singula (Grt., 1880)
9608 pulverulenta (Sm., 1891)
9609 exasperata Franc., 1941
9610 perplexa Franc., 1941
9611 damnata Franc., 1941
9612 vexata Franc., 1941
9613 daemonalis Franc., 1941

HOMOANARTA B. & Benj., 1923
9614 falcata (Neum., 1884)
9615 peralta (Barnes, 1907)
    peralto B. & McD., 1917, missp.
9616 carneola (Sm., 1891)

PSEUDANTHOECIA Sm., 1883
9617 tumida (Grt., 1880)

PHOSPHILA Hbn., 1818
    HELIOSCOTA Grt., 1895
9618 turbulenta Hbn., 1818
    arcuata (Wlk., 1857)
9619 miselioides (Gn., 1852)
    miscellus (Sm., 1902), form
    a. macerata (Sm., 1902)

MIRACAVIRA Franc., 1937
    MIRACAVERA McD., 1938,
    missp.
9620 brillians (Barnes, 1901)
9621 sylvia (Dyar, 1913)

CROPIA Wlk., 1858
    DECELEA Wlk., 1858
    LISTONIA Mösch., 1886
    LISTENIA Mösch., 1886, incorr.
    orig. spell.
    EUDIPNA; auth., not Wlk., 1856
9622 connecta (Sm., 1894)
    striata (Druce, 1898)
9623 templada (Schaus, 1906)

SPEOCROPIA Hamp., 1908
9624 fernae Benj., 1933
9625 trichoma (H.-S., 1868)

TRACHEA Ochs., 1816
    ACHATIS Billberg, 1820
    TRACHAEA Schaus, 1923, missp.
9626 delicata (Grt., 1874)
    interna (Grt., 1875)

PARATRACHEA Hamp., 1908
9627 viridescens (B. & McD., 1918), n.
    comb.

NEOPHAENIS Hamp., 1908
    NEOPHOENIS Neave, 1940,
    missp.
9628 respondens (Wlk., 1858)

FAGITANA Wlk., 1865
    PSEUDOLIMACODES Grt., 1874
9629 littera (Gn., 1852)
    lucidata Wlk., 1865
    niveicostatus (Grt., 1874)

CALLOPISTRIA Hbn., 1821
    LAGOPUS R. L., 1817, preocc. by
    Brisson, 1760
    ERIOPUS Tr., 1825
    CALOPISTRIA Steph., 1850,
    missp.
    METHORASA Moore, 1881
    HERRICHIA Grt., 1882, preocc. by
    Stgr., 1871
    EUHERRICHIA Grt., 1882
    HAPLOOLOPHUS Butler, 1891
    METATHORASA auth., missp.
9630 floridensis (Gn., 1852)
    elegantulus H.-S., 1868
    strena Grt., 1895
9631 mollissima (Gn., 1852)
    rubicunda (Wlk., 1858)
9632 granitosa (Gn., 1852)
    argentilinea Wlk., 1858
9633 cordata (Ljungh, 1825)
    monetifera (Gn., 1852), n. stat.

PHUPHENA Wlk., 1858
9634 **tura** (Druce, 1889)
9635 **obliqua** (Sm., 1900)

ACHERDOA Wlk., 1865
VARINA Neum., 1884
9636 **ferraria** Wlk., 1865
ornata (Neum., 1884)

Amphipyrini

MAGUSA Wlk., 1857
SASUNAGA Moore, 1881
CALLIXENA Saalmüller, 1891
9637 **orbifera** (Wlk., 1857)
strigifera Wlk., 1857
divaricata (Grt., 1874), form
sarpida Felder & Rogenhofer, 1874
discidens Felder & Rogenhofer, 1874
angustipennis (Mösch., 1886)
divida (Mösch., 1890)
orbiferella Strand, 1916, form
orbiferana Strand, 1916
perversa Strand, 1916

AMPHIPYRA Ochs., 1816
SCOTOPHILA Hbn., 1821
PYROPHILA Steph., 1829
PHILOPYRA Gn., 1837
9638 **pyramidoides** Gn., 1852
inornata Grt., 1864
conspersa Riley, 1871
carbonita Franc., 1941, form
9639 **tragopoginis** (Cl., 1759)
luciola (Hufn., 1766)
tetra; Haw., 1809, not F., 1787
tragopogonis auth., missp.
repressus Grt., 1871
turcomana Stgr., 1888
9640 **glabella** (Morr., 1874)
9641 **brunneoatra** Strand, 1916

PROTOPERIGEA McD., 1937
9642 **anotha** (Dyar, 1904)
9643 **posticata** (Harv., 1875)
veterata (Sm., 1894)

MICRATHETIS Hamp., 1908
MICROTHETIS Hamp., 1908, incorr. orig. spell.
9644 **triplex** (Wlk., 1857)
spilomela (Wlk., 1865)
contraria (H.-S., 1868)
conviva (Harv., 1876)
subaquila (Harv., 1878), form
minuscula (B. & McD., 1913)
pallidegrisea Strand, 1916
obscurebrunnea Strand, 1916
benjamini Draudt, 1926
obscureobrunnea McD., 1938, missp.
9645 **costiplaga** (Sm., 1908)
9646 **tecnion** Dyar, 1914

PROXENUS H.-S., 1850
9647 **miranda** (Grt., 1873)
a. **nitens** (Dyar, 1904)
9648 **mindara** B. & McD., 1913
9649 **mendosa** McD., 1927

ANORTHODES Sm., 1891
9650 **tarda** (Gn., 1852)
prima Sm., 1891
9651 **triquetra** (Grt., 1883)
9652 **indigena** (B. & Benj., 1925)

CARADRINA Ochs., 1816
CHARADRINA Agassiz, 1846, missp.
PARADRINA Boursin, 1937
9653 **morpheus** (Hufn., 1766)
pulla; Beckwith, 1794, not Esp., 1785
sepii (Hbn., 1800–03)

PLATYPERIGEA Sm., 1894
PLATYPERIGA Sm., 1894, incorr. orig. spell.
9654 **meralis** (Morr., 1875)
bilunata (Grt., 1877)
9655 **camina** Sm., 1894
a. **alpha** B. & Benj., 1926
b. **beta** B. & Benj., 1926
9656 **extima** (Wlk., 1865)
civica (Grt., 1883)
9657 **multifera** (Wlk., 1857)
fidicularia (Morr., 1874)
9658 **mona** (B. & McD., 1912)
9659 **atrostriga** (B. & McD., 1912)
9660 **distincta** (Barnes, 1928)

CRAMBODES Gn., 1852
9661 **talidiformis** Gn., 1852
conjungens (Wlk., 1858)

BALSA Wlk., 1860
GARGAZA Wlk., 1866
NOLAPHANA Grt., 1873
ASISYRA Grt., 1873
9662 **malana** (Fitch, 1856)
obliquifera Wlk., 1860
leodura (Stgr., 1887)
9663 **tristrigella** (Wlk., 1866)
zelleri (Grt., 1873)
9664 **labecula** (Grt., 1880)

SPODOPTERA Gn., 1852
LAPHYGMA Gn., 1852
PRODENIA Gn., 1852
ARIATHISA Wlk., 1865
EULAPHYGMA Butler, 1890
SPODOSOMA auth., 1966, missp.
9665 **exigua** (Hbn., 1803–08)
fulgens (Gey., 1828–32)
pygmaea (Rambur, 1834)
junceti (Zell., 1847)
cycloides (Gn., 1852)
caradrinoides (Wlk., 1856)
flavimaculata (Harv., 1876)
venosa (Butler, 1880)
sebghana (Austant, 1880)
9666 **frugiperda** (J. E. Smith, 1797)
macra (Gn., 1852)
inepta (Wlk., 1856)
plagiata (Wlk., 1856)
signifera (Wlk., 1856)
autumnalis (Riley, 1871)
fulvosa (Riley, 1871)
obscura (Riley, 1876)
9667 **praefica** (Grt., 1875)

9668 **pulchella** (H.-S., 1868)
exquisita (Mösch., 1886)
9669 **ornithogalli** (Gn., 1852)
eudiopta (Gn., 1852)
lineatella (Harv., 1875)
flavimedia (Harv., 1875)
eudioptoides (B. & Benj., 1923)
9670 **latifascia** (Wlk., 1856)
variolosa (Wlk., 1857)
cosmioides (Wlk., 1858)
9671 **dolichos** (F., 1794)
commelinae (J. E. Smith, 1797)
9672 **eridania** (Cram., 1784)
linea (F., 1794), form
phytolaccae (J. E. Smith, 1797)
putrida (Gn., 1852)
amygia (Gn., 1852)
externa (Wlk., 1856)
inquieta (Wlk., 1857)
bipunctata (Wlk., 1857)
strigifera (Wlk., 1858)
derupta (Morr., 1875)
ignobilis (Butler, 1878)
nigrofascia (Hulst, 1881)
9673 **sunia** (Gn., 1852)
albula (Wlk., 1857)
orbicularis (Wlk., 1857)
caudata (Wlk., 1865)

LEUCOCHLAENA Hamp., 1906
9674 **hipparis** (Druce, 1889)
pallens Draudt, 1925, form
colossa Draudt, 1925, form

ELAPHRIA Hbn., 1818
MONODES Gn., 1852
ALPESA Wlk., 1858
CARBONA Schaus, 1906
9675 **fuscimacula** (Grt., 1881)
9676 **nucicolora** (Gn., 1852)
unisignata (Wlk., 1856)
paginata (Morr., 1875)
clara (Harv., 1878), form
9677 **agrotina** (Gn., 1852)
trientiplaga (Wlk., 1858)
guttula (H.-S., 1868)
aduncula (Felder & Rogenhofer, 1874)
9678 **versicolor** (Grt., 1875)
9679 **chalcedonia** (Hbn., 1803–08)
arna (Gn., 1852)
expuncta (Wlk., 1857)
vincta (Wlk., 1857)
irresoluta (Wlk., 1857)
tracta (Grt., 1875)
9680 **georgei** (Moore & Rawson, 1939)
9681 **festivoides** (Gn., 1852)
varia (Wlk., 1858), form
cephalica (Butler, 1891)
albovariegata Strand, 1915
9682 **exesa** (Gn., 1852)
floridana (Wlk., 1865)
9683 **ensina** (Barnes, 1907)
obliquirena (Hamp., 1909)
9684 **grata** Hbn., 1818
orina (Gn., 1852)
rasilis (Morr., 1874)
subusta; auth., part, not Hbn., 1823

BRYOLYMNIA Hamp., 1910
9685 **viridimedia** (Sm., 1905)

9686 **semifascia** (Sm., 1900), n. comb.

GONODES Druce, 1908
GONODES Hamp., 1908 (December), preocc. by Druce, 1908 (April)
9687 **liquida** (Mösch., 1886)
violascens (Schaus, 1894)
leada (Druce, 1898)

GALGULA Gn., 1852
9688 **partita** Gn., 1852
subpartita Gn., 1852
hepara Gn., 1852, ♀
ferruginea (Wlk., 1857)
externa (Wlk., 1865)
baueri (Stgr., 1870)
vesca (Morr., 1875)
bias (Druce, 1889)
hippotamada (Druce, 1889)
mandane (Druce, 1891)
interna B. & McD., 1917, missp.

PERIGEA Gn., 1852
PERIGIA Sm., 1894, missp.
9689 **xanthioides** Gn., 1852
a. *enixa* Grt., 1875

PLATYSENTA Grt., 1874
9690 **videns** (Gn., 1852)
indigens (Wlk., 1857)
atriciliata Grt., 1874
meskei (Speyer, 1875)
a. *albipuncta* Sm., 1902
9691 **temecula** Barnes, 1905
9692 **discistriga** (Sm., 1894)
9693 **mobilis** (Wlk., 1857)
apameoides; auth., not Gn., 1852
plagiata (Wlk., 1857)
inclinata (Wlk., 1857)
subaurata (Wlk., 1865)
iole (Grt., 1877)
subaurea (Druce, 1889)
9694 **mersa** (Morr., 1875)
9695 **albolabes** (Grt., 1880)
9696 **vecors** (Gn., 1852)
remissa (Wlk., 1857)
demittens (Wlk., 1858)
luxa (Grt., 1875)
9697 **orta** (B. & McD., 1912)
9698 **concisa** (Wlk., 1856)
centralis (Wlk., 1857)
consocia (Wlk., 1857)
imbella (Wlk., 1858)
laphygmoides (Wlk., 1858)
plumbago (H.-S., 1868)
9699 **sutor** (Gn., 1852)
apameoides (Gn., 1852)
turpis (Gn., 1852)
indecisa (Wlk., 1857)
claufacta (Wlk., 1857)
clarefacta (Wlk., 1858), missp.
pauper (Wlk., 1858)
otiosa (Wlk., 1858)
detrecta (Wlk., 1865)
fabrefacta (Morr., 1874)
9700 **cervina** (Sm., 1900)
9701 **proxima** (Morr., 1876)
9702 **albigera** (Gn., 1852)
intermittens (Wlk., 1858)

9703 **hypocritica** (Dyar, 1907), extralim. ?

DRAUDTIA B. & Benj., 1926
9704 **revellata** (B. & Benj., 1924)
9705 **leucorena** (Sm., 1900)
9706 **funeralis** (Hill, 1924)
9707 **andrena** (Sm., 1911)
9708 **lunata** (B. & McD., 1916)
9709 **begallo** (Barnes, 1905)
9710 **egestis** (Sm., 1894), n. comb.
abalas (Sm., 1905), n. comb.
9711 **ignota** (B. & Benj., 1924)
9712 **morsa** (Sm., 1907)

CONDICA Wlk., 1856
MYRTALE Druce, 1891
9713 **cupentia** (Cram., 1780)
epopea (Cram., 1780)
infelix (Gn., 1852)
9714 **confederata** (Grt., 1873)

STIBAERA Wlk., 1857
9715 **curvilineata** Hamp., 1924
costiplaga; Druce, 1894, not Wlk., 1857
9716 **thyatiroides** (B. & Benj., 1924), n. comb.
habrosynoides (B. & Benj., 1924), form, n. comb.

EMARGINEA Gn., 1852
DACIRA Wlk., 1858
CYATHISSA Grt., 1881
9717 **pallida** (Sm., 1902)
9718 **percara** (Morr., 1875)
quadrate (Sm., 1906)
ochracea (Sm., 1906)
9719 **dulcinea** Dyar, 1921

OGDOCONTA Butler, 1891
9720 **cinereola** (Gn., 1852)
atomaria (Wlk., 1865)
9721 **moreno** Barnes, 1907
9722 **sexta** B. & McD., 1913
9723 **altura** Barnes, 1904
9724 **tacna** (Barnes, 1904)

STIRIODES Hamp., 1908
9725 **obtusa** (H.-S., 1854)
obtusula (Zell., 1873)
9726 **edentata** (Grt., 1883)
a. *procida* (Druce, 1889)
umbria (Druce, 1898), form
nepotica Dyar, 1912, form
9727 **perflava** (Harv., 1875)
9728 **virida** B. & McD., 1916

AZENIA Grt., 1883
9729 **implora** Grt., 1883
9730 **templetonae** Clarke, 1937

NARTHECOPHORA Sm., 1900
9731 **pulverea** Sm., 1900

LYTHRODES Sm., 1903
9732 **radiatus** Sm., 1903
9733 **venatus** Sm., 1903
9734 **tripuncta** B. & McD., 1911

HEMINOCLOA B. & Benj., 1924
9735 **mirabilis** (Neum., 1884)

BISTICA Dyar, 1912
9736 **noela** (Druce, 1892)
noela Dyar, 1912, preocc. by Druce, 1892

NEUMOEGENIA Grt., 1882
KALLITRICHIA Ottol., 1898
TRICHOCALA Dyar, 1903
9737 **poetica** Grt., 1882
smithi (Druce, 1889)
sagittalba (Ottol., 1898)
smithii (Dyar, 1903), missp.
9738 **coronides** (Druce, 1889)
pendula (Ottol., 1898)
9739 **albavena** (Ottol., 1898)

Stiriini

STIBADIUM Grt., 1874
ANTAPLAGA Grt., 1877
NEOPHAEUS Dyar, 1918
9740 **ochoa** Barnes, 1904
dollii Smith, 1908
9741 **aureolum** Hy. Edw., 1882
9742 **olvello** Barnes, 1907
9743 **unicum** B. & Benj., 1926
9744 **manti** Barnes, 1904
9745 **dimidiata** Grt., 1877
9746 **astigmatosum** Dyar, 1921
9747 **curiosum** Neum., 1884
9748 **spumosum** Grt., 1874
9749 **mavina** B. & McD., 1910
9750 **hilli** (B. & Benj., 1923)
caliente (Hill, 1923)
9751 **navium** (Harvey, 1875), mispl. ?

PLAGIOMIMICUS Grt., 1873
9752 **expallidus** Grt., 1883
9753 **triplagiatus** Sm., 1890
9754 **pityochromus** Grt., 1873
media (Morr., 1875)

POLENTA Morr., 1875
9755 **tepperi** (Morr., 1875)
richii (Grt., 1886)
richi auth., missp.
9756 **hutsoni** (Sm., 1907)
fuliginosa (Sm., 1907)
9757 **olivalis** (B. & McD., 1916)
9758 **hachita** (Barnes, 1904)
9759 **sexseriata** (Grt., 1881)
9760 **biundulalis** (Zell., 1872)
biundulata (Sm., 1893), emend.

CHRYSOECIA Hamp., 1908
CHRYSAECIA Neave, 1939, missp.
9761 **scira** (Druce, 1889)
benjamini (Hill, 1924)
9762 **requies** (Dyar, 1909)
9763 **gladiola** (Barnes, 1907)
9764 **dela** (Druce, 1894)
atrolinea (B. & McD., 1912), form, n. stat.
stigmatosa (Dyar, 1912), form
hemicrocea (Dyar, 1912), form

CIRRHOPHANUS Grt., 1872
CIRRHOPHANES Sm., 1891, missp.
9765 **dyari** Ckll., 1899
9766 **triangulifer** Grt., 1872
pretiosa (Morr., 1875)

9767 **plesioglauca** (Dyar, 1912)
    comstocki (Hill, 1924)
9768 **papago** Barnes, 1907

HOPLOLYTHRA Hamp., 1910
9769 **discistriga** (Sm., 1903)
9770 **arivaca** (Barnes, 1907)

XANTHOTHRIX Hy. Edw., 1878
    EUEDWARDSIA Grt., 1882
9771 **neumoegeni** Hy. Edw., 1881
9772 **ranunculi** Hy. Edw., 1878
    albipuncta B. & Benj., 1925, form

EULITHOSIA Hy. Edw., 1884
    PUMORA Dyar, 1918
9773 **composita** Hy. Edw., 1884

CYPTONYCHIA Druce, 1899
    CYPTONYCHIA Hamp., 1900,
    preocc. by Druce, 1899
    CHICHIMECA Hogue, 1963
9774 **thoracica** (Hy. Edw., 1884)

CHALCOPASTA Hamp., 1908
    RODRIGUESIA Dyar, 1912
9775 **territans** (Hy. Edw., 1884)
    arizona (French, 1889)
9776 **howardi** (Hy. Edw., 1877)
    ornata (Ottol., 1898)
9777 **fulgens** B. & McD., 1912
    ellica; Draudt, 1926
9778 **acema** (Druce, 1889)
9779 **koebelei** (Riley, 1893), mispl.?

BASILODES Gn., 1852
    DEOBRIGA Wlk., 1869
9780 **chrysopis** Grt., 1881
    catharops Dyar, 1911
9781 **pepita** Gn., 1852
    chrysopasa (Wlk., 1869)
9782 **inquinatus** Hogue, 1963

STIRIA Grt., 1874
9783 **blanchardi** (Hogue, 1966)
9784 **consuela** Stkr., 1900
9785 **rugifrons** Grt., 1874
9786 **dyari** Hill, 1924
9787 **sulphurea** Neum., 1882
    demaculata Strand, 1916

FALA Grt., 1875
9788 **ptychophora** Grt., 1875

CHAMAECLEA Grt., 1883
    CHAMOCLEA Hill, 1924, missp.
9789 **pernana** (Grt., 1881)
9790 **basiochrea** B. & McD., 1916

Nocloini

OSLARIA Dyar, 1904
9791 **viridifera** (Grt., 1882)
9792 **pura** (B. & McD., 1911)

NOCLOA Sm., 1906
    PHAIOECIA Dyar, 1911, n.
    syn.
9793 **plagiata** Sm., 1906
9794 **rivulosa** Sm., 1906

9795 **pallens** (Tepper, 1882)
    nesaea (Sm., 1893)
9796 **cordova** (Barnes, 1907)
9797 **nanata** (Neum., 1884)
    macula (Sm., 1891)
9798 **pilacho** (Barnes, 1904)
9799 **alcandra** (Druce, 1890)
9800 **aliaga** (Barnes, 1905)
9801 **duplicata** (Sm., 1891)

PARAMIANA B. & Benj., 1924
9802 **callaisata** A. Blanchard, 1972
9803 **smaragdina** (Neum., 1884)
9804 **marina** (Sm., 1906)
9805 **perissa** Nye, 1975
    laetabilis (Sm., 1899), preocc. by
    Zett., 1839
9806 **canoa** (Barnes, 1907)

EUAMIANA B. & Benj., 1927
9807 **contrasta** (B. & McD., 1910)
9808 **dissimilis** (B. & McD., 1910)
9809 **torniplaga** (B. & McD., 1916)

RUACODES Hamp., 1908
9810 **tela** (Sm., 1900)

PETALUMARIA Buckett & Bauer, 1968
    PETALUMA Buckett & Bauer,
    1965, preocc. by Hulst, 1888
9811 **californica** (Buckett & Bauer, 1965)

REDINGTONIA B. & McD., 1912
9812 **alba** B. & McD., 1912

Unassociated Genera

ACHYTONIX McD., 1937
9813 **epipaschia** (Grt., 1883)
    a. **nigramacula** (B. & Benj., 1926)
    b. **orae** (B. & Benj., 1926)
9814 **praeacuta** (Sm., 1894)

COSMIA Ochs., 1816
    EUSTEGNIA Hbn., 1821
    CALYMNIA Hbn., 1821
9815 **calami** (Harv., 1876)
    canescens (Behr, 1885)
    calamia McD., 1938, missp.
    orinella (Strand, 1916), form
    orinula (Strand, 1916), form

ZOTHECA Grt., 1874
9816 **tranquilla** Grt., 1874
    sambuci (Stkr., 1874)
    viridula Grt., 1878, form
    sambuci (Behr, 1885), preocc. by
    Stkr., 1874
    viridifera Sm., 1893, missp.

THURBERIPHAGA Dyar, 1920
9817 **diffusa** (Barnes, 1904)
    catalina Dyar, 1920

AMOLITA Grt., 1874
9818 **fessa** Grt., 1874
9819 **obliqua** Sm., 1903
    duplipuncta Strand, 1916, ab.
9820 **sentalis** (Kaye, 1901)
9821 **roseola** Sm., 1903

9822 **fratercula** B. & McD., 1912
9823 **delicata** B. & McD., 1912

HEMIOSLARIA B. & Benj., 1924
9824 **pima** B. & Benj., 1924

ACOPA Harv., 1875
    GRYPOTES Dyar, 1917, preocc. by
    Fieber, 1866
    ZATILPA Dyar, 1920
9825 **carina** Harv., 1875
9826 **perpallida** Grt., 1878
    incana Hy. Edw., 1882
    borealis (Dyar, 1917)
    dentifer (Dyar, 1917), form
    pura B. & L., 1922, form

NACOPA B. & Benj., 1924
9827 **bistrigata** (B. & McD., 1918)
9828 **melanderi** B. & Benj., 1927

PROTHRINAX Hamp., 1908
9829 **luteomedia** (Sm., 1907)

WALTERELLA Dyar, 1921
9830 **ocellata** (B. & McD., 1910)
    eudesmia Dyar, 1921

ESCARIA Grt., 1883
9831 **clauda** Grt., 1883
    a. **pallens** B. & Benj., 1923
9832 **homogena** McD., 1922

FOTELLA Grt., 1882
9833 **notalis** Grt., 1882
9834 **olivia** B. & McD., 1912
9835 **olivioides** B. & Benj., 1926
9836 **fragosa** (Grt., 1883)
    cervoides (B. & McD., 1912)

CEPHALOSPARGETA Mösch., 1890
9837 **elongata** Mösch., 1890

PROCHLORIDEA B. & McD., 1911
9838 **modesta** B. & McD., 1911
    a. **madonna** B. & L., 1921

ALEPTINOIDES B. & McD., 1912
9839 **ochrea** B. & McD., 1912
    haematosticta (Dyar, 1923)

ANYCTEOLA B. & Benj., 1929
9840 **fotelloides** (B. & McD., 1916)

AFOTELLA B. & Benj., 1926
9841 **cylindrica** (Grt., 1880)

FOTA Grt., 1882
9842 **armata** Grt., 1882
    medioalba B. & Benj., 1923, form
    brunneogrisea B. & Benj., 1923, form
9843 **minorata** Grt., 1882

GLOANNA Nye, 1975
    LANGONA B. & L., 1921, preocc.
    by Simon, 1901
9844 **grisescens** (B. & L., 1921)

PODAGRA Sm., 1902
9845 **crassipes** Sm., 1902

POLICOCNEMIS Benj., 1932
9846 **ungulatus** Benj., 1932

MINOFALA Sm., 1905
9847 **instans** Sm., 1905

EVIRIDEMAS B. & Benj., 1929
9848 **minuta** (B. & McD., 1910)

AXENUS Grt., 1873
9849 **arvalis** Grt., 1873
    ochraceus Hy. Edw., 1875, form
    amplus Hy. Edw., 1875, form

ANNAPHILA Grt., 1873
    PROANNAPHILA Rindge &
      Smith, 1952
9850 **danistica** Grt., 1873
9851 **hennei** Rindge & Smith, 1952
9852 **mera** Harv., 1875
    domina Hy. Edw., 1875
    a. **eremia** Rindge & Smith, 1952
9853 **pustulata** Hy. Edw., 1881
9854 **arvalis** Hy. Edw., 1875
    salicis Hy. Edw., 1881
    fletcheri (Sm., 1907)
9855 **abdita** Rindge & Smith, 1952
9856 **baueri** Rindge & Smith, 1952
9857 **astrologa** B. & McD., 1918
    divinula; B. & McD., 1912, not Grt.,
      1878
9858 **vivianae** Sala, 1964
9859 **pseudoastrologa** Sala, 1964
9860 **olgae** Sala, 1964
9861 **ida** Rindge & Smith, 1952
9862 **divinula** Grt., 1878
9863 **lithosina** Hy. Edw., 1875
    variegata Sm., 1908
9864 **miona** Sm., 1908
9865 **casta** Hy. Edw., 1890
9866 **depicta** Grt., 1873
    a. **morula** Rindge & Smith, 1952
9867 **macfarlandi** Buckett & Bauer, 1964
9868 **decia** Grt., 1875
    amicula Hy. Edw., 1875
    germana Hy. Edw., 1875
    danistica; Dyar, 1904, not Grt., 1873
9869 **diva** Grt., 1873
    yosemitensis Strand, 1912
9870 **superba** Hy. Edw., 1875
9871 **spila** Rindge & Smith, 1952
9872 **evansi** Rindge & Smith, 1952

CUCULLIINAE

Xylenini

XYLENA Ochs., 1816
    XYLITES R. L., 1817
    AXYLIA Hbn., 1821
    XYLAENA Hbn., 1822, missp.
    XYLINA Tr., 1826, missp.
    CALOCAMPA Steph., 1829
    XSYLINA Frivaldszky, 1835,
      missp.
    HYLINA Freyer, 1840, missp.
    CALLICAMPA Agassiz, 1846,
      missp.
    COLOCAMPA Stichel, 1908, missp.
9873 **nupera** (Lint., 1874)

9874 **curvimacula** (Morr., 1874)
9875 **thoracica** (Putnam-Cramer, 1886)
    brillians (Ottol., 1902)
9876 **cineritia** (Grt., 1875)
    a. **mertena** (Sm., 1909)
9877 **brucei** (Sm., 1892)

LITHOMOIA Hbn., 1821
    LITHOMIA Curt., 1838, missp.
    LITHOMOEA Hamp., 1909, missp.
9878 **solidaginis** (Hbn., 1803), extralim.
    cinerascens (Stgr., 1871)
    pallida (Tutt, 1892)
    suffusa Tutt, 1902
    a. **germana** (Morr., 1874)
    b. **morrisoni** B. & Benj., 1922
    c. **albertae** (Strand, 1916)

HOMOGLAEA Morr., 1876
    EUHARVEYA Grt., 1894
9879 **variegata** B. & McD., 1918
    ricardi (B. & Benj., 1927)
9880 **californica** (Sm., 1891)
    insinuata Sm., 1905
9881 **hircina** Morr., 1876
9882 **dives** Sm., 1907
9883 **carbonaria** (Harv., 1876)

LITHOLOMIA Grt., 1875
9884 **napaea** (Morr., 1874)
    a. **umbrifasciata** Blkmre., 1923

LITHOPHANE Hbn., 1821
    GRAPTOLITHA Hbn., 1821
    RHIZOLITHA Curt., 1830
9885 **semiusta** Grt., 1874
9886 **patefacta** (Wlk., 1858)
    niveocosta Franc., 1942, form
9887 **bethunei** (G. & R., 1868)
    duscalis Franc., 1942, form
    luteocosta Franc., 1942, form
9888 **innominata** (Sm., 1893)
    signosa; Grt., 1874 not Wlk., 1857
    illecebra Franc., 1942, normal form
9889 **petulca** Grt., 1874
9890 **ferrealis** Grt., 1874
9891 **amanda** (Sm., 1900)
    pallidior (Strand, 1916)
9892 **disposita** Morr., 1874
    argillocosta Franc., 1942, form
9893 **hemina** Grt., 1874
    lignicosta Franc., 1942, form
9894 **oriunda** Grt., 1874
    canentissima Franc., 1942, form
9895 **signosa** (Wlk., 1857)
    pallidicosta Franc., 1942, form
9896 **gausapata** Grt., 1883
9897 **contra** (B. & Benj., 1924)
9898 **longior** (Sm., 1899)
9899 **lemmeri** (B. & Benj., 1929)
9900 **subtilis** Franc., 1969
9901 **contenta** Grt., 1880
    pomona (Sm., 1889)
9902 **baileyi** Grt., 1877
9903 **vivida** (Dyar, 1910)
9904 **querquera** Grt., 1874
9905 **viridipallens** Grt., 1877
9906 **pruena** (Dyar, 1910)
9907 **puella** (Sm., 1900)
9908 **laceyi** (B. & McD., 1913)

9909 **tepida** Grt., 1874
    a. **atincta** (Sm., 1905)
9910 **antennata** (Wlk., 1858)
    cinerea (Riley, 1871)
9911 **torrida** (Sm., 1899)
9912 **pertorrida** (McD., 1942)
9913 **georgii** Grt., 1875
    a. **emarginata** (Sm., 1900)
    b. **holocinerea** (Sm., 1900)
    c. **ancilla** (Sm., 1904)
    d. **vertina** (Sm., 1904)
    e. **fletcheri** (Sm., 1904)
    f. **oregonensis** Harv., 1876
9914 **laticinerea** Grt., 1874
    a. **winnipeg** (Sm., 1900)
9915 **grotei** Riley, 1882
    cinerosa Grt., 1879, preocc. by Gn.,
      1852
9916 **unimoda** (Lint., 1878)
    a. **merceda** (Sm., 1904)
9917 **fagina** Morr., 1874
9918 **tarda** B. & Benj., 1925
9919 **tephrina** Franc., 1969
9920 **itata** (Sm., 1899)
9921 **nigrescens** (Engel, 1905)
9922 **pexata** Grt., 1874
    a. **washingtonia** Grt., 1883
9923 **dilatocula** (Sm., 1900)
9924 **atara** (Sm., 1909)
9925 **lepida** Grt., 1878
    lepida (Lint., 1878), preocc. by Grt.,
      1878
    a. **adipel** (Benj., 1936)
9926 **nasar** (Sm., 1909)
9927 **vanduzeei** (Barnes, 1928)
9928 **thaxteri** Grt., 1874
    a. **alaskensis** (Barnes, 1928)
    b. **rosetta** (Barnes, 1928)

PYREFERRA Franc., 1937
9929 **hesperidago** (Gn., 1852)
    indirecta (Wlk., 1856)
    graefiana (Grt., 1874)
    moffatiana (Grt., 1882)
9930 **citrombra** Franc., 1941
9931 **ceromatica** (Grt., 1874)
9932 **pettiti** (Grt., 1874)

EUPSILIA Hbn., 1821
    SCOPELOSOMA Curt., 1836
    MECOPTERA Gn., 1837
9933 **vinulenta** (Grt., 1864)
9933.1 **sidus** (Gn., 1852)
    walkeri (Grt., 1864)
    a. **colorado** (Sm., 1903)
9934 **cirripalea** Franc., 1952
9935 **tristigmata** (Grt., 1877)
    trisignata Fbs. & Franc., 1954,
      missp.
9936 **morrisoni** (Grt., 1874)
    castanea (Strand, 1916)
    opaca Chermock & Chermock, 1940,
      form
9937 **knowltoni** McD., 1946
9938 **fringata** (B. & McD., 1916)
9939 **devia** (Grt., 1875)

SERICAGLAEA Franc., 1941
9940 **adulta** (Gn., 1852), n. comb.
9941 **signata** (French, 1879)

XYSTOPEPLUS Franc., 1937
  JODIA; Hamp., 1906, not Hbn., 1818
9942 **rufago** (Hbn., 1818)
  honesta (Wlk., 1858)

METAXAGLAEA Franc., 1937
9943 **inulta** (Grt., 1874)
9944 **viatica** (Grt., 1874)
9945 **semitaria** Franc., 1968
9945.1 **australis** Schweitzer, 1979
9945.2 **violacea** Schweitzer, 1979

EPIGLAEA Grt., 1878
  HARPAGLAEA Hamp., 1906
9946 **decliva** (Grt., 1874)
  deleta (Grt., 1878)
9947 **apiata** (Grt., 1874)
  pastillicans (Morr., 1874)

CHAETAGLAEA Franc., 1943
9948 **cerata** Franc., 1943
9949 **tremula** (Harv., 1875)
  pastillicans; auth., not Morr., 1874
9950 **sericea** (Morr., 1874)
  venustula (Grt., 1875)

PSECTRAGLAEA Hamp., 1906
9951 **carnosa** (Grt., 1877)

EUCIRROEDIA Grt., 1875
  EUCIRRHOEDIA McD., 1938, missp.
9952 **pampina** (Gn., 1852)
  pampinella (Strand, 1916), form
  brunneoochracea (Strand, 1916), form
  a. **glenwoodi** (B. & Benj., 1922)

PSEUDOGLAEA Grt., 1876
9953 **olivata** (Harv., 1874)
  blanda (Grt., 1876)
  taedata Grt., 1876
  decepta Grt., 1881

AGROCHOLA Hbn., 1821
  ANCHOSCELIS Gn., 1839
  ANCHOCELIS Steph., 1850, missp.
  ANTHOCELIS auth., missp.
  AMATHES; Hamp., 1906, not Hbn., 1821
  LYCANADES Franc., 1937, n. syn.
9954 **purpurea** (Grt., 1874)
  crispa (Harv., 1875), rev. stat.
  fornica (Sm., 1907)
  antapica (Sm., 1907), rev. stat.
  formica Rindge, 1955, missp.
9955 **pulchella** (Sm., 1900)
9956 **lota** (Clerck, 1759) introduced, established Nfld.
  hippophaes; Rossi, 1794
  munda; Hbn., 1800–03
  americana (Morr., 1876)
  pallida (Tutt, 1892)
  suffusa (Tutt, 1892)
  rufa (Tutt, 1892)
  subdita (Warr., 1910)

SUNIRA Franc., 1950
  RUSINA; Hamp., 1906, not Steph., 1829
9957 **bicolorago** (Gn., 1852)

  spurcata (Wlk., 1857)
  ferrugineoides (Gn., 1852), form
9958 **decipiens** (Grt., 1881)
  acta (Sm., 1907)
9959 **straminea** (Sm., 1907)
9960 **verberata** (Sm., 1904)

ANATHIX Franc., 1937
9961 **ralla** (G. & R., 1868)
9962 **puta** (G. & R., 1868)
  euroa (G. & R., 1873)
  a. **dusca** (Sm., 1908)
  duscata (Sm., 1908), missp.
9963 **aggressa** (Sm., 1907)
9964 **immaculata** (Morr., 1875)

XANTHIA Ochs., 1816
  XANTHA Billberg, 1820, missp.
  CIRRHIA Hbn., 1821
  CITRIA Hbn., 1821
  MELLINIA Hbn., 1821
  EUTHEMONIA Gistl, 1848, preocc. by Steph., 1828
  TILIACEA Tutt, 1896
9965 **togata** (Esp., 1788)
  lutea (Ström, 1783), preocc.
  flavago (F., 1787), preocc.
  ochreago (Bkh., 1792), preocc.
  rubago (Don, 1801)

HILLIA Grt., 1883
  CRASIA Auriv., 1891
9966 **maida** (Dyar, 1904)
9967 **iris** (Zett., 1839)
  crasis (H.-S., 1851)
  semisigna (Wlk., 1857)
  erdmanni (Mösch., 1874)
  senescens (Grt., 1878)
  vigilans (Grt., 1878), form

FISHIA Grt., 1877
9968 **tortilis** (Grt., 1880), n. comb.
9969 **evelina** (French, 1888)
  a. **hanhami** Sm., 1909
9970 **discors** (Grt., 1881)
  vinela Sm., 1903
9971 **connecta** (Sm., 1894)
  cupola (Hamp., 1913)
9972 **yosemitae** (Grt., 1873)
  exhilarata Sm., 1903
9973 **betsia** Sm., 1905
9974 **enthea** Grt., 1877
9975 **instruta** Sm., 1910
  derelicta (Hamp., 1913)

PLATYPOLIA Grt., 1895
  EUROTYPE Hamp., 1906
9976 **anceps** (Steph., 1850)
  aplectoides (Gn., 1852), form
  confragosa (Morr., 1874)
  acutissima (Grt., 1875)
  medialis (Grt., 1876)
9977 **contadina** (Sm., 1894)
  a. **albertae** McD., 1935
9978 **loda** (Stkr., 1898)
  albiserrata (Sm., 1903)
  a. **gunderi** (B. & Benj., 1927)

XYLOTYPE Hamp., 1906
9979 **capax** (Grt., 1868)

9980 **acadia** B. & Benj., 1922, emend. McD., 1938
  arcadia B. & Benj., 1922, incorr. orig. spell.

DRYOTYPE Hamp., 1906
9981 **opina** (Grt., 1878)

MNIOTYPE Franc., 1941
  CRINO; Hamp., 1906, not Hbn., 1821
9982 **sommeri** (Lefebvre, 1836)
9983 **pallescens** McD., 1946
9984 **versuta** (Sm., 1895)
  moilena (Stkr., 1898)
9985 **miniota** (Sm., 1908)
9986 **fervida** (Sm., 1908)
9987 **ducta** (Grt., 1878)
9988 **tenera** (Sm., 1900)

SUTYNA Todd, 1958
  ANYTUS Grt., 1873, preocc. by Stål, 1865
  AMYTUS Grünberg, 1910, missp.
9989 **privata** (Wlk., 1857)
  monstrata (Wlk., 1857)
  sculpta (Grt., 1873)
  a. **plana** (Grt., 1882)
9990 **profunda** (Sm., 1900)
  a. **teltowa** (Sm., 1910)
  b. **obscura** (Sm., 1900)
9991 **tenuilinea** (Sm., 1903)

PACHYPOLIA Grt., 1874
9992 **atricornis** Grt., 1874

BRACHYLOMIA Hamp., 1906
  CLEOCERIS; auth., not Bdv., 1836
  BOMBYCIA; auth., not Hbn., 1822
  ITEOPHAGA Boursin, 1965
9993 **populi** (Stkr., 1898)
  albidior Strand, 1916, form
  contracta Strand, 1916, form
9994 **elda** (French, 1887), n. comb.
9995 **rectifascia** (Sm., 1891)
9996 **curvifascia** (Sm., 1891), n. comb.
9997 **thula** (Stkr., 1898), n. comb.
9998 **algens** (Grt., 1878), n. comb.
9999 **discinigra** (Wlk., 1856)
10000 **onychina** (Gn., 1852)

LOMILYSIS Franc., 1937
10001 **discolor** (Sm., 1904)

EPIDEMAS Sm., 1894
10002 **cinera** Sm., 1894
10003 **obscura** Sm., 1903
10004 **melanographa** Hamp., 1906

Feraliini

FERALIA Grt., 1874
  MOMOPHANA Grt., 1875
  ARTHROCHLORA Grt., 1875
  MOMAPHANA Hamp., 1906, missp.
  ARTHRACHLORA Hamp., 1906, missp.
10005 **jocosa** (Gn., 1852)
  furtiva Sm., 1909, form
  jocosides Strand, 1916, form

10006 **deceptiva** McD., 1920
10007 **major** Sm., 1890
10008 **comstocki** (Grt., 1874)
    comstocci Hamp., 1906, missp.
    a. **columbiana** Sm., 1903
10009 **februalis** Grt., 1874
10010 **meadowsi** Buckett, 1968

Psaphidini

BRACHIONYCHA Hbn., 1819
    ASTEROSCOPUS Bdv., 1828
    PETASIA Steph., 1829
    BRACHIONYX Meigen, 1832,
      missp.
    BRACHYONYCHA Agassiz, 1846,
      missp.
    SELENOSCOPUS Heinemann, 1859
    BRACHYONIX Berio, 1966, missp.
10011 **borealis** (Sm., 1899)

EUTOLYPE Grt., 1874
10012 **electilis** (Morr., 1875), n. comb.
    depilis (Grt., 1881), form, n. syn., n.
      stat.
    bombyciformis Sm., 1892
10013 **grandis** Sm., 1898
10014 **rolandi** Grt., 1874
    vernalis (Morr., 1874)
10015 **damalis** (Grt., 1879)

COPIPANOLIS Grt., 1874
10016 **styracis** (Gn., 1852)
    cubilis Grt., 1874, form
    stigma Sm., 1890, form
    borealis Sm., 1892, form
    fasciata Sm., 1892, form

PSEUDOCOPIVALERIA Buckett &
    Bauer, 1966
10017 **sonoma** (McD., 1941)
10018 **anaverta** Buckett & Bauer, 1966

PSAPHIDA Wlk., 1865
    DICOPIS Grt., 1874
    PSAPHIDIA Dyar, 1901, missp.
10019 **resumens** Wlk., 1865
    viridiscens (Wlk., 1865)
    muralis (Grt., 1874)
10020 **thaxteriana** (Grt., 1874)

COPIVALERIA Grt., 1883
10021 **grotei** (Morr., 1874)

APSAPHIDA Franc., 1973
10022 **eremna** Franc., 1973

VIRIDEMAS Sm., 1908
10023 **galena** Sm., 1908

Oncocnemidini

SUPRALATHOSEA B. & Benj., 1924
10024 **baboquivariensis** B. & Benj., 1924

PROVIA B. & McD., 1910
10025 **argentata** B. & McD., 1910

PLEROMELLA Dyar, 1921
10026 **opter** Dyar, 1921

PLEROMELLOIDA Nye, 1975
    PLEROMA Sm., 1891, preocc. by
      Sollas, 1888
10027 **conserta** (Grt., 1881)
    apposita (Sm., 1892)
10028 **obliquata** (Sm., 1891)
    a. **smithi** (B. & Benj., 1922)
10029 **bonuscula** (Sm., 1898)
10030 **arizonata** (B. & Benj., 1922)
10031 **cinerea** (Sm., 1904)

NEOGALEA Hamp., 1906
10032 **esula** (Druce, 1889)
    braziliensis Hamp., 1906

CATABENA Wlk., 1865
    TURBULA Wlk., 1869
    ADIPSOPHANES Grt., 1873
10033 **lineolata** Wlk., 1865
    petraea (Wlk., 1869)
    miscellus (Grt., 1873)
10034 **sagittata** B. & McD., 1913
10035 **pronuba** B. & McD., 1916
    nanuscula (Dyar, 1922)
10036 **vitrina** (Wlk., 1857)
10037 **terminella** (Grt., 1883)
    candida (Sm., 1900)
10038 **divisa** (H.-S., 1868)

OXYCNEMIS Grt., 1882
10039 **advena** Grt., 1882
    pacifica (Hy. Edw., 1884)
    baboquavaria Sm., 1907
10040 **orbicularis** B. & McD., 1912
10041 **subsimplex** Dyar, 1904
10042 **franclemonti** A. Blanchard, 1968
10043 **gracillima** (Grt., 1881)
    yuma Sm., 1907
10044 **acuna** Barnes, 1907
    adusta Sm., 1907
10045 **gustis** Sm., 1907
10046 **fusimacula** Sm., 1902
10047 **erratica** B. & McD., 1913
10048 **grandimacula** B. & McD., 1910

LEUCOCNEMIS Hamp., 1908
10049 **perfundis** (Sm., 1894)
10050 **nivalis** (Sm., 1894)
    silveroides (Hill, 1924), n. syn.
10051 **obscurella** B. & McD., 1916
10052 **variabilis** B. & McD., 1918
    alba B. & Benj., 1923, form

TRISTYLA Sm., 1893
10053 **alboplagiata** Sm., 1893

HEMISTILBIA B. & Benj., 1929
10054 **apposita** (B. & McD., 1918)

APHARETRA Grt., 1901
10055 **dentata** (Grt., 1875)
10056 **pyralis** (Sm., 1895)
10057 **purpurea** McD., 1940
10058 **californiae** McD., 1946

HOMOHADENA Grt., 1873
10059 **badistriga** (Grt., 1872)
    a. **rayata** Sm., 1908
    b. **tenuistriga** McD., 1940
10060 **induta** Harv., 1874

10061 **inconstans** Grt., 1883
10062 **stabilis** Sm., 1895
10063 **rustica** B. & McD., 1911
10064 **incomitata** Harv., 1875
10065 **infixa** (Wlk., 1856)
    kappa Grt., 1874
    retroversa Morr., 1874
    a. **dinalda** Sm., 1908
10066 **fifia** Dyar, 1904

ADITA Grt., 1874
10067 **chionanthi** (J. E. Smith, 1797)

HOMONCOCNEMIS Hamp., 1906
10068 **fortis** (Grt., 1880)
    vorax (Behr, 1884)
    a. **picina** (Grt., 1880)

ONCOCNEMIS Led., 1853
    PHORNACISA Wlk., 1862
    COPIHADENA Morr., 1875
    METAHADENA Morr., 1876
    PHARNACISA Fbs., 1954, missp.
10069 **mirificalis** Grt., 1879
10070 **dayi** Grt., 1873
10071 **euta** Sm., 1903
10072 **hayesi** Grt., 1873
10073 **regina** Sm., 1902
10074 **corusca** Sm., 1899
    ate (Dyar, 1921)
10075 **kelloggi** (Hy. Edw., 1875)
    exemplaris Sm., 1894
10076 **albifasciata** Hamp., 1906
    fasciata Sm., 1888, preocc. by Hy.
      Edw., 1886
10077 **melantho** Sm., 1899
10078 **sandaraca** Buckett & Bauer, 1967
10079 **pudorata** Sm., 1893
10080 **tenuifascia** Sm., 1888
10081 **parvanigra** Blkmre., 1923
10082 **lepipoloides** McD., 1922
10083 **glennyi** Grt., 1873
10084 **phairi** McD., 1927
10085 **terminalis** Sm., 1888
10086 **linda** B. & McD., 1913
10087 **benjamini** Linds., 1923
    punctilinea B. & L., 1922, preocc. by
      Hamp., 1906
    lindseyi Draudt, 1925
10088 **deceptiva** B. & L., 1922
10089 **basifugens** (Dyar, 1914)
    tinkhami McD., 1933
10090 **modesta** McD., 1933
10091 **balteata** Sm., 1902
10092 **iricolor** Sm., 1888
10093 **levis** Grt., 1880
10094 **sanina** Sm., 1910
10095 **simplex** Sm., 1888
10096 **augustus** Harv., 1875
10097 **nita** Sm., 1910
10098 **meadiana** Morr., 1875
10099 **saundersiana** Grt., 1876
10100 **fasciata** (Hy. Edw., 1886)
10101 **occata** (Grt., 1875)
10102 **pernotata** Grt., 1883
10103 **polingii** Barnes, 1904
10104 **homogena** Grt., 1877
    a. **pallida** Barnes, 1928
10105 **heterogena** A. Blanchard, 1972
10106 **arizonensis** Barnes, 1928

10107 **melalutea** Sm., 1899
10108 **viriditincta** Sm., 1894
10109 **singularis** B. & McD., 1912
10110 **wilsonensis** Hill, 1924
10111 **primula** B. & McD., 1918
10112 **flagrantis** Sm., 1893
10113 **chorda** (Grt., 1880)
    refecta Sm., 1894
    a. **extremis** Sm., 1890
10114 **rosea** Sm., 1903
10115 **bakeri** Dyar, 1905
    baceri Hamp., 1906, missp.
10116 **youngi** McD., 1922
10117 **simplicia** Sm., 1903
10118 **columbia** McD., 1922
10119 **punctilinea** Hamp., 1906
10120 **deserta** (Sm., 1891)
10121 **barnesii** Sm., 1899
10122 **umbrifascia** Sm., 1894
10123 **piffardi** (Wlk., 1862)
    atrifasciata (Morr., 1876)
10124 **cibalis** Grt., 1880
    a. **canadicola** Strand, 1916
10125 **laticosta** Dyar, 1904
10126 **lacticollis** Sm., 1908
    laticollis Rindge, 1955, missp.
10127 **nigrocaput** Sm., 1892
10128 **atricollaris** (Harv., 1875)
10129 **cottami** A. Blanchard, 1972
10130 **figurata** (Harv., 1875)
    a. **pallidior** Barnes, 1928
10131 **minor** Barnes, 1928
10132 **ragani** Barnes, 1928
10133 **semicollaris** Sm., 1909
10134 **astrigata** B. & McD., 1912
10135 **riparia** Morr., 1875
    a. **major** Grt., 1881
    b. **aqualis** Grt., 1881
    c. **manitoba** McD., 1941
    d. **subterminalis** McD., 1941
    e. **deserticola** McD., 1941
10136 **obscurata** B. & McD., 1912
10137 **intruda** Sm., 1910
10138 **ciliata** Sm., 1900
10139 **curvicollis** Grt., 1883
10140 **chandleri** Grt., 1873
    poliochroa Hamp., 1906
10141 **colorado** Sm., 1893
    chandleri; Hamp., 1906, not Grt.,
    1873
10142 **sagittata** B. & McD., 1916
10143 **mackiei** B. & Benj., 1924
10144 **ibapahensis** B. & Benj., 1924
10145 **extranea** Sm., 1892
10146 **utahensis** B. & Benj., 1924
10147 **tetrops** Dyar, 1904
10148 **griseicollis** Grt., 1882
    gerdis Sm., 1910
10149 **sectilis** Sm., 1894
10150 **sectiloides** (B. & McD., 1913)
    subtilis (B. & McD., 1913)
10151 **dunbari** (Harv., 1876)
    definita (B. & McD., 1912)
10152 **poliafascies** (Dyar, 1910)
    barbara (B. & McD., 1915)
10153 **toddi** A. Blanchard, 1968

LEPIPOLYS Gn., 1852
10154 **perscripta** Gn., 1852
10155 **behrensi** (Grt., 1874)

SYMPISTIS Hbn. 1823
10156 **melaleuca** (Thunb., 1791)
    leucoptera (Esp., 1798)
    moesta (Hbn., 1803–08)
    bicycla (Pack., 1867)
10157 **lapponica** (Thunb., 1791)
    amissa (Dup., 1836)
    tenebricosa (Mösch., 1877)
10158 **wilsoni** B. & Benj., 1924
10159 **zetterstedti** (Stgr., 1857)
    sibirica Alphéraky, 1895
    a. **labradoris** (Stgr., 1901)
10160 **kolthoffi** (Auriv., 1890)
10161 **besla** (Skin. & Mengel, 1892)
10162 **funesta** (Paykull, 1793)
    funebris (Hbn., 1808–09)
    a. **cocklei** (Dyar, 1904)

STYLOPODA Sm., 1891
10163 **cephalica** Sm., 1891
10164 **aterrima** (Grt., 1879)
10165 **groteana** (Dyar, 1903)
10166 **modestella** (B. & McD., 1918)
10167 **anxia** Sm., 1908
10168 **sexpunctata** (B. & McD., 1916)

COPANARTA Grt., 1895
10169 **aurea** (Grt., 1879)
10170 **nigerrima** (Sm., 1903)

PSEUDACONTIA Sm., 1883
10171 **louisa** Sm., 1908
10172 **crustaria** (Morr., 1875)
10173 **cansa** Sm., 1908

TRIOCNEMIS Grt., 1881
10174 **saporis** Grt., 1881

CRIMONA Sm., 1902
10175 **pallimedia** Sm., 1902

CERAPODA Sm., 1893
    CEROPODA Sm., 1894, missp.
10176 **stylata** Sm., 1893
    a. **arida** B. & Benj., 1923

CALOPHASIA Steph., 1829
    CLEOPHANA Bdv., 1833
    RHABDOPHANA Sodoffsky, 1837
10177 **lunula** (Hufn., 1766), introduced
    biological control, established
    linariae (D. & S., 1775)

BEHRENSIA Grt., 1875
10178 **conchiformis** Grt., 1875
    a. **suffusa** Buckett, 1964
10179 **bicolor** McD., 1941

Culliini

LATHOSEA Grt., 1881
10180 **pulla** Grt., 1881
    pullata Grt., 1890
10181 **dammersi** McD., 1935
10182 **spaldingi** B. & Benj., 1922, emend.
    spauldingi B. & Benj., 1922, incorr.
    orig. spell.

RANCORA Sm., 1892
10183 **strigata** Sm., 1892

10184 **serraticornis** (Lint., 1874)
    matricariae (Stkr., 1874)
10185 **comstocki** McD., 1937
10186 **brucei** Sm., 1903
10187 **solidaginis** (Stkr., 1874)
10187.1 **albicinerea** Sm., 1903
10187.2 **albida** (Sm., 1894)
10187.3 **ketchikana** B. & Benj., 1922

CUCULLIA Schr., 1802
    EUDERAEA Hbn., 1821
    EUCALIMIA Hbn., 1821
    CALLAENIA Hbn., 1821
    ARGYRITIS Hbn., 1821
    TRIBUNOPHORA Hbn., 1822
    CALLAINIA Hbn., 1826, missp.
    LOPHIA Sodoffsky, 1837
    ARGYROGALEA Hamp., 1906
10188 **dentilinea** (Sm., 1899)
10189 **obtusa** Sm., 1909
10190 **speyeri** Lint., 1874
    a. **dorsalis** Sm., 1892
10191 **laetifica** Lint., 1875
    cita Grt., 1883
    hartmanni French, 1888
10192 **alfarata** Stkr., 1898
    phila Sm., 1908
10193 **minor** B. & McD., 1913
10194 **intermedia** Speyer, 1870
    a. **cinderella** Sm., 1892
10195 **similaris** Sm., 1892
10196 **lilacina** Schaus, 1898
    agua Barnes, 1904
10197 **florea** Gn., 1852
    a. **obscurior** Sm., 1892
10198 **postera** Gn., 1852
    indicta Sm., 1904, n. syn.
10199 **omissa** Dod, 1916
10200 **asteroides** Gn., 1852
10201 **montanae** Grt., 1882
10201.1 **eucaena** Dyar, 1919, n. record
10202 **convexipennis** G. & R., 1868
10203 **oribac** Barnes, 1904
    aribac B. & McD., 1917, missp.
10204 **arizona** Sm., 1905
    strigata Schaus, 1898, preocc. by
      Sm., 1892
    perstrigata Hamp., 1906

NYCTEROPHAETA Sm., 1882
    (Febr.)
    EPINYCTIS Grt., 1882 (Apr.)
10205 **luna** (Morr., 1875)
    magdalena Hlst., 1882
    notatella (Grt., 1882)

COPICUCULLIA Sm., 1894
    COPICULLIA Sm., 1909, missp.
10206 **antipoda** (Stkr., 1878)
    propinqua Sm., 1894
    a. **adopitna** B. & Benj., 1927
10207 **basipuncta** B. & McD., 1918
10208 **mcdunnoughi** Henne, 1940
10209 **eulepis** (Grt., 1876)
    bistriga (Sm., 1892)
    mala Sm., 1908
10210 **cucullioides** B. & Benj., 1923
10211 **jemezensis** Dyar, 1922
10212 **incresa** Sm., 1910
10213 **heinrichi** B. & Benj., 1924

10214 **astigma** Sm., 1894
10215 **luteodisca** Sm., 1909

OPSIGALEA Hamp., 1906
10216 **blanchardi** Todd, 1966

EMARIANNIA Benj., 1933
10217 **cucullidea** Benj., 1933

HADENINAE

Hadenini

SPARKIA Nye, 1975
   CEA Grt., 1883, preocc. by Haliday,
     1837
10218 **immacula** (Grt., 1883)
   immaculata auth., missp.

TRICHOCOSMIA Grt., 1883
10219 **inornata** Grt., 1883
   demacula Strand, 1912, form
10220 **drasteroides** (Sm., 1903)

HADENELLA Grt., 1883
10221 **pergentilis** Grt., 1883

DISCESTRA Hamp. 1905
   SALACIA Boie, 1839, preocc. by
     Lamouroux, 1816
   DICESTRA McD., 1937, missp.
10222 **chartaria** (Grt., 1873)
10223 **trifolii** (Hufn., 1766)
   chenopodii (F., 1787)
   infraina (Haw., 1809)
   inquieta (Wlk., 1857)
   glaucovaria (Wlk., 1860)
   major (Speyer, 1875)
   a. **albifusa** (Wlk., 1857), n. comb.
10224 **mutata** (Dod, 1913), n. comb.
10225 **fulgora** (B. & McD., 1918), n. comb.
10226 **castrae** (B. & McD., 1912), n. comb.
   a. **ultra** (B. & Benj., 1924), n.
     comb.
10227 **hamata** (McD., 1930), n. comb.
10228 **oregonica** (Grt., 1881), n. comb.
   a. **morana** (Sm., 1910), n. comb.
   b. **columbica** (McD., 1930), n. comb.
   c. **montanica** (McD., 1930), n.
     comb.
10229 **alta** (B. & Benj., 1924), n. comb.
10230 **subalbida** (B. & Benj., 1924), n.
   comb.
10231 **obesula** (Sm., 1904), n. comb.
   a. **ortruda** (Sm., 1910)
10232 **farnhami** (Grt., 1873), n. comb.
10233 **crotchi** (Grt., 1880), n. comb.
   a. **fusculenta** (Sm., 1891), n. comb.
10234 **oaklandiae** (McD., 1937), n. comb.
10235 **chunka** (Sm., 1910), n. comb.
10236 **projecta** (McD., 1938), n. comb.

TRUDESTRA McD., 1937
10237 **hadeniformis** (Sm., 1894)

SCOTOGRAMMA Hy. Edw., 1887
10238 **submarina** (Grt., 1883)
10239 **fervida** B. & McD., 1912
   a. **proxima** B. & Benj., 1924
   b. **campestris** McD., 1943

10240 **densa** Sm., 1893
10241 **stretchi** Hy. Edw., 1887
10242 **harnardi** B. & Benj., 1924
10243 **addenda** B. & Benj., 1924
10244 **ptilodonta** (Grt., 1883)
   a. **nevada** B. & McD., 1912
   b. **albescens** McD., 1943
10245 **megaera** Sm., 1899
10246 **hirsuta** McD., 1938
10247 **yakima** (Sm., 1900)
10248 **orida** (Sm., 1903)
10249 **gatei** (Sm., 1910)
10250 **fieldi** B. & Benj., 1927
   wrighti B. & Benj., 1927, form
10251 **deffessa** (Grt., 1880)
10252 **inconcinna** Sm., 1888, mispl.?

TRIDEPIA McD., 1937
10253 **nova** (Sm., 1903)

TRICHOCLEA Grt., 1883
   TRICHCLEA Sm., 1903, missp.
10254 **postica** Sm., 1891
10255 **edwardsii** Sm., 1888
   edvardsi Hamp., 1905, missp.
   a. **deserticola** Hill, 1924
10256 **fuscolutea** (Sm., 1892)
10257 **florida** (Sm., 1900)
10258 **antica** (Sm., 1891)
10259 **decepta** Grt., 1883
   paupera; Druce, 1889, part, not
     Wlk., 1858
10260 **mojave** Benj., 1932
10261 **uscripta** (Sm., 1891)
10262 **ruisa** Fbs., 1913
10263 **artesta** (Sm., 1903)
10264 **vindemialis** (Gn., 1852)

SIDERIDIS Hbn., 1821
10265 **rosea** (Harv., 1874)
10266 **congermana** (Morr., 1874)
10267 **palmillo** (Barnes, 1907)
10268 **maryx** (Gn., 1852)
   rubefacta (Morr., 1874)
   vindemialis (Grt., 1875)

ADMETOVIS Grt., 1873
10269 **oxymorus** Grt., 1873
10270 **similaris** Barnes, 1904

MAMESTRA Ochs., 1816
   AETHRIA Hbn., 1821, preocc. by
     Hbn., 1819
   BARATHRA Hbn., 1821
   MAMISTRA Sodoffsky, 1837,
     emend.
   COPIMAMESTRA Grt., 1883
10271 **configurata** Wlk., 1856
   occidenta (Grt., 1883)
10272 **curialis** (Sm., 1888)

POLIA Ochs., 1816
   POLIA Bdv., 1828, preocc. by
     Ochs., 1816
   APLECTA Gn., 1838
   MAMESTRA; auth., not Ochs.,
     1816
10273 **discalis** (Grt., 1877)
10274 **piniae** Buckett & Bauer, 1967
10275 **nimbosa** (Gn., 1852)

   a. **mystica** (Sm., 1898)
   b. **mysticoides** (B. & Benj., 1924)
10276 **imbrifera** (Gn., 1852)
10277 **rogenhoferi** (Mösch., 1870)
10278 **leomegra** (Sm., 1908)
   a. **carbonifera** (Hamp., 1908), n.
     syn.
10279 **richardsoni** (Curt., 1834)
   septentrionis (Wlk., 1857)
   feildeni (McLachlan, 1878)
   a. **lanuginosa** (Sm., 1900)
   b. **tamsi** (Benj., 1933)
   c. **squara** (Sm., 1908)
10280 **purpurissata** (Grt., 1864)
   juncimacula (Sm., 1883), n. stat.
   crydina (Dyar, 1904), n. stat.
   languida (Sm., 1893), n. syn.
   a. **apurpura** (B. & McD., 1913)
10281 **nugatis** (Sm., 1898)
   tufa (Sm., 1905), n. stat.
10282 **noverca** (Grt., 1878)
10283 **vauorbicularis** Sm., 1902
10284 **delecta** B. & McD., 1916
10285 **tuana** (Sm., 1906)
10286 **nipana** (Sm., 1910)
10287 **montara** (Sm., 1910)
10288 **detracta** (Wlk., 1857)
   claviplena (Grt., 1873)
   a. **neoterica** (Sm., 1898)
10289 **goodelli** (Grt., 1875)
   a. **acutermina** (Sm., 1904)
10290 **obscura** (Sm., 1888)
10291 **latex** (Gn., 1852)
   demissa (Wlk., 1857)

MELANCHRA Hbn., 1820
   CERAMICA Gn., 1852, n. syn.
10292 **adjuncta** (Gn., 1852), n. comb.
   declarata (Wlk., 1856)
   benjamini (Lemmer, 1937), form
10293 **picta** (Harr., 1841), n. comb.
   exusta (Gn., 1852), n. comb.
   contraria (Wlk., 1856), n. comb.
10294 **pulverulenta** (Sm., 1888), n. comb.
10295 **assimilis** (Morr., 1874), n. comb.

LACANOBIA Billberg, 1820
   DIATARAXIA Hbn., 1821
   PEUCEPHILA Hamp., 1909
10296 **nevadae** (Grt., 1876), n. comb.
   canadensis (Sm., 1888), n. stat.
10297 **atlantica** (Grt., 1874), n. comb.
   discolor (Speyer, 1875), n. comb.
10298 **radix** (Wlk., 1857), n. comb.
   dimmocki (Grt., 1875), n. comb.
10299 **subjuncta** (G. & R., 1868), n. comb.
   a. **eleanora** (B. & McD., 1918), n.
     comb.
10300 **grandis** (Gn., 1852), n. comb.
   libera (Wlk., 1856), n. comb.
10301 **lutra** (Gn., 1852), n. comb.
   lubens (Grt., 1875), n. comb.
   rufula (Morr., 1875), n. comb.
   a. **glaucopis** (Hamp., 1905), n.
     comb.
10302 **rugosa** (Morr., 1875), n. comb.
10303 **tacoma** (Stkr., 1900), n. comb.
10304 **legitima** (Grt., 1864), n. comb.
10305 **dodii** (Sm., 1904), n. comb.
10306 **beani** (Grt., 1877), n. comb.

10307 **lilacina** (Harv., 1874), n. comb.
    illebefacta (Morr., 1874), form, n. comb.
    luski (B. & McD., 1913), n. stat., n. comb.
10308 **liquida** (Grt., 1881), n. comb.
    a. **meodana** (Sm., 1910), n. comb., n. syn.
10309 **prodeniformis** (Sm., 1888), n. comb.

PAPESTRA Sukhareva, 1972
10310 **quadrata** (Sm., 1891), n. comb.
    a. **ingravis** (Sm., 1895), n. comb.
10311 **biren** (Goeze, 1781)
    glauca (Hbn., 1808–09)
    bombycina; auth., not Hufn., 1766
    frustrata (McD., 1946), n. comb., n. syn.
10312 **cristifera** (Wlk., 1858), n. comb.
    impolita (Morr., 1874), n. comb.
10313 **brenda** (B. & McD., 1916), n. comb.
10314 **invalida** (Sm., 1891), n. comb.

ANARTOMIMA Boursin, 1952
    PSEUDANARTA Kozhanchikov, 1947, preocc. by Grt., 1878
10315 **secedens** (Wlk., 1858), n. comb.

HADENA Schr., 1802
    DIANTHOECIA Bdv., 1834
    DIANTHAECIA Gn., 1838, missp.
    ADENA Agassiz, 1846, missp.
    DIANTHECIA Taylor, 1855, missp.
    DIAUTHOECIA Mösch., 1864, missp.
10316 **ectypa** (Morr., 1875)
    bella (Grt., 1883)

ANEPIA Hamp., 1918
    EPIA Hbn., 1821, preocc. by Hbn., 1820
10317 **capsularis** (Gn., 1852)
    propulsa (Wlk., 1857)
10318 **minorata** (Sm., 1888)
10319 **amabilis** (B. & McD., 1918)
10320 **jola** (B. & Benj., 1924)
10321 **ectrapela** (Sm., 1898)
10322 **circumvadis** (Sm., 1902)
10323 **plumasata** Buckett & Bauer, 1967

HECATERA Gn., 1852
    ASTRAPETIS; auth., not Hbn., 1821
10324 **sutrina** (Grt., 1881)

MISELIA Ochs., 1816
    HARMODIA Hbn., 1820
10325 **glaciata** (Grt., 1882)
10326 **variolata** (Sm., 1888)

ANARTA Ochs., 1816
    CHARELIA Sodoffsky, 1837
10327 **melanopa** (Thunb., 1791)
    alpicola (Quenzel, 1802)
    rupestralis (Hbn., 1796–99)
    vidua (Hbn., 1803–08)
    tristis (Hbn., 1808–09)
    rupestris (Hbn., 1818–19)
    nigrolunata Pack., 1867

    wistroemi Lampa, 1885
    a. **laerta** Sm., 1903
10328 **sierrae** B. & McD., 1916
    a. **laertidia** B. & McD., 1916
10329 **mimuli** Behr, 1885
10330 **etacta** Sm., 1900
10331 **magna** B. & Benj., 1924
10332 **cordigera** (Thunb., 1788)
    albirena (Hbn., 1800–03)
    luteola G. & R., 1865
    suffusa Tutt, 1892
    aethiops Hoffmann, 1893
    carbonaria Christoph, 1893
10333 **myrtilli** (L., 1761)
    albirena; Haw., 1809, not Hbn., 1800–03
    alpina Raetzer, 1890
    rufescens Tutt, 1892
    anglica Culot, 1915

LASIESTRA Hamp., 1905
10334 **staudingeri** (Auriv., 1891)
    schoenherri; Stgr., 1861, not Zett., 1839
10335 **quadrilunata** (Grt., 1874)
10336 **leucocycla** (Stgr., 1857)
    a. **moeschleri** (Stgr., 1901)
    b. **hampa** (Sm., 1908)
    c. **flanda** (Sm., 1908)
    d. **albertensis** (McD., 1925)
    e. **subfumosa** (Gibson, 1920)
10337 **coloradensis** Richards, 1943
10338 **lagganata** (B. & Benj., 1924)
10339 **impingens** (Wlk., 1857)
    nivaria (Grt., 1875)
    perpura (Morr., 1875)
    a. **curta** (Morr., 1875)
    b. **phaea** (Hamp., 1905)
10340 **discolor** (Sm., 1899)
10341 **phoca** (Mösch., 1864)
10342 **luteola** (Sm., 1893)
10343 **klotsi** Richards, 1943
10344 **promulsa** (Morr., 1875)
10345 **macleani** (McD., 1927)
10346 **uniformis** (Sm., 1893)
10347 **infuscata** (Sm., 1899)
10348 **poca** (B. & Benj., 1923), n. stat.
10349 **preblei** (Benj., 1933), n. stat.
10350 **dolosa** (B. & Benj., 1923)
10351 **mimula** (Grt., 1883)

LASIONYCTA Auriv., 1892
    LASCIONYCTA Hill, 1927, missp.
10352 **perplexa** (Sm., 1888)
10353 **alberta** B. & Benj., 1923
10354 **marloffi** (Dyar, 1922)
10355 **albinuda** (Sm., 1903)
10356 **subdita** (Mösch., 1860)
10357 **membrosa** (Morr., 1875)
10358 **subfuscula** (Grt., 1874)
10359 **benjamini** Hill, 1927
10360 **mutilata** (Sm., 1898)
10361 **rainieri** (Sm., 1900)
10362 **conjugata** (Sm., 1899)
10363 **sedilis** (Sm., 1899)
10364 **arietis** (Grt., 1879)
10365 **wyatti** (B. & Benj., 1926)
10366 **insolens** (Grt., 1874)
    earina (Harv., 1874)
10367 **ochracea** (Riley, 1892)

LACINIPOLIA McD., 1937
10368 **meditata** (Grt., 1873)
    a. **columbia** (Sm., 1888)
10369 **determinata** (Sm., 1891)
10370 **lustralis** (Grt., 1875)
    cervina (Sm., 1898)
    a. **suffusa** (Sm., 1888)
10371 **cuneata** (Grt., 1873)
    a. **gertana** (Sm., 1913)
    basirufa (Strand, 1917)
    rubicunda (Strand, 1917)
10372 **anguina** (Grt., 1881)
    a. **larissa** (Sm., 1895)
10373 **incurva** (Sm., 1888)
10374 **longiclava** (Sm., 1891)
10375 **gnata** (Grt., 1882)
10376 **agnata** (Sm., 1905)
10377 **prognata** McD., 1940
10378 **luteimacula** (B. & Benj., 1925)
10379 **umbrosa** (Sm., 1888)
    intentata (Sm., 1898)
10380 **vittula** (Grt., 1882)
10381 **naevia** (Sm., 1898)
    griseata (Sm., 1900)
10382 **stenotis** (Hamp., 1905)
10383 **palilis** (Harv., 1875)
10384 **uliginosa** (Sm., 1905)
    bicolor (B. & McD., 1913)
10385 **canities** (Hamp., 1905)
10386 **brachiolum** (Harv., 1876)
10387 **selama** (Stkr., 1898)
10388 **rubrifusa** (Hamp., 1905)
10389 **roseosuffusa** (Sm., 1900)
10390 **falsa** (Grt., 1880), n. comb.
    micta (Hamp., 1918), n. comb.
10391 **francisca** (Sm., 1910)
10392 **leucogramma** (Grt., 1873), n. comb.
10393 **teligera** (Morr., 1875), n. stat.
    imbuna (Sm., 1905), n. syn.
10394 **vicina** (Grt., 1874)
    a. **acutipennis** (Grt., 1880)
    b. **sareta** (Sm., 1906)
10395 **pensilis** (Grt., 1874)
    indistincta (Strand, 1917)
    a. **doira** (Stkr., 1898)
    ascula (Sm., 1905)
10396 **basiplaga** (Sm., 1905)
10397 **renigera** (Steph., 1829)
    herbimacula (Gn., 1852)
    infecta (Wlk., 1857)
10398 **stricta** (Wlk., 1865)
    ferrea (Grt., 1881)
    a. **cinnabarina** (Grt., 1874)
    b. **tenisca** (Sm., 1910)
    c. **papka** (B. & Benj., 1929)
    kappa (B. & Benj., 1925), preocc. by Hbn., 1808–09
10399 **circumcincta** (Sm., 1891)
10400 **spiculosa** (Grt., 1883)
10401 **lepidula** (Sm., 1888)
10402 **perta** (Druce, 1889), n. comb., n. record
10403 **erecta** (Wlk., 1857)
    constipata (Wlk., 1857)
    innexa (Grt., 1874)
10404 **perfragilis** (Dyar, 1923), n. comb., n. record
10405 **lorea** (Gn., 1852)
    ligata (Wlk., 1860)
    dodgei (Morr., 1875)

10406 **olivacea** (Morr., 1874)
    obscurior (Sm., 1888)
    a. **lucina** (Sm., 1901)
    b. **megarena** (Sm., 1901)
    c. **altua** (Sm., 1901)
    d. **petita** (Sm., 1901)
    e. **vaumedia** (Sm., 1888)
10407 **davena** (Sm., 1901)
10408 **comis** (Grt., 1876)
    obnigra (Sm., 1901)
10409 **rectilinea** (Sm., 1888)
10410 **lunolacta** (Sm., 1903)
10411 **laudabilis** (Gn., 1852)
    indicans (Wlk., 1857)
    mediosuffusa (Strand, 1917), form
    rufoirrorata (Strand, 1917), form
10412 **marinitincta** (Harv., 1875)
    a. **appendicula** McD., 1937
10413 **explicata** McD., 1937
10414 **implicata** McD., 1937
10415 **strigicollis** (Wallgr., 1860)
10416 **illaudabilis** (Grt., 1875)
    dilatata (Sm., 1900), n. syn.
    a. **alboguttata** (Grt., 1877)
    b. **restora** (Sm., 1910)
10417 **tricornuta** McD., 1937
10418 **runica** (Hamp., 1918)
10419 **viridifera** McD., 1937
10420 **consimilis** McD., 1937
10421 **buscki** (B. & Benj., 1927)
10422 **quadrilineata** (Grt., 1873)
    cinereoviridis (Strand, 1917)
10423 **patalis** (Grt., 1873)
    a. **fletcheri (Grt., 1888)**
10424 **parvula** (H.-S., 1868)
    lustralis; Grossb., 1917, not Grt.,
    1875

TRICHOCERAPODA Benj., 1932
10425 **comstocki** Benj., 1932
10426 **strigata** (Sm., 1891)
10427 **oblita** (Grt., 1877)
    insertans (Sm., 1890)
    deserta (Grinnell, 1912)
    arrosta (Dyar, 1921)

DARGIDA Wlk., 1856
EUPSEPHOPAECTES Grt., 1873
EUPSEPHOPACTES Hamp.,
    1905, missp.
10428 **procincta** (Grt., 1873)
10429 **graminivora** Wlk., 1856, emend.
    Hamp., 1905
    grammivora Wlk., 1856, incorr. orig.
    spell.

FARONTA Sm., 1908
PSEUDOLEUCANIA McD., 1937,
    preocc. by Stgr., 1899
PROTOLEUCANIA McD., 1937
10430 **quadrannulata** (Morr., 1876)
10431 **diffusa** (Wlk., 1856)
    albilinea; auth., not Hbn., 1821
    moderata (Wlk., 1856)
    harveyi (Grt., 1873)
    a. **obscurior** (Sm., 1902)
    b. **limitata** (Sm., 1902)
    c. **neptis** (Sm., 1902)
10432 **terrapictalis** Buckett, 1969
10433 **tetera** (Sm., 1902)

10434 **rubripennis** (G. & R., 1870)
10435 **aleada** Sm., 1908

ALETIA Hbn., 1821
HYPHILARE Hbn., 1821
10436 **oxygala** (Grt., 1881)
    minorata (Sm., 1894)
    rubripallens (Sm., 1902), form
    a. **luteopallens** (Sm., 1902)
10437 **yukonensis** (Hamp., 1911), emend.
    McD., 1938
    yuconensis (Hamp., 1911), incorr.
    orig. spell.

PSEUDALETIA Franc., 1951
10438 **unipuncta** (Haw., 1809)
    extranea (Gn., 1852)
10438.1 **sequax** Franc., 1951, n. record

LEUCANIA Ochs., 1816
DONOCHLORA Sodoffsky, 1837
LEUCADIA Sodoffsky, 1837
BOROLIA Moore, 1881
10439 **extincta** Gn., 1852
    texana (Morr., 1875)
    ligata (Grt., 1875)
    a. **flabilis** (Grt., 1881)
    rimosa (Grt., 1882)
10440 **linita** Gn., 1852
    amygdalina (Harv., 1878)
    punctata (Strand, 1917)
10441 **farcta** (Grt., 1881)
    palliseca Sm., 1902
    oregona Sm., 1902, n. syn.
    roseola Sm., 1894, form
10442 **anteoclara** Sm., 1902
    calgariana Sm., 1902, form
10443 **februalis** (Hill, 1924)
10444 **phragmatidicola** Gn., 1852
10445 **linda** Franc., 1952
10446 **multilinea** Wlk., 1856
    solita Wlk., 1856
10446.1 **lapidaria** (Grt., 1876), rev. stat.
10447 **commoides** Gn., 1852
10448 **comma** (L., 1761)
    turbida (Hbn., 1800–03)
    congener (Hbn., 1800–03)
    suffusa Tutt, 1888
    nigrofasciata Hamp., 1894
    rhodocomma Püngler, 1900
10449 **insueta** Gn., 1852
    adonea (Grt., 1874)
    a. **dia** Grt., 1879
    b. **megadia** Sm., 1902
    c. **heterodoxa** Sm., 1894
    mimica Stkr., 1899
10450 **incognita** (B. & McD., 1918)
10451 **oaxacana** Schaus, 1898
    complicata Stkr., 1898, n. syn.
10452 **imperfecta** Sm., 1894
10453 **stolata** Sm., 1894
10454 **latiuscula** H.-S., 1868
    subpuncta (Harv., 1875)
10455 **scirpicola** Gn., 1852
    calpota Sm., 1908
    pendens Sm., 1905
    juncicola Gn., 1852
10456 **adjuta** (Grt., 1874)
10457 **infatuans** Franc., 1972
10458 **humidicola** Gn., 1852

10459 **inermis** (Fbs., 1936)
10460 **callidior** (Fbs., 1936)
10461 **ursula** (Fbs., 1936)
10462 **pseudargyria** Gn., 1852
    callida (Grt., 1882), form
    derufata (Strand, 1917), form
10463 **pilipalpis** (Grt., 1877)

PERIGONICA Sm., 1890
10464 **tertia** Dyar, 1903
10465 **fermata** Sm., 1911
10466 **fulminans** Sm., 1890
10467 **eldana** Sm., 1911
10468 **angulata** Sm., 1890
10469 **pectinata** (Sm., 1888)
    a. **punctilinea** Sm., 1906

ACERRA Grt., 1874
10470 **normalis** Grt., 1874

STRETCHIA Hy. Edw., 1874
10471 **plusiaeformis** Hy. Edw., 1874
    a. **coloradicola** Strand, 1917
10472 **pictipennis** McD., 1949
10473 **muricina** (Grt., 1876)
10474 **pacifica** McD., 1949
10475 **inferior** (Sm., 1888)
10476 **prima** Sm., 1891

ORTHOSIA Ochs., 1816
ORTHOA Billberg, 1820
MONIMA Hbn., 1821
CUPHANOA Hbn., 1821
SEMIOPHORA Steph., 1829
TAENIOCAMPA Gn., 1852
CYPHONOA Agassiz, 1846, missp.
ANCATA Căpuşe, 1958
ANACTA Kristensen, 1966, missp.
10477 **erythrolita** (Grt., 1879)
    apicata (Sm., 1911), form
    acutangula (Sm., 1911), form
    erythrolitoides (Strand, 1917), form
10478 **pulchella** (Harv., 1876)
    addenda (Sm., 1890), form
    orbiculata (Sm., 1911), form
    palomarensis (Hill, 1924)
    a. **achsha** (Dyar, 1904)
    algula (Sm., 1911), form
    b. **hepatica** (B. & McD., 1911)
10479 **transparens** (Grt., 1882)
    hamifera Grt., 1888
    fringata (Sm., 1908)
10480 **praeses** (Grt., 1879)
    saleppa (Sm., 1907), form
    stigmata (B. & Benj., 1924)
10481 **mys** (Dyar, 1903)
    a. **caloramica** (B. & McD., 1911)
    agravens (B. & McD., 1911)
10482 **ferrigera** (Sm., 1894)
    strigatteria (Hill, 1923)
    puncticostata (Dyar, 1921), form
10483 **johnstoni** McD., 1943
10484 **terminata** (Sm., 1888)
10485 **behrensiana** (Grt., 1875), n. comb.
10486 **macona** (Sm., 1908)
10487 **rubescens** (Wlk., 1865)
    rubrescens auth., missp.
    venata (Sm., 1890)
10488 **garmani** (Grt., 1879)
10489 **arthrolita** (Harv., 1875)

10490 **revicta** (Morr., 1876)
    subterminata (Sm., 1888)
    subterminalis (Sm., 1890), missp.
10491 **alurina** (Sm., 1902)
10492 **desperata** (Sm., 1891), n. comb., n. stat.
10493 **segregata** (Sm., 1893), n. comb.
    gussata (Sm., 1895), n. comb.
    negussa (Sm., 1900), form, n. comb.
    a. **plicata** (Sm., 1898), n. comb.
10494 **pacifica** (Harv., 1874)
    infrapicta (Strand, 1917)
10495 **hibisci** (Gn., 1852)
    alia; auth., not Gn., 1852
    confluens (Morr., 1874)
    insciens Wlk., 1857
    a. **brucei** (Sm., 1910)
      malora (Sm., 1910)
    b. **nubilata** (Sm., 1910)
    c. **quinquefasciata** (Sm., 1909)
      latirena (Dod, 1910)
      proba (Sm., 1910)
      inherita (Sm., 1910)
      inflava (Sm., 1910)
10496 **flaviannula** (Sm., 1899)
10497 **annulimacula** (Sm., 1891)
10498 **tenuimacula** (B. & McD., 1913)
10499 **mediomacula** B. & Benj., 1924
10500 **nongenerica** B. & Benj., 1924

CROCIGRAPHA Grt., 1875
10501 **normani** (Grt., 1874)

HIMELLA Grt., 1875
10502 **intractata** (Morr., 1874)
    fidelis Grt., 1875

EGIRA Dup., 1845
    XYLOMYGES Gn., 1852
    XYLOMIGES auth., missp.
    XYLOMANIA Hamp., 1905
10503 **baueri** (Buckett, 1968), n. comb.
10504 **variabilis** (Sm., 1891), n. comb.
10505 **hiemalis** (Grt., 1874), n. comb.
    californica (Stkr., 1874), n. comb.
10506 **simplex** (Wlk., 1865), n. comb.
    pallidior (Sm., 1899), n. comb.
10507 **vanduzeei** (B. & Benj., 1926), n. comb.
10508 **crucialis** (Harv., 1875), n. comb.
    a. **peritalis** (Sm., 1892), n. comb.
10509 **cognata** (Sm., 1894), n. comb.
    a. **minorata** (B. & McD., 1918), n. comb.
10510 **februalis** (B. & McD., 1918), n. comb.
10511 **curialis** (Grt., 1873), n. comb.
    a. **indurata** (Sm., 1894), n. comb.
    b. **nicalis** (Sm., 1909), n. comb.
    c. **tantiva** (Sm., 1909), n. comb.
    d. **argus** (Sm., 1909), n. comb.
10512 **candida** (Sm., 1894), n. comb.
10513 **dolosa** (Grt., 1880), n. comb.
10514 **rubrica** (Harv., 1878), n. comb.
    a. **rubricoides** (B. & Benj., 1924), n. comb.
    b. **mustelina** (Sm., 1911), n. comb.
    c. **pulchella** (Sm., 1894), n. comb.
10515 **perlubens** (Grt., 1881), n. comb.
    subapicalis (Sm., 1888), n. comb.
10516 **purpurea** (B. & McD., 1910), n. comb.

10517 **alternans** (Wlk., 1857), n. comb.
    tabulata (Grt., 1878), n. comb.

ACHATIA Hbn., 1813
    ASTRAPETIS Hbn., 1821
    ACHALIA Hbn., 1825, missp.
    ACHATEA Curt., 1826, missp.
    ASTRAPETES Agassiz, 1846, emend.
10518 **distincta** Hbn., 1813
    vitis (French, 1879)

MORRISONIA Grt., 1874
10519 **mucens** (Hbn., 1827–31)
    multifaria (Wlk., 1857)
    spoliata (Wlk., 1857)
    sectilis (Gn., 1852), form
    rileyana Sm., 1890
    sectilana Strand, 1917
10520 **evicta** (Grt., 1873)
    vomerina (Grt., 1873), form
    infidelis Grt., 1879, form
10521 **confusa** (Hbn., 1827–31)
    infructuosa (Wlk., 1857)

CERAPTERYX Curt., 1833
10522 **graminis** (L., 1758)
    tricuspis (Esp., 1786)
    gramineus (Haw., 1803), missp.
    hibernicus Curt., 1833
    albineura (Bdv., 1837)
    albipuncta (Lampa, 1884)
    brunnea (Lampa, 1884)
    rufa (Tutt, 1889)
    rufocosta (Tutt, 1889)
    ochrea (Tutt, 1889)
    pallida (Tutt, 1889)
    obsoleta (Tutt, 1889)

THOLERA Hbn., 1821
    NEURONIA Hbn., 1821
    CHARAEAS Steph., 1829
    EPINEURONIA Rebel, 1901
10523 **americana** (Sm., 1894)

NEPHELODES Gn., 1852
    MONOSCA Wlk., 1869
    MONOSTOLA Alphéraky, 1892
10524 **minians** Gn., 1852
    emmedonia; auth., not Cram., 1780
    violans Gn., 1852, form
    rubeolans Gn., 1852
    subdolens (Wlk., 1856)
    expansa (Wlk., 1857)
    subnotata (Wlk., 1869)
    a. **tertialis** Sm., 1903
    b. **pectinata** Sm., 1900
10525 **demaculata** B. & McD., 1918
10526 **mendica** B. & L., 1921
10527 **adusta** Buckett, 1973
10528 **carminata** (Sm., 1890)

Eriopygini

ANHIMELLA McD., 1943
10529 **perbrunnea** (Grt., 1879)
10530 **contrahens** (Wlk., 1860)
    thecata (Morr., 1875)
    purpurascens (Strand, 1917)
    a. **conar** (Stkr., 1898)

    quadristigma (Sm., 1900)
    infidelis (Dyar, 1904)
10531 **pacifica** McD., 1943

HOMORTHODES McD., 1943
10532 **furfurata** (Grt., 1875)
    peredia (Grt., 1883)
    a. **uniformis** (Sm., 1888)
    b. **lindseyi** (Benj., 1922)
10533 **communis** (Dyar, 1904)
    affurata (Hamp., 1905)
    brunneosuffusa (Strand, 1917)
10534 **fractura** (Sm., 1906)
    a. **mecrona** (Sm., 1908)
10535 **discreta** (B. & McD., 1916)
10536 **dubia** (B. & McD., 1912)
10537 **mania** (Stkr., 1899)
10538 **rubritincta** McD., 1943
10539 **hanhami** (B. & McD., 1911)
    a. **semicarnea** (B. & McD., 1918)
10540 **carneola** McD., 1943
10541 **reliqua** (Sm., 1899)
    gracea (Sperry, 1938)
10542 **rectiflava** (Sm., 1908)
10543 **flosca** (Sm., 1906)
10544 **perturba** McD., 1943
10545 **gigantoides** (B. & McD., 1912), n. comb.

PROTORTHODES McD., 1943
10546 **curtica** (Sm., 1890)
    a. **bostura** (Sm., 1908)
10547 **utahensis** (Sm., 1888)
10548 **akalus** (Stkr., 1898)
10549 **eureka** (B. & Benj., 1927)
10550 **daviesi** (B. & Benj., 1927)
10551 **indra** (Sm., 1906)
10552 **incincta** (Morr., 1874)
10553 **saturnus** (Stkr., 1900)
10554 **smithii** (Dyar, 1904), n. comb.
10555 **argentoppida** McD., 1943
10556 **perforata** (Grt., 1883)
10557 **rufula** (Grt., 1874)
10558 **antennata** (B. & McD., 1912)
10559 **alfkeni** (Grt., 1895)
    perplexa (Grt., 1895)
    latens (Sm., 1908)
    occluna (Sm., 1909)
10560 **texana** (Sm., 1900)
    a. **consors** (Sm., 1900)
10561 **variabilis** (B. & McD., 1912)
10562 **mulina** (Schaus, 1894)
    pseudochroma (Dyar, 1913)
10563 **oviduca** (Gn., 1852)
    capsella (Grt., 1875)
10564 **lindrothi** Krogerus, 1954
10565 **orobia** (Harv., 1876)
10566 **melanopis** (Hamp., 1905)
    a. **coloradensis** (Strand, 1917)

ULOLONCHE Sm., 1888
10567 **culea** (Gn., 1852)
    modifica (Morr., 1874)
10568 **consopita** (Grt., 1881)
10569 **modesta** (Morr., 1874)
10570 **fasciata** Sm., 1888
10571 **marloffi** (B. & Benj., 1924)
10572 **dilecta** (Hy. Edw., 1885)
10573 **disticha** (Morr., 1875)
10574 **orbiculata** (Sm., 1891)
10575 **niveiguttata** (Grt., 1873)

HYPEREPIA B. & L., 1922
10576 **jugifera** (Dyar, 1920)
19577 **pi** B. & L., 1922

PSEUDORTHODES Morr., 1874
10578 **vecors** (Gn., 1852)
    enervis (Gn., 1852)
    prodeuns (Wlk., 1856)
    velata (Wlk., 1860)
    togata (Wlk., 1865)
    nitens (Grt., 1883)
    purpureobrunnea (Strand, 1917)
    a. **griseocincta** (Harv., 1874)
10579 **calceolaris** (Stkr., 1900)
10580 **imora** (Stkr., 1898)
10581 **keela** (Sm., 1908)
19582 **irrorata** (Sm., 1888)
10583 **puerilis** (Grt., 1874)
10584 **virgula** (Grt., 1883)

ORTHODES Gn., 1852
    ORTHODEA Sm., 1906, missp.
10585 **crenulata** (Butler, 1890)
10586 **furtiva** McD., 1943
10587 **cynica** Gn., 1852
    nimia Gn., 1852
    candens Gn., 1852
    tecta Wlk., 1865
10588 **adiastola** Franc., 1976
10589 **bolteri** (Sm., 1900)
    a. **gigas** Sm., 1906

SYNORTHODES Franc., 1976
10590 **auriginea** Franc., 1976
10591 **typhedana** Franc., 1976

HEXORTHODES McD., 1943
10592 **serrata** (Sm., 1900)
    dubiosa (B. & McD., 1913)
10593 **inconspicua** (Grt., 1883)
    pectinicornis (Sm., 1906)
10594 **jocosa** (B. & McD., 1916)
10595 **planalis** (Grt., 1883)
10596 **agrotiformis** (Grt., 1881)
10597 **trifascia** (Sm., 1891)
10598 **euxoiformis** (B. & McD., 1913)
10599 **alamosa** (Barnes, 1904)
10600 **hueco** (Barnes, 1904)
10601 **accurata** (Hy. Edw., 1882)
 1602 **senatoria** (Sm., 1900)
10603 **catalina** (B. & McD., 1912)
10604 **optima** (Dyar, 1920), n. comb., n.
    record

ERIOPYGA Gn., 1852
10605 **crista** (Wlk., 1856)
    dissimilis (B. & McD., 1910)

URSOGASTRA Sm., 1906
10606 **lunata** Sm., 1906

ZOSTEROPODA Grt., 1874
10607 **hirtipes** Grt., 1874
10608 **clementei** Meadows, 1942

NELEUCANIA Sm., 1902
10609 **suavis** (B. & McD., 1912)
10610 **patricia** (Grt., 1880)
10611 **bicolorata** (Grt., 1881)
    a. **citronella** Sm., 1902
10612 **niveicosta** Sm., 1902

10613 **praegracilis** (Grt., 1877)
    gracillima (Grt., 1895)

TRICHORTHOSIA Grt., 1883
10614 **ferricola** (Sm., 1905)
10615 **diplogramma** (Schaus, 1903), n. comb.
    albidior (B. & McD., 1910), n. comb.
    inquisita (Dyar, 1920), n. comb.
10616 **parallela** (Grt., 1883)
    terminatissima (Dyar, 1904)
10617 **tristis** (B. & McD., 1910)

TRICHOPOLIA Grt., 1883
    EUPOLIA Sm., 1894, preocc. by
    Hulbrecht, 1887
10618 **dentatella** Grt., 1883
    obtusa (Sm., 1888)
    licentiosa (Sm., 1894)
10619 **suspicionis** B. & Benj., 1920

TRICHAGROTIS McD., 1929
10620 **spinosa** (B. & McD., 1912)

TRICHOFELTIA McD., 1929
    TRICHOFELTIA B. & Benj., 1929
    (Sept.), preocc. by McD., 1929
    (April)
10621 **circumdata** (Grt., 1883)

MIMOBARATHRA B. & McD., 1915
10622 **antonito** (Barnes, 1907)

MIODERA Sm., 1908
10623 **stigmata** Sm., 1908
10624 **eureka** B. & Benj., 1927

ENGELHARDTIA B. & Benj., 1923
10625 **ursina** (Sm., 1898)

TRICHOLITA Grt., 1875
10626 **elsinora** (Barnes, 1904)
    erebus Sm., 1911
10627 **signata** (Wlk., 1860)
    semiaperta (Morr., 1874)
    igna (B. & Benj., 1927), form
    a. **semitropicae** (B. & Benj., 1927)
10628 **notata** Stkr., 1898
    syrissa Stkr., 1898
    ota (B. & Benj., 1927), form
10629 **baranca** Barnes, 1905
10630 **fistula** Harv., 1878
    ulamora Sm., 1911
10631 **chipeta** Barnes, 1904
    a. **endiva** Sm., 1911
    vespera (B. & L., 1922), form
10632 **bisulca** (Grt., 1881)
10633 **lutina** (Sm., 1902)
    velutina (Sm., 1900), preocc. by
    Evers., 1846

LOPHOCERAMICA Dyar, 1908
10634 **artega** (Barnes, 1907)
    pallicauda (Sm., 1908)
    eriopygoides B. & Benj., 1926, form

HYDROECIODES Hamp., 1905
10635 **juvenalis** (Grt., 1881), n. comb.
10636 **repleta** (Bird, 1911), n. comb.
10637 **serrata** (Grt., 1880), n. comb.
10638 **ochrimacula** (B. & McD., 1913), n.
    comb., n. stat.

10639 **auripurpura** (A. Blanchard, 1968), n.
    comb.

Glottulini

XANTHOPASTIS Hbn., 1821
    EUTHISANOTIA Hbn., 1831
    PHILOCHRYSA Grt., 1863
10640 **timais** (Cram., 1782)
    amaryllidis (Sepp, 1832–40)
    regnatrix (Grt., 1864)
    antillium Dyar, 1913
    moctezuma Dyar, 1913
    molinoi Dyar, 1919

NOCTUINAE

Agrotini

AGROTIS Ochs., 1816
    SCOTIA Hbn., 1821
    AGRONOMA Hbn., 1821
    AGRONOMAE Hbn., 1821, incorr.
    orig. spell.
    GEORYX Hbn., 1821
    PSAMMOPHILA Steph., 1850,
    preocc. by Brown, 1827
    POROSAGROTIS Sm., 1890
    PARORAGROTIS Dod, 1910,
    missp.
10641 **vetusta** Wlk., 1865
    muraenula G. & R., 1868
    a. **catenuloides** (Sm., 1910)
    catenula; auth., not Grt., 1879
    b. **mutata** (B. & Benj., 1924)
10642 **daedalus** (Sm., 1890)
10643 **dolli** Grt., 1882
10644 **mollis** Wlk., 1857
    fernaldi Morr., 1876
10645 **orthogonia** Morr., 1876
    a. **delorata** (Sm., 1908)
    b. **duae** (B. & Benj., 1927)
10646 **orthogonoides** McD., 1946
10647 **patula** Wlk., 1857
    septentrionalis Mösch., 1862
10648 **gladiaria** Morr., 1874
    morrisoniana Morr., 1875 (Jan.)
    morrisoniana Riley, 1875 (Mar.),
    preocc. by Morr., 1875
10649 **kingi** McD., 1932
10650 **robustior** (Sm., 1899)
10651 **venerabilis** Wlk., 1857
    a. **arida** (Ckll., 1913)
10652 **vancouverensis** Grt., 1873
    agilis Grt., 1873
    hortulana Morr., 1876
    a. **semiclarata** Grt., 1881
    b. **dentilinea** Sm., 1890
10653 **gravis** Grt., 1874
    vapularis Grt., 1876
10654 **buchholzi** (B. & Benj., 1929)
10655 **musa** (Sm., 1910)
10656 **atha** Stkr., 1898
10657 **aeneipennis** Grt., 1876
10658 **stigmosa** Morr., 1874, rev. stat.
    clodiana Grt., 1881
10659 **volubilis** Harv., 1874
    a. **fumipennis** McD., 1932
10660 **obliqua** (Sm., 1903)
10661 **malefida** Gn., 1852
    inspinosa Gn., 1852

consueta Wlk., 1857
   submucosa H.-S., 1868
10662 **apicalis** H.-S., 1868, n. record
10663 **ipsilon** (Hufn., 1766)
   suffusa (D. & S., 1775)
   ypsilon (Rottemburg, 1776), missp.
   idonea (Cram., 1780)
   spinula; Esp., 1786, not D. & S.,
     1775
   spinifera (Villers, 1789)
   telifera Harr., 1841
10664 **subterranea** (F., 1794)
   annexa Tr., 1825
   anteposita Gn., 1852
   lutescens (Blanchard, 1852)
   decernens Wlk., 1857
   interferens Wlk., 1858
   lutea (Druce, 1889)
   interposita Maassen, 1890
10665 **repleta** Wlk., 1857
10666 **manifesta** Morr., 1875
   impingens (Dyar, 1920)
10667 **triphaenoides** (Dyar, 1912)
   orbipuncta (B. & McD., 1916)
10668 **haesitans** Wlk., 1857

ONYCHAGROTIS Hamp., 1903
10669 **rileyana** (Morr., 1874)

FELTIA Wlk., 1856
10670 **jaculifera** (Gn., 1852), rev. stat.
   ducens Wlk., 1856
   radiata Sm., 1891
   subgothica; auth., not Haw., 1809
10671 **hudsoni** Sm., 1903
10672 **evanidalis** (Grt., 1878)
10673 **californiae** McD., 1939
10674 **subgothica** (Haw., 1809)
   jaculifera; auth., not Gn., 1852
10675 **tricosa** (Lint., 1874), rev. stat.
10676 **herilis** (Grt., 1873)
10677 **pectinicornis** Sm., 1890
10678 **subpallida** McD., 1932
10679 **edentata** Sm., 1902
10680 **geniculata** G. & R., 1868

COPABLEPHARON Harv., 1878
10681 **grande** (Stkr., 1878)
   subflavidens Grt., 1882
10682 **canarianum** McD., 1932
10683 **contrastum** McD., 1932
10684 **serratum** McD., 1932
10685 **serraticorne** A. Blanchard, 1976
10686 **gillaspyi** A. Blanchard, 1976
10687 **viridisparsum** Dod, 1916
10688 **hopfingeri** Franc., 1954
10689 **longipenne** Grt., 1882
10690 **absidum** (Harv., 1875)
   absida B. & McD., 1917, missp.
10691 **sanctaemonicae** Dyar, 1904
10692 **album** (Harv., 1876)
10693 **albisericeum** A. Blanchard, 1976

EUCOPTOCNEMIS Grt., 1874
10694 **fimbriaris** (Gn., 1852)
   obvia (Wlk., 1858)
   sordida Grt., 1893, form
10695 **tripars** (Wlk., 1856)
   worthingtoni (Grt., 1880)
10696 **dapsilis** (Grt., 1882)

MANRUTA Sm., 1903
10697 **elingua** Sm., 1903

TRICHOSILIA Hamp., 1918
10698 **acarnea** (Sm., 1905)

PRORAGROTIS McD., 1929
10699 **longidens** (Sm., 1890)
10700 **dubitata** McD., 1933

MESEMBRAGROTIS B. & Benj., 1927
10701 **ruckesi** B. & Benj., 1927

EUXOA Hbn., 1821
by J. DONALD LAFONTAINE [nos.
   10702–10868]
   METAXYJA Hbn., 1821
   EXARNIS Hbn., 1821
   BROTIS Hbn., 1921
   TELMIA Hbn., 1821
   MIMETES Hbn., 1821, preocc. by
     Eschscholtz, 1818
   METAXYA Wlk., 1857, missp.
   PLEONECTOPODA Grt., 1873
   CARNEADES Grt., 1883
   CHORIZAGROTIS Sm., 1890
   OROSAGROTIS Hamp., 1903
   METAXYIA Hamp., 1903, missp.
   MIMETIS Hamp., 1903, missp.
   PROSAGROTIS Warr., missp.
   MESOEUXOA Corti, 1932
   AGROTIPHILA auth.
   LONGIVESICA Hdwk., 1970

LONGIVESICA Hdwk., 1970,
   subgenus
10702 **divergens** (Wlk., 1857)
   versipellis (Grt., 1875)
   fusimacula (Sm., 1902)
   abar (Stkr., 1899)
   factoris (Sm., 1900)
10703 **sinelinea** Hdwk., 1965
10704 **edictalis** (Sm., 1893)
10705 **messoria** (Harr., 1841)
   spissa (Gn., 1852)
   inextricata (Wlk., 1865)
   indirecta (Wlk., 1865)
   displiciens (Wlk., 1865)
   expulsa (Wlk., 1865)
   ordinata (Wlk., 1865)
   reticens (Wlk., 1865)
   cochranis (Riley, 1867)
   cochrani auth.
   repentis (G. & R., 1868)
   friabilis (Grt., 1875)
   atrifera (Grt., 1878)
   territorialis (Sm., 1900)
   fulda (Sm., 1900)
   pindar (Sm., 1900)
   inordita (B. & Benj., 1927)

PLEONECTOPODA Grt., 1873,
   subgenus
10706 **dissona** (Mösch., 1860)
   opipara (Morr., 1874)
   labradoriensis (Staud., 1881)
   solitaria (Sm., 1888), quest. syn.
10707 **westermanni** (Staud., 1857)
   polaris (B.-H., 1910)
10708 **extranea** (Sm., 1888)
   quinquelinea (Sm., 1890)

10709 **vallus** (Sm., 1900)
   luteomaculata Hdwk., 1966
10710 **luteosita** (Sm., 1900)
10711 **churchillensis** (McD., 1932)
10712 **macleani** McD., 1927
10713 **chimoensis** Hdwk., 1966
10714 **quebecensis** (Sm., 1900)
   moxa Sm., 1907
   quinta Sm., 1908
10715 **scandens** (Riley, 1869)
   fulminans (Grt., 1895)
   elata (Sm., 1898)
10716 **aurulenta** (Sm., 1888)
   aurulentoides (Strand, 1916)
10717 **trifasciata** (Sm., 1888)
10718 **lewisi** (Grt., 1873)
   colata (Grt., 1881)
10719 **juliae** Hdwk., 1968
10720 **altens** McD., 1946
10721 **austrina** Hdwk., 1968
10722 **cryptica** Hdwk., 1968
10723 **tristicula** (Morr., 1876)
   nesilens Sm., 1903
10724 **vetusta** (Wlk., 1865)
   tetra (Wlk., 1865)
   euroides (Grt., 1875)
   perpura (Morr., 1874)
10725 **fuscigera** (Grt., 1874)
   feniseca (Harv., 1875)
   leucopterota McD., 1946, n.
     syn.
10726 **atomaris** (Sm., 1890)
   a. *detesta* (Sm., 1893)
   choris; Sm., 1890
   b. *esta* Sm., 1906
10727 **pleuritica** (Grt., 1876)
   confracta (Sm., 1890)
10728 **pestula** Sm., 1904
   petruska McD., 1932
10729 **simona** McD., 1932
   pleuriticoides Benj., 1936

CHORIZAGROTIS Sm., 1890,
   subgenus
10730 **drewseni** (Staud., 1857)
   thanatologia (Dyar, 1904)
   sordida (Sm., 1908)
   boretha (Sm., 1908)
   perfida Dod, 1916
   pseudovitta Boursin, 1959
10731 **auxiliaris** (Grt., 1873)
   introferens (Grt., 1875)
   agrestis (Grt., 1877)
   flexilis (Sm., 1890)
   tegularis Strand, 1916, ab.,
     infrasubsp.
   tegularis (Draudt, 1924)
   montanus (Cook, 1930)
10732 **inconcinna** (Harv., 1875)
   mercenaria (Grt., 1878)
   sorror (Sm., 1888)
10733 **terrealis** (Grt., 1883)

EUXOA Hbn., 1821, subgenus
10734 **condita** Lafontaine, 1975
10735 **shasta** Lafontaine, 1975
10736 **biformata** Sm., 1910
10737 **intermontana** Lafontaine, 1975
10738 **mimallonis** (Grt., 1873)
   gagates (Grt., 1875)
   rufipennis (Grt., 1875)

caenis (Grt., 1879)
muscosa (Grt., 1883)
lenola Sm., 1910
10739 **septentrionalis** (Wlk., 1865)
incubita (Sm., 1900)
10740 **vernalis** Lafontaine, 1976
10741 **olivia** (Morr., 1876)
lacunosa (Grt., 1878)
segregata (Sm., 1894)
enteridis (Sm., 1900)
vanidica (Sm., 1900)
anacosta Sm., 1905
zembla Sm., 1905
fieldii Dyar, 1908
10742 **terrena** (Sm., 1900)
lagganae (Sm., 1900)
10743 **antica** Lafontaine, 1974
10744 **serricornis** (Sm., 1888)
itodes (Sm., 1900), n. syn.
epictata Sm., 1907, n. syn.
10745 **tocoyae** (Sm., 1900)
sonoma McD., 1941, n. syn.
10746 **scotogrammoides** McD., 1932
10747 **annulipes** (Sm., 1890)
10748 **oncocnemoides** (B. & Benj., 1927)
10749 **intrita** (Morr., 1874)
strigilis (Grt., 1876)
alticola (Sm., 1890), n. syn.
titubatis (Sm., 1894)
reuda (Stkr., 1898)
10750 **rufula** (Sm., 1888)
a. **basiflava** (Sm., 1890)
compressipennis (Sm., 1900), n. syn.
intrusa (Sm., 1900), n. syn.
10751 **silens** (Grt., 1875)
10752 **pimensis** B. & McD., 1910
10753 **immixta** (Grt., 1881)
tetrica (Sm., 1888), n. syn., quest. syn.
knoxvillea McD., 1937, n. syn.
10754 **simulata** McD., 1946
10755 **declarata** (Wlk., 1865)
decolor (Morr., 1874)
spectanda (Sm., 1890)
a. **californica** Hdwk., 1973
10756 **campestris** (Grt., 1875)
10757 **rockburnei** Hdwk., 1973
10758 **flavidens** (Sm., 1888)
intricata Dyar, 1920, n. syn.
10759 **punctigera** (Wlk., 1865)
pastoralis (Grt., 1875)
10760 **aurantiaca** Lafontaine, 1975
10761 **stygialis** B. & McD., 1912
10762 **cana** Lafontaine, 1975
10763 **spumata** McD., 1940
10764 **stigmatalis** (Sm., 1900)
10765 **pallipennis** (Sm., 1888)
alcesta Sm., 1905
10766 **misturata** (Sm., 1890), Morr. mss.
perturbata (Sm., 1890)
candida (Sm., 1894)
gian (Stkr., 1898)
falerina (Sm., 1900)
vertesta Sm., 1910
10767 **melana** Lafontaine, 1975
10768 **atristrigata** (Sm., 1890)
collocata (Sm., 1894)
10769 **nevada** (Sm., 1900)
floramina (Sm., 1905)
10770 **cinereopallida** (Sm., 1903)

10771 **serotina** Lafontaine, 1976
10772 **camalpa** (Dyar, 1912)
clavigera (Dyar, 1922)
a. **manca** Benj., 1936
10773 **maderensis** Lafontaine, 1976
10774 **mitis** (Sm., 1894)
colla (Stkr., 1898), n. syn.
ura Sm., 1905, n. syn.
uramina Sm., 1905, n. syn.
10775 **luctuosa** Lafontaine, 1976
10776 **aequalis** (Harv., 1876)
10777 **acornis** (Sm., 1895)
megastigma (Sm., 1900)
testula (Sm., 1900)
a. **alko** (Stkr., 1899)
naevulus (Sm., 1900)
termessus (Sm., 1900)
sessilis (Sm., 1900)
10778 **conjuncta** (Sm., 1890)
10779 **cona** (Stkr., 1898)
sotnia Sm., 1905
10780 **comosa** (Morr., 1876)
atropulverea (Sm., 1900), n. syn.
ternaria (Sm., 1900), n. syn.
masoni Ckll., 1905, n. syn.
a. **lutulenta** (Sm., 1895), n. stat.
vulpina (Sm., 1890), n. syn.
latebra Benj., 1936, n. syn.
luteotincta McD., 1940, n. syn.
johnstoni McD., 1946, n. syn.
b. **annir** (Stkr., 1898), n. stat.
dakota (Sm., 1900), n. syn.
c. **altera** McD., 1940, n. stat.
d. **ontario** (Sm., 1900), n. stat.
vestitura Sm., 1905
10781 **fumalis** (Grt., 1873)
permunda (Morr., 1874)
10782 **lucida** B. & McD., 1912
10783 **incallida** (Sm., 1890)
10784 **lineifrons** (Sm., 1890)
audentis (Sm., 1894)
10785 **infausta** (Wlk., 1865)
brunneigera (Grt., 1876), n. syn.
rubefactalis (Grt., 1881), n. syn.
sponsa (Sm., 1888), n. syn.
numa (Stkr., 1898)
relaxa (Sm., 1900), n. syn.
holoberba (Sm., 1900), n. syn.
10786 **satis** (Harv., 1876)
micronyx (Grt., 1878), n. syn.
perfusca (Grt., 1883), n. syn.
cocklei Sm., 1908, n. syn.
10787 **loya** (Sm., 1900)
monteclara (Sm., 1906), n. syn.
obscura Hill, 1924, n. syn.
a. **excogita** (Sm., 1900), n. stat.
10788 **bicollaris** (Grt., 1878)
abnormis (Sm., 1890), n. syn.
10789 **inyoca** Benj., 1936
10790 **selenis** (Sm., 1900)
claromonta Sm., 1906
10791 **piniae** Buckett & Bauer, 1964
10792 **satiens** (Sm., 1890)
10793 **scholastica** McD., 1920
10794 **setonia** McD., 1927
10795 **pluralis** (Grt., 1878)
10796 **bifasciata** (Sm., 1888)
lowensis Benj., 1936, n. syn.
bisagittifera Benj., 1936, n. syn.
10797 **cinnabarina** B. & McD., 1918

10798 **basalis** (Grt., 1879)
10799 **permixta** McD., 1940
coloradensis Strand, 1916, rev. stat., infrasubsp. name
10800 **nostra** (Sm., 1890)
10801 **ochrogaster** (Gn., 1852)
insignata (Walker, 1857), preocc. by Wlk., 1857
illata (Wlk., 1857)
subsignata (Wlk., 1865)
cinereomacula (Morr., 1874)
turris (Grt., 1875)
gularis (Grt., 1875)
10802 **wirima** Hdwk., 1965
10803 **velleripennis** (Grt., 1874)
10804 **plagigera** (Morr., 1874)
10805 **tessellata** (Harr., 1841)
insulsa (Wlk., 1856)
insignata (Wlk., 1857)
perlentans (Wlk., 1857)
maizi (Fitch, 1864)
atropurpurea (Grt., 1877)
tesselloides (Grt., 1882)
finis (Sm., 1888)
remota (Sm., 1890), n. syn.
flaviscapula (Sm., 1900)
noctuiformis (Sm., 1900), n. syn.
neotelis (Sm., 1900)
atrofusca (Sm., 1900), n. syn.
objurgata (Sm., 1900)
cariosa (Sm., 1900)
nordica (Sm., 1900)
caesia (Sm., 1900), n. syn.
acutifrons (Sm., 1900)
laminis (Sm., 1900)
focina (Sm., 1903)
marinensis McD., 1941, n. syn.
10806 **henrietta** (Sm., 1900)
delicata B. & McD., 1912, n. syn.
adusta B. & McD., 1912
10807 **albipennis** (Grt., 1876)
verticalis (Grt., 1880)
nigripennis (Grt., 1881)
malis (Sm., 1900)
bialba Sm., 1905
indensa Sm., 1910
10808 **lillooet** McD., 1927
10809 **catenula** (Grt., 1879)
contagionis (Sm., 1900)
lindseyi Blkmre., 1923, n. syn.
10810 **violaris** (G. & R., 1868)
10811 **siccata** (Sm., 1893)
placida B. & McD., 1912, n. syn.
10812 **bostoniensis** (Grt., 1874)
10813 **medialis** (Sm., 1888)
kerrvillei (Sm., 1900)
poncha Sm., 1910
truva Sm., 1910
discilinea Dyar, 1918, n. syn.
rufosuffusata McD., 1940, n. syn.
10814 **ustulata** Lafontaine, 1976
10815 **sculptilis** (Harv., 1875)
xyliniformis (Sm., 1890)
10816 **perexcellens** (Grt., 1875)
excellens (Grt., 1875), preocc. by Staud., 1867
infelix (Sm., 1890)
10817 **obeliscoides** (Gn., 1852)
sexatilis (Grt., 1873)
infusa (Sm., 1902)

10818 **oberfoelli** Hdwk., 1973
10819 **choris** (Harv., 1876)
    cogitans (Sm., 1890)
    achor (Stkr., 1899)
10820 **hollemani** (Grt., 1874)
10821 **xasta** B. & McD., 1910
10822 **andera** Sm., 1910
10823 **cincta** B. & Benj., 1924
10824 **brevipennis** (Sm., 1888)
    brevistriga Sm., 1910
    angulirena Sm., 1910
10825 **costata** (Grt., 1876)
10826 **idahoensis** (Grt., 1878)
    furtiva (Sm., 1890)
10827 **clausa** McD., 1923
10828 **foeminalis** (Sm., 1900)
10829 **laetificans** (Sm., 1894)
    masculina (Sm., 1903)
10830 **quadridentata** (G. & R., 1865)
    pugionis (Sm., 1900)
    a. **flutea** Sm., 1910
10831 **niveilinea** (Grt., 1882)
    rabiata Sm., 1910, n. syn.
10832 **agema** (Stkr., 1899)
10833 **olivalis** (Grt., 1879)
    mcdunnoughi Cook, 1930
10834 **oblongistigma** (Sm., 1888)
10835 **melura** McD., 1932
10836 **dargo** (Stkr., 1898)
    rumatana (Sm., 1903)
10837 **unica** McD., 1940
10838 **detersa** (Wlk., 1856)
    pitychrous (Grt., 1873)
    a. **personata** (Grt., 1876)
    azif (Stkr., 1898)
10839 **cicatricosa** (G. & R., 1865)
    neomexicana (Sm., 1890)
    ducalis (Sm., 1907)
    teplia Sm., 1910
10840 **recula** (Harv., 1876)
10841 **citricolor** (Grt., 1880)
    postmedialis Strand, 1916, ab., infrasubsp.
    postmedialis Draudt, 1924
10842 **tronella** (Sm., 1903)
10843 **teleboa** (Sm., 1890)
    pedalis (Sm., 1890), n. syn.
    recticincta (Sm., 1895), n. syn.
10844 **mercedes** B. & McD., 1912
10845 **difformis** (Sm., 1900)
10846 **murdocki** (Sm., 1890)
10847 **moerens** (Grt., 1883)
    luteola (Sm., 1888)
10848 **latro** (B. & Benj., 1927)
10849 **dodi** McD., 1923
10850 **infracta** (Morr., 1875)
10851 **redimicula** (Morr., 1874)
10852 **auripennis** Lafontaine, 1974
10853 **arizonensis** Lafontaine, 1974
10854 **servita** (Sm., 1895)
    a. **novangliae** McD., 1950, n. stat.
10855 **munis** (Grt., 1879)
    sublatis (Grt., 1880)
    rena (Sm., 1890), n. syn.
    dolens (Sm., 1908), n. syn.
    cervinea Sm., 1910, n. syn.

OROSAGROTIS Hamp., 1903, subgenus
10856 **montana** (Morr., 1875)
    rigida (Sm., 1891)

10857 **taura** Sm., 1905
10858 **cooki** McD., 1925
    orbicularis (Sm., 1888), preocc. by
    Wlk., 1865
10859 **macrodentata** Hdwk., 1965
10860 **perolivalis** (Sm., 1905)
10861 **ridingsiana** (Grt., 1875)
    maimes (Sm., 1903)
10862 **aberrans** McD., 1932
10863 **manitobana** McD., 1925
10864 **flavicollis** (Sm., 1888)
10865 **perpolita** (Morr., 1876)
    exculta (Sm., 1900)
    criddlei Sm., 1908
10866 **incognita** (Sm., 1893)
10867 **wilsoni** (Grt., 1873)
    specialis (Grt., 1874)
10868 **riversii** (Dyar, 1903)

Aniclini

LOXAGROTIS McD., 1929
10869 **grotei** Franc. & Todd, n. name
    apicalis (Grt., 1881), preocc. by
    H.-S., 1868
10870 **acclivis** (Morr., 1875)
    reclivis (Dyar, 1907)
    a. **opaca** (Harv., 1875)
10871 **proclivis** (Sm., 1888)
    oaxacana (Schaus, 1898)
10972 **neoclivis** (B. & Benj., 1924)
10873 **salina** (Barnes, 1904)
    arabella (Dyar, 1910)
10874 **socorro** (Barnes, 1904)
10875 **pampolycala** (Dyar, 1912)
10876 **capnota** (Sm., 1908)
10877 **timbor** (Dyar, 1919), n. record
10878 **albicosta** (Sm., 1888)
10879 **serano** (Sm., 1910)
10880 **kyune** (Barnes, 1904)

RICHIA Grt., 1887
10881 **chortalis** (Harv., 1875)
    aratrix (Harv., 1875), form
10882 **parentalis** (Grt., 1879)
    decipiens (Grt., 1879), form
10883 **distichoides** (Grt., 1883)
10884 **lobato** (Barnes, 1904)
    lobata auth., missp.
10885 **hahama** (Dyar, 1919), rev. stat.
10886 **larga** (Sm., 1908), n. comb.
10887 **cyminopristes** (Dyar, 1912), n. comb.,
    n. record
10888 **pyrsogramma** (Dyar, 1916), n. comb.

PSEUDORTHOSIA Grt., 1874
10889 **variabilis** Grt., 1874
    a. **pallidior** Ckll., 1906

PSEUDOSEPTIS McD., 1929
10890 **grandipennis** (Grt., 1883)
    jalapa (Hamp., 1903), quest. syn.

OCHROPLEURA Hbn., 1821
10891 **plecta** (L., 1761)
    vicaria Wlk., 1856
    unimacula (Stgr., 1859)
    costalis Moore, 1867
    anderssoni (Lampa, 1885)
    glaucimacula (Graeser, 1888)

    ignota Swinhoe, 1889
    plectella (Strand, 1916)

PROTOGYGIA McD., 1929
10892 **lagena** (Grt., 1875)
10893 **enalaga** McD., 1932
10894 **querula** (Dod, 1915)
10895 **comstocki** McD., 1934
10896 **polingi** (B. & Benj., 1922)
10897 **epipsilioides** (B. & Benj., 1926)
10898 **milleri** (Grt., 1876)
10899 **biclavis** (Grt., 1879)
    demutabilis (Sm., 1893)
10900 **elevata** (Sm., 1891)
    terrifica (Sm., 1893)

EUAGROTIS McD., 1929
10901 **lubricans** (Gn., 1852)
    associans (Wlk., 1858)
    spreta (Sm., 1902)
10902 **forbesi** Franc., 1952
10903 **illapsa** (Wlk., 1857)
10904 **beata** (Grt., 1883)
10905 **exuberans** (Sm., 1898)
10906 **bairdii** (Sm., 1908)
10907 **simplicia** (Morr., 1874)
    simplaria (Morr., 1875), missp.
10908 **digna** (Morr., 1875)
    nigrovittata (Grt., 1876)
10909 **tenuescens** (Sm., 1890)
10910 **tepperi** (Sm., 1888)
    atricincta (Sm., 1895)

ANICLA Grt., 1874
    ANIDA Grt., 1875, missp.
    ANCIOLA Neave, 1939, missp.
10911 **infecta** (Ochs., 1816)
    praecox; Hbn., 1803–08, not L., 1758
    incivis (Gn., 1852)
    praecipua (Wlk., 1869)
    bartolemica (Wallgr., 1871)
    alabamae (Grt., 1874)
    pauper (Butler, 1878)
    incisa (Gundlach, 1881)
    mulina (Mösch., 1886)
    mahalpa Schaus, 1898
10912 **cemolia** Franc., 1967

CRASSIVESICA Hdwk., 1970
10913 **bocha** (Morr., 1874)
    brochus (Morr., 1875), missp.

HEMIEUXOA McD., 1929
10914 **rudens** (Harv., 1875)
    pellucidalis (Grt., 1882), form

PERIDROMA Hbn., 1821
10915 **saucia** (Hbn., 1803–08)
    margaritosa (Haw., 1809)
    majuscula (Haw., 1809)
    aegua (Hbn., 1809–13)
    stictica (Blanchard, 1854)
    infuscata (Blanchard, 1854)
    impacta (Wlk., 1856)
    intecta (Wlk., 1856)
    ambrosioedes (Wlk., 1857)
    angulifera (Wallgr., 1860)
    ortonii (Pack., 1869)
    ochronota (Hamp., 1903)
    fuscobrunnea (Strand, 1916)

porphyrea; Edelsten, 1939, not D. &
   S., 1775

DIARSIA Hbn., 1821
10916 **calgary** (Sm., 1898)
10917 **rubifera** (Grt., 1875)
   cynica (Sm., 1898)
   perversa (Strand, 1916)
   a. **perumbrosa** (Dyar, 1904)
      umbrosa (Dyar, 1904), preocc. by
      Hbn., 1809–13
10918 **dislocata** (Sm., 1904)
10919 **jucunda** (Wlk., 1857)
   perconflua (Grt., 1876), form
   eriensis (Grt., 1878), form
   hospitalis (Grt., 1882), form, n.
      syn.
10920 **esurialis** (Grt., 1881)
   a. **uclueleti** (B. & Benj., 1929)
10921 **rosaria** (Grt., 1878)
   pseudorosaria Hdwk., 1950
10922 **freemani** Hdwk., 1950, rev. status

Noctuini

PROTEXARNIS McD., 1929
10923 **balanitis** (Grt., 1873)

ACTEBIA Steph., 1829
   HAPALIA Hbn., 1821, preocc. by
      Hbn., 1818
   ACTOBIA Agassiz, 1846, emend.
10924 **fennica** (Tauscher, 1806)
   intracta (Wlk., 1857)

RHYACIA Hbn., 1821
   EPIPSILIA Hbn., 1821
   EPISILIA Hamp., 1903, missp.
10925 **quadrangula** (Zett., 1839)
   rava (H.-S., 1850)
   umbratus (Pack., 1867)
   a. **clemens** (Sm., 1890)

SPAELOTIS Bdv., 1840
10926 **clandestina** (Harr., 1862)
   unicolor (Wlk., 1856)
   nigriceps (Wlk., 1865)
10927 **havilae** (Grt., 1880)

GRAPHIPHORA Ochs., 1816
   GRAPHOPHORA Agassiz, 1846,
      emend.
   PSEUDOSPAELOTIS McD., 1929
10928 **haruspica** (Grt., 1875)
   unimacula (Morr., 1874), preocc. by
      Stgr., 1859
   grandis (Speyer, 1875)
   a. **sierrae** (Harv., 1876)
   b. **inopinatus** (Sm., 1898)

EUROIS Hbn., 1821
   EURHOIS Agassiz, 1846, emend.
10929 **occulta** (L., 1758)
   passetii (Thierry-Mieg, 1886)
   a. **implicata** (Lefebvre, 1836)
10930 **astricta** Morr., 1874
   a. **subjugata** Dyar, 1904
   b. **elenae** (B. & Benj., 1927)
10931 **nigra** (Sm., 1892)
   a. **argni** (B. & Benj., 1927)

10932 **docilis** (Grt., 1881)
   ingeniculata (Sm., 1890)
10933 **praefixa** (Morr., 1875)

BARROVIA B. & McD., 1916
10934 **fasciata** (Skin., 1902)

PARABARROVIA Gibson, 1920
10935 **keelei** Gibson, 1920

PACHNOBIA Gn., 1852
10936 **tecta** (Hbn., 1803–08)
   carnea; auth., not Thunb., 1788
   a. **roosta** Sm., 1903
      tectoides (Corti, 1926)
10937 **wockei** (Mösch., 1862)
10938 **scropulana** (Morr., 1874)
   moeschleri (B.-H., 1910)
10939 **okakensis** (Pack., 1867)
   cinerea (Stgr., 1871)
10940 **morandi** Benj., 1934

XESTIA Hbn., 1818
   AMATHES Hbn., 1821
   MEGASEMA Hbn., 1821
   LYTAEA Steph., 1829
   SEGETIA Steph., 1829
10941 **bolteri** (Sm., 1898), n. comb.
10942 **adela** Franc., 1980
   c-nigrum; auth., part
10942.1 **dolosa** Franc., 1980
   c-nigrum; auth., part
10943 **normaniana** (Grt., 1874), n. comb.
   triangulum var. A (Gn., 1852)
   triangulum; Grt., 1874, not Hufn.,
      1766
   obtusa (Speyer, 1875)
10944 **smithii** (Snell., 1896), n. comb.
   baja; auth., not F., 1787
10945 **trumani** (Sm., 1903), n. comb. (equal
      xanthographa D. & S., 1775?)
10946 **conchis** (Grt., 1879), n. comb.
10947 **oblata** (Morr., 1875), n. comb.
   hilliana (Harv., 1878)
   a. **streckeri** (B. & Benj., 1927)
10948 **substrigata** (Sm., 1895), n. comb.
10949 **cinerascens** (Sm., 1891), n. comb.
10950 **bicarnea** (Gn., 1852), n. comb.
   plagiata (Wlk., 1865)
10951 **tenuicula** (Morr., 1874), n. comb.
   treati (Grt., 1875)
10952 **flavotincta** (Sm., 1892), n. comb.
   a. **evelynae** (B. & Benj., 1927)
10953 **scaramangoides** (B. & Benj., 1926),
      n. comb. (equal form of *flavotincta*?)
10954 **collaris** (G. & R., 1868), n. comb.
10955 **badinodis** (Grt., 1874), n. comb.
10956 **bollii** (Grt., 1881), n. comb.
   hilaris (Grt., 1880), preocc. by
      Freyer, 1838

ANOMOGYNA Stgr., 1871
   PTEROSCIA Morr., 1874
   PLATAGROTIS Sm., 1890
   ANOMAGYNA Benj., 1933 missp.
10957 **atrata** (Morr., 1874)
   a. **yukona** (McD., 1921)
   b. **ursae** McD., 1940
10958 **fabulosa** Fgn., 1965
   sincera; auth., not H.-S., 1851

10959 **speciosa** (Hbn., 1809–13)
   a. **arctica** (Zett., 1839)
   b. **mixta** (Wlk., 1856)
      livalis (Sm., 1910)
10960 **apropitia** Benj., 1933
10961 **aklavicensis** Benj., 1933
10962 **perquiritata** (Morr., 1874)
   baileyana (Grt., 1880)
   a. **partita** McD., 1921
   b. **beddeki** (Hamp., 1913)
      beddeci (Hamp., 1913), incorr.
      orig. spell.
   c. **clarkei** Benj., 1934
10963 **laetabilis** (Zett., 1839)
10964 **homogena** McD., 1921
   a. **conditoides** Benj., 1934
10965 **imperita** (Hbn., 1827–31)
   comparata (Mösch., 1862)
   saxigena (Morr., 1874)
   a. **discitincta** (Wlk., 1856)
      arufa (Sm., 1905)
10966 **mallochi** Benj., 1933
   arufoides Benj., 1933, form
   a. **stejnigeri** Benj., 1933
10967 **elimata** (Gn., 1852)
   grisatra (Sm., 1907), rev. stat.
10968 **badicollis** (Grt., 1873), rev. stat.
10969 **dilucida** (Morr., 1875), rev. stat.
   janualis (Grt., 1878), rev. stat.
10970 **youngii** (Sm., 1902)
   a. **atoma** (Sm., 1907)
10971 **mustelina** (Sm., 1900)
   occidens (Hamp., 1913)
10972 **infimatis** (Grt., 1880)
   dernarius (Sm., 1907), rev. stat.
10973 **vernilis** (Grt., 1879)
   filiis (Sm., 1907)

SETAGROTIS Sm., 1890
10974 **planifrons** (Sm., 1890)
10975 **cinereicollis** (Grt., 1876)
   pallidicollis (Grt., 1880)
   congrua (Sm., 1890)
   a. **vocalis** (Grt., 1879)
      invenusta (Grt., 1883), form
10976 **radiatus** (Sm., 1900)
   radiola (Hamp., 1903)
10977 **atrifrons** (Grt., 1873)
10978 **piscipellis** (Grt., 1878)
   a. **amia** (Dyar, 1907)
   b. **exculpatrix** (Dyar, 1907)
   c. **corrodera** (Sm., 1907)
   d. **cinibarina** (Hill, 1924)
10979 **fortiter** (B. & McD., 1918)

AGROTIPHILA Grt., 1875
   AGROTIMORPHA B. & Benj., 1929
10980 **staudingeri** (Mösch., 1862)
10981 **maculata** Sm., 1893
   maculate Sm., 1894, missp.
10982 **colorado** Sm., 1891
   montana; Grt., 1875, not Morr., 1875

EPIPSILIAMORPHA B. & Benj.,
   1929
10983 **alaskae** (Grt., 1876)
10984 **aequaeva** Benj., 1934

ARCHANARTA B. & Benj., 1929
10985 **quieta** (Hbn., 1809–13)

schoenherri (Zett., 1839)
    a. **constricta** (Wlk., 1857)
      rigida (Wlk., 1857)
10986 **bryanti** Benj., 1933
10987 **acraea** Benj., 1934

EUGRAPHE Hbn., 1821
  COENOPHILA Steph., 1850
  AMMOGROTIS Stgr., 1895
10988 **subrosea** (Steph., 1829), extralim.
    subcaerulea (Stgr., 1871)
    latefasciata (Hoyningen-Huene, 1901)
    kieferi (Rebel, 1912)
    decipiens (Warnecke, 1924)
    rubifera (Warnecke, 1933)
    a. **opacifrons** (Grt., 1878)

ADELPHAGROTIS Sm., 1890
10989 **stellaris** (Grt., 1880)
10990 **quarta** (Grt., 1881)
10991 **indeterminata** (Wlk., 1865)
    washingtoniensis (Grt., 1881)
    a. **innotabilis** (Grt., 1874)

PARADIARSIA McD., 1929
10992 **littoralis** (Pack., 1867)
    a. **pectinata** (Grt., 1874)
      ferruginoides (Sm., 1890)

HEMIPACHNOBIA McD., 1929
10993 **subporphyrea** (Wlk., 1858)
    a. **monochromatea** (Morr., 1874)

CERASTIS Ochs., 1816
  CERASTIA Steph., 1850, emend.
  SORA Heinemann, 1859, preocc. by Wlk., 1859
  MATUTA Grt., 1874, preocc. by Weber, 1795
  GYPSITEA Tams, 1939
10994 **tenebrifera** (Wlk., 1865)
    catherina (Grt., 1874)
    manifestolabes (Morr., 1874)

METALEPSIS Grt., 1875
10995 **cornuta** (Grt., 1874)
10996 **salicarum** (Wlk., 1857)
    claviformis (Morr., 1874)
    orilliana (Grt., 1875)
10997 **fishii** (Grt., 1878)
    heinrichi (B. & Benj., 1929)

CHOEPHORA G. & R., 1868
10998 **fungorum** G. & R., 1868

APLECTOIDES Butler, 1878
10999 **condita** (Gn., 1852)
    trabalis (Grt., 1877)

ANAPLECTOIDES McD., 1929
11000 **prasina** (D. & S., 1775)
    herbacea (Gn., 1852)
11001 **pressus** (Grt., 1874)
    a. **fales** (Sm., 1905)
    b. **discolor** (Sm., 1905)
    c. **abbea** (Sm., 1908)
11002 **brunneomedia** McD., 1946
    intermedia (Fbs., 1954), *laps. cal.*

CHERSOTIS Bdv., 1840
11003 **juncta** (Grt., 1878)
    a. **patefacta** (Sm., 1895)

PROTOLAMPRA McD., 1929
11004 **rufipectus** (Morr., 1875)
11005 **hero** (Morr., 1876)
11006 **brunneicollis** (Grt., 1865)

EUERETAGROTIS Sm., 1890
11007 **sigmoides** (Gn., 1852)
11008 **perattenta** (Grt., 1876)
    a. **inattenta** (Sm., 1903)
11009 **attenta** (Grt., 1874)

HEPTAGROTIS McD., 1929
11010 **phyllophora** (Grt., 1874)

HEMIGRAPHIPHORA McD., 1929
11011 **plebeia** (Sm., 1898)
    a. **bajoides** (B. & Benj., 1929)

CRYPTOCALA Benj., 1921
11012 **acadiensis** (Bethune, 1870)
    gilvipennis (Grt., 1874)

ABAGROTIS Sm., 1890
11013 **erratica** (Sm., 1890)
    a. **ornata** Sm., 1903
11014 **kirkwoodi** Buckett, 1968
11015 **alcandola** (Sm., 1908)
    tristis B. & McD., 1912
11016 **vittifrons** (Grt., 1864)
11017 **bimarginalis** (Grt., 1883)
11018 **trigona** (Sm., 1893)
    sambo (Sm., 1908)
11019 **mirabilis** (Grt., 1879)
11020 **rubricundis** Buckett, 1968
11021 **striata** Buckett, 1968
11022 **glenni** Buckett, 1968
11023 **hennei** Buckett, 1968
11024 **nefascia** (Sm., 1908)
    negascia (Sm., 1908), incorr. orig. spell.
    forbesi (Benj., 1921)
11025 **alampeta** Franc., 1967
11026 **denticulata** McD., 1946
11027 **barnesi** (Benj., 1921)
    nevadensis (Benj., 1921)
11028 **baueri** McD., 1949
11029 **alternata** (Grt., 1864)
    alternatella (Strand, 1916)
    uniformis (Strand, 1916)
11030 **turbulenta** McD., 1927
11031 **crumbi** Franc., 1955
    a. **benjamini** Franc., 1955
11032 **variata** (Grt., 1876)
    varix (Grt., 1876)
    orbis (Grt., 1876)
    orbitis (Strand, 1916)
11033 **scopeops** (Dyar, 1904)
11034 **tecatensis** Buckett, 1968
11035 **discoidalis** (Grt., 1876)
11036 **pulchrata** (Blkmre., 1925)
11037 **apposita** (Grt., 1878)
11038 **nanalis** (Grt., 1881)
    mantalini (Sm., 1894)
11039 **duanca** (Sm., 1908)
11040 **reedi** Buckett, 1969

11041 **placida** (Grt., 1876)
    minimalis (Grt., 1879)
11042 **dodi** McD., 1927

RHYNCHAGROTIS Sm., 1890
11043 **cupida** (Grt., 1865)
    velata (Wlk., 1865)
11044 **adulta** (Gn., 1852), n. comb.
    brunneipennis (Grt., 1875), n. syn. & n. stat.
11045 **anchocelioides** (Gn., 1852)
11046 **belfragei** Sm., 1890
11047 **exertistigma** (Morr., 1874)
    formalis (Grt., 1874), form
    observabilis (Grt., 1875), form
    carissima (Harv., 1875), form
    cupidissima (Grt., 1875), form
    laetula (Grt., 1876)
    distracta Sm., 1890
    falcula (Grt., 1876), form
    emarginata (Grt., 1876), form
    inelegans (Sm., 1890), form
    faculana (Strand, 1916)
    morrisonistigma (Grt., 1876), form
    binominalis (Sm., 1888)
    crenulata (Sm., 1888)
    niger Sm., 1903, form
    meta Sm., 1903, form
    faculoides (Strand, 1916), form
    faculella (Strand, 1916), form
11048 **insularis** (Grt., 1876)
    confusa (Sm., 1888), form

PRONOCTUA Sm., 1894
11049 **typica** Sm., 1894
    peabodyae (Dyar, 1903), form
11050 **pyrophiloides** (Harv., 1876)

Ufeini

UFEUS Grt., 1873
11051 **satyricus** Grt., 1873
    a. **sagittarius** Grt., 1883
11052 **plicatus** Grt., 1873
    barometricus (Goossens, 1881)
    a. **electra** Sm., 1908
    b. **hulsti** Sm., 1908
11053 **faunus** Stkr., 1898
11054 **unicolor** Grt., 1878
    a. **coloradica** Strand, 1916

HELIOTHINAE

DERRIMA Wlk., 1858
  PHILOMMA Grt., 1864
11055 **stellata** Wlk., 1858
    henrietta (Grt., 1864)
    cinocentralis Strand, 1912

MICROHELIA Hamp., 1910
11056 **angelica** (Sm., 1900)
11057 **restrictalis** (Sm., 1900)
    immacula Strand, 1912

HELIOTHODES Hamp., 1910
11058 **diminutivus** (Grt., 1873)
    suffusana Strand, 1912
    macromacula Strand, 1912
    bifida Strand, 1912
11059 **joaquin** McD., 1946

11060 **fasciatus** (Hy. Edw., 1875)
    sabulosa (Sm., 1906)

BAPTARMA Sm., 1904
11061 **felicita** Sm., 1904

EUTRICOPIS Morr., 1875
11062 **nexilis** Morr., 1875
    elaborata (Hy. Edw., 1881)
    a. **subcolorata** B. & McD., 1918

PYRRHIA Hbn., 1821
11063 **umbra** (Hufn., 1766)
    rutilago (D. & S., 1775)
    marginata (F., 1775)
    conspicua (Bkh., 1792)
    umbrago (Esp., 1798)
    marginago (Haw., 1809)
    tibetana (Moore, 1878)
    vexilliger (Christoph, 1893)
    aconiti Höltzermann, 1902
11064 **exprimens** (Wlk., 1857)
    angulata Grt., 1874
    stilla Grt., 1879

RHODOECIA Hamp., 1910
11065 **aurantiago** (Gn., 1852)
    illiterata (Grt., 1875)
    differta (Morr., 1875)
    illinoisensis (French, 1879)

ERYTHROECIA Hamp., 1910
11066 **suavis** (Hy. Edw., 1884)
11067 **hebardi** Skin., 1917

HELIOTHIS Ochs., 1816

    HELICOVERPA Hdwk., 1965,
    subgenus
11068 **zea** (Boddie, 1850)
    obsoleta (F., 1793), technically
      preocc. by F., 1775
    umbrosus Grt., 1862
    ochracea Ckll., 1889
    armigera; auth., not Hbn., 1803–08
11069 **stombleri** Okumura & Bauer, 1969

    HELIOTHIS Ochs., 1816, subgenus
    HELIOTHIS Hbn., 1806, suppr.
      (ICZN Op. 97 & 278)
    HELIOTIS Lefebvre, 1827,
      missp.
    HELIOTHISA Meigen, 1832,
      emend.
    HELIOTIS Sodoffsky, 1837,
      emend.
    CHLORIDEA Duncan and
      [Westwood], 1841
    ASPILA Gn., 1842, preocc. by
      Steph., 1834
    TIMORA; auth., not Wlk., 1856
    HELIOCHEILUS Grt., 1865
    RHODOSEA Grt., 1883
    DISOCNEMIS Grt., 1883
    DYSOCNEMIS Grt., 1883,
      emend.
    CANTHYLIDIA Butler, 1886
    HELIOTHRIS Olliff, 1890, missp.
11070 **subflexus** Gn., 1852
11071 **virescens** (F., 1777)

rhexiae (J. E. Smith, 1797)
    viridescens (Wlk., 1857)
    prasina (Wlk., 1857)
    spectanda (Stkr., 1876)
11072 **phloxiphagus** G. & R., 1867
    phlogophagus G. & R., 1870, emend.
    interjacens Grt., 1880
    acesias Felder & Rogenhofer, 1874,
      form
    luteitinctus Grt., 1875
11073 **turbatus** (Wlk., 1858)
    albidentina (Wlk., 1865)
    lupatus Grt., 1875
11074 **paradoxus** (Grt., 1865)
    hyperfusca (Strand, 1916)
11075 **julia** (Grt., 1883)
    grandis (Druce, 1890)
11076 **toralis** (Grt., 1881)
    rosario (Barnes, 1904)
    unicolor (Walter, 1928)
11077 **ononis** (F., 1787)
    ononis (D. & S., 1775), nom. nud.
    ononidis Gn., 1852, missp.
    septentrionalis (Hy. Edw., 1884)
    intensiva (Warr., 1911)
    lugubris Klemensiewicz, 1912
11078 **oregonicus** (Hy. Edw., 1875)
11079 **proruptus** Grt., 1873
    venustus (Hy. Edw., 1875)
    fimbria (Williams, 1905)
11080 **belladona** (Hy. Edw., 1881)
    anartoides (Strand, 1915)
11081 **borealis** (Hamp., 1903)

PROTOSCHINIA Hdwk., 1970
11082 **scutosa** (F., 1787)
    nuchalis (Grt., 1878)

SCHINIA Hbn., 1818
    EUCLIDIA Hbn., 1808, suppr.
      (ICZN Op. 789)
    MELICLEPTRIA Hbn., 1823
    ANTHOECIA Bdv., 1840
    ALARIA Duncan and [Westwood],
      1841
    TRYPANA Gn., 1841
    ORIA Gn., 1852, preocc. by Hbn.,
      1821
    RHODOPHORA Gn., 1852
    TAMILA Gn., 1852
    EULEUCYPTERA Grt., 1865
    LYGRANTHOECIA G. & R., 1873
    TRICOPIS Grt., 1874
    OXYLOS Grt., 1875
    HELIOPHANA Grt., 1875
    ADONISEA Grt., 1875
    PIPPONA Harv., 1875
    RHODODIPSA Grt., 1877
    PORRIMA Grt., 1877
    BESSULA Grt., 1881
    DASYPOUDAEA Sm., 1883
    PSEUDOTAMILA Sm., 1883
    TRILEUCA Grt., 1883
    EUPANYCHIS Grt., 1890
    CANIDIA Grt., 1890
    TRICHOSELLUS Grt., 1890
    INCITA Grt., 1895
    PALADA Sm., 1900
    CHLOROCLEPTRIA Hamp., 1903
    TRILENCA Neave, 1940, missp.

11083 **villosa** (Grt., 1864)
    pauxillus (Grt., 1875)
    a. **sexata** (Sm., 1906)
    b. **subatra** (Sm., 1906)
    c. **intermontana** Hdwk., 1958
11084 **triolata** (Sm., 1906)
11085 **steenensis** Hdwk., 1973
11086 **graefiana** (Tepper, 1882)
11087 **vacciniae** (Hy. Edw., 1875)
    vanella (Grt., 1879)
11088 **sueta** (Grt., 1873)
    a. **californica** (Grt., 1873)
    b. **sierrae** Hdwk., 1958
    c. **martini** Hdwk., 1958
    d. **aetheria** (B. & McD., 1912)
11089 **aurantiaca** (Hy. Edw., 1881)
    californica (Hamp., 1903)
    a. **tenuimargo** (B. & McD., 1913)
11090 **amaryllis** (Sm., 1891)
11091 **perminuta** (Hy. Edw., 1881)
    dubitans (Tepper, 1883)
11092 **roseitincta** (Harv., 1875)
    exalta (Hy. Edw., 1885)
11093 **antonio** (Sm., 1906)
11094 **carminatra** (Sm., 1903)
11095 **indiana** (Sm., 1908)
11096 **scarletina** (Sm., 1900)
11097 **pulchripennis** (Grt., 1874)
    languida (Hy. Edw., 1881)
11098 **scissa** (Grt., 1876)
11099 **scissoides** (Benj., 1936)
11100 **avemensis** (Dyar, 1904)
11101 **dobla** (Sm., 1906)
11102 **honesta** (Grt., 1881)
    kasloa (Sm., 1903)
11103 **persimilis** (Grt., 1873)
    flavidenta (Sm., 1906)
    coloradica (Strand, 1916)
11104 **spinosae** (Gn., 1852)
    hirtella (G. & R., 1866)
    camina (Sm., 1906)
11105 **bina** (Gn., 1852)
    meskeana (Grt., 1875)
    fastidiosa (Stkr., 1876)
    rufimedia (Grt., 1880)
11106 **volupia** (Fitch, 1868)
    volupides (Strand, 1916)
11107 **fulleri** (McElvare, 1961)
11108 **masoni** (Sm., 1896)
11109 **aden** (Stkr., 1898)
11110 **septentrionalis** (Wlk., 1858)
    atrites (Grt., 1864)
    brevis (Grt., 1865)
11111 **approximata** Stkr., 1898
11112 **sordida** Sm., 1883
11113 **petulans** (Hy. Edw., 1885)
11114 **ar** Stkr., 1898
11115 **siren** (Stkr., 1876)
11116 **tuberculum** (Hbn., 1827–31)
    dorsilutea (Wlk., 1857)
11117 **lynx** (Gn., 1852)
11118 **obscurata** Stkr., 1898
11119 **tanena** Stkr, 1898
11120 **parmeliana** (Hy. Edw., 1882)
11121 **concinna** Sm., 1891
11122 **bicuspida** Sm., 1891
11123 **labe** Stkr., 1898
11124 **errans** Sm., 1883
11125 **biforma** Sm., 1906
11126 **inclara** (Stkr., 1876)

11127 **ultima** (Stkr., 1876)
11128 **arcigera** (Gn., 1852)
    arcifera (Gn., 1852), emend.
    spraguei (Grt., 1864)
    a. **ferricasta** Sm., 1906
      ferricosta McD., 1938, missp.
11129 **limbalis** (Grt., 1875)
11130 **olivacea** Sm., 1906
11131 **mortua** (Grt., 1865)
    packardii (Grt., 1865)
    nobilis (Grt., 1865)
11132 **jaguarina** (Gn., 1852)
    demaculata Strand, 1916
11133 **edwardsii** (Sm., 1906)
11134 **cupes** (Grt., 1875)
    crotchii (Hy. Edw., 1875)
    navarra Dyar, 1914
    a. **deserticola** B. & McD., 1916
11135 **rivulosa** (Gn., 1852), rev. stat.
    marginata (Haw., 1809), preocc. by
      F., 1798
    contracta (Wlk., 1858)
    designata (Wlk., 1865)
    constricta (Hy. Edw., 1882), form
11136 **hanga** Stkr., 1898
11137 **nubila** (Stkr., 1876)
11138 **dolosa** Stkr., 1898
11139 **lora** Stkr., 1898
11140 **saturata** (Grt., 1874)
    rubiginosa (Stkr, 1876)
11141 **thoreaui** (G. & R., 1870)
11142 **bifascia** Hbn., 1818
    divergens (Wlk., 1858)
    imperspicua (Stkr., 1876)
    digitalis Sm., 1891
11143 **sara** Sm., 1907
11144 **baueri** McElvare, 1951
11145 **ernesta** Sm., 1907
11146 **buta** Sm., 1907
11147 **gracilenta** Hbn., 1818
    gracilis (Hbn., 1808), suppr. (ICZN
      Op. 789)
11148 **oleagina** Morr., 1875
11149 **trifascia** Hbn., 1818
    lineata (Wlk., 1858)
11150 **accessa** Sm., 1906
    viridens (Walter, 1928)
11151 **sexplagiata** Sm., 1891
    pyraloides Stkr., 1898
11152 **tobia** Sm., 1906
11153 **velaris** (Grt., 1878)
    ochreifascia Sm., 1891
11154 **lanul** (Stkr., 1878)
11155 **miniana** (Grt., 1881)
11156 **pallicincta** Sm., 1906
11157 **biundulata** Sm., 1891
11158 **espea** Sm., 1908

11159 **intrabilis** Sm., 1893
11160 **ligeae** Sm., 1893
11161 **jaegeri** (G. H. Sperry, 1940)
11162 **simplex** Sm., 1891
11163 **felicitata** (Sm., 1894)
    imperialis (B. & McD., 1911)
11164 **florida** (Gn., 1852)
11165 **neglecta** Stkr., 1898
11166 **regia** (Stkr., 1876)
11167 **niveicosta** (Sm., 1906)
    melliflua Dyar, 1921
11168 **gaurae** (J. E. Smith, 1797)
    matutina (Hbn., 1827–31)
11169 **mitis** (Grt., 1873)
    obliquata (Sm., 1891)
11170 **illustra** Sm., 1906
11171 **gloriosa** (Stkr., 1878)
11172 **terrifica** B. & McD., 1918
11173 **sanguinea** (Gey., 1832)
    carminosa Neum., 1884
11174 **lucens** (Morr., 1875)
    luxuriosa (Grt., 1882)
11175 **meadi** (Grt., 1873)
11176 **zuni** (McElvare, 1950)
11177 **nundina** (Drury, 1773)
    nigrirena (Haw., 1809)
11178 **arefacta** (Hy. Edw., 1885)
11179 **tertia** (Grt., 1874)
    a. **megarena** Sm., 1883
11180 **argentifascia** B. & McD., 1912
11181 **albafascia** Sm., 1883
    a. **erosa** Sm., 1906
11182 **brunnea** B. & McD., 1913
11183 **diffusa** Sm., 1891
11184 **walsinghami** (Hy. Edw.,
    1881)
11185 **crenilinea** Sm., 1891
11186 **balba** (Grt., 1881)
    a. **brucei** Sm., 1891
11187 **coercita** (Grt., 1881)
11188 **unimacula** Sm., 1891
    coolidgei Hill, 1924
11189 **alensa** Sm., 1906
11190 **separata** (Grt., 1879)
11191 **velutina** B. & McD., 1912
11192 **cumatilis** (Grt., 1865)
    sulmala (Stkr., 1878)
    sulmula auth., missp.
11193 **hulstia** Tepper, 1883
11194 **tenuescens** (Grt., 1883)
11195 **reniformis** Sm., 1900
11196 **obliqua** Sm., 1883
11197 **oculata** Sm., 1900
    macroptica Sm., 1906
11198 **alencis** (Harvey, 1875)
    aleucis auth., missp.
11199 **chrysella** (Grt., 1874)

    conchula (Felder & Rogenhofer,
      1872?, 1874)
11200 **ciliata** Sm., 1900
11201 **bimatris** (Harv., 1875)
11202 **carolinensis** (B. & McD., 1911)
11203 **luxa** (Grt., 1881)
11204 **citrinella** (G. & R., 1870)

    THYREION Sm., 1891
11205 **rosea** Sm., 1891
    stena Sm., 1906
11206 **snowi** (Grt., 1875)

    HELIOLONCHE Grt., 1873
    HELIOSEA Grt., 1875
11207 **modicella** Grt., 1873
11208 **carolus** McD., 1936
11209 **celeris** (Grt., 1873)
    a. **melicleptrioides** (Benj., 1936)
11210 **pictipennis** (Grt., 1875)
    defasciata (Benj., 1936)
11211 **cresina** (Sm., 1906)

    MELAPORPHYRIA Grt., 1874
11212 **immortua** Grt., 1874

Grotellini

    GROTELLA Harv., 1875
11213 **septempunctata** Harv., 1875
11214 **harveyi** B. & Benj., 1922
11215 **sampita** Barnes, 1907
11216 **blanca** Barnes, 1904
11217 **dis** Grt., 1883
11218 **parvipuncta** B. & McD., 1912
11219 **stretchi** B. & Benj., 1922
11220 **vagans** B. & Benj., 1922
11221 **binda** Barnes, 1907
11222 **tricolor** Barnes, 1904
11223 **soror** B. & McD., 1912
11224 **margueritaria** A. Blanchard, 1968
11225 **grisescens** (B. & McD., 1910)
11226 **olivacea** B. & McD., 1911
11227 **citronella** B. & McD., 1916
11228 **vauriae** McElvare, 1950
11229 **blanchardi** McElvare, 1966

    HEMIGROTELLA B. & McD., 1918
11230 **argenteostriata** B. & McD., 1918

    NEOGROTELLA B. & Benj., 1922
11231 **confusa** B. & Benj., 1922
11232 **spaldingi** (B. & McD., 1913)
    spauldingi (B. & McD., 1913), incorr.
      orig. spell.
11233 **macdunnoughi** B. & Benj., 1922

INDEX

258